ATLAS OF THE 2020 ELECTIONS

EDITED BY Robert H. Watrel
Kimberly Johnson Maier
Ryan Weichelt
Fiona M. Davidson
John Heppen
Erin H. Fouberg
J. Clark Archer
Richard L. Morrill
Fred M. Shelley
Kenneth C. Martis

CARTOGRAPHY BY Ryan Weichelt, J. Clark Archer, and Robert H. Watrel

ROWMAN & LITTLEFIELD
Lanham • Boulder • New York • London

We would like to dedicate this atlas to our friend and colleague, Ronald "Ron" J. Johnston, who passed away in May 2020. He was instrumental in putting electoral geography on the map! He will be dearly missed.

Published by Rowman & Littlefield
An imprint of The Rowman & Littlefield Publishing Group, Inc.
4501 Forbes Boulevard, Suite 200, Lanham, Maryland 20706
www.rowman.com

86-90 Paul Street, London EC2A 4NE

British Library Cataloguing in Publication Information available

Library of Congress Cataloging-in-Publication Data

Names: Watrel, Robert H., editor, cartographer. | Archer, J. Clark, editor, cartographer. | Davidson, Fiona M., editor. | Fouberg, Erin Hogan, editor. | Heppen, John, editor. | Maier, Kimberly Johnson, editor. | Martis, Kenneth C., editor. | Morrill, Richard L., 1934- editor. | Shelley, Fred M., 1952- editor. | Weichelt, Ryan Douglas, editor, cartographer.
Title: Atlas of the 2020 elections / edited by Robert H. Watrel, J. Clark Archer, Fiona M. Davidson, Erin H. Fouberg, John Heppen, Kimberly Johnson Maier, Kenneth C. Martis, Richard Morrill, Fred M. Shelley, Ryan Weichelt ; cartography by Ryan Weichelt, J. Clark Archer, and Robert H. Watrel.
Description: Lanham : Rowman & Littlefield, [2022] | Includes bibliographical references and index.
Identifiers: LCCN 2021047819 (print) | LCCN 2021047820 (ebook) | ISBN 9781538151976 (hardcover) | ISBN 9781538151983 (epub)
Subjects: LCSH: Presidents—United States—Election—2020—Maps. | Elections—United States—Maps. | Elections—United States—Statistics—Maps. | Voting—United States—Maps. | Political campaigns—United States—Maps.
Classification: LCC G1201.F9 A8 2022 (print) | LCC G1201.F9 (ebook) | DDC 324.973/09330223—dc23/eng/20211203
LC record available at https://lccn.loc.gov/2021047819
LC ebook record available at https://lccn.loc.gov/2021047820

∞™ The paper used in this publication meets the minimum requirements of American National Standard for Information Sciences—Permanence of Paper for Printed Library Materials, ANSI/NISO Z39.48-1992.

CONTENTS

FIGURES AND TABLES

FIGURES

TABLES

PREFACE

The *Atlas of the 2020 Elections* offers a spatial perspective of the presidential primaries and campaigns of the major-party candidates leading up to the presidential election on Tuesday, November 3, 2020, and the events that followed in the United States.

The editors span several generations, which brought an incredible wealth of political and geographic knowledge to our final product. While the COVID-19 pandemic prevented us from meeting in person at the American Association of Geographers' Annual Meeting and isolated many of us to our homes, technology allowed us to meet more frequently. Through biweekly Zoom meetings, we discussed the significance of the 2020 election and the events that followed—often as they happened. We looked forward to our nuanced political geography discussions every other week. Our frequent meetings enabled us to navigate the isolation we experienced and kept us connected and talking with each other about topics we find richly interesting. As a result, the level of collaboration in the *Atlas of the 2020 Elections* was higher than in previous versions, and we hope the higher degree of collaboration is reflected here.

Robert H. Watrel and Kimberly Johnson Maier

1 INTRODUCTION

ROBERT H. WATREL AND KIMBERLY JOHNSON MAIER

Democratic nominee Joseph R. Biden won the 2020 presidential election by carrying twenty-seven states and the Second Congressional District of Nebraska (which allocates its electoral votes by congressional district), giving him 306 of the 538 votes in the Electoral College. Biden won approximately seven million more popular votes than Donald J. Trump, breaking Barack H. Obama's previous record by winning 81.2 million votes, the most votes ever cast for a candidate in US history. Turnout was so high in the 2020 election that Trump also broke Obama's record, winning 74.2 million votes—almost five million more than Obama's record 69.5 million votes. The 2020 election came down to a few key swing and tipping-point states including Arizona (Biden won by 10,457 votes: 49.36 percent to Trump's 49.06 percent), Georgia (Biden won by 11,779 votes: 49.47 percent to Trump's 49.24 percent), Wisconsin (Biden won by 20,682 votes: 49.45 percent to Trump's 48.82 percent), and Pennsylvania (Biden won by 80,555 votes: 50.01 percent compared to Trump's 48.84 percent). Surprisingly, Georgia cast their electoral votes for a Democratic candidate for the first time since 1980. Arizona also cast their electoral votes for a Democratic candidate for the first time since 1996.

It is important to look at elections through the geographic lens as it showcases spatial and temporal changes of territorial constituencies, providing insights into the dynamics of American political processes and outcomes at different scales. Geographically, the pattern of support for Biden and Trump is like patterns that have been around since the early 1990s. However, with Georgia, Biden was the first Democrat to carry a southeastern state other than Florida for the first time since 2008. Like the 2016 election, Biden won the Northeast and Pacific Coast, while Trump won the Southeast, the Great Plains, and the Mountain West regions. However, unlike the 2016 election, the "Democratic blue wall" states of Pennsylvania, Michigan, and Wisconsin held for Biden. He also made inroads in the Mountain West by winning Arizona. Ultimately, Biden flipped five states and one congressional district from Republican in 2016 to Democratic in 2020.

An analysis of the county-level map confirms the continued sharp divide between rural and urban. The 2020 county scale map looks quite similar to the last several elections, with rural Republican areas mapped in red and urban Democratic areas mapped in blue. With more rural than urban counties in the United States, Trump won 2,560 counties while Biden won only 471 counties. While the county scale map looks like a Trump victory, counties do not vote—people do. A large portion of the counties that Biden won were urban areas with millions of more voters.

THE 2020 ELECTION

At the time of the November 2020 election, President Trump had been impeached, COVID-19 cases were climbing, more than 237,000 Americans had died of COVID, and the vaccine was still weeks away. After the election, the sitting US president refused to accept the election results and vigorously touted voter fraud accusations, which resulted in numerous recounts and failed court cases. Trump finally conceded the election on January 7, 2021, the day after the January 6 insurrection of the Capitol.

The contributors to this atlas analyze the primaries leading up to the 2020 US presidential election, the campaigns, and the US presidential election from a political geography standpoint. Since Trump ran virtually unopposed, analysis of the primaries focuses on the Democratic primary. The Democratic primary was a diverse pool of candidates, including six women and the first openly gay candidate for president. On the hundredth anniversary of women's suffrage, the United States saw the election of the first female vice president, Kamala Harris.

The geography of the Democratic primary votes reveals a slight shift of support away from Democratic Socialist candidate Bernie Sanders toward the more moderate Democrat, Joe Biden, when compared to the 2016 election. Unlike previous elections, third-party candidates received minimal support. Early in the Democratic primaries, Pete Buttigieg won Iowa and Bernie Sanders won several other states. However, it became clear following the South Carolina primary and his ten-state win on Super Tuesday that Biden was the clear favorite for the Democratic nomination. Sanders suspended his presidential bid on April 8, 2020.

The COVID-19 pandemic dramatically altered how campaigns and elections were conducted. For example, some candidates opted for virtual, small-scale, and socially distanced rallies, while others tried to maintain the pre-pandemic status quo. Conventions were also significantly altered. Both the Republican and Democratic Conventions were postponed and, in some cases, had locations changed. In the end, the Democratic Convention included both a face-to-face and virtual component. The Republican Convention was originally slotted for Charlotte, North Carolina, but the Republican National Committee looked to move it to Jacksonville, Florida, to avoid COVID restrictions in North Carolina. Eventually, the Republican Convention was held in a combination of locations, including Charlotte and the White House in Washington, DC, in part because of the rising COVID cases in Florida during summer 2020 (the first wave peak in Florida was mid-July).

To accommodate voting during the pandemic, many states allowed more opportunities for early, mail-in, and absentee ballots during the general election. As a result, the 2020 presidential election saw the largest percentage of voter turnout since 1960 and the largest number of votes cast in US history. The partisan climate of the election also motivated voter participation. Comparing voter participation in 2016 and 2020, the Republican votes increased from 63 million to 74.2 million and Democratic votes grew from 65.8 million to 81.2 million.

The November 2020 election also included races for 35 Senate seats, 2 of which were special elections (Arizona and Georgia) and 435 seats in the House of Representatives. In the Senate the Democrats gained 3 seats, which gave the party a 50-50 split and a majority on party-line votes when Vice President Harris would cast the deciding vote. In the House of Representatives, the Republicans gained fourteen seats, narrowing the Democratic majority to just nine, with two current vacancies. In the end, Democrats controlled both the Senate and the House. As shown by maps in this atlas, like the presidential vote, most urban districts elected Democrats and most rural districts elected Republicans.

ORGANIZATION OF THE ATLAS

The atlas is divided into eleven chapters; each, except for this introduction, contains maps and accompanying narratives. Contributors to the atlas created and analyzed more than 150 maps of voting patterns and demographics at the national, state, county, and local precinct scales.

Chapter 2 addresses the primary election process leading up to the nominations of Trump and Biden by the Republican and Democratic parties, respectively. While Trump was the incumbent candidate, he did not run unopposed, but none of the other challengers posed much of a threat. Trump became the official Republican candidate to accept his nomination at the White House, while the Republican National Convention took place in Charlotte, North Carolina. In the Democratic primary, Joseph Biden was one of twenty-eight major candidates who vied to become the Democratic party nominee. After March 3, 2020, Super Tuesday, it was clear that Biden was well on his way to becoming the Democratic nominee. This chapter also discusses the impact COVID-19 had on state primary elections.

Chapter 3 focuses on the electoral process and the campaign. Each party raised and spent millions of dollars from campaign donors, focusing on areas where they believed they could make a difference in the Electoral College outcome. The chapter also analyzes campaign stops during the last twelve days of the campaign, and geographic language in the 2020 debates. The maps in this chapter reveal clear geographic variations.

Chapter 4 assesses the outcome of the election. Of interest are election results mapped at the county level. Maps in this chapter identify the voting patterns of the major and third parties, the Electoral College, the popular vote by county, and comparisons of the state and national level results with past elections. When comparing the 2016 to 2020 elections, maps reveal that much of the increase in Democratic votes came from the Mountain West region (see map on atlas front cover), while Republicans increased support near the southern border with Mexico. The remainder of this chapter discusses voter turnout, metropolitan, micropolitan, and rural voting patterns.

In chapter 5, we examine the 2020 election at the regional level. Seven regions are discussed within this chapter, including the Northeast, Appalachia, the Southeast, the Midwest, the Great Plains, the Mountain West, and the Pacific Coast states. The overall regional patterns were similar to the 2016 election, with the Northeast and Pacific Coast regions remaining heavily Democratic, while the Great Plains, Appalachia, and the Southeast remain predominantly Republican—although there were pockets of Democratic support in urban areas, the Black Belt, and Native American reservations.

Chapter 6 examines the distribution of demographic indicators, including measures of ethnicity, gender, religion, occupation, and population growth and decline. The essays in this chapter use maps to discuss possible associations of population subgroups, including Black voters and women, to provide unique interpretations about the 2020 election.

Chapter 7 focuses on the congressional elections, party affiliation, and specific roll-call votes in the 116th Congress. Roll-call votes include those pertaining to Russian sanctions, emergency supplemental appropriation for assistance and security at the southern border, and the US, Mexico, and Canada Trade Agreement. Additional entries in the chapter focus on specific congressional races in states including Arizona, Maine, Montana, Colorado, and Texas.

In chapter 8, contributors highlight selected campaigns for statewide elections and referenda. Analyses of gubernatorial elections include Montana and North Carolina. Furthermore, essays focus on pertinent race issues, such as the impact of the George Floyd murder on voter turnout in Minnesota, and the removal of the Confederate flag from the Mississippi state flag. Additionally, contributors analyze votes on marijuana legalization and the minimum wage.

Chapter 9 examines postelection events, including the Senate runoff elections in Georgia, the events of January 6, 2021, and the 2020 Census and subsequent reapportionment.

Chapter 10 concludes the atlas with some final thoughts on the 2020 election. Here, contributors discuss how the current geography of voting and changing demographic patterns may impact the 2024 presidential election.

Finally, chapter 11 discusses technical issues, including data collection, analysis, and mapping methods and techniques.

2 PRIMARY ELECTIONS

THE 2020 PRESIDENTIAL PRIMARY CAMPAIGNS

FRED M. SHELLEY AND HEATHER HOLLEN

Going into the 2020 election, Republican President Donald Trump was highly popular among the Republican rank and file, and he faced only nominal opposition in winning his party's nomination. Thus, this chapter focuses on the Democratic primary elections that led to the selection of Joseph Biden and Kamala Harris as the Democratic nominees for president and vice president, respectively.

In 2016, the Democratic Party experienced strong tensions between its moderate and progressive wings. This tension was evident in the contest for the party's nomination between the eventual nominee, Hillary Clinton, and her main challenger, Bernie Sanders. The tension and animosity between the two factions of the party continued after Trump defeated Clinton narrowly in the 2016 general election. Each side blamed the other for Clinton's loss. In the 2018 midterm elections, the Democrats picked up forty-one seats in the House of Representatives. While Trump remained popular among Republicans, public opinion polls showed that his approval rating among all voters hovered between 40 percent and 45 percent during his time in office. Thus, many Democratic leaders believed that the general election could be won by a strong challenger.

In 2018, the divide between moderates and progressives intensified with the election of progressive young Democratic representatives such as Alexandria Ocasio-Cortez of New York, Deb Haaland of New Mexico, Ayanna Pressley of Massachusetts, Rashida Tlaib of Michigan, and Ilhan Omar of Minnesota. Going into the 2020 campaign, this tension in the Democratic Party was represented by the divide in support for moderate primary candidates such as Biden, Pete Buttigieg, Amy Klobuchar, and Andrew Yang and progressive primary candidates such as Sanders, Harris, Elizabeth Warren, Cory Booker, and Julián Castro.

CHANGES TO THE NOMINATION PROCESS

Before the 2020 Democratic primaries began, the Democratic Party made several changes to the primary election process. Some of these changes were made in response to the continuing divisiveness and acrimony that had emerged within the party during, and after, the 2016 primary and general election campaigns.

In 2016, the party's national convention included 4,767 voting delegates. These included 716 unpledged delegates, who were often referred to as superdelegates. These superdelegates included all Democratic members of the US Senate and the US House of Representatives, all Democratic state governors, and elected members of the Democratic National Committee (DNC). In contrast to the pledged delegates who were chosen in state primaries or caucuses, the superdelegates were free to vote as they saw fit.

More than 90 percent of the superdelegates supported Hillary Clinton in 2016, whereas only 55 percent of the pledged delegates supported Clinton and 45 percent supported Bernie Sanders. Given this disparity, many Sanders supporters demanded that superdelegates be given reduced roles in selecting the party's nominees. In response, the DNC revised its rules such that superdelegates would not be allowed to cast decisive votes on the convention's first ballot. Meanwhile, several states that had held caucuses to select delegates in 2016 and earlier years switched from caucuses to primary elections in 2020. This change was significant in that Sanders had been much more successful in caucuses in 2016 whereas Clinton had done better in primaries.

THE CANDIDATES

The sheer number of Democratic candidates vying for the presidential nomination in 2020 demonstrated the broad spectrum of political views within the Democratic Party. In total, there were twenty-eight candidates who tossed their figurative hats in the ring. The diversity among the group was unprecedented—female, openly gay, former military service members, people of color, tech millionaires and billionaires, and a wide range of ages from under forty to late seventies. The twenty-eight candidates (listed in alphabetical order) included Senator Michael Bennet of Colorado, former Vice President Joseph Biden of Delaware, former New York City Mayor Michael Bloomberg, Senator Cory Booker of New Jersey, Governor Steve Bullock of Montana, South Bend, Indiana Mayor Pete Buttigieg, former Secretary of Housing and Urban Development Julián Castro of Texas, New York City Mayor Bill de Blasio, former Representative John Delaney of Maryland, Representative Tulsi Gabbard of Hawaii, Senator Kirsten Gillibrand of New York, former Senator Michael Gravel of Alaska, Senator Kamala Harris of California, former Governor John Hickenlooper of Colorado, Governor Jay Inslee of Washington, Senator Amy Klobuchar of Minnesota, Mayor Wayne Messam of Miramar, Florida, Representative Seth Moulton of Massachusetts, former Representative Beto O'Rourke of Texas, former State Senator Richard Ojeda of West Virginia, former Governor Deval Patrick of Massachusetts, Representative Tim Ryan of Ohio, Senator Bernie Sanders of Vermont, former Representative Joe Sestak of Pennsylvania, hedge fund investor Tom Steyer of California, Representative Eric Swalwell of California, Senator Elizabeth Warren of Massachusetts, self-help lecturer and New Age author Marianne Williamson of California, and tech executive Andrew Yang of New York (figure 2.1).

Most public opinion polls taken in 2019 showed Biden with the largest amount of support among potential Democratic voters. This broad level of support was attributed to Biden's name recognition, his service as vice president under the highly popular President Barack Obama, and his high level of support among non-white Democrats. Sanders, who also had high name recognition following his 2016 run, remained the favorite of Democratic voters who identified as progressives.

CRITERIA FOR INCLUSION IN THE DEBATES

The DNC scheduled eleven debates among the candidates. The first six debates took place in 2019 with one each in June, July, September, October, November, and December. The remaining five debates took place in early 2020 before the Iowa, New Hampshire, Nevada, South Carolina, and Super Tuesday primaries and caucuses (figure 2.2).

Given the unprecedented size of the field, the DNC spelled out criteria for inclusion in each debate. For the first two debates, the criteria required a candidate to receive 1 percent (or higher) support in more than three national or early state polls—Iowa, Nevada, New Hampshire, and/or South Carolina. Only polling data released between January 1, 2019, and fourteen days prior to the debate would be recognized. Polling data from the same organization would only be accepted if the polls were conducted in different geographic regions. If a candidate failed to meet the polling criterion, she or he could still qualify for a debate by meeting a fundraising threshold. This required a candidate to provide verifiable evidence that they obtained donations from at least 65,000 unique donors and that these donors represented at least 200 unique donors per state in at least 20 states. Given the large candidate pool, each debate was capped at twenty total candidates (ten per night). If more than twenty candidates qualified, then the candidates would be prioritized based on those who reached both polling and fundraising thresholds, those with the highest polling averages, and those with the most contributions.

The polling and fundraising criteria increased for the next two debates. The polling criterion increased to 2 percent, and the fundraising numbers increased to at least 130,000 unique donors with at least 400 unique donors per state in 20 states. For the fifth debate in November, the polling criterion required the candidate to receive 5 percent support in at least two single state polls in Iowa, Nevada, New Hampshire, and/or South

FIGURE 2.1

2020 Leading Democratic Presidential Candidates' Hometowns

FIGURE 2.2

2019 - 2020 Democratic Presidential Primary Debates Dates and Locations

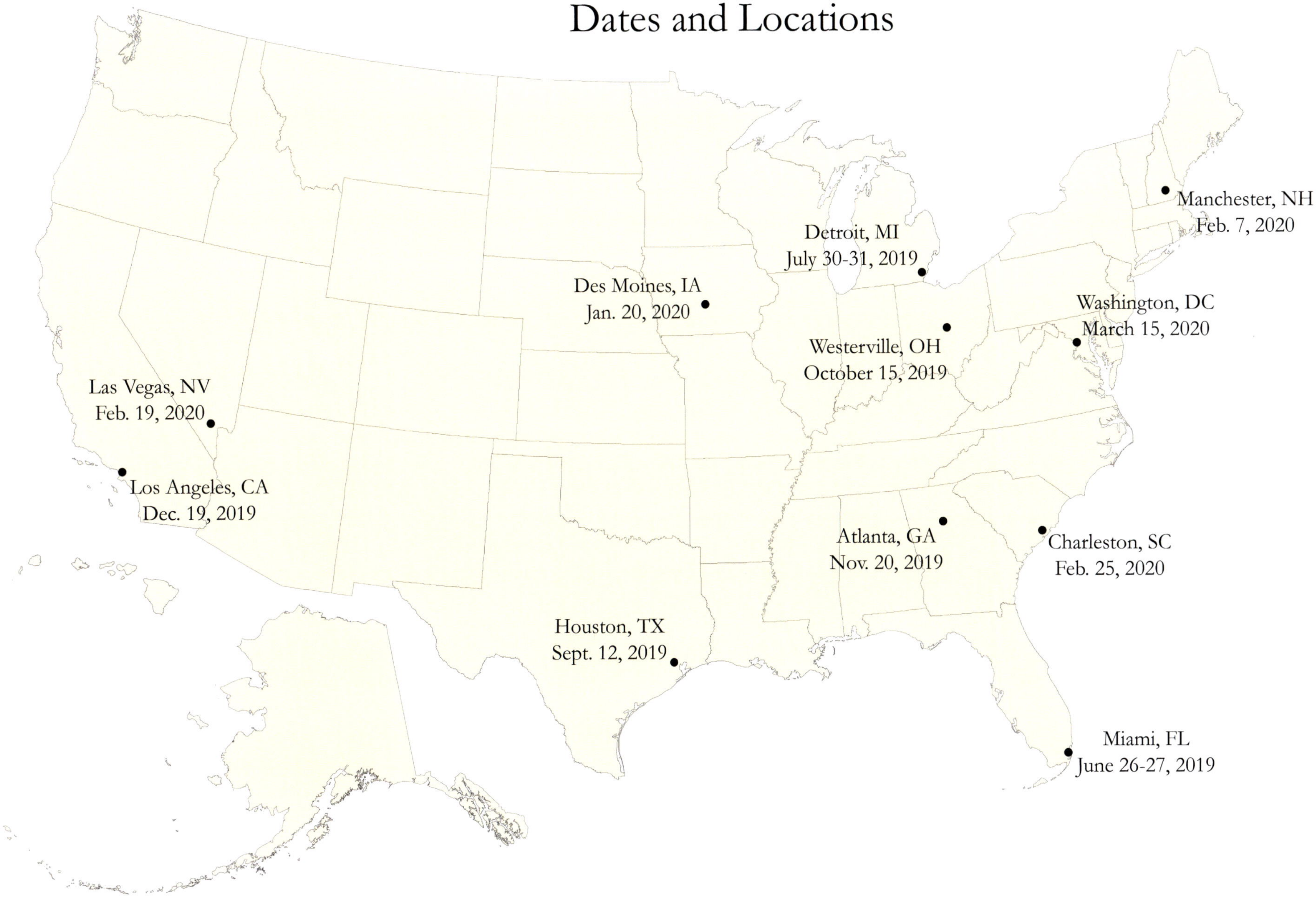

Carolina. Both polls could come from the same organization and geographic area, but they had to be released between September 13, 2019, and a week before the November debate. Unlike the previous debates, the candidates were *also* required to meet a fundraising threshold of at least 165,000 unique donors with at least 600 unique donors per state in 20 states. For the sixth debate in December, the previous polling criteria mentioned for the fifth debate increased to 4 percent and 6 percent, respectively. Qualifying polling data had to be released between October 16, 2019, and December 12, 2019. The fundraising threshold increased to 200,000 unique donors with at least 800 unique donors per state in 20 states.

For the seventh debate in January before the Iowa precinct caucuses, the polling requirement increased to 5 percent and 7 percent, respectively. Qualifying polling data had to be released between November 14, 2019, and January 10, 2020. The fundraising threshold increased to 225,000 unique donors with at least 1,000 unique donors per state in 20 states. On January 17, 2020, the DNC released qualifying criteria for the eighth debate to take place before the New Hampshire primary. The polling and fundraising criteria did not change from the seventh debate. However, candidates could also qualify for the eighth debate by meeting a delegate threshold, consisting of obtaining at least one pledged delegate from Iowa to the Democratic National Convention.

For the ninth debate in mid-February, the fundraising criteria used in previous debates were eliminated. Candidates could qualify by meeting delegate or polling criteria. A candidate could qualify if they obtained at least one pledged delegate from Iowa or New Hampshire to the Democratic National Convention. A candidate could also qualify by meeting one of two polling requirements—receiving 10 percent support in at least four national, South Carolina, and/or Nevada polls, or receiving 12 percent support in at least two polls from Nevada and/or South Carolina. Some progressives and some of the other candidates were critical of the DNC's decision to remove the fundraising criterion at this point because this decision was seen as favoring the very wealthy Bloomberg, who had the resources to self-fund his campaign without relying on outside donors. Bloomberg drew significant interest after announcing his candidacy. However, observers regarded his performance in the two debates in which he participated as poor, and he subsequently withdrew in favor of Biden.

For the tenth debate in late February, candidates could qualify by obtaining at least one pledged delegate from Iowa, New Hampshire, or Nevada to the Democratic National Convention. For the eleventh and final debate in March, candidates were required to receive at least 20 percent of the pledged delegates awarded through March 15, 2020. Overall, the increasingly stringent criteria helped to winnow down the field. Twenty of the twenty-eight announced candidates qualified for the first debate, whereas only Biden and Sanders qualified for the final debate.

DEBATES AND DROPOUTS

With twenty candidates having qualified for the first two debates, the DNC recognized that a single debate among all twenty of them would be unwieldy and would give each candidate only a small amount of speaking time. In response, the DNC decided to divide the candidates into two groups of ten. Each candidate was assigned at random to one of the two groups. The first pair of debates took place on June 26 and 27, 2019. The June 27 debate was notable for a sharp exchange between Biden and Harris over Biden's past opposition to school busing for racial balance, and over Biden's comments that he had worked with fellow senators who were known to have been segregationists during their Senate careers. A second set of debates, with the same qualification criteria and with the candidates reassigned to groups at random once again, took place on July 30 and 31. Before these debates, Swalwell dropped out of the race and was replaced in the debates by Bullock. After the July 30–31 debates, Gillibrand, Gravel, Hickenlooper, Inslee, and Moulton dropped out of the race.

The next four debates included one debate each in September, October, November, and December. The four ensuing debates, with more stringent qualification criteria as described earlier, took place in January and February, with one in each of the first four primary states. The South Carolina debate on February 25 also preceded the Super Tuesday primaries on March 3. The final debate took place on March 15.

Over the course of the fall, Bullock, Castro, de Blasio, Harris, Messam, O'Rourke, Ryan, and Williamson left the race. Thus, by January 2020 only eleven of the original candidates remained, along with Bloomberg and Patrick, who had entered the race in the fall. Of these thirteen candidates, only Biden, Buttigieg, Klobuchar, Sanders, Steyer, and Warren qualified for the January debate. These six candidates, along with Bloomberg (who qualified for the ninth debate, which took place after the New Hampshire primary) and Gabbard (who did not qualify for any debates after November), stayed in the race until after the first four primaries.

THE IOWA PRECINCT CAUCUSES

As in past years, the nomination process began with contests in four states—Iowa, New Hampshire, South Carolina, and Nevada. The Iowa precinct caucuses, which led the nomination process, took place on Monday, February 3; 1,678 local precinct caucuses took place throughout the state. An additional eighty-seven "satellite" caucuses also took place to accommodate registered Iowa Democrats who, for a variety of reasons, could not participate in their local precinct caucuses. These included caucuses at work-related sites and on college campuses. Other "satellite" caucuses were held to accommodate accessibility needs, to accommodate voters who did not speak English, and to accommodate Iowans who were abroad or away from the state on caucus night. Participants in these "satellite" caucuses could not participate in their local precinct caucuses.

At each local caucus, participants selected delegates to countywide conventions on a proportional basis, with the number of delegates for each candidate proportional to the number of votes that that candidate had received in that local caucus. Delegates pledged to a particular candidate could be elected to the county-level convention only if his or her candidate received at least 15 percent of the votes of caucus attendees. Precinct residents could vote only if they attended their caucuses in person—absentee balloting was not allowed. Once the initial vote was taken, those voters who chose candidates who received less than 15 percent of the original votes could realign themselves with another candidate who did receive more than 15 percent of these votes.

Immediately prior to the caucuses, aggregations of public opinion polls showed Sanders in the lead with 22.6 percent of the vote, followed by Biden with 19.4 percent, Buttigieg with 15.9 percent, Warren with 15.2 percent, and Klobuchar with 10.3 percent. After the votes were counted after realignment, however, Sanders won a narrow plurality with 26.5 percent of the final vote. Buttigieg finished second with 25.1 percent, followed by 20.3 percent for Warren, 13.7 percent for Biden, and 12.2 percent for Klobuchar. Thus Sanders, Buttigieg, and Warren exceeded expectations, whereas Biden fell considerably short of expectations.

Statewide, Buttigieg carried fifty-seven of Iowa's ninety-nine counties, whereas Sanders carried only nineteen (figure 2.3). Sanders's support was more concentrated in urbanized areas and in cities, whereas Buttigieg's support was spread more evenly around the state. Warren carried Johnson County, which includes Iowa City and the University of Iowa and is recognized as the most liberal and progressive county in the state. She also carried Sioux County in northwestern Iowa, but Sioux County has a relatively small population and is very heavily Republican. Biden and Klobuchar won relatively few counties, with Klobuchar's support concentrated in northern Iowa near the border with Klobuchar's home state of Minnesota. Four of the six Iowa counties (which include the five most populous counties in the state) that Biden carried in the general election were carried by the more progressive candidates, Sanders and Warren, in the precinct caucuses. Exit polls showed that Sanders did slightly better than Buttigieg among men and did significantly better among voters aged under forty-five, whereas Buttigieg was more popular among older voters.

After the caucuses ended on the evening of February 3, significant delays in reporting the results took place. Many of these delays occurred because precinct chairs were expected to report their precincts' results using a newly implemented app intended to record the results. As it turned out, some of the precinct chairs had not been trained on how to use the app. Some of those chairs who did not know how to use the app attempted to report the results in their precincts by telephone, but a long backlog of telephone calls forced some to stay on hold for a long time to get through to report them.

Although some observers feared that the system had been hacked, investigation showed that this was unlikely to have occurred. However, there were numerous errors and inconsistencies in reporting the results. As a result of these issues, complete and accurate results were not released for several days. In the meantime, both the Buttigieg and Sanders campaigns declared victory. The two finished in a virtual tie for numbers of delegates for county conventions. The delays and errors associated with compiling and reporting the results led to intensified criticism of the caucus procedure, which many already viewed as biased against people who could not attend their precinct caucuses.

ON TO NEW HAMPSHIRE

After the Iowa caucuses, the candidates moved on to New Hampshire to campaign for votes in the New Hampshire primary, which took place on Tuesday, February 11. The aggregation of public

FIGURE 2.3

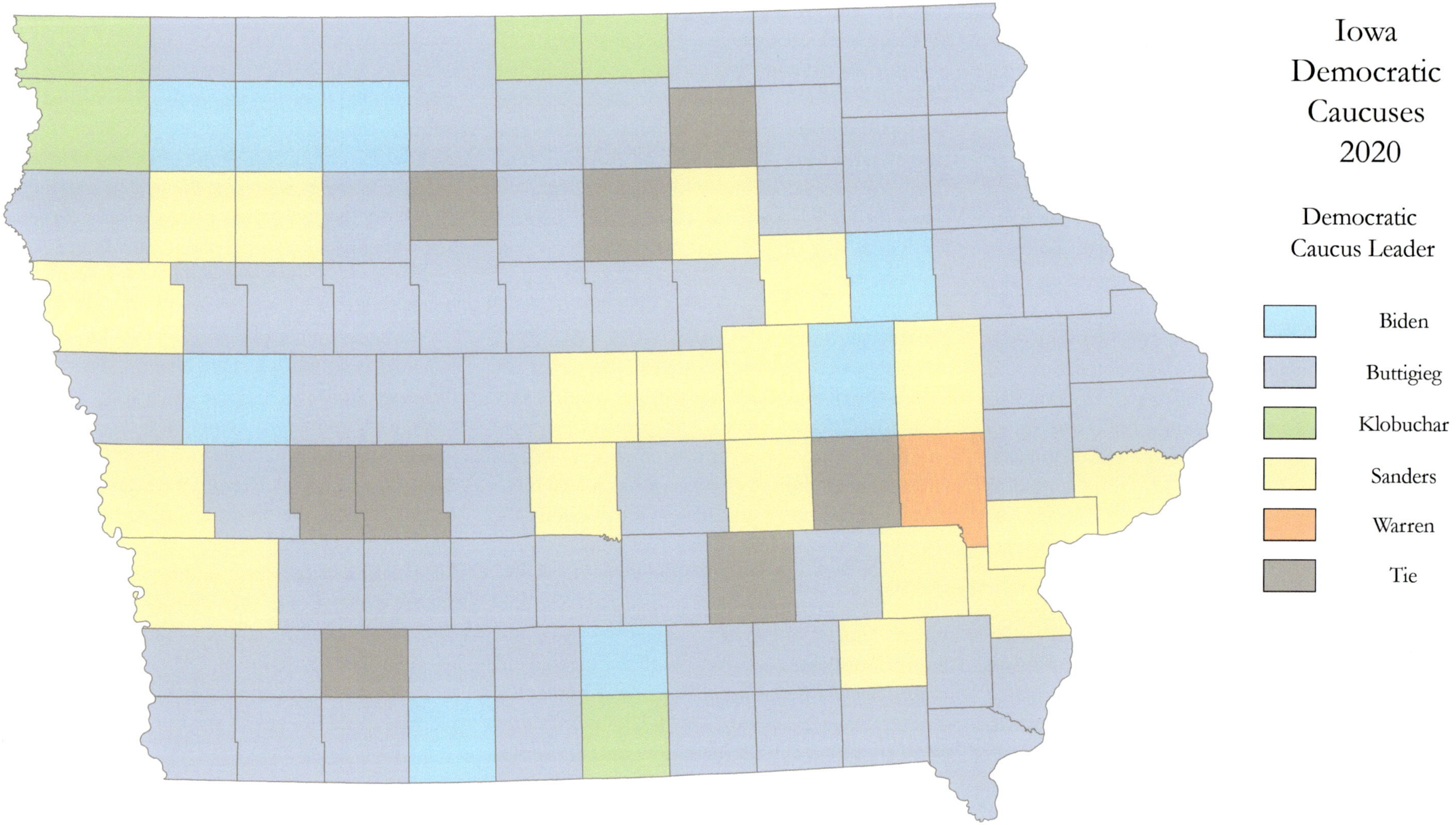
Iowa
Democratic
Caucuses
2020
Democratic
Caucus Leader
Biden
Buttigieg
Klobuchar
Sanders
Warren
Tie

FIGURE 2.4

New Hampshire Democratic Primary 2020

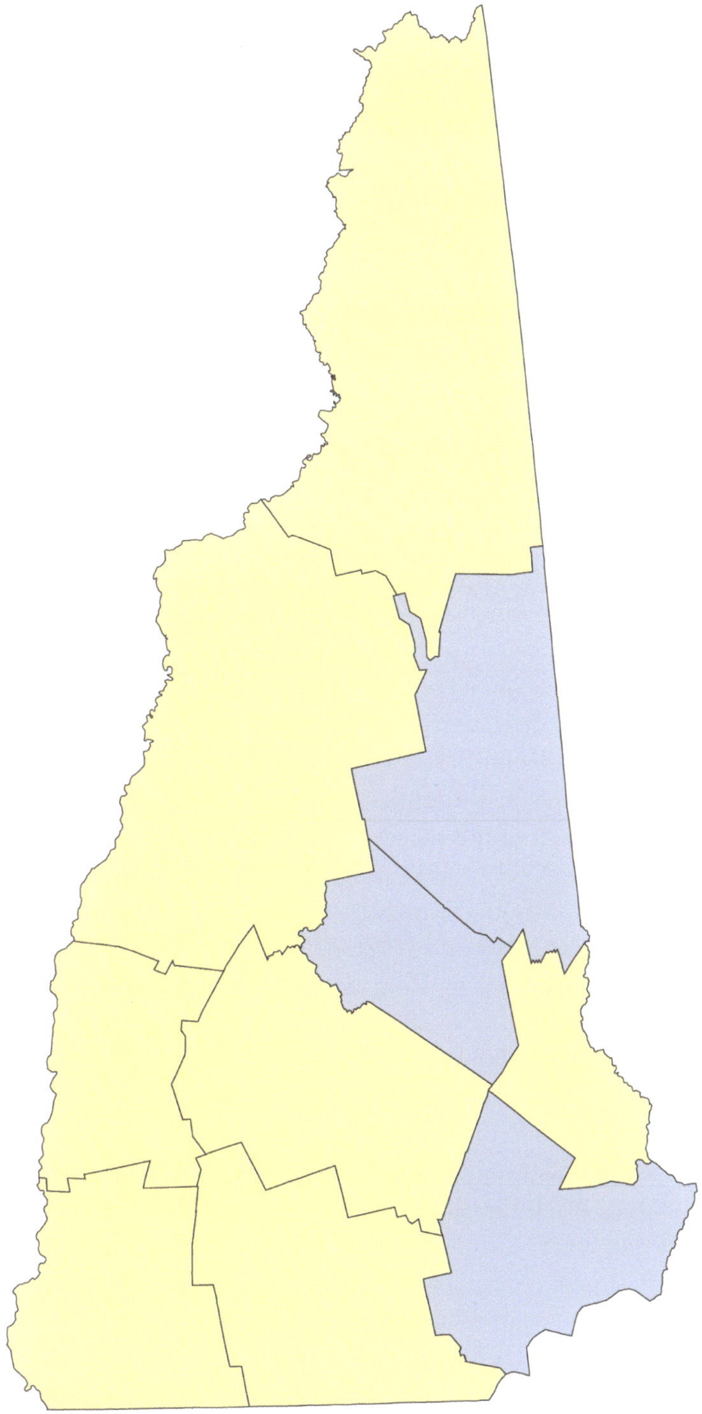

Democratic Primary Leader

Buttigieg

Sanders

opinion polls conducted shortly before the primary showed Sanders in the lead with 27.3 percent of the predicted vote followed by Buttigieg with 21.3 percent, Warren with 12.2 percent, Biden with 11.7 percent, and Klobuchar with 10.8 percent. In the primary itself, with 25.6 percent of the vote Sanders edged out Buttigieg, who finished second with 24.3 percent. Klobuchar finished in third place with 19.7 percent of the vote, followed by Warren with 9.2 percent and Biden with 8.4 percent.

The result was disappointing to the Sanders campaign because Sanders represents neighboring Vermont in the Senate and because he had won more than 60 percent of the vote in the New Hampshire primary in 2016. Klobuchar's third-place finish was a surprise, because her actual vote percentage was nearly twice as high as had been predicted. Both Warren and Biden underperformed expectations, however, lending some to doubt the long-run viability of their campaigns. The result was particularly damaging to Warren, who represented neighboring Massachusetts in the Senate. The other candidates finished far behind the first five, and Bennet, Patrick, and Yang withdrew from the race shortly after the New Hampshire primary.

Sanders carried seven of New Hampshire's ten counties, whereas Buttigieg carried the other three (figure 2.4). As in Iowa, Sanders did better in urban areas. Sanders won 31.1 percent of the vote in the state's largest city of Manchester as compared to 20.5 percent for Buttigieg. Sanders also carried the five next-largest cities in the state, including Nashua, Concord, Derry, Dover, and Rochester. Exit polls from the New Hampshire primary were consistent with those taken in Iowa. Sanders did best among younger voters and progressives, whereas Buttigieg was more successful with moderate and middle-aged voters. Klobuchar polled best among voters aged sixty-five or over.

NEVADA AND SOUTH CAROLINA

Precinct caucuses in Nevada took place on February 22, using the realignment procedure that had been used in Iowa. After realignment, Sanders finished in first place with 40.5 percent of the vote. Biden finished second with 18.9 percent, followed by Buttigieg with 17.3 percent. Sanders carried eleven of Nevada's sixteen counties along with the independent city of Carson City (figure 2.5). Sanders carried Clark County, which includes Las Vegas and contains more than half of the state's population, with 49 percent of the vote. The five counties carried by Buttigieg are all sparsely populated rural counties. Some difficulties and irregularities in calculating and reporting the results took place, reinforcing the criticism of the caucus process that had occurred in Iowa a few weeks earlier.

The primary in South Carolina, which took place on February 29, proved pivotal to Biden's eventual nomination. In contrast to Iowa and New Hampshire with their relatively small minority populations, 27 percent of South Carolina's residents are Black. And because a large majority of white South Carolinians are Republicans, more than half of the state's Democrats are Black. Thus, a strong showing among Black Democrats would be key to victory in the South Carolina primary. Three days before the South Carolina primary, Representative Jim Clyburn gave Biden a ringing endorsement. Clyburn, who is Black, was first elected to the House of Representatives in 1993. He was Majority Whip and the third-ranking Democrat in the House. Clyburn's endorsement gave the Biden campaign a significant boost.

According to public opinion survey aggregations, Biden was favored with 38.0 percent of the predicted vote. In the primary itself, Biden exceeded expectations with 48.7 percent of the vote. Sanders finished second with 19.8 percent, and Steyer, who had invested heavily in South Carolina, came in third. The exit poll showed that Biden won 61 percent among Black voters, suggesting that Clyburn's endorsement shortly before the primary may have boosted Biden's overall percentage. According to the exit poll, 70 percent of those voters who had not made up their minds before Clyburn's endorsement voted for Biden.

In winning nearly half of the vote, Biden carried all of South Carolina's forty-six counties. He won more than 50 percent of the vote in twenty-nine counties, and he exceeded 65 percent in six of them, all of which have majority-Black populations (figure 2.6). He won less than 40 percent of the vote in only four counties, including the three predominately white counties in far northwestern South Carolina and including Lexington County, which contains affluent suburbs of the state capital of Columbia and whose population is only about 12 percent Black.

Biden's win in South Carolina gave his campaign significant momentum going into the Super Tuesday primaries three days later. After the South Carolina primary, the moderate candidates Buttigieg, Klobuchar, and Steyer withdrew from the race. Buttigieg and Klobuchar formally endorsed Biden, further encouraging moderate voters in future primaries to support Biden rather than to divide their votes among more than one moderate candidate. Their names remained on the ballot,

FIGURE 2.5

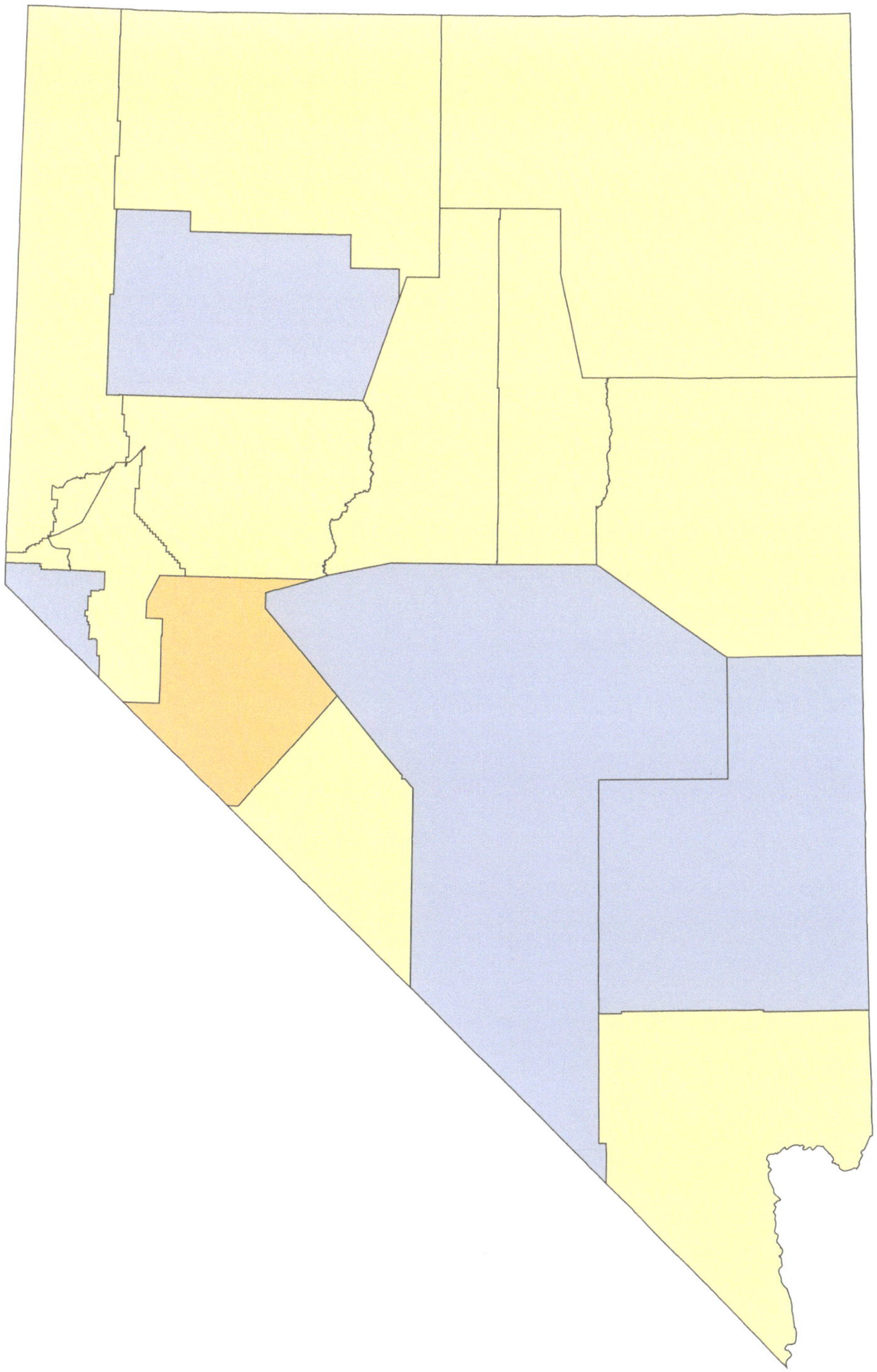

Nevada Democratic Primary 2020

Democratic Primary Leader

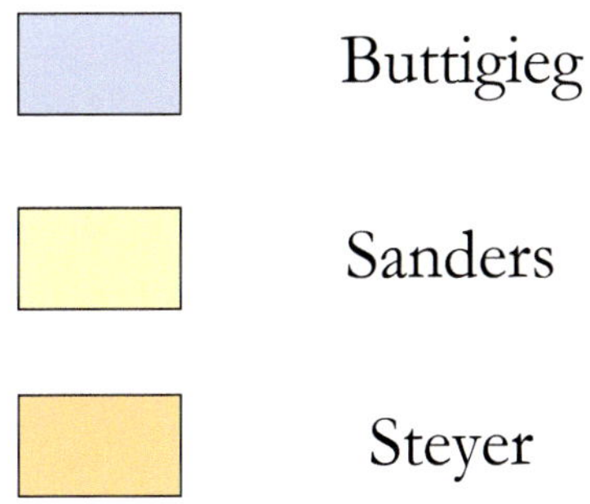

FIGURE 2.6

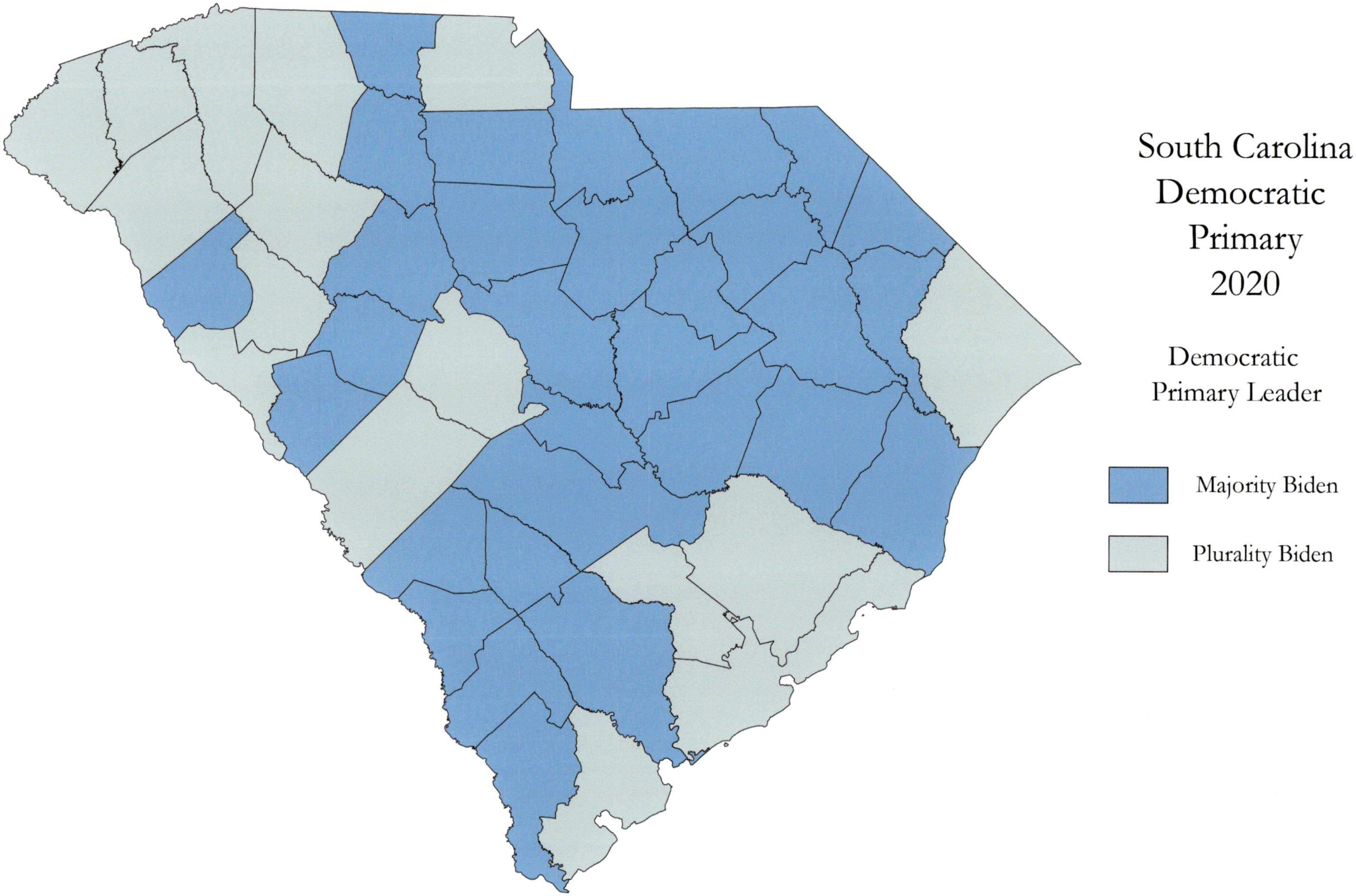
South Carolina Democratic Primary 2020
Democratic Primary Leader
Majority Biden
Plurality Biden

however, and some voters had already cast their ballots for them via early or absentee voting before they announced their withdrawals.

SUPER TUESDAY

Three days after the South Carolina primary, the Super Tuesday primaries took place on Tuesday, March 3. On that day, fourteen states held primary elections (figure 2.7). More than a third of the pledged delegates to the Democratic National Convention came from one of these fourteen states.

The term Super Tuesday refers to the large number of states that hold their primaries simultaneously on the same day. The original idea behind Super Tuesday came from the South during the 1980s. At that time, Democratic Party leaders in some southern states felt that their region was being ignored or overlooked in the party's selection of its nominee. These party leaders agreed to hold their states' primaries on the same day to increase the region's influence on the choice. In 1988, nine southern states held their primaries on March 8 of that year. Although southern party leaders hoped that these primaries would benefit a moderate candidate, it turned out that the vote was split among several viable candidates. The nomination went eventually to the more liberal Michael Dukakis, who lost the general election subsequently to George H. W. Bush. Later, however, success in Super Tuesday primaries helped the moderate Bill Clinton in 1992, Al Gore in 2000, and Hillary Clinton in 2016.

Historically, California had held its primaries near the end of the primary season. However, by that point the nomination had generally been decided, and therefore California had no influence over the choice of the party's nominee. Accordingly, California decided to hold its primary on Super Tuesday also. Thus, two of the largest states participated in Super Tuesday.

When the votes were counted, Biden won ten of the fourteen Super Tuesday primaries, whereas Sanders won California, Colorado, Utah, and his home state of Vermont. Biden finished second in each of these four states. His strongest states were Alabama, in which he won 63 percent of the vote, and Virginia, in which he won 53 percent of the vote. He also exceeded 40 percent in Arkansas, North Carolina, and Tennessee. Biden carried a large majority of counties in these and other southern states. He won every county in Alabama and lost only Washington County, which contains Fayetteville and the University of Arkansas, to Sanders in that state. He also carried every county in Oklahoma, all but three counties in Tennessee, and all but four counties in North Carolina.

County-level analysis illustrates regional breakdowns within some of these states. In Texas, Biden defeated Sanders by a margin of 34.6 percent to 29.9 percent. Although his statewide margin of victory was relatively small, Biden won a large majority of Texas's 254 counties (figure 2.8). His support was strongest in rural northern and eastern Texas. He also did very well in Harris County (Houston), Dallas County (Dallas), and Tarrant County (Fort Worth), all of which have significant minority populations in contrast to the largest cities in Iowa and New Hampshire, whose minority populations are relatively small. However, Sanders swept heavily Latino South Texas. He carried most of the counties along the Red River that separates Texas and Mexico, from El Paso in the west to Brownsville along the Gulf of Mexico. Sanders also won Travis County, which contains the city of Austin and the University of Texas at Austin and is recognized generally as the most liberal county in the state. He won neighboring and rapidly growing Hays and Comal counties as well.

Biden bested Sanders in Massachusetts by a larger margin of 33.4 percent to 26.6 percent. This result was unexpected because public opinion polls had predicted that Sanders would win the primary. Biden carried eleven of the Bay State's fourteen counties, whereas Sanders carried Franklin, Hampshire, and Suffolk counties (figure 2.9). Hampshire County includes many people associated with the University of Massachusetts–Amherst, Smith College, Mount Holyoke College, Amherst College, and Hampshire College. These colleges contribute strongly to Hampshire County's liberal politics. Suffolk County includes the cities of Boston and Cambridge. It too is a very liberal county. It was the only county in Massachusetts in which Warren, who had been a professor of law at Harvard University in Cambridge, finished second and relegated Biden to third place. Biden, meanwhile, enjoyed substantial support in other areas of the state, including the cities of Worcester and Springfield, which contain larger working-class populations.

One of Sanders's strongest Super Tuesday states was Colorado. Here, Sanders defeated Biden by a margin of 37.0 percent to 24.6 percent, with Bloomberg (18.5 percent) and Warren (17.6 percent) trailing them. Sanders carried fifty-five of the state's sixty-four counties while Biden carried seven counties and Bloomberg carried two (figure 2.10). The counties carried by Biden or Bloomberg are sparsely populated rural counties, and those in the eastern, Great Plains region of Colorado are heavily Republican. Sanders did especially well, however,

FIGURE 2.7

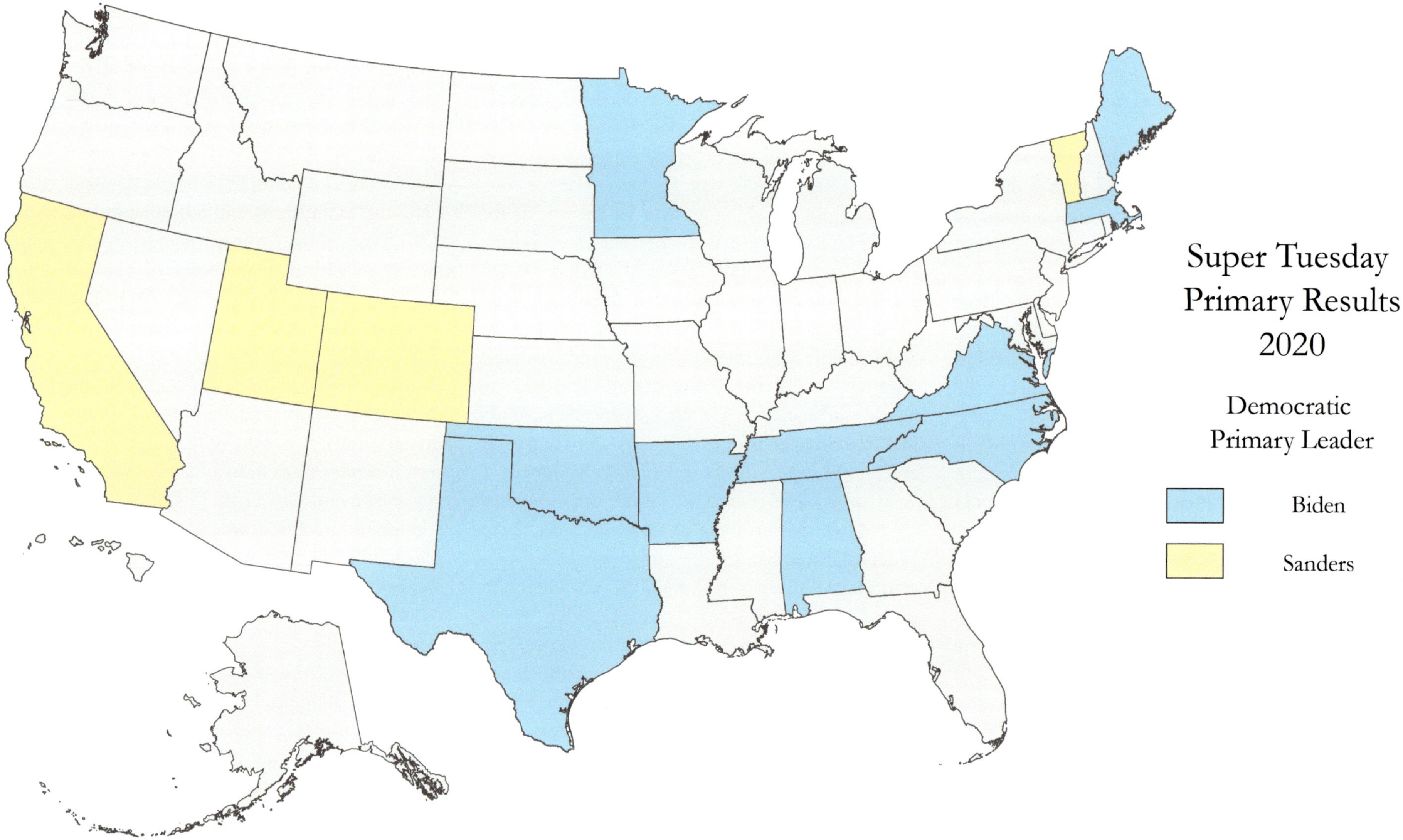
Super Tuesday
Primary Results
2020
Democratic
Primary Leader
Biden
Sanders

FIGURE 2.8

Texas Democratic Primary 2020

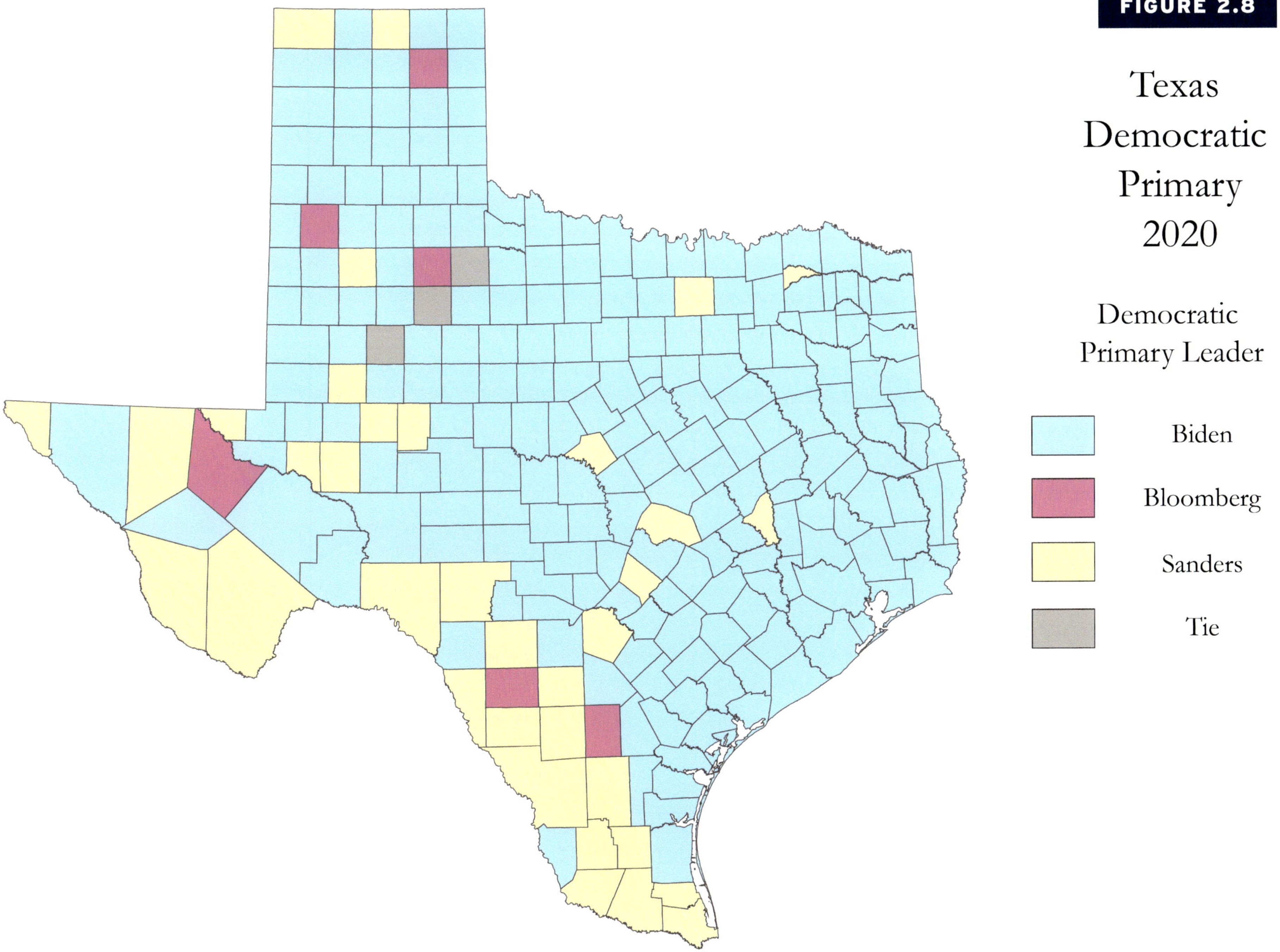

FIGURE 2.9

Massachusetts Democratic Primary 2020

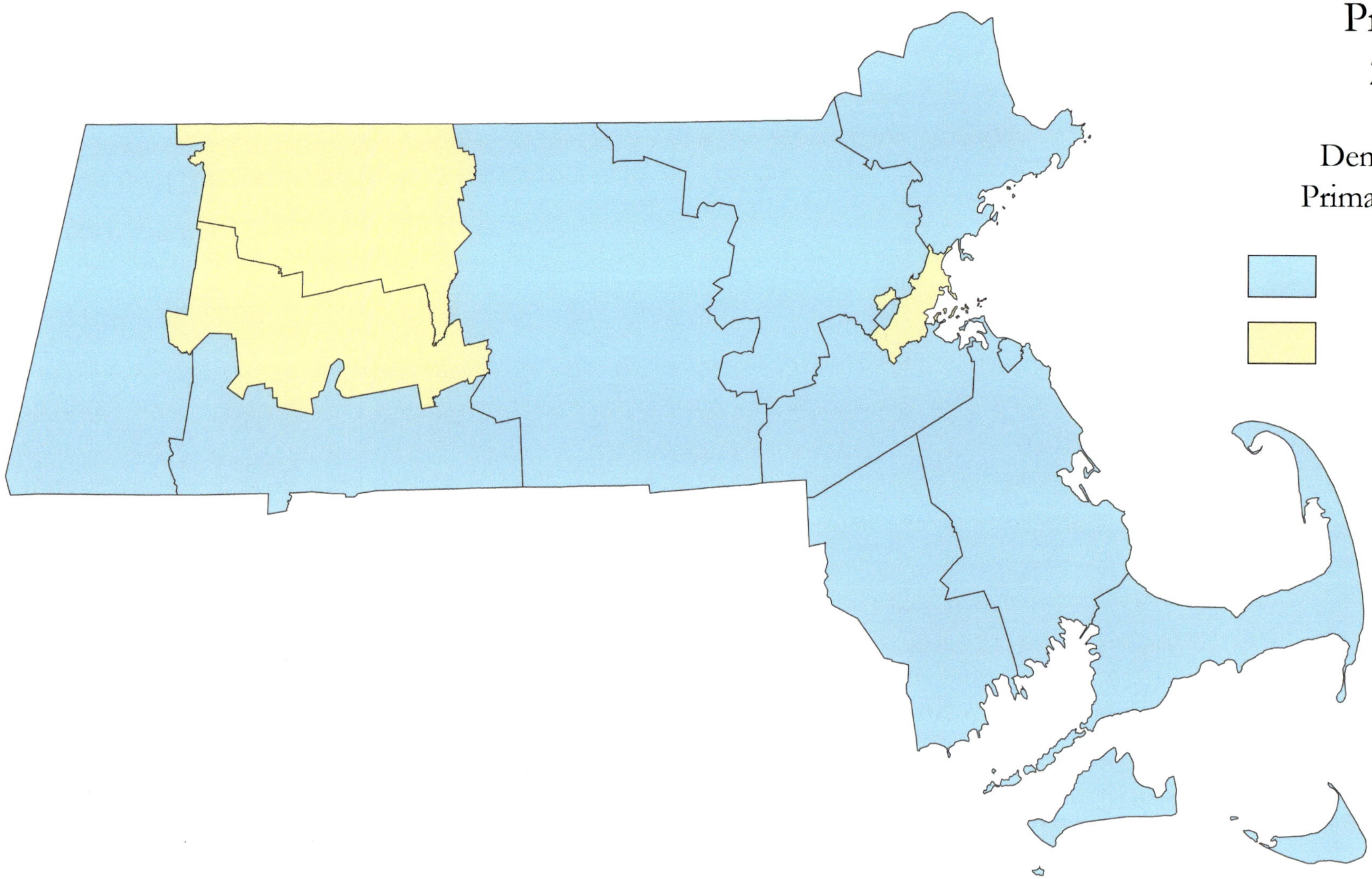

FIGURE 2.10

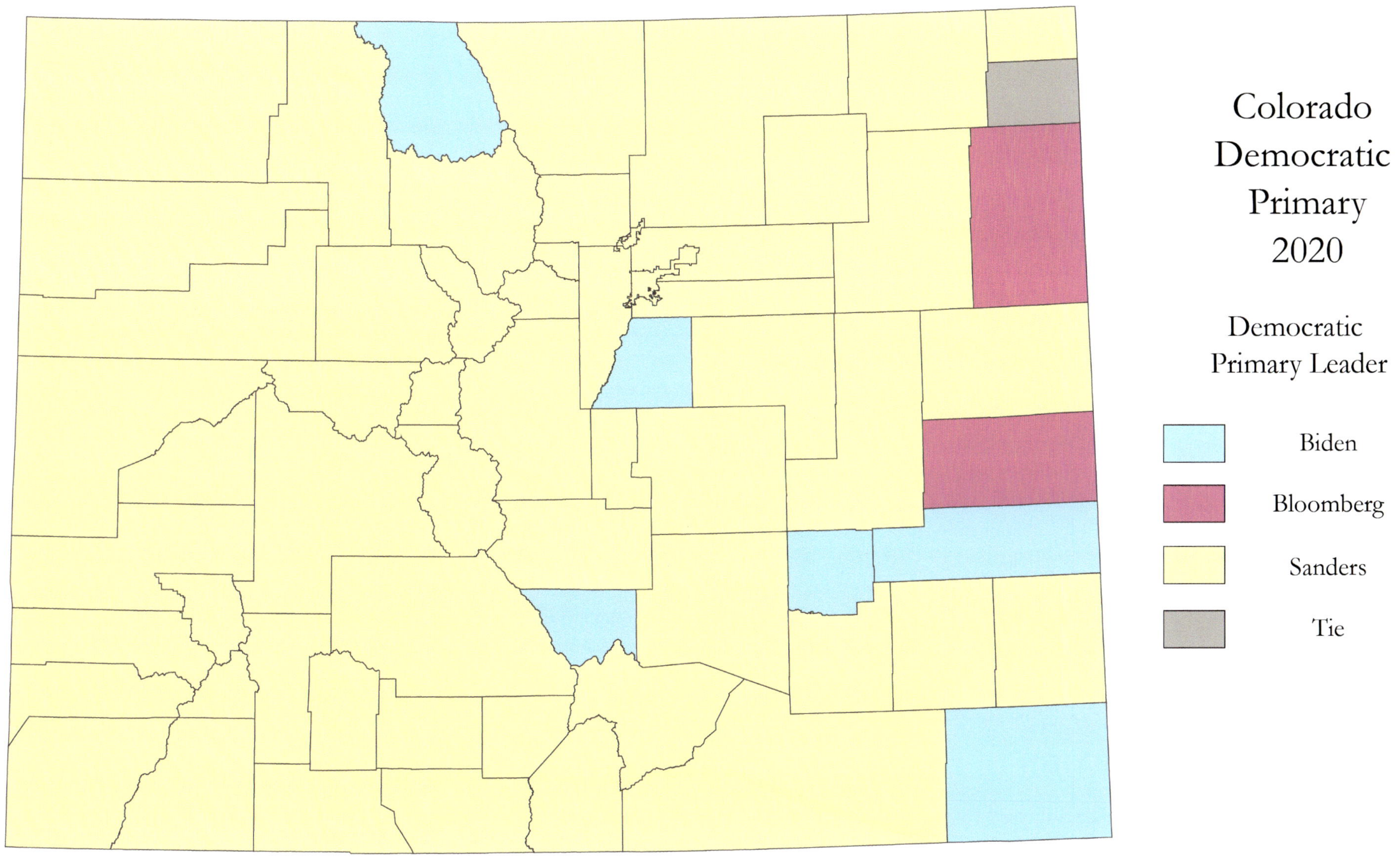

Colorado Democratic Primary 2020
Democratic Primary Leader
Biden
Bloomberg
Sanders
Tie

in the counties located on the Front Range, which contain 80 percent of the state's population and include the cities of Denver, Boulder, Colorado Springs, and Fort Collins.

After Super Tuesday, it was generally recognized that Biden was well on his way to the party's nomination. Warren withdrew after the Super Tuesday primaries, leaving Biden and Sanders as the major candidates remaining in the field. (Gabbard remained in the race but with little support outside of her home state of Hawaii.)

On March 10, the states of Idaho, Michigan, Mississippi, Missouri, and Washington held their primaries. Biden cemented his lead by winning all five of them. North Dakota held caucuses, which Sanders won, but this result had little impact on the nationwide outcome. Arizona, Florida, and Illinois followed suit a week later, and Biden continued his momentum by winning them. Biden clinched the nomination in early June, by which time he had amassed more than half of the pledged delegates (figure 2.11). Before Biden clinched the nomination, however, the COVID-19 pandemic that began early in 2020 was becoming a serious public health issue that would affect the rest of the campaign up through the general election in November.

THE COVID-19 PANDEMIC AFFECTS THE CAMPAIGN

A virus causing coronavirus disease triggered by severe acute respiratory syndrome was identified in China in December 2019 and named COVID-19 after its year of identification. The COVID-19 virus spread throughout China and then throughout the world in January and February. The first COVID-19 case in the United States was confirmed on January 20, with the first American fatality associated with the virus taking place on February 6. The virus spread quickly across the United States. By early March, at the time of the Super Tuesday primaries, more than a thousand Americans had been diagnosed with COVID-19. The number of confirmed cases reached one hundred thousand by the end of March, five million by the end of July, and more than twenty million by the end of 2020. The overall death toll reached one hundred thousand by the end of May, two hundred thousand by the end of September, and nearly four hundred thousand by the end of December.

In efforts to slow the pandemic, public health officials encouraged Americans to wear masks covering the mouth and nose in public, to stay at least six feet apart from other persons, and to avoid congregating in restaurants, bars, churches, and other public places. Some states, counties, and cities mandated the use of masks, while others refused to do so. Such mask mandates became a highly politicized issue, with Democrats much more likely to support mask mandates and to wear masks in public places, even if mask mandates were not in place. However, many Republicans opposed mask mandates, in part because they regarded such mandates as infringing on their civil liberties. As well, some skeptics regarded the numbers of confirmed cases and fatalities as overstated and inaccurate.

The rapid spread of COVID-19 during and after March had major effects on the later primaries and the national conventions. Twelve states postponed their primaries (figure 2.12). Additionally, four states replaced in-person voting with voting by mail for their primaries. The governors of Kentucky, New York, New Jersey, Delaware, and Connecticut issued executive orders allowing or implementing absentee voting for their states' primaries. Other states took legislative action to delay primary elections and focus on absentee or mail-in voting. The 2020 elections marked a shift in the ways in which voters cast their ballots, with many choosing to use absentee or mail-in voting for the first time. The increased use of absentee ballots and mail-in voting would prove to be a significant political issue after the 2020 general election, when Trump and other Republicans claimed that irregularities in these processes cheated him of victory.

For individuals who chose to vote in person for postponed primaries, the experience was unlike any election in recent history. Most voting sites required poll workers and voters to wear face masks. Many sites also put markers six feet apart on the ground indicating where voters should stand in line. There was an unprecedented level of sanitization, with pens being cleaned between voters and hand sanitizer stations throughout voting locations. Some polling places were forced to shut down because of a lack of poll workers, while others faced concerns about how to keep their poll workers safe. Many polling places were unable to turn away voters who had a fever or presented symptoms.

The pandemic also impacted the ways in which candidates of both parties interacted with voters. Large rallies with thousands of people crowded into stadiums or concert halls became a thing of the past, at least for the Democratic Party. Once Biden and Trump were confirmed as the Democratic and Republican nominees respectively, each candidate chose to interact with voters in his own way. Trump held his first major rally following the outbreak of COVID-19 in Tulsa, Oklahoma, on June 20. The rally took place in

FIGURE 2.11

Democratic Primary Winner by State, 2020

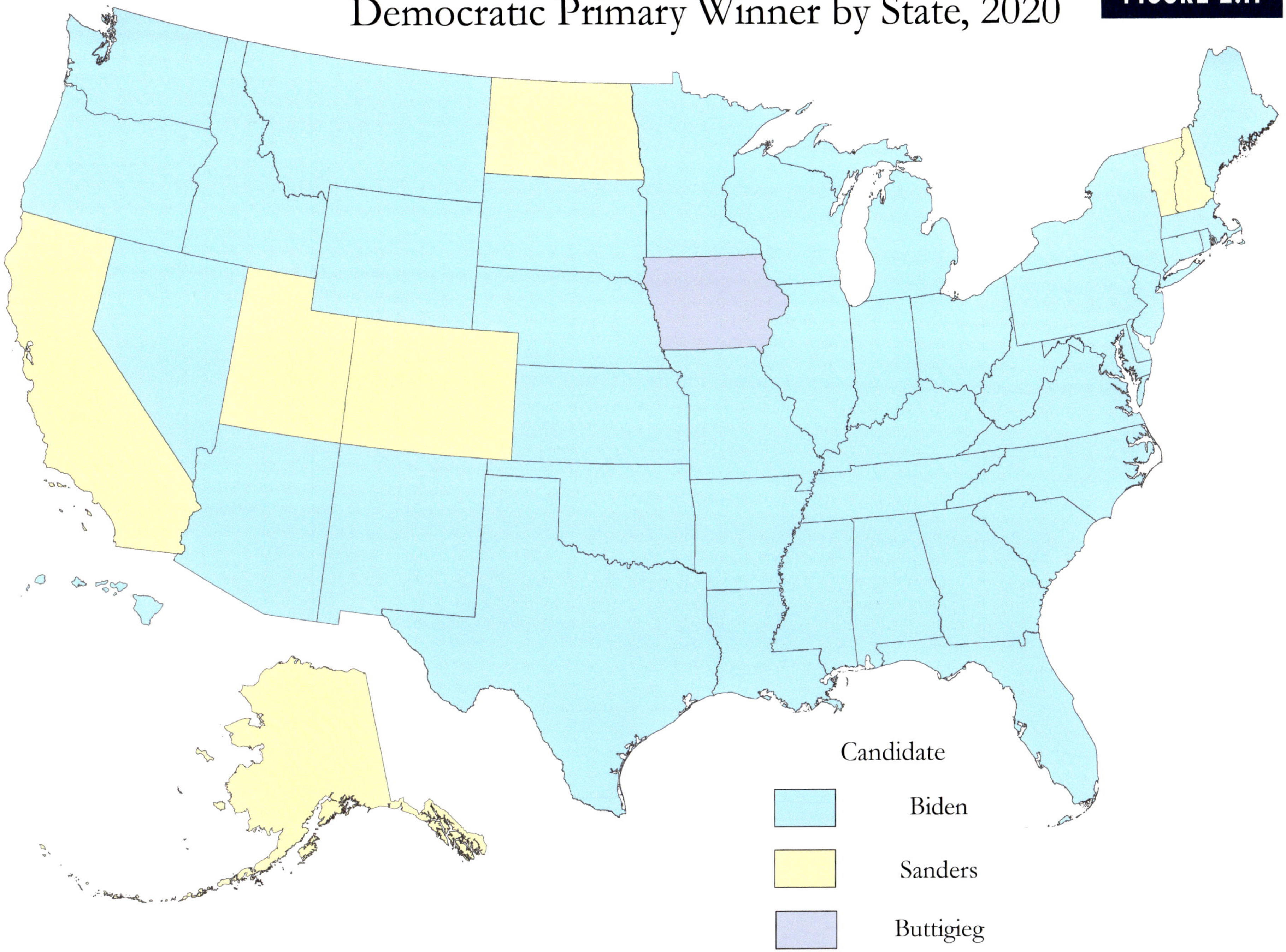

FIGURE 2.12

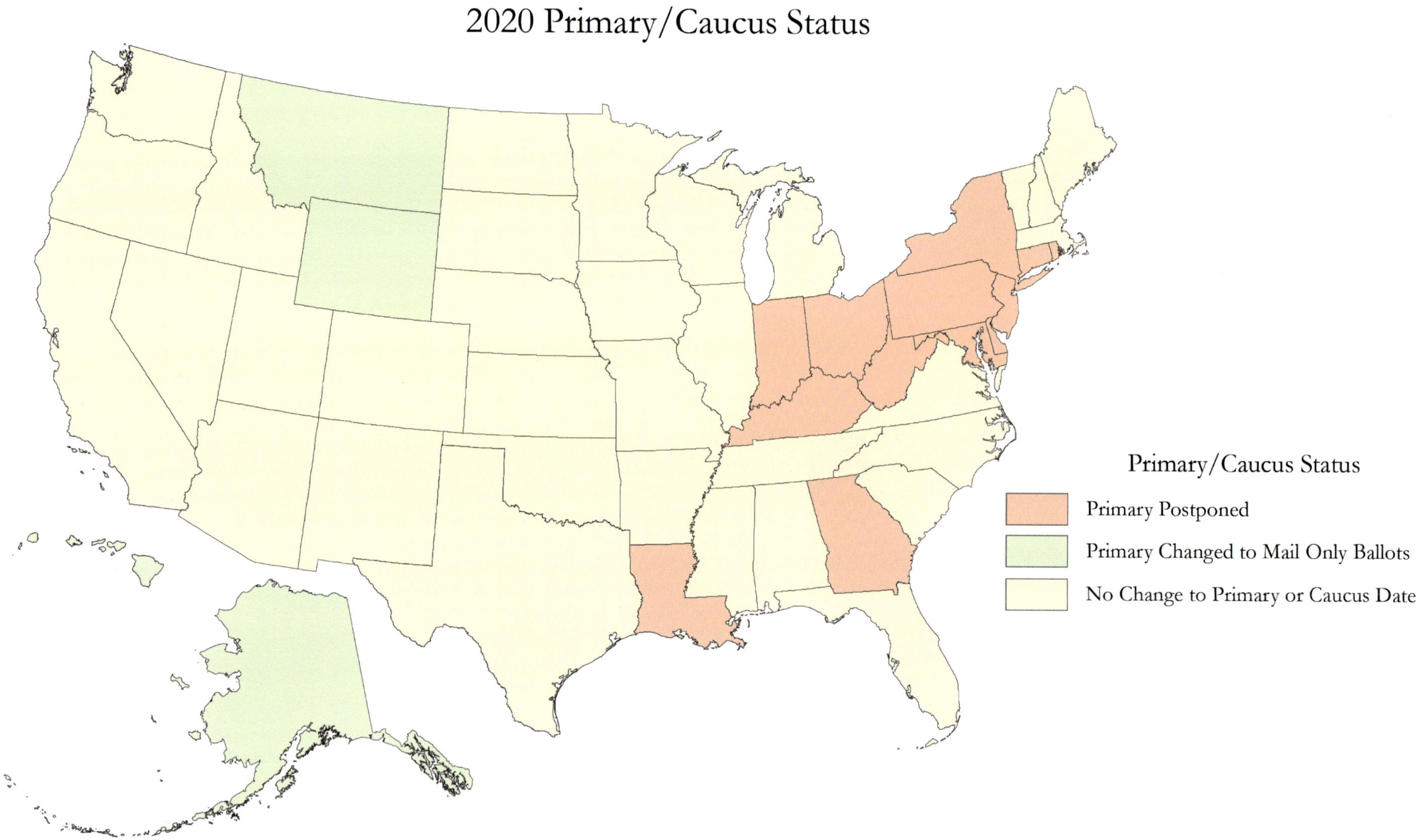
2020 Primary/Caucus Status
Primary/Caucus Status
Primary Postponed
Primary Changed to Mail Only Ballots
No Change to Primary or Caucus Date

a large arena used normally for basketball games and concerts. Many of the attendees did not wear masks, and a subsequent spike in COVID-19 cases in Tulsa was attributed in part to the rally. Among the fatalities was Herman Cain, who had been a candidate for the Republican nomination in 2012 and who attended the Tulsa rally without a mask. Trump also held online tele-rallies where supporters could listen to a live audio stream and access the recording afterward. Biden chose a different route, requiring social distancing and masks at his post-pandemic events. His campaign held socially distanced rallies that often took place in parking lots where supporters could remain in their cars.

SELECTING A DEMOCRATIC RUNNING MATE

Once Joe Biden had wrapped up his party's nomination, all eyes were on him to make a unique and glass-shattering choice as his vice presidential pick. Hopes were high that he would choose a woman, and especially a woman of color. On March 15, Biden announced that he was committed to selecting a woman as his running mate.

Biden's announcement generated considerable speculation about who might be chosen for the vice presidential nomination. Senators Harris, Klobuchar, and Warren were all considered, although Klobuchar announced in June that she did not wish to be considered and urged Biden to choose a woman of color. Other women reported to have been under serious consideration (again listed in alphabetical order) included Stacey Abrams, who had lost a close election for Governor of Georgia in 2018; Senator Tammy Baldwin of Wisconsin; Representative Karen Bass of California; Mayor Keisha Lance Bottoms of Atlanta, Georgia; Representative Val Demings of Florida; Senator Tammy Duckworth of Illinois; Governor Michelle Lujan Grisham of New Mexico; former US Ambassador to the United Nations Susan Rice of the District of Columbia; and Governor Gretchen Whitmer of Michigan. Many of these potential nominees are women of color. Abrams, Bass, Bottoms, Demings, Harris, and Rice are Black; Lujan Grisham is Latina; Duckworth is of partial Asian ancestry; and Harris has South Asian as well as African ancestry.

On August 11, Biden announced that Harris would be his running mate. Having launched a presidential bid herself, Harris chose to support Biden after she dropped out in December of 2019. Harris was a senator from California with a background as a prosecutor, district attorney, and attorney general of California. Harris was the only Black woman serving in the Senate at the time that she was chosen as Biden's running mate. The selection of Harris was widely supported by most Democrats, although a few progressives expressed skepticism based on Harris's background as a prosecutor.

VIRTUAL CONVENTIONS

Due to the pandemic, the Democratic National Convention was postponed from July to mid-August. The event was significantly scaled down, with delegates voting remotely and a largely virtual audience. This format provided a unique opportunity for the party to reimagine how the convention included voters and their voices from across the country. Speakers including Biden, Harris, several of the other potential nominees who had competed during the primaries, and other prominent Democrats such as Obama made their speeches and remarks remotely. A few prominent Republicans who had endorsed Biden, including former Governor John Kasich of Ohio, also spoke remotely to the virtual convention.

The voting process used to formalize Biden's nomination was also done remotely in a unique manner. Democratic Party leaders in each state prepared videos showcasing their states and announcing the numbers of delegate votes cast for each candidate from their states. These efforts were regarded by observers as successful and more meaningful than the traditional process by which delegates from each state read that state's results from the convention floor. Of particular interest were the videos produced in Alabama, Nebraska, and New Mexico. The Alabama video emphasized the state's history during the civil rights movement in the 1960s. Nebraska's video highlighted the state's rural character and the plight of family farmers, and New Mexico called attention to the state's large Native American and Latino populations.

The Republican National Convention had been scheduled for August 24–27 in Charlotte, North Carolina. Due to North Carolina's statewide restrictions limiting large crowds to prevent the spread of COVID-19, Trump announced in June that he would deliver his acceptance speech in Jacksonville, Florida. As COVID-19 cases continued to rise in Florida, Trump decided to change the location again and give his acceptance speech in front of a large crowd at the White House in Washington, DC. The official business of the Republican National Convention took place in Charlotte in a significantly scaled-down manner.

3 THE CAMPAIGN

THE 2020 CAMPAIGN

JOHN HEPPEN

Joe Biden secured enough delegates to be the Democratic nominee on June 6, 2020. By that date it was clear that two major issues would dominate the campaign—the societal and economic impacts of COVID-19 and the Black Lives Matter (BLM) movement accelerated by the death of George Floyd and others at the hands of the police. COVID-19, in addition to its impacts on public health, led to the worst economic crisis since the Great Depression, with unemployment hovering in the teens all spring and summer. After a break due to COVID-19, Trump held his first in-person rally at Tulsa, Oklahoma, on June 20. The original date of June 19 was changed because it fell on Juneteenth and Tulsa was the scene of the worst racial massacre of Black people in the country's history in 1921. Trump's event drew about 6,400 people in an arena that could have held over 19,000. The crowd was not required to wear masks or socially distance themselves. In his speech Trump used racist language to describe COVID-19, made calls for law and order, and attacked Muslim Representative Ilhan Omar of Minneapolis. Former Republican candidate for president in 2012 Herman Cain, who attended the rally, tested positive for COVID-19 on June 29 and died from it.

Trump faced many obstacles besides COVID-19. The economic slowdown, Wall Street crash, and general unease and conflict in society over the BLM movement and racism in the country hindered him. He was only the third president to ever be impeached. Like Andrew Johnson and Bill Clinton, he was not convicted in the Senate on February 5, 2020. Trump was accused of soliciting foreign involvement in his favor for his presidential campaign. His almost daily appearances on televised briefs on the coronavirus in March and April with Dr. Anthony Fauci and Dr. Deborah Birx frequently did more harm than good. He was prone to exaggerations and factual misstatements, and appeared in conflict with medical advice and expertise. Trump feuded with governors in Michigan and New York over state-mandated lockdowns, and he routinely downplayed the dangers of COVID-19 and blamed China for COVID-19. Those statements about China fueled prejudice against Asian Americans. Trump signed the CARES Act on March 27, which supported extended unemployment benefits, loans to small businesses, direct aid, and direct cash payments to individuals and families earning modest incomes. Millions of households received the checks in the mail and/or a letter from President Trump making sure to associate him with the relief checks. The economy did see historically low rates of unemployment and high levels of stock market gains before COVID-19, but the massive layoffs and stay-at-home orders throughout the country made those days seem like a distant memory to many voters.

Compounding his problems, former cabinet members including Defense, State, National Security, and Chief of Staff made statements questioning Trump's fitness for office, which was unprecedented. Trump based his reelection on a campaign of law and order, attacks on Joe Biden, stopping illegal immigration, support for the police and military, nationalism, social and cultural conservativism, attacking the Left as socialists, and his belief that the best economy in the world before COVID-19 was going to come roaring back. A Fox News Poll in conducted in June showed Biden with a lead of 12 percent nationally (Blanton 2020). Trump also faced opposition from a group of prominent Republicans who created the Lincoln Project in December 2019. The goal of the Lincoln Project was to prevent the reelection of Donald Trump because they regarded him as an apostasy to Republican Party values and not true to the legacy of Abraham Lincoln. Trump's rhetoric, fitness for office, foreign policy, moral character, general behavior, honesty, tweets, commitment to democracy, commitment to racial and gender equality, and suspected use of the office for his personal gain were prime issues for his opponents. Despite these questions, Trump did have unequivocal support from a large segment of the population among white voters and Christian conservatives, as well as Wall Street and Main Street Republicans.

Biden, though riding high in the polls in June, faced challenges. He was best known as Barack Obama's loyal vice president and Delaware senator before. Biden had failed in previous runs for the presidency. He is older than Trump by a few years and would be the oldest person to ever hold the presidency. He ran on a campaign based on bringing back normalcy, bipartisanship, confronting racism, following scientific advice on COVID-19, extending economic aid, and a fairer and more balanced post–COVID-19 economy with a more active role for the federal government. Trump routinely called him "Sleepy Joe" and questioned his stamina and mental awareness to be president. He was accused of sexual assault by former staffer Tara Reade in March but the allegations and heat from those accusations had faded by June with Biden making strong denials. His son Hunter Biden was accused of using his father's influence in his role on the board of directors of a Ukrainian energy company. Biden also faced suspicion from the left wing of the party given his past votes in the Senate on busing and the 1990s crime and welfare reform bills. There was concern that Bernie Sanders's most ardent supporters would not vote for Biden in the general election (Sunkara 2020). Biden based his campaign on minority voters, women voters, suburban voters tired of Trump, and those most directly hurt medically and economically by COVID-19.

Biden avoided the traditional in-person rallies. Biden generally had drive-in events where he would make a speech and supporters would watch socially distanced in their automobiles. He also campaigned extensively from his home in Delaware via television and webcasting. Typical was an event he held at the Minnesota State Fairgrounds on Friday, October 30, where invited supporters parked, and he made a speech from a stage. His campaign practiced all protocols to stop the spread of COVID-19. Trump on the contrary on September 18 in Bemidji, Minnesota, held a traditional rally with crowds in person with no social distancing and very few people wearing masks. Sixteen cases of COVID-19 were traced to the event by the Minnesota Department of Health (Vigdor 2020). Trump continued to have campaign events with thousands of people even after surviving COVID-19 and being hospitalized himself in early October.

The first debate held September 29 in Cleveland, Ohio, was universally described as disastrous and an embarrassment to the country with Trump's constant interruptions and attacks and the loss of control by the moderator, Chris Wallace of Fox News. Biden called Trump the worst president in history. Trump's refusal to explicitly denounce the white supremacist group the Proud Boys was condemned. He claimed to not know who they were and told them to "stand back and stand by." The awfulness of the debate, though, did not alter the course of the race as Biden continued to hold a rather large lead in national polls. A poll conducted by CNN from October 1 to October 4 after the debate gave Biden a fourteen-point lead (Stevens 2020).

The vice presidential debate was held on October 7 at the University of Utah in Salt Lake City. California Senator Kamala Harris was the first non-white woman nominated for the vice presidency by a major party. She had battled Biden during the primary season but became his running mate to solidify support from women, Black voters, and Asian American voters. Vice President Pence of Indiana remained on the ticket to maintain support from conservative Christians. Both vice presidential choices were there to cement support from each party's most loyal constituents. The debate was not nearly as raucous as the first debate. Nothing dramatic from the debate changed the nature of the race; the vice presidential debates in this century have been relatively uneventful.

The second debate was canceled on October 9; Trump suffered from COVID-19 and was hospitalized from October 2 to the fifth. Trump refused to participate in a virtual or online debate as a makeup. Instead, both candidates held town halls at the same time on Thursday, October 15. Trump was broadcast on NBC and Biden was on ABC. Biden's event achieved higher ratings with 1.6 million more viewers (Grynbaum and Koblin 2000). The final debate was held October 22 in Nashville, Tennessee, and was considered more civil with much less of the rancor of the first debate. Polling data conducted after the debate gave Biden a victory in the debate (Elbeshbishi 2020). As in 2016, the debates and the behavior of the candidates seemed to do little to move the polls or change voters' opinions.

During the last twelve days of the campaign, Trump made thirty-seven visits over thirteen states and concentrated his visits in Michigan, Pennsylvania, and Wisconsin (figure 3.7). It was to those states with their 77,774 votes and the forty-six electoral votes that pundits credited Trump's 2016 victory. He made nine stops in the Quaker State and six in the Great Lakes State in 2020. Biden in this last twelve days visited those three states in addition to six others for a total of only nine states and twenty-one visits, with eight in Pennsylvania alone and two in tiny Delaware (figure 3.6). Trump's strategy to campaign like 2016 more than Biden did not pay dividends. Perhaps Biden's more cautious strategy illustrated to voters a greater concern regarding the COVID-19 crisis versus the more denying Trump.

Much of the voting was done by mail via absentee or early voting throughout the country due to COVID-19 concerns. There was no announced winner on election night or early the next morning due to the high volume of votes that could not be counted until after the polls closed. On November 7, 2020, Biden was declared president-elect by major news organizations after days of counting ballots. The call of Pennsylvania for Biden after days of counting mail-in ballots gave Biden at the time 273 electoral votes with four states remaining and three of them with Biden in the lead. He eventually won 306 electoral votes with Georgia being razor-thin in his favor. The final national popular vote was Biden with 51.26 percent to Trump's 47.80 percent. Most polls had Biden with an eight-to-ten-point lead going into Election Day, so the polls missed it again, but this time the polls predicted the winner. By November 16, Biden was universally recognized by all mainstream media and Fox News as the president-elect despite Trump refusing to concede, although he did tweet on November 15 that Biden had "won" but quickly took back that claim. Trump made claims that he won on Election Day and that millions of illegal or fraudulent ballots by mail was why he lost. Despite the events of January 6, Joe Biden was inaugurated as the forty-sixth president of the United States on January 20, 2021, ending a campaign like no other.

REFERENCES

Blanton, D. 2020. Fox News Poll: Biden Widens Lead over Trump; Republicans Enthusiastic, but Fear Motivates Dems. June 18, https://www.foxnews.com/politics/fox-news-poll-biden-widens-lead-over-trump-republicans-enthusiastic-but-fear-motivates-dems (last accessed June 7, 2021).

Elbeshbishi, S. 2020. Majority of Voters Say Biden Won Second Debate, Poll Finds. *USA Today*. October 25, https://www.usatoday.com/story/news/politics/elections/2020/10/24/biden-beat-trump-second-presidential-debate-majority-says-poll/6025522002/ (last accessed June 7, 2021).

Grynbaum, M. R., and J. Koblin. 2020. Biden Beats Trump in Ratings Battle of the Network Town Halls. *New York Times*. October 26, https://www.nytimes.com/2020/10/16/business/media/biden-trump-town-hall-ratings.html (last accessed June 7, 2021).

Stevens, M. 2020. A CNN Poll Taken after the Debate Finds Biden Leading Trump by 16 Points Nationally. *New York Times*. October 6, https://www.nytimes.com/2020/10/05/us/elections/cnn-poll-biden-trump.html (last accessed June 7, 2021).

Sunkara, B. 2020. You've Probably Heard Socialists Won't Vote for Biden. *New York Times*. May 28, https://www.nytimes.com/2020/05/28/opinion/socialists-biden-trump.html (last accessed June 7, 2021).

Vigdor, N. 2020. Minnesota Connects Two Dozen Coronavirus Cases with Campaign Events, Most of Them Visits by Trump. *New York Times*. October 13, https://www.nytimes.com/2020/10/13/us/elections/minnesota-connects-two-dozen-coronavirus-cases-with-campaign-events-most-of-them-visits-by-trump.html (last accessed June 7, 2021).

DONORSHEDS: INDIVIDUAL CONTRIBUTIONS TO BIDEN AND TRUMP

CARL T. DAHLMAN

An estimated $5.7 billion was spent on the 2020 presidential race. This total includes nearly $1 billion spent by outside interests, including corporations and special interests. Some of this outside spending comes from "dark money" that entered politics following the Supreme Court's 2010 *Citizens United* decision. The identities of dark money donors are not reported to the Federal Election Commission (FEC). Outside spending now rivals traditional campaign contributions made by US citizens through donations to specific candidates or political action committees. The Center for Responsive Politics estimates that about $300 million was spent by outside groups—including super political action committees (PACs) and dark money—against each party's presidential candidate. It is revealing that about $387 million was spent by outside groups in favor of Biden's election but only $53 million was spent in favor of Trump.

Individual donations are nonetheless a vital source of campaign funding even though they are capped by law at $2,800 per person to any one candidate. All donations over $200 require full disclosure of the donor's name and address to the FEC. In the 2020 election cycle, direct individual donations to all presidential candidates, in both primary and general elections, totaled about $2 billion. Overall, the race cost the candidates approximately $4 billion, about two and a half times more than the sums spent in 2016. Making up the difference between individual contributions and total expenditures was money transferred from PACs or contributions by candidates to their own campaigns. Two Democratic primary candidates broke records in the amounts they gave to their own campaigns: Michael Bloomberg spent $1.1 billion, and Tom Steyer spent $342 million. Bloomberg, in fact, spent more of his own money than Biden raised in total from donors.

Individual donations can tell us something about the level of support and enthusiasm for candidates. In the 2020 election cycle, more than four million individual donors gave Joe Biden $823 million, the largest such sum among all primary and general election presidential candidates. This was a 30 percent increase over individual contributions to the Democratic nominee in 2016. Biden received the larger share of these contributions in the form of large donations: 61 percent of all funds donated to his campaign were in amounts of $200 or more. In 2020, women were 45 percent of all individual donors, an 8 percent increase over 2016, and they formed more of Biden's donor base (48 percent) than Trump's (37 percent). Combined with contributions and transfers from other committees, Biden was the first candidate in history to break the billion-dollar mark in campaign financing, having raised a total of $1.074 billion from donors.

Donald Trump raised $458 million in individual contributions from about two million donors. This represented about 55 percent of the money Biden received and less than one-third of what all Democratic candidates received in total. There were also fewer than half as many individual donors to Trump than to Biden. As the incumbent, Trump had no significant fundraising challenges from other Republicans. Trump received 49 percent of individual contributions to his campaign in amounts smaller than $200, a larger share than Biden but down considerably from what Trump raised in 2016. Combined with contributions and transfers from other committees, Trump's campaign raised a total of $744 million for his reelection campaign.

Mapping "donorsheds" of where campaign contributors live reveals a pattern similar to a map of the US population weighted by income. The most important donorsheds are the wealthier neighborhoods of major metropolitan areas. These donors are a key part of national political campaigns, yet they represent a small fraction of the population and an even smaller cross-section of American communities. Donors giving at least $200 to any federal election in the 2020 election cycle were only 1.8 percent of the US adult population. This was nearly three times the rate of the 2016 election, however, a level of public interest reflected in record-breaking voter turnout in 2020.

What is, perhaps, more interesting is to compare the donorsheds of the parties' presidential candidates. Using FEC data, it is possible to build maps of itemized contributions at the state level (figure 3.1). More detailed data mapping by metropolitan area and zip code reveals that while

FIGURE 3.1

Donorsheds: Reported Campaign Contributions January 1, 2019 to December 31, 2020

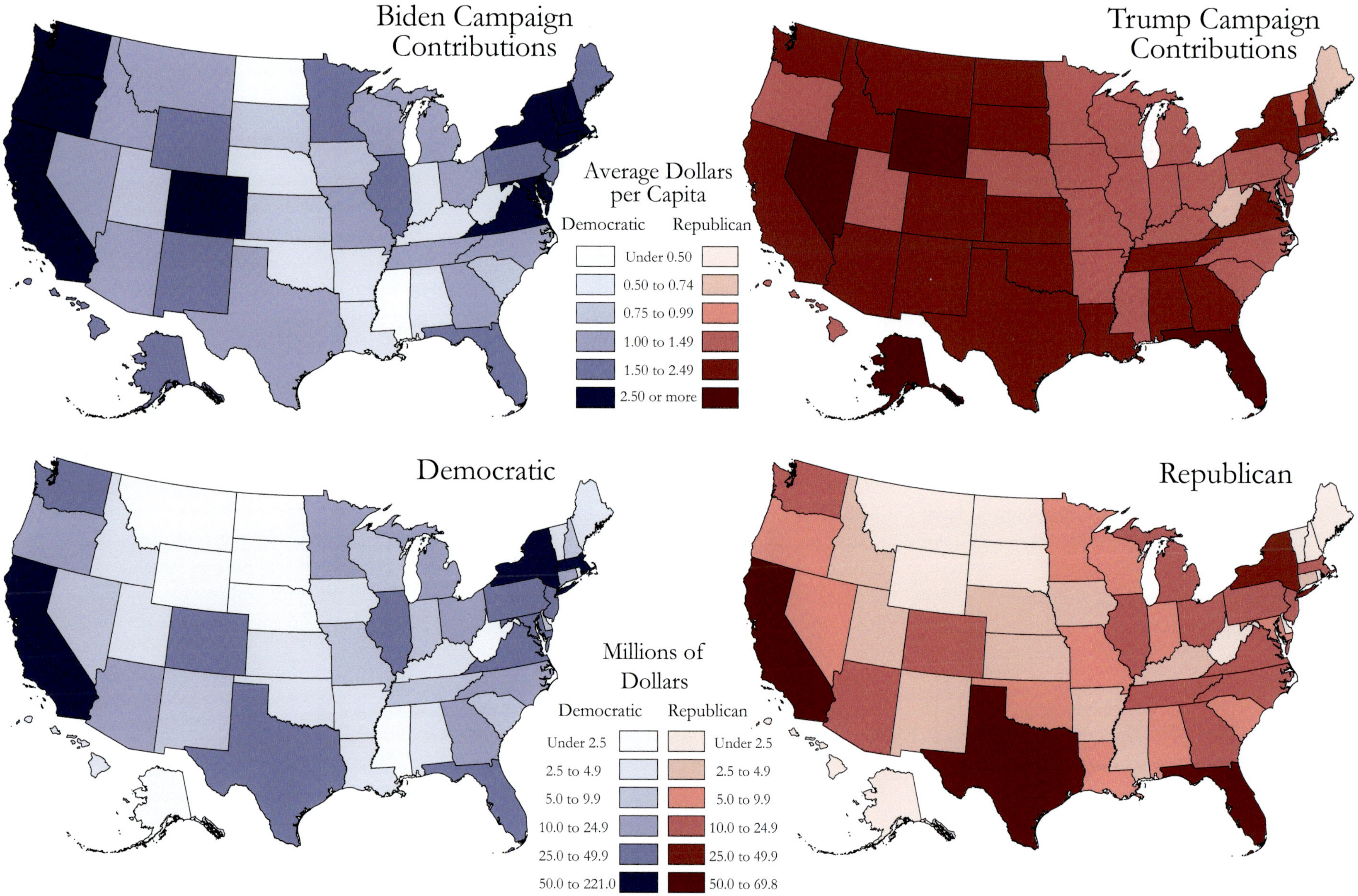

candidates seem worlds apart politically, their largest donors often live quite near one another. Smaller donors, on the other hand, are spread across the country and are important as a whole for presidential campaigns, but individually they do not attract special attention from candidates. Fundraising dinners in Silicon Valley or Manhattan can raise more money than mailers to a thousand farming communities.

At the state level, donors in California, Texas, New York, and Florida contributed the largest sums to both presidential candidates in 2020. The donorsheds for Biden and Trump contrasted more sharply at the local level, however. Repeating a pattern from earlier elections, donations to conservatives generally come from suburban and rural areas while liberals do better in urban areas and wealthy suburbs. This largely reflects voting patterns, although the largest donors to either party often live in the same exclusive neighborhoods.

California provided the largest state totals for donations to both Biden and Trump campaigns, albeit in much different amounts. Californians donated $184 million directly to Biden while Trump received only $85 million. The difference was even greater when comparing the total donations given to the candidates from individuals and PACs, combined. Biden raised $305 million in the Sunshine State, more than three times what Trump collected there. Texas donations totaled less than half those of California and split for Trump $72 million to Biden's $45 million. Trump was able to raise an additional $13 million from Texas PACs. Biden's success in New York easily overcame Trump's totals in Texas and Florida, however. In New York, Biden raised a total of $221 million from individuals and PACs, outperforming Trump in his home state.

Biden led Trump in fundraising across key metropolitan areas. Biden dominated the coastal metropolitan regions and his top five sources for individual contributions were Washington, DC ($67 million), New York ($63 million), Los Angeles ($48 million), Boston ($38 million), and San Francisco ($38 million). Biden raised an addition $377 million from PACs in these five cities and in Chicago he outperformed Trump by about three to one. By contrast, Trump's major donorsheds produced smaller yields: Los Angeles ($16 million), Houston ($14 million), Dallas ($12 million), Atlanta ($12 million), and Phoenix ($12 million). PAC activity increased his total take in Las Vegas ($98 million), West Palm Beach ($35 million), and Washington, DC ($34 million).

The staggering amounts of money that entered the 2020 presidential campaign were joined by another $4 billion in the House and Senate campaigns that year. While individual contributions reflect authentic citizen participation in the political process, the limitless spending now open to America's wealthiest families and corporations raises serious questions about whether access and influence have been taken away from the many in favor of the few. Since the Supreme Court in *Buckley v. Valeo* (1976) ruled that campaign money is free speech, regulation has been made harder. The FEC, established after Watergate, has been increasingly stymied by partisan dysfunction. Congressional efforts at reversing these trends, such as H.R. 1 "For the People Act," seem unlikely to gain support among lawmakers who benefit from the current campaign finance system.

REFERENCES

Center for Responsive Politics. 2021. 2020 Presidential Race [including data on individual candidates, donor geography, and donor demographics]. Washington, DC. https://www.opensecrets.org/2020-presidential-race/ (last accessed May 25, 2021).

Evers-Hillstrom, K. 2020. Donors, Big and Small, Propel Biden to Victory. Washington, DC: Center for Responsive Politics. https://www.opensecrets.org/news/2020/11/biden-wins/ (last accessed May 25, 2021).

Federal Election Commission. 2020. *Presidential Candidate Data Summary Tables*. Washington, DC. https://www.fec.gov/campaign-finance-data/presidential-candidate-data-summary-tables/?year=2020&segment=24 (last accessed May 25, 2021).

CAMPAIGN EXPENDITURES IN THE 2020 PRESIDENTIAL ELECTION

CHRIS MAIER

With each election, presidential contests continue to get more expensive, and the 2020 presidential election, between incumbent Republican President Donald Trump and his Democratic challenger and former Vice President Joe Biden, was no exception. What is interesting about campaign expenditures in the 2020 presidential election was the change in narrative that occurred around spending.

Early in 2020, President Trump's campaign was flush with cash. This was noted by President Trump's campaign manager, Brad Parscale, who, in early May of 2020, referred to the campaign as the Death Star that just needed to be unleashed (Gabbatt 2020). While President Trump started the 2020 election cycle on solid financial grounds, the same cannot be said for Biden. After a fourth-place finish in the Iowa caucuses and fifth-place finish in the New Hampshire primary, the first two states to vote in the 2020 Democratic presidential primary, Biden's campaign was on rocky ground. Reports, after the significant defeat in New Hampshire, questioned whether the campaign had the financial resources to continue (Schwartz 2020; Viser, Wootson, and Scherer 2020). Biden's campaign, however, persevered, made it to the South Carolina primary, and came away with a dominant victory, which propelled him to a successful day on Super Tuesday and the Democratic nomination. As the 2020 campaign transitioned to the general election, the financial picture for the Biden campaign improved and became a strength as the former vice president eventually outspent the incumbent president.

During the 2020 presidential election, President Donald Trump and former Vice President Joe Biden spent over $650 million during the primary season and another $1.2 billion in the general election. In total, the two candidates spent 1,866,095,890 on the 2020 presidential election (Federal Election Commission, Biden Trump 2020; Federal Election Commission, Trump 2020). The calculated spending totals in this section do not include each candidate's itemized deductions, nor do they include outside spending from organizations such as super PACs.

President Trump spent roughly $313 million during the 2020 Republican primaries (Federal Election Commission, Trump 2020). Figure 3.2 shows the expenditures for Trump's primary campaign. As the incumbent, Trump ran a different primary campaign than his Democratic opponent, Joe Biden. While Trump did not run unopposed in the primary, none of the other candidates posed much of a threat to the Trump campaign, which allowed the president to direct his attention to the general election. While much of Trump's primary spending was in Virginia, he spent money in forty-nine of fifty states, ranging from a low of $2,000 in Vermont to a high of $206,076,877 in Virginia. Roughly 65 percent of Trump's primary spending occurred in Virginia, which was home to his campaign headquarters in Arlington. South Dakota was the only state that President Trump's campaign did not spend money in during the primary season.

During the 2020 Democratic presidential primaries and caucuses, Biden's campaign spent $337,584,294, which placed him just ahead of President Trump as the campaigns moved into the general election (Federal Election Commission, Biden 2020). Figure 3.2 shows the expenditures for Biden's primary campaign. Biden faced a much more competitive primary; he faced strong competitors in Senator Bernie Sanders, Senator Elizabeth Warren, Senator Amy Klobuchar, and former mayor Pete Buttigieg. He also faced two wealthy businessmen in Michael Bloomberg and Tom Steyer. Biden's campaign spent roughly a third of his money ($121,888,336) in Georgia, which was home to two competitive senate races as well. Georgia is also home to Media Buying & Analytics LLC, which is a company located in Atlanta that received many large expenditures from the Biden campaign. In addition to Georgia, Biden spent significant money in Pennsylvania, home to his campaign headquarters and a vital swing state in the general election, as well as traditional Democratic strongholds in New York and California. Biden spent money in all fifty states, ranging from $5,513 in South Dakota to $121,888,336 in Georgia.

Figure 3.2 shows the expenditures for President Trump's 2020 general election campaign, which totaled $421,176,546. Unlike the primary season, President Trump's campaign spent

FIGURE 3.2

Reported Primary and General Election Expenditures

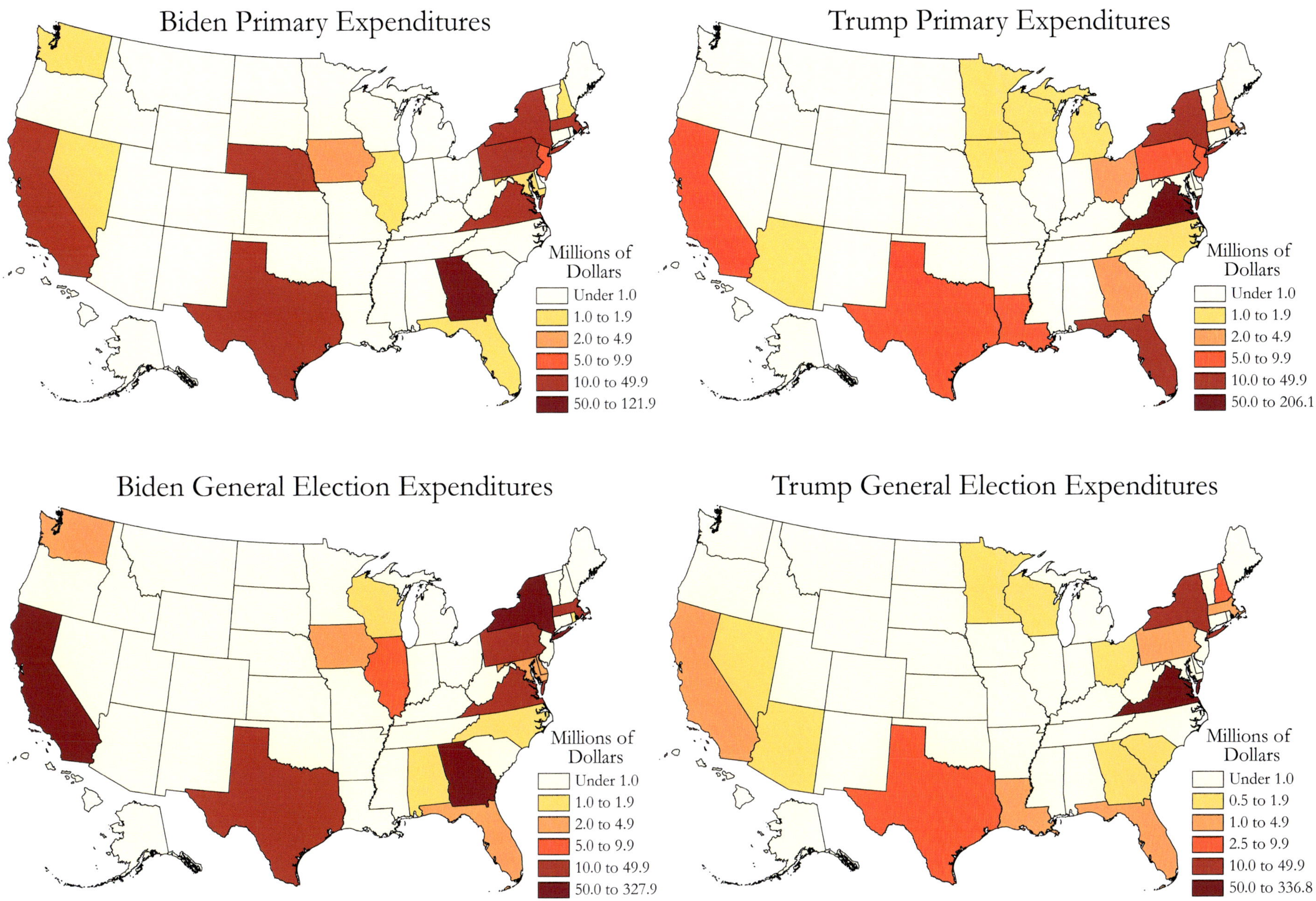

money in all fifty states, with the most money ($336,839,271) spent in Virginia. In addition to Virginia, President Trump's campaign spent significantly in Washington, DC, which hosted part of the Republican National Convention, and New York. Another geographic pattern seen in President Trump's campaign expenditures are the large outlays in "battleground" or "toss-up" states, including Arizona, Florida, Georgia, Nevada, New Hampshire, Pennsylvania, Texas, and Wisconsin (Federal Election Commission, Trump 2020).

Figure 3.2 shows Joe Biden's expenditures for the 2020 general election campaign, which totaled $794,235,709. With plenty of money to spend, Vice President Biden spent money in all fifty states. The geographic pattern created by Biden's largest expenditures in the general election was similar to the primary season, with significant money spent in Georgia ($327,877,201), Washington, DC, New York, and California. As was the case with Trump, Biden focused his spending on battleground states, including Florida, Iowa, North Carolina, Pennsylvania, Texas, and Wisconsin (Federal Election Commission, Biden 2020). For Biden, it was these battleground states, especially Arizona, Georgia, Pennsylvania, and Wisconsin, that swung the election for Joe Biden.

REFERENCES

Federal Election Commission, Biden. 2020. Operating Expenditures. https://www.fec.gov/data/committee/C00703975/?cycle=2020 (last accessed June 24, 2021).

Federal Election Commission, Trump. 2020. Operating Expenditures. https://www.fec.gov/data/candidate/P80001571/?cycle=2020&election_full=true (last accessed June 24, 2021).

Gabbatt, Adam. 2020. Trump's "Juggernaut" Reelection Bid Like the Death Star, Campaign Boss Boasts. *The Guardian*. May 7.

Schwartz, Brian. 2020. Some Joe Biden Donors Are Panicking after New Hampshire Primary Flop, But Others Remain Hopeful. *CNBC*. https://www.cnbc.com/2020/02/11/biden-donors-divided-on-his-prospects-after-new-hampshire-primary-loss.html (last accessed June 23, 2021).

Viser, M., C. R. Wooton Jr., and M. Scherer. 2020. His Campaign on the Line, Joe Biden Goes Missing in New Hampshire. *Washington Post*. February 6.

PRESIDENTIAL CANDIDATE CAMPAIGN TRAVEL DURING THE 2020 PRESIDENTIAL ELECTION CAMPAIGN

J. CLARK ARCHER AND JILL A. ARCHER

The US Constitution requires a new federal contest every four years to select electors who in turn cast ballots to fill the highest executive offices of president and vice president through an indirect Electoral College process. Since the mid-nineteenth century, members of the Electoral College have been popularly elected in each state. But the overall competition is still a "federal" election and not a "national" election. There have been hundreds of amendments proposed to modify or abolish the Electoral College, but every one has failed save for the Twelfth Amendment ratified in 1804 to require separate electoral votes for president and vice president.

The fact that electors are popularly elected within each state and then their electoral votes are summed over all states together has profound consequences for the conduct of a presidential election campaign. In a national election, one more voter's approval would be equally sought after, regardless of where within the territory of the United States that vote might be cast. But in an indirect federal election, one more voter's approval can be far more valuable in one state than in another state. Most states follow a "winner-take-all" rule that allocates all their electoral votes to the candidate who receives the most popular votes, whether a majority or only a plurality. The only modern exceptions are Maine and Nebraska, which split electoral votes between congressional districts.

The state-level maps in figure 3.3 illustrate the most important numerical quantities relevant to campaign planning for each new federal presidential election. These are the numbers of electoral votes apportioned to each state after the most recent US Census of Population, the state-level electoral vote winners in each state in the most recent presidential election, and the numbers of popular votes that separated winning from losing candidates in each state in that election. These quantities, of course, are widely known at the beginning of each new election cycle. Polls and other later information add surface texture to the process, but the underlying bedrock factors are the apportionment of electoral votes and the popular and electoral outcomes of the preceding presidential election. The relevance of the preceding election was unusually great in 2020, because the incumbent Republican President Donald Trump had lost the national popular contest in 2016 by more than 2.8 million votes but won a fragmented Electoral College tally by 304 electoral votes for Trump to 227 electoral votes for Hillary Clinton and 7 "faithless" electoral votes for other candidates (Watrel et al. 2018). The outcome in 2016 printed a very large boldface exclamation point on the difference between a national election and a federal election.

The top map in figure 3.3 shows the numbers of electoral votes held by each state after the 2010 Census of Population, and the popular vote leader in each state in the 2016 election (ignoring faithless electors). The bottom map in figure 3.3 shows states classed by the difference in the numbers of popular votes for Clinton and Trump expressed as percentages of the total votes cast in each state in the 2016 election. The categories are 15 percent or greater advantages for Clinton or Trump, 5–14 percent advantages for Clinton or Trump, and no more than a 5 percent advantage for either candidate. Darker shades of blue or red used for the 15 percent or greater categories represent "very safe states," medium shades of blue or red represent "somewhat competitive states," and a contrasting green is used for the less than 5 percent difference category to highlight "tossup" or "battleground states."

A bit of a harbinger of irregularities and inconsistencies during the Trump administration was the finding reported in the *Atlas of the 2016 Elections* that there were comparatively strong correlations between county-level changes in the major-party vote split from 2012 to 2016 and percentages of adults with only a high school education versus those with a BA degree or above education (Watrel et al. 2018, 142–62). Previous survey evidence had indicated slightly greater support among college graduates for Republican Romney than for Democrat Obama in the 2012 election (Archer et al. 2014, 160–63). But from 2012 to 2016 vote split changes in counties with higher percentages of adults with only a high school education favored Republican Trump while vote split changes in counties with higher percentages of adults with a BA degree or above educational attainment favored Democrat Clinton. As discussed more fully in a

FIGURE 3.3

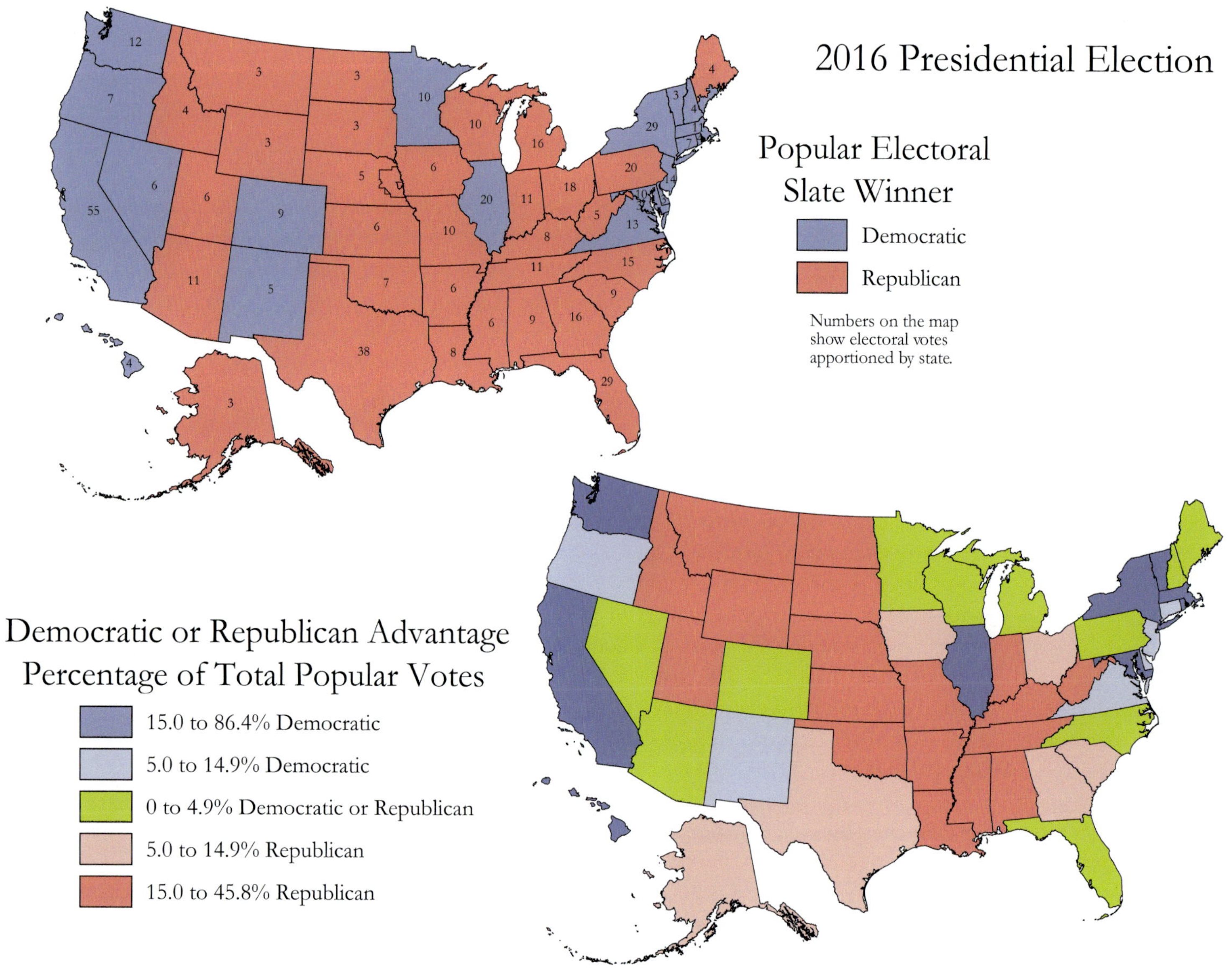

2016 Presidential Election
Popular Electoral Slate Winner
Democratic
Republican
Numbers on the map show electoral votes apportioned by state.
Democratic or Republican Advantage Percentage of Total Popular Votes
15.0 to 86.4% Democratic
5.0 to 14.9% Democratic
0 to 4.9% Democratic or Republican
5.0 to 14.9% Republican
15.0 to 45.8% Republican

later section of this atlas, the county-level correlation between percent voter support for Trump in the 2020 election and percent of adults with a BA degree or above was found to have a strongly negative value of -0.537, which is one of the largest values found for any demographic variable. Trump Republicans thus seem quite different from the Lincoln Republicans who enacted the Morrill Act initiating federal support for higher education in 1862. The "fake news" alternative reality promoted by Trump to cover his failings and shortcomings evidently repelled more highly educated voters even more strongly in 2020 than in 2016.

Another unanticipated influence on the 2020 election campaign was the COVID-19 pandemic. *Washington Post* reporter Bob Woodward reported that Donald Trump was alerted during a daily briefing on January 28, 2020, to the emergence of a mysterious pneumonia-like virus in Wuhan, China, and warned by National Security Advisor Robert O'Brien that "This will be the biggest national security threat you face in your presidency" (Woodward 2020, xiii). Trump later told Woodward (2020, 384) in an interview on July 8, 2020, that "we were riding high. The election was over. I was going to win easily. And all of a sudden we got hit with the China virus." A great deal of fake news was generated to try to divert the influence of the virus away from the election. There was not even agreement about whether wearing a mask to try to slow the rate of spread of a highly infectious airborne virus was "American" or "Un-American" or "Democratic" or "Republican," although a partisan divide did develop with Democratic candidates tending to be more cautious than Republican candidates on the matter. Indeed, Democratic rivals Bernie Sanders and Joe Biden both canceled new in-person primary campaign appearances after March 10, even though the Democratic presidential nomination was still undecided. Polling data at the time indicated about 75 percent of Democrats regarded COVID-19 as a "real threat," compared with only 40 percent of Republicans (Wikipedia 2021).

Pandemic concerns were great enough for the White House, albeit grudgingly, to caution against "unnecessary gatherings of more than ten people" while at the same time issuing very mixed and contradictory messages about mask wearing. Various state and municipal governments issued "stay-at-home" orders, and many schools and colleges shifted to "remote" rather than "in-person" instruction.

The cumulative number of COVID-19 cases across the United States surpassed one million by the end of April 2020 and exceeded 9.1 million by the time of the US presidential election on November 3, 2020. The map in figure 3.4 shows the cumulative rate of COVID-19 cases per 10,000 population by county from January 21 to November 3, 2020. This period spans the primary elections, the nominating conventions, and the general election campaign during the 2020 presidential election cycle. The map is based on county-level COVID-19 case data from the *New York Times* (2021), and population data from the US Bureau of Census (2019).

The virus initially diffused hierarchically, so people in metropolitan areas along the coasts were more likely to become exposed earlier. Clusters of infections appeared near major international airports in New York, Los Angeles, Dallas, Atlanta, Seattle, and Washington, DC, during the winter and spring quarters of 2020. Smaller metropolitan and micropolitan areas showed rising infection rates later in the spring or early summer, and some interior rural areas remained uninfected before late summer or early fall. But there were many seeming exceptions to this pattern, including a locally televised case in May involving a teenage boy who had not traveled outside his small rural community in northeastern Nebraska. Indeed, while more populous places usually had higher total COVID-19 caseloads, the average rates of infection were slightly lower in metropolitan counties than in micropolitan or rural counties by the time of the election. Using GIS techniques, it was found that by November 3 the mean COVID-19 cumulative case rate was 275.9 per 10,000 people in metropolitan counties, versus 310.3 in micropolitan counties, and 308.1 in rural counties.

The COVID-19 pandemic began to interfere with the 2020 presidential campaign even before the end of the nomination period. The Trump administration's bungling efforts to try to deflect public attention away from the pandemic promoted a large field of competitors to enter the race on the Democratic side. But social distancing and "stay-at-home" guidelines and orders issued by several state and local governments confounded campaign efforts by Democratic candidates competing for their party's presidential nomination. The televised Democratic primary debate in Washington, DC, on March 15, for example, was held with no audience present. And several states, including Georgia, Kentucky, Louisiana, Maryland, and Ohio, delayed primary elections originally scheduled for March or April, to May, June, or even July.

Recent election history had played a role in each major party's choice of a convention site before the onset of the pandemic. Democratic officials had been horrified by Donald Trump's breach of the assumed "blue wall" of traditional Democratic strength in the industrial Midwest

FIGURE 3.4

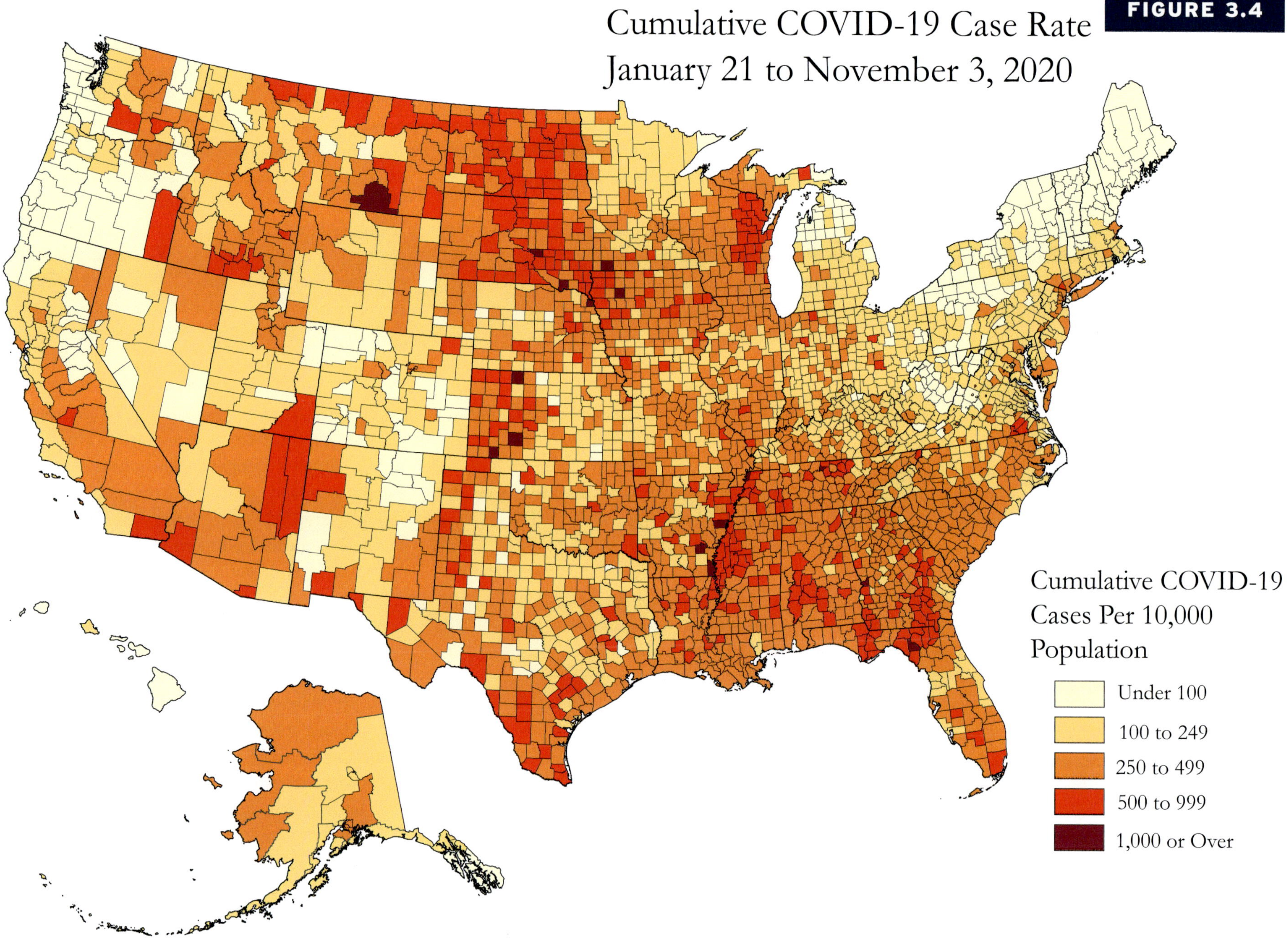
Cumulative COVID-19 Case Rate
January 21 to November 3, 2020
Cumulative COVID-19
Cases Per 10,000
Population
Under 100
100 to 249
250 to 499
500 to 999
1,000 or Over

when Trump won the electoral votes of Michigan and Wisconsin by very slim pluralities in the 2016 general election. In hindsight, Democratic critics noted that the Clinton-Kane campaign had slighted the industrial Midwest in their campaign activities while the Trump-Pence campaign had unexpectedly targeted this area for campaign visits and other campaign resources four years earlier (Watrel et al. 2018, 36–42). In response, the Democratic National Committee chose to hold the 2020 Democratic National Convention in Milwaukee, Wisconsin, in an effort to boost the chances of the eventual Democratic presidential nominee in Wisconsin and the rest of the industrial Midwest and thus avoid a repeat of the debacle of 2016. The Republican National Committee chose Charlotte, North Carolina, over other contending sites for the Republican National Convention for very similar reasons, to try to stem potential pro-Democratic trends in the Tar Heel State. But as it turned out, the expanding COVID-19 pandemic disrupted the conventions of both major parties.

The Democratic National Convention originally scheduled for July 13–16 in Milwaukee, Wisconsin, was delayed by one month until August 17–20 due to the pandemic. And very few of the convention proceedings were conducted in person at the convention site but rather most were held online according to a hastily devised remote-virtual format that allowed presenters and delegates to avoid traveling to the convention site. One of the more innovative features of the Democratic National Convention was that each state's nomination vote was memorably cast remotely via video links from picturesque locations across the United States. And Joe Biden's presidential nomination acceptance speech was delivered remotely to convention delegates and to the larger national television audience from his home studio in Wilmington, Delaware. Indeed, Biden gave specific emphasis in his remarks to following Centers for Disease Control and Prevention (CDC) guidelines for mask-wearing and social distancing during the Democratic National Convention.

The Republican National Convention also was disrupted by the pandemic. Donald Trump's aversion to public mask-wearing and social distancing put plans for the Republican National Convention at risk when the state of North Carolina insisted that mask-wearing and social distancing guidelines would need to be followed at the Republican National Convention in Charlotte. Trump was reportedly enraged by the state's enforcement of antivirus measures, and parts of the convention were proposed to be moved to a secondary site located in Jacksonville, Florida. But growing evidence of an increasingly rapid deterioration of COVID-19 conditions in Florida very shortly caused the plans for the Jacksonville relocation to be scrapped. No other potential relocation site emerged, and Trump finally reluctantly agreed to return the Republican National Convention back to Charlotte, where it was held under a much scaled-down mixed in-person/virtual program over a three-day period from August 24 to August 27. Many commentators remarked that the revised Republican National Convention appeared much less professional and much more disorganized than the revised Democratic National Convention. The so-called Trump base was retained, but there was little evidence of a larger "post-convention bounce" for his campaign.

There usually is a lull after the major-party conventions but before the customary start of the general election campaign on Labor Day, which fell on September 2 in 2020. Four years earlier Democrat Hillary Clinton and Republican Donald Trump had received their party's presidential nominations before the end of July and the general election was not held until Tuesday, November 7 in 2016. But the pandemic-delayed late August conventions in 2020 abbreviated the post-convention campaign leading up to Election Day on November 3, 2020, by more than a month compared with the preceding election cycle. Polls conducted in late August and early September 2020 gave Biden a lead of roughly 5 to 7 percentage points nationally. But Hillary Clinton had entered the general election contest ahead in the polls in 2016 as well, albeit by smaller margins. So Republican partisans had reasons to worry about Trump's reelection prospects, but Democratic partisans had reasons to fear that a "hidden" Trump advantage might reappear before Election Day.

In 2016 Republican candidates Trump and Pence had out-hustled Democratic candidates Clinton and Kane by visiting nearly twice as many states (Watrel et al. 2018, 42). And this history seemed to be repeated in 2020, when Democratic candidates Biden and Harris began the general election campaign under intentionally restricted travel schedules. Their posture was that the COVID-19 epidemic made campaign travel too risky and advisedly to be avoided under social distancing guidelines in effect in many states and localities.

Trump and Pence held no such views and followed vigorous campaign travel schedules, at least until President Donald Trump and First Lady Melania Trump tested positive for the coronavirus on October 2, 2020. This was just days after the first presidential debate between Trump and Biden in Cleveland, Ohio, at Case Western Reserve University on September 29, prompting fears that Biden might have been exposed to the virus by Trump. Somehow Biden continued to test negative for the COVID-19 virus. But this is another example of the role of the pandemic in the 2020 election. So too is the fact that the first debate had originally been scheduled to be held at Notre Dame University in South Bend, Indiana, but worries about spreading the COVID-19 virus on campus had caused Notre Dame University to withdraw their offer, leading to the selection of the Cleveland venue instead. Trump did not resume active campaign travel again until October 12, a week after he had been given a successful experimental antiviral treatment only available in very limited quantities and then released from Walter Reed Hospital.

Biden continued to follow a virtual campaign from his home in Wilmington, Delaware, until after the last presidential debate was held on October 22 between Biden and Trump at Belmont University in Nashville, Tennessee. Another debate between Trump and Biden had been planned for October 15 in Miami, Florida, but Trump was still technically under COVID-19 isolation restrictions and refused to participate in a virtual format rather than an in-person confrontation so that this debate was canceled.

States receiving presidential and vice presidential campaign stops by Democratic nominees Biden and Harris and by Republican nominees Trump and Pence from September 1 to November 3, 2020, are shown on maps in figure 3.5. The data for these maps were painstakingly collected by the authors daily from televised news reports (CNN, CBS, PBS, ABC, and NBC), and from daily newspapers in print and online forms (*Lincoln Journal Star*, *New York Times*, *Omaha World-Herald*, *USA Today*, and *Washington Post*). Careful efforts were made to exclude what were virtual rather than in-person events of various kinds. Since candidate time is one of the scarcest of all resources in an election campaign, it is very plausible to infer that these campaign stops are revealing indicators of the strategies and priorities of each campaign organization. Campaign fundraising is possible, but candidate time-raising is not. Dispatching campaign surrogates including spouses, children, friends, and others does offer a bit of flexibility to candidate time constraints, but observations made during the 2020 campaign suggest that there are few meaningful differences between the travel of surrogates and the travel of candidates, except that surrogates are sometimes dispatched to less important destinations. Indeed, the simple correlation on the Democratic side between Biden stops and Biden surrogate stops including by Barack Obama and Jill Biden was found to be +0.68, compared on the Republican side with a correlation of +0.90 between Trump stops and Trump surrogate stops including by Donald Trump Junior and Melania Trump.

The maps and some statistical correlations make several inferences very apparent. First, most states received very little if any attention from either major-party presidential candidate during the general election campaign in 2020. Indeed, thirty of the fifty states received no general election campaign visits at all from either Biden or Trump. Second, the campaign travel patterns of the Democratic and the Republican presidential candidates were very similar, as shown on the maps and reinforced by a calculated correlation coefficient of +0.85 between Biden's and Trump's state-level campaign stop patterns. The campaign stop patterns of the vice presidential candidates were slightly less similar, with a calculated correlation coefficient of +0.68 found between Harris's and Pence's state-level campaign stop patterns. Third, Trump did make over one-third more campaign stops than Biden from September 1 to November 3, 2020: ninety-two stops by Trump versus seventy stops by Biden.

While the candidates and their campaign organizations didn't agree about much, they implicitly agreed that potential voters in some states were much more important than potential voters in other states and targeted their campaign travel plans accordingly. Comparing the maps of state-level campaign stops with the earlier shown map of differences between Clinton and Trump election support in 2016 it is clear that both campaigns agreed that states with more than about a 5 percent difference either way in 2016 were to be avoided and routed around except under special circumstances involving such matters as candidate's home states. Stops in California, Delaware, Indiana, and New York illustrate home-state attractions, though California and New York are also prime campaign fundraising sources.

The campaign travel activities of Biden and Trump after their last debate on October 22 are especially revealing. The map in figure 3.6 shows the localities that Biden visited during the last twelve days of the campaign, from October 23 to November 3. While Biden had previously campaigned mainly in a virtual mode by delivering speeches from his home office in Wilmington, Delaware, before the final debate, he became much more active after the final debate, by holding more than a dozen in-person rallies during the last twelve days of the general election campaign. But unlike Trump's rallies, Biden's rallies were held outside in drive-in theaters or similar venues where attendees were required to remain seated in their

FIGURE 3.5

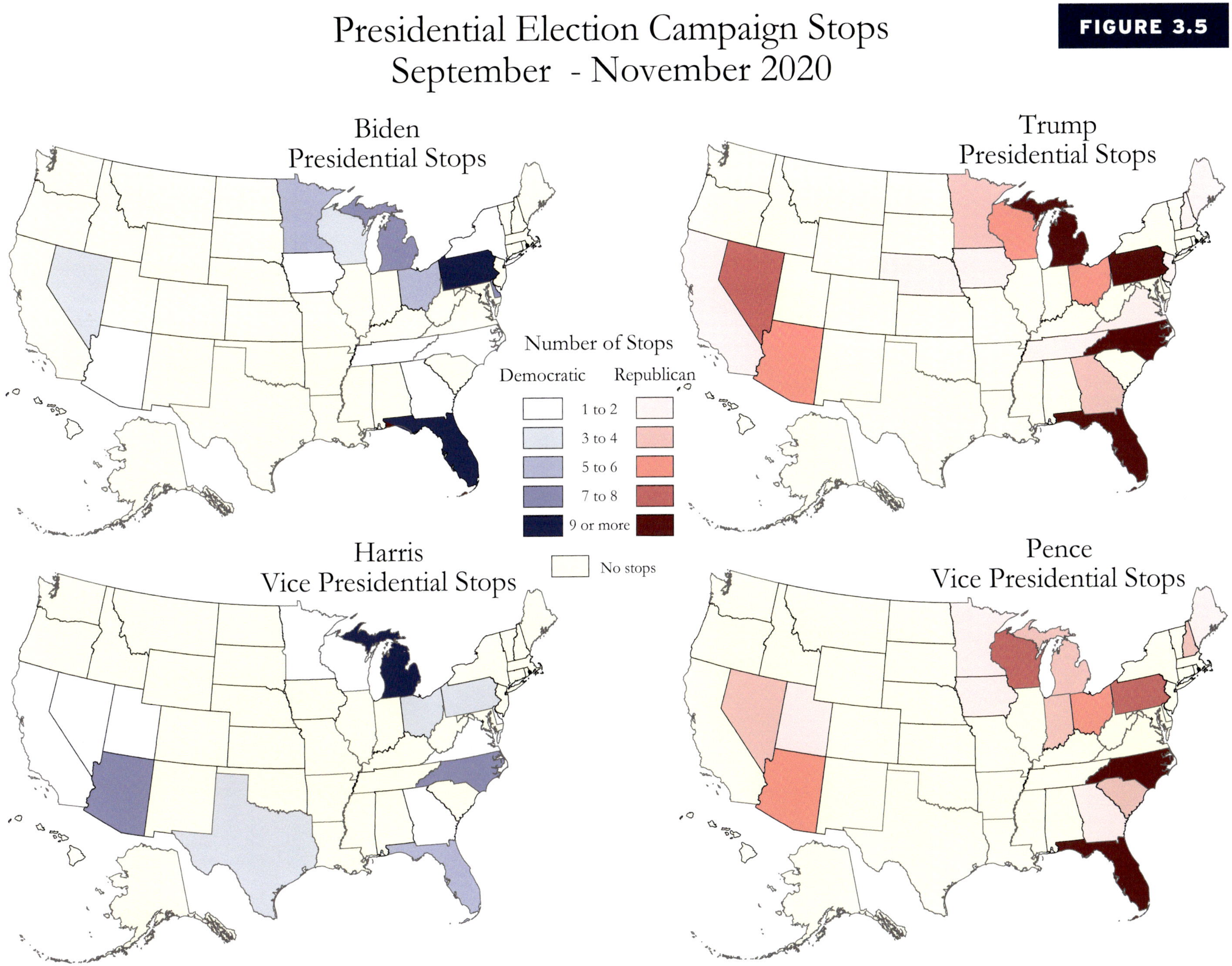
Presidential Election Campaign Stops
September - November 2020
Biden
Presidential Stops
Trump
Presidential Stops
Harris
Vice Presidential Stops
Pence
Vice Presidential Stops
Number of Stops
Democratic
Republican
1 to 2
3 to 4
5 to 6
7 to 8
9 or more
No stops

FIGURE 3.6

Biden Visits during the Last Twelve Days of the Campaign
October 23 through November 3, 2020

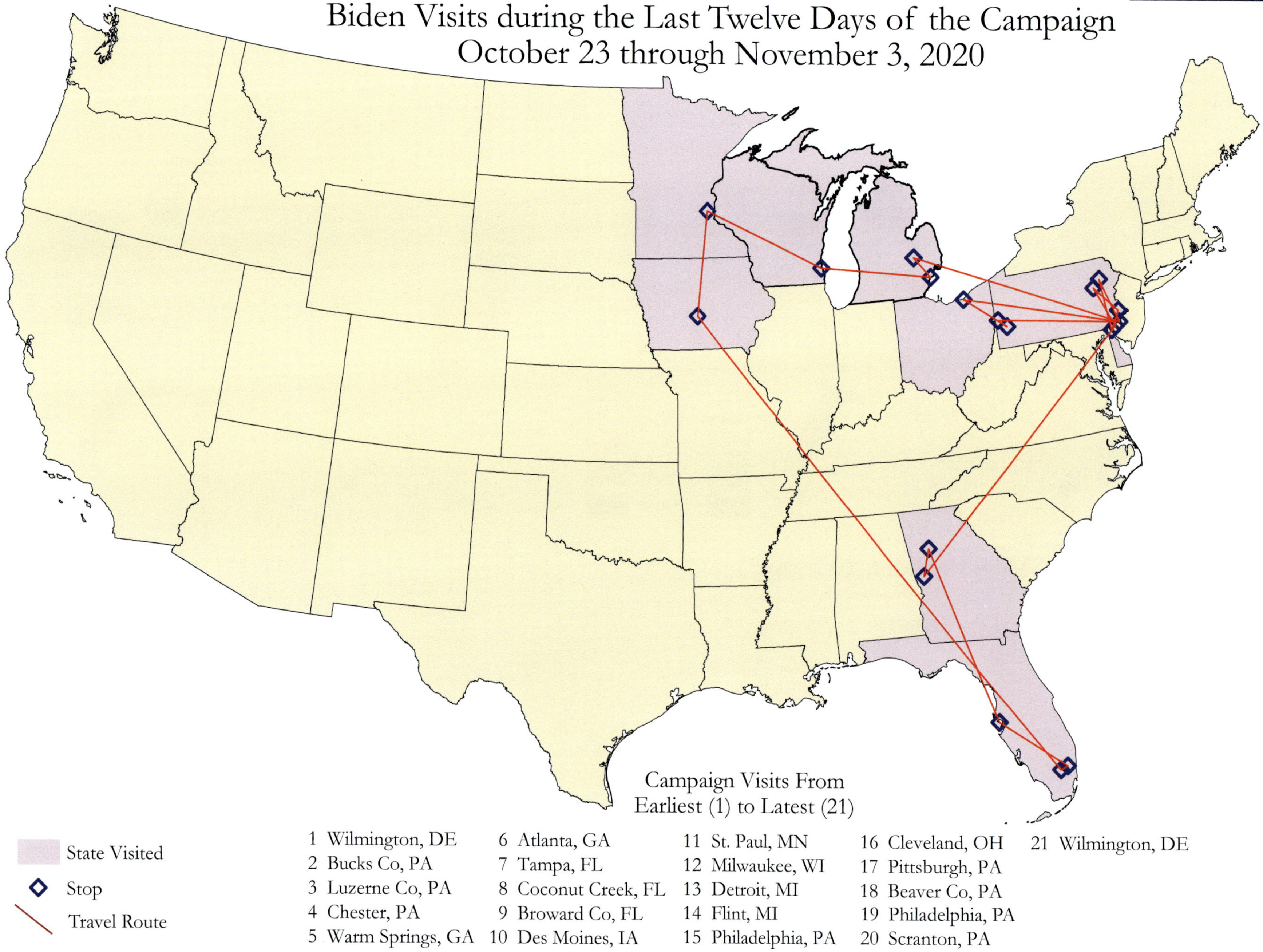

FIGURE 3.7

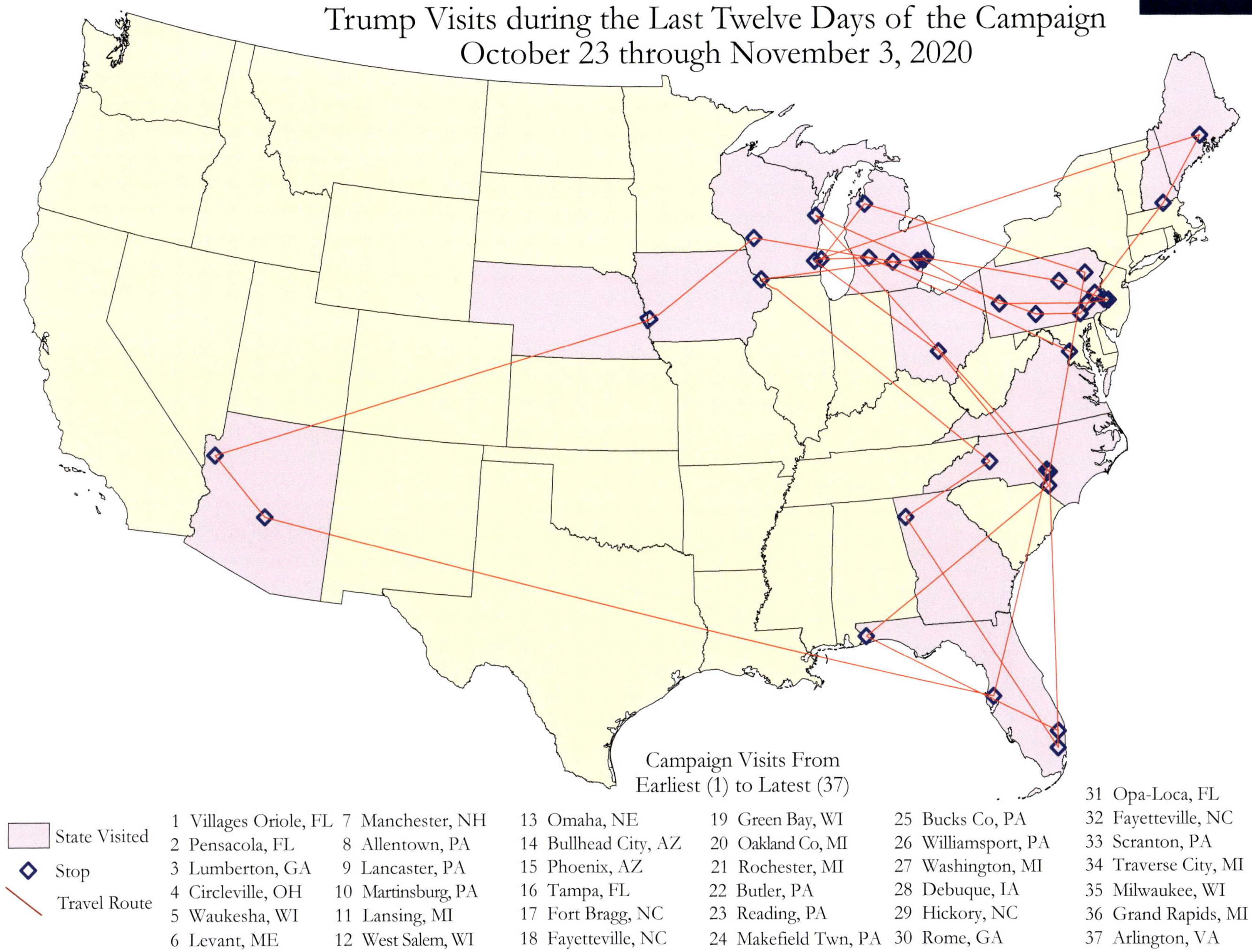

Trump Visits during the Last Twelve Days of the Campaign
October 23 through November 3, 2020
State Visited
Stop
Travel Route
Campaign Visits From
Earliest (1) to Latest (37)
1 Villages Oriole, FL
2 Pensacola, FL
3 Lumberton, GA
4 Circleville, OH
5 Waukesha, WI
6 Levant, ME
7 Manchester, NH
8 Allentown, PA
9 Lancaster, PA
10 Martinsburg, PA
11 Lansing, MI
12 West Salem, WI
13 Omaha, NE
14 Bullhead City, AZ
15 Phoenix, AZ
16 Tampa, FL
17 Fort Bragg, NC
18 Fayetteville, NC
19 Green Bay, WI
20 Oakland Co, MI
21 Rochester, MI
22 Butler, PA
23 Reading, PA
24 Makefield Twn, PA
25 Bucks Co, PA
26 Williamsport, PA
27 Washington, MI
28 Debuque, IA
29 Hickory, NC
30 Rome, GA
31 Opa-Loca, FL
32 Fayetteville, NC
33 Scranton, PA
34 Traverse City, MI
35 Milwaukee, WI
36 Grand Rapids, MI
37 Arlington, VA

vehicles and to wear masks to minimize the risks of COVID-19 transmission. In obvious recollection of Clinton's damaging neglect of the industrial Midwest in 2016, Biden traveled to campaign stops in Iowa, Michigan, Minnesota, Ohio, and Wisconsin during the final period of the campaign. And, perhaps in response to rising poll indicators, he also visited Georgia even though Georgia had delivered very few Democratic electoral votes since native Georgian Jimmy Carter had won the White House from Plains, Georgia, in 1976.

The map in figure 3.7 shows the places where Donald Trump campaigned after the final debate, from October 23 to Election Day on November 3, 2020. The similarities with the map of Biden's campaign stops are obvious, with similar concentrations of stops in several states in the industrial Midwest, in Florida, and in Pennsylvania on both maps. But Trump was again more highly traveled than Biden, making thirty-seven stops from October 23 to November 3, or nearly twice as many stops as Biden's twenty-one stops during the same period. Interestingly, Trump's travels carried him to Maine and Nebraska, which are the only two states that do not allocate electoral votes on a "winner-take-all" basis but rather split electoral votes by congressional district. By that point, the Trump campaign organization was increasingly worried that Trump would again lose the national popular vote but not repeat an electoral vote win a second time.

According to Bloomberg News (2020), during the last week of the campaign the Trump organization pulled nearly all its rapidly diminishing advertising dollars out of Florida and transferred them to just the four battleground states of Michigan, Minnesota, Ohio, and Pennsylvania. Trump campaign polling apparently had shown him behind in all these states except Ohio, and he needed to try to squeak out an Electoral College victory with less than even a plurality of the national popular vote. By late in the 2020 election campaign, it was looking very, very unlikely that the national popular and electoral outcomes would be divergent once again in 2020.

Trump's campaign visits to congressional districts in Maine and Nebraska shined rays of illumination toward what presidential election campaigns might be like if there were no Electoral College. But of course there is a Federal Electoral College, and this time around the Republican Trump-Pence ticket's 46.9 percent of the national popular vote translated to 232 electoral votes while the Democratic Biden-Harris ticket's 51.4 percent of the national popular vote translated to 306 electoral votes. Trump left the White House, and Biden moved in.

REFERENCES

Archer, J. Clark, R. H. Watrel, F. Davidson, E. H. Fouberg, K. C. Martis, R. L. Morrill, F. M. Shelley, and G. R. Webster, eds. 2014. *Atlas of the 2012 Elections*. Lanham, MD: Rowman & Littlefield.

Bloomberg News. 2020. Trump Pulls Florida Ads. https://www.bloomberg.com/news/articles/2020-10-27/trump-pulls-florida-ads-as-cash-poor-campaign-enters-final-week (last accessed October 27, 2020).

New York Times. 2021. County Level Coronavirus (COVID-19) Data. https://github.com/nytimes/covid-19-data#readme (last accessed April 2, 2021).

US Bureau of Census. 2019. CO-EST2019-all data: Annual Resident Population Estimates, Estimated Components of Resident Population Change, and Rates of the Components of Resident Population Change for States and Counties: April 1, 2010 to July 1, 2019. File: 7/1/2019 County Population Estimates Source: US Census Bureau, Population Division Release Date: March 2020. http://census.gov (last accessed June 24, 2020).

Watrel, Robert H., R. Weichelt, F. Davidson, J. Heppen, E. H. Fouberg, J. C. Archer, R. L. Morrill, F. M. Shelley, and K. C. Martis, eds. 2018. *Atlas of the 2016 Elections*. Lanham, MD: Rowman & Littlefield.

Wikipedia. 2021. COVID-19 Pandemic in the United States. https://en.wikipedia.org/wiki/COVID-19_pandemic_in_the_United_States (last accessed April 14, 2021).

Woodward, Bob. 2020. *RAGE*. New York: Simon and Schuster.

GEOGRAPHIC LANGUAGE AND THE 2020 PRESIDENTIAL DEBATES

MATTHEW BALENTINE AND GERALD R. WEBSTER

The institution of regular, face-to-face presidential debates began in 1976, although the first televised debates took place in 1960. Since 1976, presidential candidates representing the two major parties are expected to participate in public debates in the run-up to the election. Campaign activities are inherently geographic (see other chapters), and as regular and integral parts of the campaigning process, debates are no exception. The policy topics broached in debates are often compartmentalized by media discourse and debate moderators. Past moderators have structured debate sections and entire debates around one of two geographic policy domains: foreign versus domestic. Additionally, geography and political strategy potentially intersect in the debate process. There is evidence to suggest that candidates consider two modes of deployment of geographic language for strategic political communication (Balentine and Webster 2018). Evidence suggests that incumbent candidates tend to tout foreign place and policy knowledge that highlights prior experience, while challengers tend to emphasize domestic issues.

Another aspect of debate language's geographic and strategic connections emerges considering the US electoral system. The Electoral College (EC) encourages campaign efforts primarily in the most competitive states, colloquially referred to as "battleground states" or "swing states." The EC system places a premium on these states; they often determine the outcome of the presidential race.

Donald Trump's 2020 candidacy included a litany of aberrations, and many are traceable to the 2016 campaign. Most notable of these unprecedented events occurred during the final 2016 debate, in which he refused to confirm whether he would accept the election's outcome. This arguably foreshadowed the 2021 insurrection at the US Capitol, which was largely fueled by Trump's continued baseless claims that his 2020 electoral loss was due to fraud.

Like other aspects of the 2020 election, the debates were influenced by the COVID-19 pandemic. Donald Trump contracted COVID-19 after the first debate. The second scheduled debate was canceled due to Trump's refusal to debate Biden remotely. In lieu of the canceled second debate, each candidate participated in separate town hall events televised at the same time. Biden's town hall was broadcast on ABC and Trump's town hall was broadcast on NBC.

The 2020 debates also included an unprecedented discourse that highlights the intersection of geographic language and political strategic communication. While beyond our methodology, we share selected anecdotes from the debates. First, Trump explicitly mobilizes "place" along partisan lines to criticize Biden. There are multiple instances where he alludes to "Democratic run" cities and states. This spatially grounding partisan language is unique in the debate record because of its overtness and blatant divisiveness. The only previous instance includes exchanges between Texas Senator Ted Cruz and President Trump during the 2016 Republican Party primaries. In the run-up to the Iowa Caucus, Cruz attacked Trump's "New York values." Following his caucus victory in Iowa, the electoral importance of New York in defeating Donald Trump became apparent, and Cruz rolled back his comments on New York by providing geographic and class specificity. This was likely to prevent alienating the entire state of New York and all of New York City's boroughs.

> As I travel the country here in Iowa, New Hampshire, South Carolina, Nevada, everyone knows what New York values are. It's the values of the elite liberals that have done enormous damage to New York and they're [*sic*] a bunch of cops and firemen and hard-working men and women in the great state of New York who are fed up with the out-of-touch values of Manhattan. (Saenz 2016)

Trump's spatial grounding of partisanship came in the first debate, when he suggested a partisan plan to use COVID-19 restrictions to influence the presidential election. His claim leans upon the spatialization of partisanship for evidence (see table 3.1). There are multiple examples where Trump attacked Democratic-led cities, states, and areas. The spatialization of partisanship was not exclusive to specific places. The spatial grounding

TABLE 3.1. Anecdotal Examples of Spatialization of Partisanship

Candidate [Time]	First Debate—September 29, 2020, in Ohio
Trump [05:15]	Wait a minute, Joe. Let me shut you down for a second, Joe, just for one second. He wants to shut down the country. We just went through it. We had to, because we didn't know anything about the disease. Now we've found that elderly people with heart problems and diabetes and different problems are very, very vulnerable. We learned a lot. Young children aren't, even younger people aren't. We've learned a lot, but he wants to shut it down. More people will be hurt by continuing. *If you look at Pennsylvania, if you look at certain states that have been shut down, they have Democrat governors, all, one of the reasons they shut down is because they want to keep it shut down until after the election on November 3rd.*
Trump [10:31]	*A lot of people, between drugs and alcohol and depression, when you start shutting it down, you take a look at what's happening at some of your Democrat-run states where they have these tough shutdowns. And I'm telling you it's because they don't want to open it. One of them came out last week, you saw that, "Oh, we're going to open up on November 9th." Why November 9th? Because it's after the election.* They think they're hurting us by keeping them closed. They're hurting people. People know what to do. They can social distance. They can wash their hands, they can wear masks. They can do whatever they want, but they got to open these states up.
Trump [54:04]	I don't think you have any law enforcement. You can't even say the word law enforcement. Because if you say those words, you're going to lose all of your radical left supporters. And why aren't you saying those words, Joe? Why don't you say the words law enforcement? Because you know what*? If called us [sic] in Portland, we would put out that fire in a half an hour. But they won't do it, because they're run by radical left Democrats. If you look at Chicago, if you look at any place you want to look, Seattle, they heard we were coming in the following day and they put up their hands and we got back Seattle. Minneapolis, we got it back, Joe, because we believe in law and order, but you don't. The top ten cities and just about the top forty cities are run by Democrats, and in many cases radical left.* And they've got you wrapped around their finger, Joe, to a point where you don't want to say anything about law and order. And I'll tell you what, the people of this country want and demand law and order and you're afraid to even say it.
Candidate Exchange [59:08]	President Donald J. Trump: During the Obama-Biden administration, there was tremendous division. There was hatred. You look at Ferguson, you look at, or you go to very... many places, look at Oakland. Look what happened in Oakland. Look what happened in Baltimore. Look what happened.... Frankly, it was more violent than what I'm even seeing now. Vice President Joe Biden: Oh, my Lord. President Donald J. Trump: But the reason— Vice President Joe Biden: This is ridiculous. President Donald J. Trump: *is that the Democrats that run these cities—* Vice President Joe Biden: Absolutely ridiculous.

Candidate Exchange [60:29]	President Donald J. Trump: *But if he ever got to run this country and they ran it the way he would want to run it, we would have by the way our suburbs would be gone. And you would see problems like you've never seen before.* Vice President Joe Biden: *He wouldn't know a suburb unless you took a wrong turn.* President Donald J. Trump: *Oh, I know suburbs so much that you.* Chris Wallace: Gentlemen wait a minute. Vice President Joe Biden: *I was raised in the suburbs. This is not 1950. All these dog whistles and racism don't work anymore. Suburbs are by and large integrated.* There's many people today driving their kids to soccer practice and/or Black and white and Hispanic in the same car as there have been any time in the past, *what really is a threat to the suburbs and their safety is his failure to deal with COVID. They're dying in the suburbs. His failure to deal with the environment, they're being flooded, they're being burned out because his refusal to do anything. That's why the suburbs are in trouble.*
Second Debate—October 22 in Tennessee	
Trump [22:09]	Thank you, and I appreciate that. Look, all he does is talk about shutdowns. But forget about him. *His Democrat governors, Cuomo in New York, you look at what's going on in California, you look at Pennsylvania, North Carolina, Democrats, Democrats all, they're shut down so tight and they're dying.* They're dying. And he supports all these people. All he talks about is shutdowns. No, we're not going to shut down. And we have to open our schools. And it's like, as an example, I have a young son, he also tested positive. By the time I spoke to the doctor the second time, he was fine. It just went away. Young people, I guess, it's their immune system.
Biden [25:15]	Take a look at what New York has done in terms of turning the curve down, in terms of the number of people dying. *And I don't look at this in terms of the way he does, blue states and red states. They're all the United States. And look at the states that are having such a spike in the coronavirus. They're the red states, they're the states in the Midwest, they're the states in the Upper Midwest.* That's where the spike is occurring significantly. But they're all Americans. They're all Americans. And what we have to do is say, wear these masks, number one. Make sure we get the help that the businesses need. That money's already been passed to do that. It's been out there since the beginning of the summer, and nothing's happened.
Biden [62.02]	. . . *I'm running as a proud Democrat, but I'm going to be an American president. I don't see red states and blue states.* What I see is American United States. And folks, every single state out there finds themself in trouble. They're going to start laying off, whether they're red or blue, cops, firefighters, first responders, teachers, because they have to balance their budget. And the founders were smart. They allowed the federal government to deficit spend to compensate for the United States of America.
Trump Town Hall—October 15 in Florida	
Trump [18:06]	And frankly, you want to know something? I denounce Antifa, and I denounce these people on the left that are burning down our cities, *that are run by Democrats who don't know what they're doing–*
Exchange [18:54]	Savannah Guthrie: They believe that it is a Satanic cult run by the deep state. President Trump: . . . study the subject. I'll tell you what I do know about. I know about Antifa, and I know about the radical left, and I know how violent they are and how vicious they are. *And I know how they are burning down cities run by Democrats, not run by Republicans.*

Source: Collected by authors.

Note: Emphasis added by authors.

of partisanship also implicated broader urban forms, as both candidates sparred over their familiarity with, and interest in, suburban America.

Table 3.2 lists the nine swing states in the 2020 presidential election. To evaluate the strategic use of geographic language related to swing states, the candidates' combined battleground state mentions are compared to the total domestic references. Nine states qualified as swing states. Excluding New York and Delaware, the respective home states of Donald Trump and Joe Biden, 19 percent of the contiguous states qualified as swing states (9/47 = 19 percent). Candidate mentions of swing states, as a proportion of all state mentions, exceeding the 19 percent threshold may suggests a strategic intent by the debater and their campaigns to court swing state voters in their debate and town hall appearances.

TABLE 3.2. 2020 Swing States

Arizona
Florida
Georgia
Michigan
Nevada
North Carolina
Pennsylvania
Wisconsin
Texas

Source: Calculated by authors, states meeting the following criteria were classified as swing states: 2020 margin of victory ≤ 6%; mean of state polls before the first debate ≤ 6%; and inclusion in the RealClearPolitics.com list of "toss-up" states.

In all three events the candidates exceeded the 19 percent threshold. The 19 percent mark is the rate at which swing state mentions could be attributable to chance. In two events both candidates' swing state reference proportions were three times higher than the 19 percent threshold.

As previously noted, the 2020 cycle included several unprecedented developments. The communicative engagement with and about swing states during the 2020 debates was additionally unprecedented. While the 2020 cycle, and previous cycles, have provided some support for the theory that candidates strategically mention swing states, all the examples familiar to the authors have reflected self-acclaims related to swing states, or simply the "name-dropping" of swing states. Previous examples suggest candidates use these references in positive or neutral ways to seemingly appeal to voters in those states. However, the second debate saw President Trump deploy geographic language linked to the swing state of Pennsylvania, to attack Biden (table 3.3). Biden has touted a close connection to Pennsylvania, and this was reflected in the debates by his multiple references to Scranton, his birthplace. Incidentally, candidates tend to overemphasize their home states, and states where they have conducted meaningful political work, as those places are necessarily part of their personal narratives, experiences, and accomplishments. However, Biden's birth state of Pennsylvania was not only important as a part of his personal narrative; it was strategically important for both candidates. This is reflected in the fact that Pennsylvania ranked in the top two positions in both candidates' lists of top domestic references (table 3.4). Its strategic importance is also reflected in its competitive history. Pennsylvania's margins of victory in the last two presidential contests were less than 2 percent, and its average margin of victory in 1976–2020 is 5.16 percent. Trump attacked Biden by directly calling into question his Pennsylvania credentials.

> Donald Trump: And he doesn't come from Scranton. He lived there for a short period of time before he even knew it.
>
> Kristen Welker: Okay. We're going to move on to the next question.
>
> Donald Trump: And he left. And the people of Pennsylvania will show you that.
>
> Kristen Welker: Let me move on to my next question, gentlemen.
>
> Donald Trump: They understand.

The home states of the two candidates, New York and Delaware, and the critical swing state of Pennsylvania, were either the first or second most mentioned domestic places by the two candidates (table 3.4 and figure 3.8). The greater amount and diversity of domestic places mentioned by Trump is surprising. As noted earlier, incumbents tend to highlight foreign place names to emphasize their experience (Balentine and Webster 2018). This was not the case with respect to President Trump. It is possible that his emphasis on spatially partisan language aimed at large metropolitan areas may account for this. Additionally, this approach might be connected to the Trump campaign's "law and order" discourse. While Biden seemed to utilize domestic language in a measured way, he did not completely omit references to Rust Belt states as Hillary Clinton did in 2016. Biden had fewer

TABLE 3.3. Swing States References in the 2020 Debates as a Percentage of All Domestic References

	Biden Percentage	Trump Percentage
Debate 1	33	30
Town Hall	57	60
Debate 2	75	68

Note: Only specific references at or below the state level were included in the analysis. Data were aggregated to the state level. References to Washington, DC, and candidates' home states were excluded from the analysis.

TABLE 3.4. Candidates' Top Domestic References

Trump's Top Domestic References		Biden's Top Domestic References	
Pennsylvania	17	Delaware	16
New York	15	Pennsylvania	7
California	9	Florida	2
Ohio	9	New York	2
Florida	7	Ohio	2
Michigan	6	California	1
Illinois	3	Iowa	1
Oregon	3	Michigan	1
South Carolina	3	Texas	1
Texas	3		
Virginia	3		
North Carolina	2		
Oklahoma	2		
Washington	2		
Alabama	1		
Arizona	1		
Maryland	1		
Minnesota	1		
Missouri	1		
New Jersey	1		
Vermont	1		
West Virginia	1		

domestic references overall and fewer swing state mentions, but of those, he included historically important swing states in the Rust Belt.

Each candidate's most referenced foreign countries suggested similar foreign policy concerns and reflected most issues that had been highlighted in national media (table 3.5). Russia and China accounted for both candidates' top two most frequently mentioned foreign countries, likely because of the shadow of Russian efforts toward influencing the 2016 election, concerns that the Russians would repeat the effort in 2020, and the COVID-19 pandemic that originated in Wuhan, China (figure 3.8).

The ratio between foreign and domestic reference may indicate a candidate's efforts to focus on a particular geographic policy or issue domain (i.e., foreign or domestic; see table 3.6). A value of 1.0 would suggest a balanced representation of the geographic policy domains. Both candidates' totaled foreign/domestic (F/D) ratios were greater than 1.0, suggesting an overall emphasis on the foreign policy domain. This is to be expected; both candidates brought with them previous foreign experience from their time in the executive branch. While both were greater than 1.0, Biden's ratio was substantially higher. This may be indicative of two aspects of the debates. First, it may reflect Biden's efforts to highlight the fact that his administration would represent a return to normalcy for domestic affairs along with the reestablishment of cooperation with US allies that were marginalized during Trump's term as president.

Second, the large difference in F/D ratio might be explained by the logistical failures of the Trump administration to plan for, and alleviate, the domestic impacts of COVID-19. The economic gains during President Trump's term

FIGURE 3.8

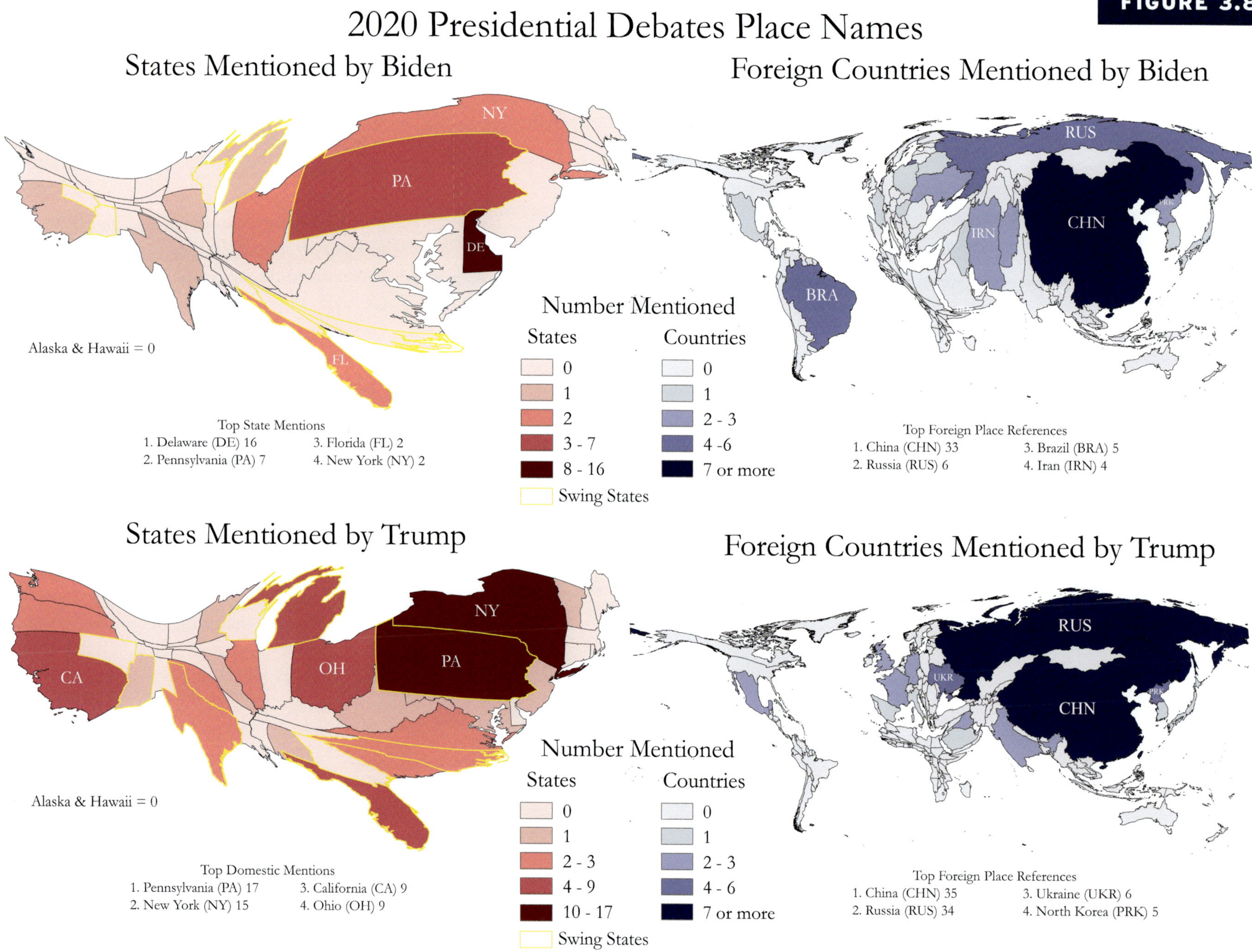

2020 Presidential Debates Place Names
States Mentioned by Biden
NY
PA
DE
FL
Alaska & Hawaii = 0
Top State Mentions
1. Delaware (DE) 16
2. Pennsylvania (PA) 7
3. Florida (FL) 2
4. New York (NY) 2
Foreign Countries Mentioned by Biden
RUS
CHN
PRK
IRN
BRA
Top Foreign Place References
1. China (CHN) 33
2. Russia (RUS) 6
3. Brazil (BRA) 5
4. Iran (IRN) 4
Number Mentioned
States
0
1
2
3 - 7
8 - 16
Swing States
Countries
0
1
2 - 3
4 -6
7 or more
States Mentioned by Trump
NY
PA
OH
CA
Alaska & Hawaii = 0
Top Domestic Mentions
1. Pennsylvania (PA) 17
2. New York (NY) 15
3. California (CA) 9
4. Ohio (OH) 9
Foreign Countries Mentioned by Trump
RUS
UKR
PRK
CHN
Top Foreign Place References
1. China (CHN) 35
2. Russia (RUS) 34
3. Ukraine (UKR) 6
4. North Korea (PRK) 5
Number Mentioned
States
0
1
2 - 3
4 - 9
10 - 17
Swing States
Countries
0
1
2 - 3
4 - 6
7 or more

TABLE 3.5. Candidates' Top Foreign References

Trump's Top Foreign References		Biden's Top Foreign References	
China	35	China	33
Russia	34	Russia	6
Ukraine	6	Brazil	5
North Korea	5	Iran	4
France	3	North Korea	4
India	3	Ukraine	4
Mexico	3	Afghanistan	3
United Kingdom	3	Mexico	2
Germany	2	South Korea	2
Iraq	2	Belarus	1
Italy	1	France	1
Japan	1	Hungary	1
Saudi Arabia	1	Iraq	1
South Korea	1	Israel	1
Spain	1	Japan	1
		Poland	1

TABLE 3.6. Candidates' Ratio of Foreign-to-Domestic References

	Trump			Biden		
	Foreign	Domestic	F/D	Foreign	Domestic	F/D
Debate 1 Ohio	25	38	0.66	10	8	1.25
Town Hall	21	28	0.75	21	16	1.31
Debate 2 Nashville	65	32	2.03	51	13	3.92
Total	111	98	1.13	82	37	2.22

rapidly evaporated and were potentially forgotten by voters. Attempts to appeal to voters' concerns in the domestic policy sphere, or remind them of past economic gains, would lower Trump's F/D ratio. Additionally, the Biden campaign may have considered Trump's recent domestic failures so substantial as to not require a rebuttal, allowing Biden the latitude to focus on foreign policy.

REFERENCES

Balentine, M., and G. Webster. 2018. Place Names as a Form of Strategic Political Communication: An Analysis of Geographic Language Used in U.S. Presidential Debates from 1976–2012. In B. Stanley and R. Kehrein, eds., *The Changing Global Map of Languages, 1961–1980*. New York: Springer.

Saenz, A. 2016. Ted Cruz Credits Attack on Donald Trump's "New York Values" in Iowa Win. *ABC News*. https://abcnews.go.com/Politics/ted-cruz-credits-attack-donald-trumps-york-values/story?id=36658796 (last accessed June 19, 2021).

NEWSPAPER ENDORSEMENTS OF PRESIDENTIAL CANDIDATES IN THE 2020 ELECTION

DANIEL A. MCGOWIN AND GERALD R. WEBSTER

During the 2020 presidential campaign, thirty newspapers owned by the McClatchy Company did not endorse a candidate for president. According to an internal memo, in order to make a presidential endorsement the editorial board of each newspaper had to conduct interviews with both President Donald Trump and former Vice President Joe Biden (Smith 2020). The extreme challenge of fulfilling the requirement meant that papers such as the *Miami Herald*, *Kansas City Star*, and *Charlotte Observer* could not make an endorsement.

Although part of the reasoning behind the move to require interviews was an attempt to orient papers more toward local news, it may also reflect the lack of influence such endorsements may have on persuading voters to support a particular candidate. This was evident in the 2016 election cycle when only 6 percent of all newspaper endorsements went to Mr. Trump. Although former Secretary Hillary Clinton won the popular vote, given her loss in the Electoral College it is no surprise that some now question the rationale behind making endorsement at the national level (Editorial Board, *Dallas Morning News* 2020; Hiner 2020; Sonenshine 2020).

The timing of many endorsements is also problematic, especially given the rise of early voting. As there is no set date for newspapers to make endorsements, editorial boards may wait until late in the year to announce their selection. Of the top hundred newspapers by circulation, forty-two announced their endorsement within a month of the election, and thirteen did so within two weeks of Election Day. This is even more striking given that only fifty-nine of the top hundred papers in circulation made any public announcement on an endorsement.

Despite a recent shift away from the tradition of making presidential contest candidate endorsements, newspaper endorsements remain an important aspect of presidential elections. The trend in the 2020 election cycle is reminiscent of the pattern that emerged in the 2016 campaign, with a majority of the major newspapers in the United States endorsing the Democratic candidate (figure 3.9). Of the fifty-nine top circulation newspapers making statements on an endorsement, forty-six endorsed Mr. Biden and only six endorsed President Trump; the remaining seven made a statement on not endorsing a candidate. Although seventeen papers made an endorsement that differed from 2016, most were changes either to a no endorsement position in 2020, or away from a non-endorsement stance in 2016. Only one major paper, the *Spokesman-Review* in Spokane, Washington, switched from one party to another by endorsing President Trump in 2020 after supporting Secretary Clinton in 2016.

In terms of newspaper circulation, Mr. Biden received endorsements from thirty papers with a circulation over one hundred thousand readers and fourteen with a circulation over two hundred thousand. The average circulation among papers endorsing Mr. Biden was 207,182. For Mr. Trump, the average—the *New York Post* and the *Las Vegas Review-Journal*—have a circulation over one hundred thousand readers.

In total, 167 different daily newspapers released a statement on presidential endorsements. Of that number, 71.9 percent endorsed Joe Biden while only 7.8 percent endorsed Donald Trump. However, nearly 20 percent of those newspapers made no endorsement. And, in an article titled "America, We Deserve Better," the *Concord Monitor* in New Hampshire took an "anti-endorsement" position by specifically not endorsing President Trump (Editorial Board, *Concord Monitor* 2020).

In terms of the no endorsement positions, many justified such a stance by focusing on local news and races rather than national ones (Hiner 2020; Smith 2020). Some dailies felt that given the contentious, partisan environment surrounding the current state of politics in the United States, it was counterproductive to make such endorsements (Van Brimmer 2020; Van Nostrand 2020). Others felt that it was not for them to tell voters for whom to vote, instead opting to lay out the strengths and weaknesses of each candidate and leaving the choice up to the readers (Love 2018). In the case of the *Salt Lake Tribune*, the paper became a tax-exempt 501(c)(3) nonprofit and because of that "the rules forbid [the paper from] supporting any candidate for public office" (Pyle 2020). The *Detroit News*, which supported Libertarian Gary Johnson in 2016, felt that neither major candidate

FIGURE 3.9

2020 Presidential Endorsements of Top 100 Newspapers by Circulation: Biden, Trump, and Non-Endorsements

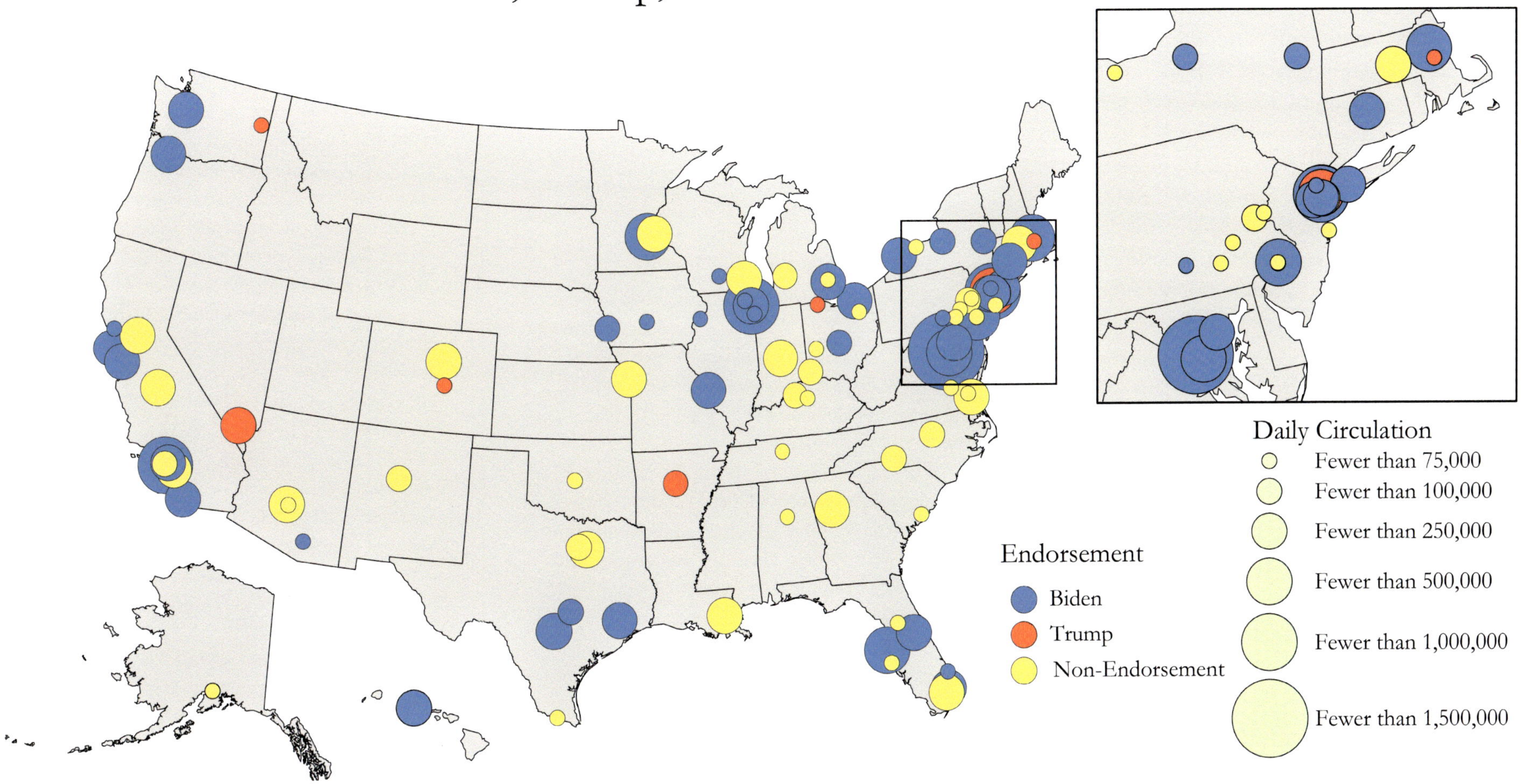

was worthy of an endorsement (Editorial Board, *Detroit News* 2020).

Of greater interest is the change in precedent among some publications. *USA Today*, which in 2016 broke from tradition and made an "anti-Trump" endorsement, made its first ever endorsement by backing Joe Biden. In closing out its endorsement, the paper's editorial board argued that "this extraordinary moment in the history of our nation requires an extraordinary response. With his plans, his personnel picks, his experience and his humanity, Joe Biden can help lead the United States out of this morass and into the future" (Editorial Board, *USA Today* 2020). Champaign, Illinois's *News-Gazette* endorsed a Democratic candidate for president for the first time in its history, arguing that Joe Biden was a better alternative than a Republican candidate who did not espouse "the traditional Republican values that Midwesterners cherish" (Editorial Board, *The News-Gazette* 2020). The magazine *Scientific American* endorsed its first candidate ever, endorsing Mr. Biden predominantly because of what the editors viewed as Mr. Trump's attacks on science (Editorial Board, *Scientific American* 2020). The *Pittsburgh Post-Gazette* endorsed its first Republican candidate since 1972, stating that Mr. Trump's record of getting things done outweigh his personality faults (Editorial Board, *Pittsburgh Post-Gazette* 2020).

Perhaps most telling is how some newspapers ultimately informed voters to "vote with your conscience." This message was stated even if a paper made an explicit endorsement. However, Shippensburg University's paper *The Slate* took that statement as its official position, encouraging readers to take part in "a hallmark of American democracy." The editors close out their "endorsement" by imploring readers to "educate yourself and vote your conscience. Your vote matters" (Editorial Board, *The Slate* 2020).

REFERENCES

Editorial Board, *Concord Monitor*. 2020. Editorial: America, We Deserve Better. October 25, https://www.concordmonitor.com/Election-2020-36919524 (last accessed March 29, 2021).

Editorial Board, *Dallas Morning News*. 2020. We Won't Recommend a Candidate for President in 2020 and Here's Why. February 16, https://www.dallasnews.com/opinion/we-recommend/2020/02/16/we-wont-recommend-a-candidate-for-president-in-2020-and-heres-why/ (last accessed March 27, 2021).

Editorial Board, *The Detroit News*. 2020. Editorial: For President, We Can't Lend Our Name to Men Whose Values We Don't Share. October 28, https://www.detroitnews.com/story/opinion/editorials/2020/10/29/presidential-endorsement-withheld-detroit-news-fourth-time/3745346001/ (last accessed March 30, 2021).

Editorial Board, *The News-Gazette*. 2020. Editorial: Our Choice for President. October 11, https://www.news-gazette.com/opinion/editorials/editorial-our-choice-for-president/article_b6b66ad7-e444-5ea0-8a90-ae1b9e539e1b.html (last accessed March 30, 2021).

Editorial Board, *Pittsburgh Post-Gazette*. 2020. The Man and the Record. October 31, https://www.post-gazette.com/opinion/editorials/2020/10/31/editorial-donald-trump-joe-biden-mike-pence-kamala-harris-presidential-candidate-endorsement/stories/202010310021 (last accessed March 30, 2021).

Editorial Board, *Scientific American*. 2020. *Scientific American* Endorses Joe Biden. October 1, https://www.scientificamerican.com/article/scientific-american-endorses-joe-biden1/ (last accessed March 30, 2021).

Editorial Board, *The Slate*. 2020. The Slate Speaks: Exercise Your Right to Vote, Register to Vote Before It's Too Late. September 29, https://www.theslateonline.com/article/2020/09/the-slate-speaks-exercise-your-right-to-vote-register-to-vote-before-its-too-late (last accessed March 29, 2021).

Editorial Board, *USA Today*. 2020. Elect Joe Biden. Reject Donald Trump. October 20, https://www.usatoday.com/in-depth/opinion/todaysdebate/2020/10/20/elect-joe-biden-reject-donald-trump-editorials-debates/5919435002/ (last accessed March 30, 2021).

Hiner, J. 2020. Letter from the Editor: MLive Is Not Endorsing a Candidate in the Presidential Race, and Here's Why. October 15, https://www.mlive.com/news/2020/10/letter-from-the-editor-mlive-is-not-endorsing-a-candidate-in-the-presidential-race-and-heres-why.html (last accessed March 29, 2021).

Love, B. 2018. Rethinking the Newspaper Endorsement. *The Cincinnati Enquirer*, October 26, https://www.cincinnati.com/story/opinion/2018/10/26/enquirer-editor-beryl-love-writes-newspaper-editorial-board-endorsement-policy/1772082002/?fbclid=IwAR0OZ3BwVMUI-7LvYrp020cxlxYOQKkcrDQH7UZUdVPTb5gu9yXz_bB5XgY (last accessed March 29, 2021).

Pyle, G. 2020. George Pyle: The Flaw in Chris and Spencer's Excellent Adventure. *Salt Lake Tribune*, October 23, https://www.sltrib.com/opinion/commentary/2020/10/23/george-pyle-flaw-chris/ (last accessed March 30, 2021).

Smith, B. 2020. McClatchy's 30 Newspapers Probably Won't Be Endorsing This Year. *New York Times*, September 17, https://www.nytimes.com/2020/09/17/us/elections/mcclatchys-30-newspapers-probably-wont-be-endorsing-this-year.html (last accessed October 7, 2020).

Sonenshine, T. D. 2020. It's Time for Newspapers to Stop Endorsing Presidential Candidates. *The Hill*, September 14, https://thehill.com/opinion/campaign/516200-its-time-for-newspapers-to-stop

-endorsing-presidential-candidates?rl=1 (last accessed October 7, 2020).

Van Brimmer, A. 2020. Editors: Why the *Savannah Morning News* Isn't Endorsing Candidates in 2020 Election. *Savannah Morning News*, October 24, https://www.savannahnow.com/story/opinion/editorials/2020/10/24/editors-why-savannah-morning-news-isnrsquot-endorsing-candidates-in-2020-election/42884777/ (last accessed March 30, 2021).

Van Nostrand, J. 2020. Editor's Notebook: Why the *Tribune* Isn't Endorsing Candidates This Election Cycle. *Columbia Daily Tribune*, October 31, https://www.columbiatribune.com/story/news/columns/2020/10/31/editors-notebook-why-columbia-daily-tribune-isnt-endorsing-candidates/6105935002/ (last accessed March 30, 2021).

BLACK LIVES MATTER PROTESTS, 2020

KENNETH FRENCH

The historic Black Lives Matter (BLM) marches in 2020 were the biggest political protests recently in terms of social scope and geographic reach. The Black Lives Matter movement was started in 2013 by Opal Tometi, Patrisse Cullors, and Alicia Garza in response to the acquittal of George Zimmerman over the killing of Trayvon Martin in Sandford, Florida (Lebron 2017). Since 2013, BLM protests have followed the shooting and/or killing of African Americans by hate crime or by the police. Past protests have followed the police killings of Eric Garner (New York City), Michael Brown (Ferguson, Missouri), Tamir Rice (Cleveland, Ohio), Walter Scott (North Charleston, South Carolina), Alton Sterling (Baton Rouge, Louisiana), Philando Castile (St. Paul, Minnesota), Stephon Clark (Sacramento, California), and Botham Jean (Dallas, Texas), to name a few. However, Black Lives Matter demonstrations in 2020 (figure 3.10) were more numerous and wide ranging than previous protests. The explanations for the geographic distribution of BLM protests come down to local shootings of African Americans and the seminal moment when George Floyd was killed in Minneapolis.

A reason for many Georgia protests was the killing of Ahmaud Arbery—he was pursued and killed while jogging by three White residents of Brunswick, Georgia. The three men were charged with false imprisonment, aggravated assault, murder, and federal hate crimes (Bowman 2021). In Louisville, Kentucky, most of the BLM demonstrations in 2020 were in response to the killing of Breonna Taylor in her own bed by Louisville Metro Police during a "no-knock" warrant. The warrant was for Ms. Taylor's ex-boyfriend, who was said to receive drug packages at her address; no drugs were found in the apartment (Carrega and Ghebremedhin 2020). The shooting of Jacob Blake in Kenosha, Wisconsin, by Kenosha Police on August 23, 2020, ignited rioting in the city and counterprotests turned deadly. Two Black Lives Matter protestors were shot and killed by Kyle Rittenhouse of Antioch, Illinois; he was seen as a patriot by many conservatives and his legal defense fund has raised millions of dollars from donations (*Chicago Sun Times* Staff 2020). While there are several local protests after African Americans were shot or killed in 2020, the main spark to national and worldwide outrage occurred after people watched George Floyd gasp for his last breath on camera.

On May 25, 2020, George Floyd was arrested on suspicion of using a counterfeit $20 bill at a Minneapolis store. After his arrest and handcuffing, Mr. Floyd complained about not being able to breathe. He was restrained on the ground by Minneapolis police officers; Officer Derek Chauvin was seen putting his knee over the neck of George Floyd for several minutes. Even with Mr. Floyd and bystanders pleading for Officer Chauvin to get off, Mr. Floyd went motionless and later died (Hill et al. 2021). Officer Chauvin was found guilty of second-degree unintentional murder, third-degree murder, and second-degree manslaughter in April 2021. The outrage over George Floyd's death led to BLM protesting police brutality and systemic racism worldwide. The peak BLM protest was on June 6, 2020, in which half a million people protested in 550 places in America (Buchanan, Bui, and Patel 2020). In the Minneapolis–St. Paul metropolitan area, outrage and grief turned into rioting and looting as an angry crowd burned buildings (Romo 2020). Black Lives Matter protests, as the figure shows, also spread into major metropolitan areas across the United States. The Washington, DC, BLM protests became a flashpoint—demonstrators focused protesting at the White House, home to President Donald Trump and his pro-police, hardline rhetoric that stated it was okay for police to injure suspects. On June 1, 2020, President Trump ordered law enforcement to use tear gas and riot control tactics on peaceful BLM protesters to clear a path from the White House to Lafayette Park and St. John's Church for a photo op (Gjelten 2020). BLM protesters were also threatened by military helicopters hovering at low altitude at night. In response, the mayor of Washington, DC, renamed the two-block pedestrian section of Sixteenth Street as Black Lives Matter Plaza with the words "Black Lives Matter" painted on the street. The main difference between the Black Lives Matter protests in 2020 compared to the past were the geographic distribution of protests in predominately White small towns and in conservative States. Three hundred people

FIGURE 3.10

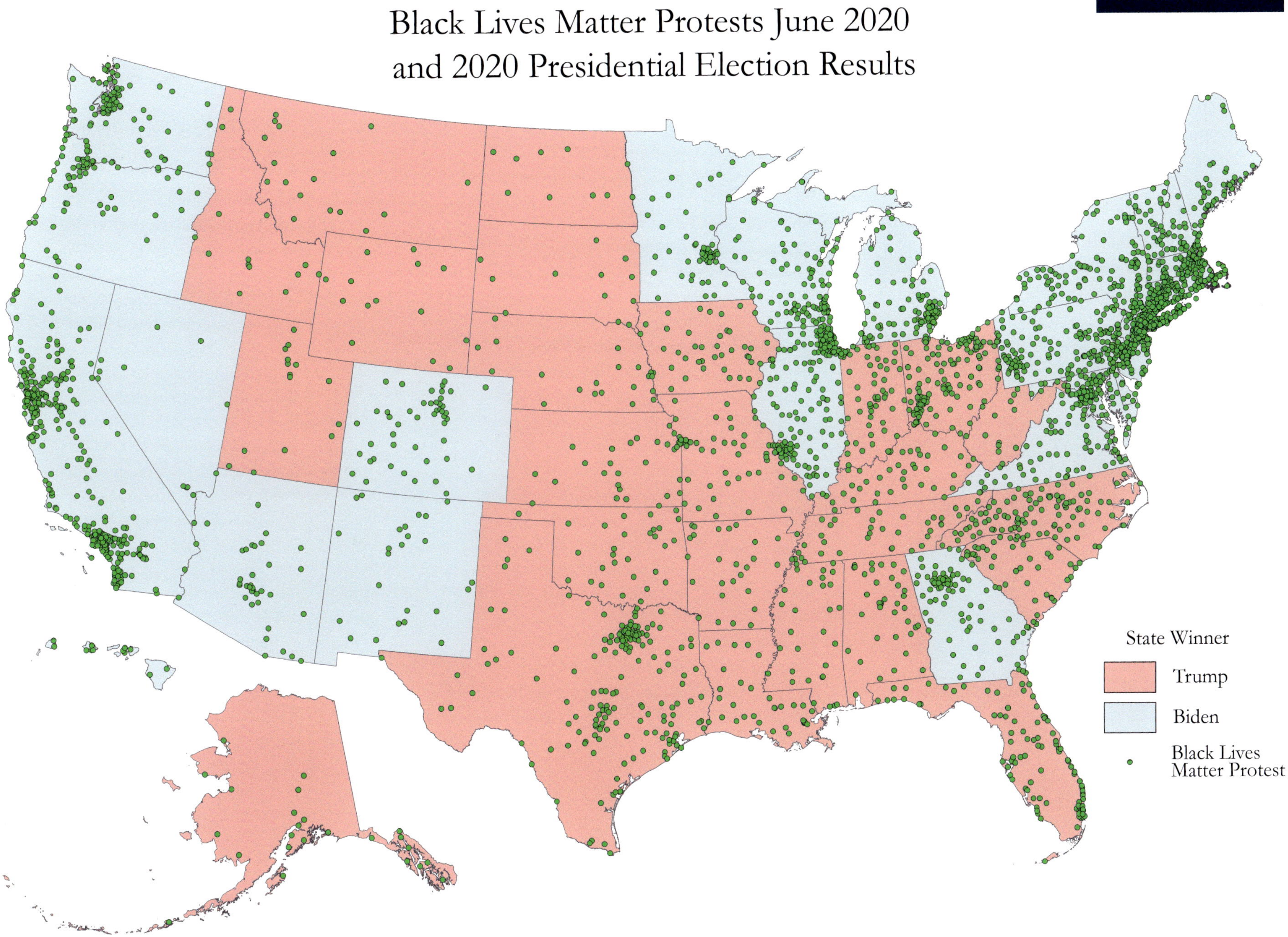
Black Lives Matter Protests June 2020
and 2020 Presidential Election Results
State Winner
Trump
Biden
Black Lives
Matter Protest

demonstrated in Norfolk, Nebraska (24,000 population) and 250 protested in Farmington, New Mexico (44,000 population; Schulte 2020). Black Lives Matter marches also took place in the small towns of Fairmont, West Virginia, Lynn, Massachusetts, and Havre, Montana (Putnam, Chenoweth, and Pressman 2020).

The critics of the Black Lives Matter movement include the counterprotest of pro-police Blue Lives Matter and those who were more upset over the rioting and looting than the killing of African Americans extrajudicially. On May 29, 2020, President Trump tweeted in response to the BLM protests: "when the looting starts, the shooting starts," a call to the bigoted Miami Police Chief Walter Headly quote in 1967 to justify the police shooting of African American protesters (Sprunt 2020). The killing of several African Americans in 2020 led to protests and counterprotests in a highly politically charged climate. Regardless of political pressure and counterprotests, the historical Black Lives Matter protests in 2020 indicated that many Americans were in favor of ending police brutality and systemic racism in the criminal justice system.

REFERENCES

Bowman, E. 2021. 3 Men Indicted on Federal Hate Crimes Charges in Ahmaud Arbery Killing. *National Public Radio*. April 28, https://www.npr.org/2021/04/28/991722733/3-men-indicted-on-federal-hate-crime-charges-in-ahmaud-arbery-killing (last accessed April 30, 2021).

Buchanan, L., Q. Bui, and J. Patel. 2020. Black Lives Matter May Be the Largest Movement in U.S. History. *New York Times*. July 3, https://www.nytimes.com/interactive/2020/07/03/us/george-floyd-protests-crowd-size.html (last accessed April 30, 2021).

Carrega, C. and S. Ghebremedhin. 2020. Timeline: Inside the Investigation of Breonna Taylor's Killing and Its Aftermath. *ABC News*. November 17, https://abcnews.go.com/US/timeline-inside-investigation-breonna-taylors-killing-aftermath/story?id=71217247 (last accessed April 29, 2021).

Chicago Sun Times Staff. 2020. Kyle Rittenhouse Charged in Shooting Deaths of 2, Wounding of 1 during Kenosha Unrest. *Chicago Sun Times*. December 30, https://chicago.suntimes.com/news/2020/12/30/22206292/kyle-rittenhouse-kenosha-wisconsin (last accessed April 29, 2021).

Gjelten, T. 2020. Peaceful Protesters Tear-Gassed to Clear Way for Trump Church Photo-Op. *National Public Radio*. June 1, https://www.npr.org/2020/06/01/867532070/trumps-unannounced-church-visit-angers-church-officials (last accessed April 30, 2021).

Hill, E., A. Tiefenthaler, C. Triebert, D. Jordan, H. Willis, and R. Stein. 2021. How George Floyd Was Killed in Police Custody. *New York Times*. May 31, https://www.nytimes.com/2020/05/31/us/george-floyd-investigation.html (last accessed April 28, 2021).

Lebron, C. 2017. *The Making of Black Lives Matter: A Brief History of an Idea*. Oxford: Oxford University Press.

Putnam, L., E. Chenoweth, and J. Pressman. 2020. The Floyd Protests Are the Broadest in U.S. History—and Are Spreading to white, Small-Town America. *Washington Post*. June 6, https://www.washington-post.com/politics/2020/06/06/floyd-protests-are-broadest-us-history-are-spreading-white-small-town-america/ (last accessed April 30, 2021).

Romo, V. 2020. Photos: Protests Over George Floyd's Death Escalate into Violence, Destruction. *National Public Radio*. May 28, https://www.npr.org/sections/pictureshow/2020/05/28/864559449/photos-protests-over-george-floyds-death-grow-violent-it-is-how-we-express-pain (last accessed April 29, 2021).

Schulte, G. 2020. George Floyd Protests Spread to Smaller, Mostly White Towns. *ABC News*. June 7, https://abcnews.go.com/US/wireStory/george-floyd-protests-spread-smaller-white-towns-71122457 (last accessed April 30, 2021).

Sprunt, B. 2020. The History Behind When the Looting Starts, the Shooting Starts. *National Public Radio*. May 29, https://www.npr.org/2020/05/29/864818368/the-history-behind-when-the-looting-starts-the-shooting-starts (last accessed May 2, 2021).

4 OUTCOMES

RESULTS OF THE 2020 PRESIDENTIAL ELECTION AT THE STATE AND COUNTY LEVELS

JOHN HEPPEN, FRED M. SHELLEY, AND RICHARD L. MORRILL

Democrat and former Vice President Joseph Biden defeated incumbent Donald J. Trump at the polls by about seven million votes and in the Electoral College 306 to 232 electoral votes and took twenty-five states and the District of Columbia (figures 4.1 and 4.2). Although Biden won most of the electoral votes of Maine, Maine law grants an electoral vote to the candidate carrying each House of Representatives district and Trump carried Maine's Second District. Thus, Maine's electoral votes were split, with three going to Biden and one going to Trump. Nebraska has similar laws; Trump won the popular vote, but Biden captured the electoral votes of the Omaha-based district. Nebraska's electoral vote was split 4 to 1 in favor of Trump.

At the state level, the pattern of popular vote support for Trump and Biden paralleled patterns of support for Republican and Democratic candidates in recent elections, with Biden carrying the Northeast and the Pacific Coast states while Trump carried the South, the Great Plains, and most of the Rocky Mountain states. The Midwest was split. Biden was successful in that he recaptured states that had flipped from Obama to Trump in Michigan, Wisconsin, and Pennsylvania. In 2016, Trump carried them by narrow margins—10,704 popular votes in Michigan, 44,292 in Pennsylvania, and 22,748 in Wisconsin. In 2020 Biden won Michigan by 154,188 votes, Pennsylvania by 80,555 votes, and Wisconsin by 20,682. In addition to those states, Biden's victory was augmented by snatching Arizona and Georgia. Arizona was won by just 10, 457 votes and Georgia was won by a scant 12, 670. Biden failed to recapture Ohio, Iowa, and Florida.

Presidential elections are won and lost at the state level in the Electoral College, and examination of state-level results blurs vital additional insights that can be discerned by examining county-level results. A map showing which candidate earned the most votes in each county illustrates the sharp divide between urban and rural places (figure 4.3). The map illustrates the great rural and urban divide in the country. This gap was evident in previous elections but intensified in the 2016 election and again in 2020 as rural voters supported Trump strongly. Trump dominated the largely white and rural counties throughout the country. However, Biden did carry many rural counties containing large minority populations. Examples include Black majority counties in the Mississippi Delta, central Alabama, and northeastern North Carolina. Plus, he won heavily Hispanic counties in South Texas and counties containing American Indian reservations.

Otherwise, Biden's support was confined largely to metropolitan areas. He won Illinois and Minnesota by carrying the Chicago and Twin Cities metropolitan areas respectively while losing

FIGURE 4.1

2020 Presidential Election

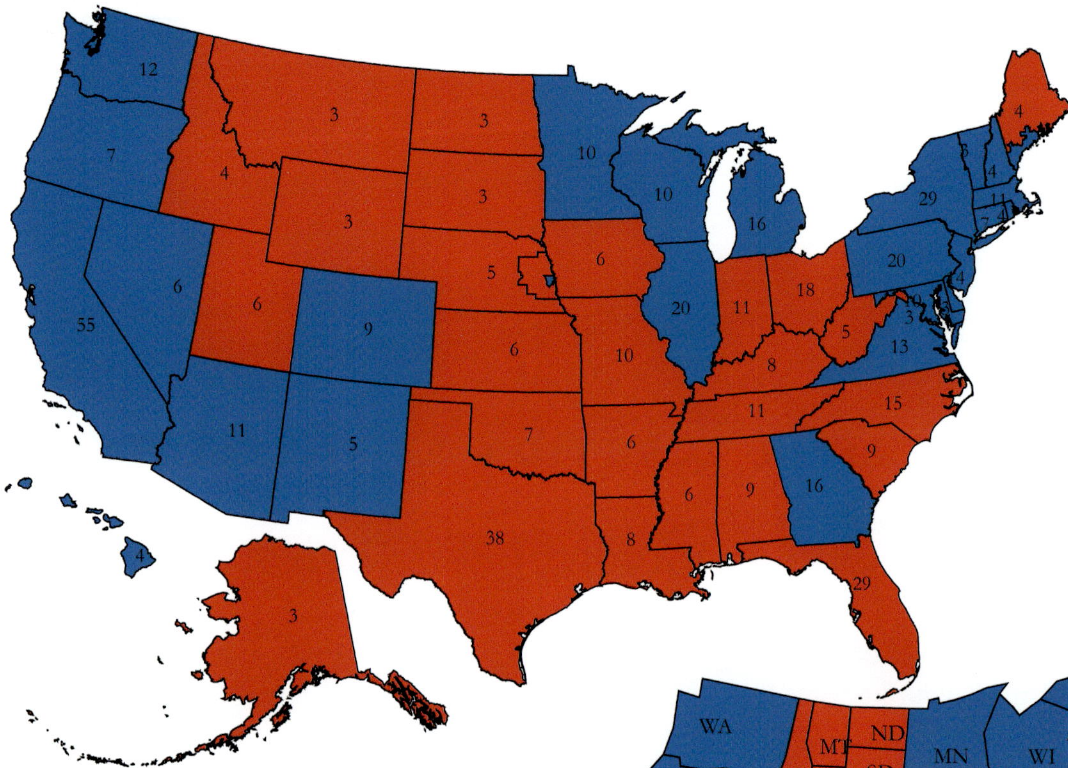

Popular Electoral Slate Winner

Democratic

Republican

Numbers on the map show electoral votes apportioned by state.

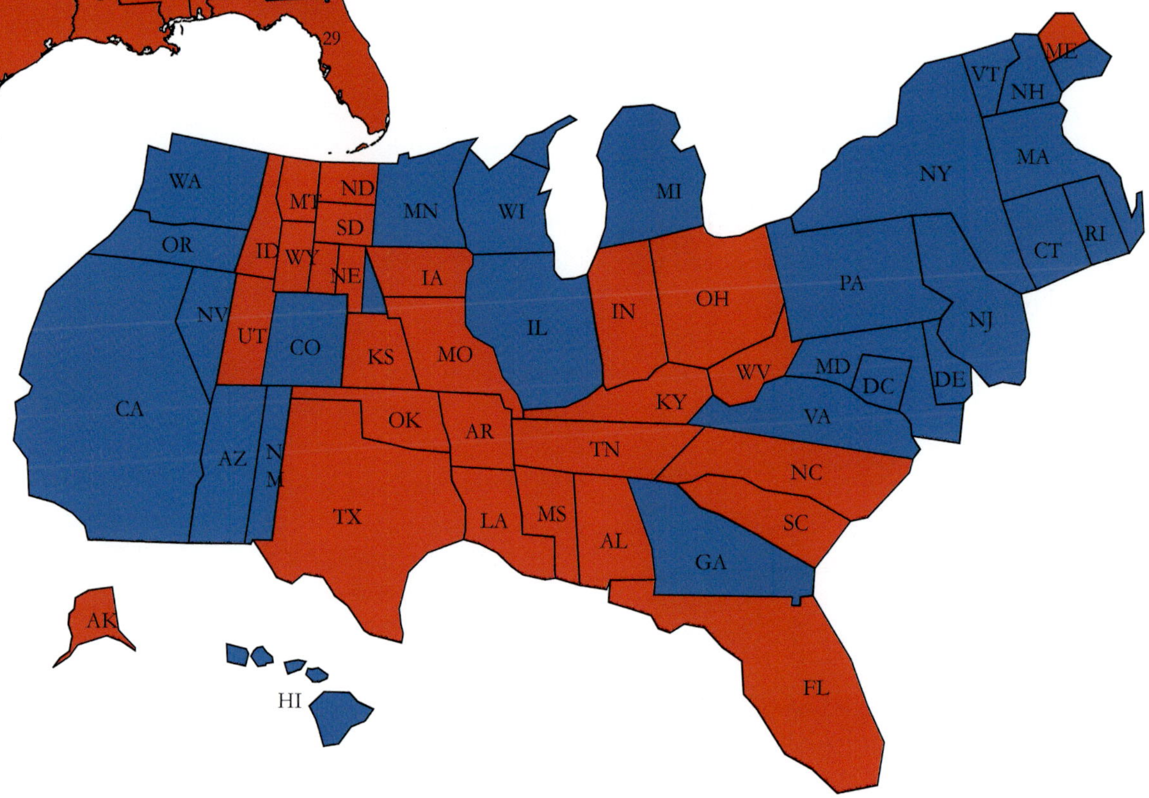

Electoral College Vote

Biden (306)

Trump (232)

FIGURE 4.2

Percentage of Popular Vote
2020 Presidential Election

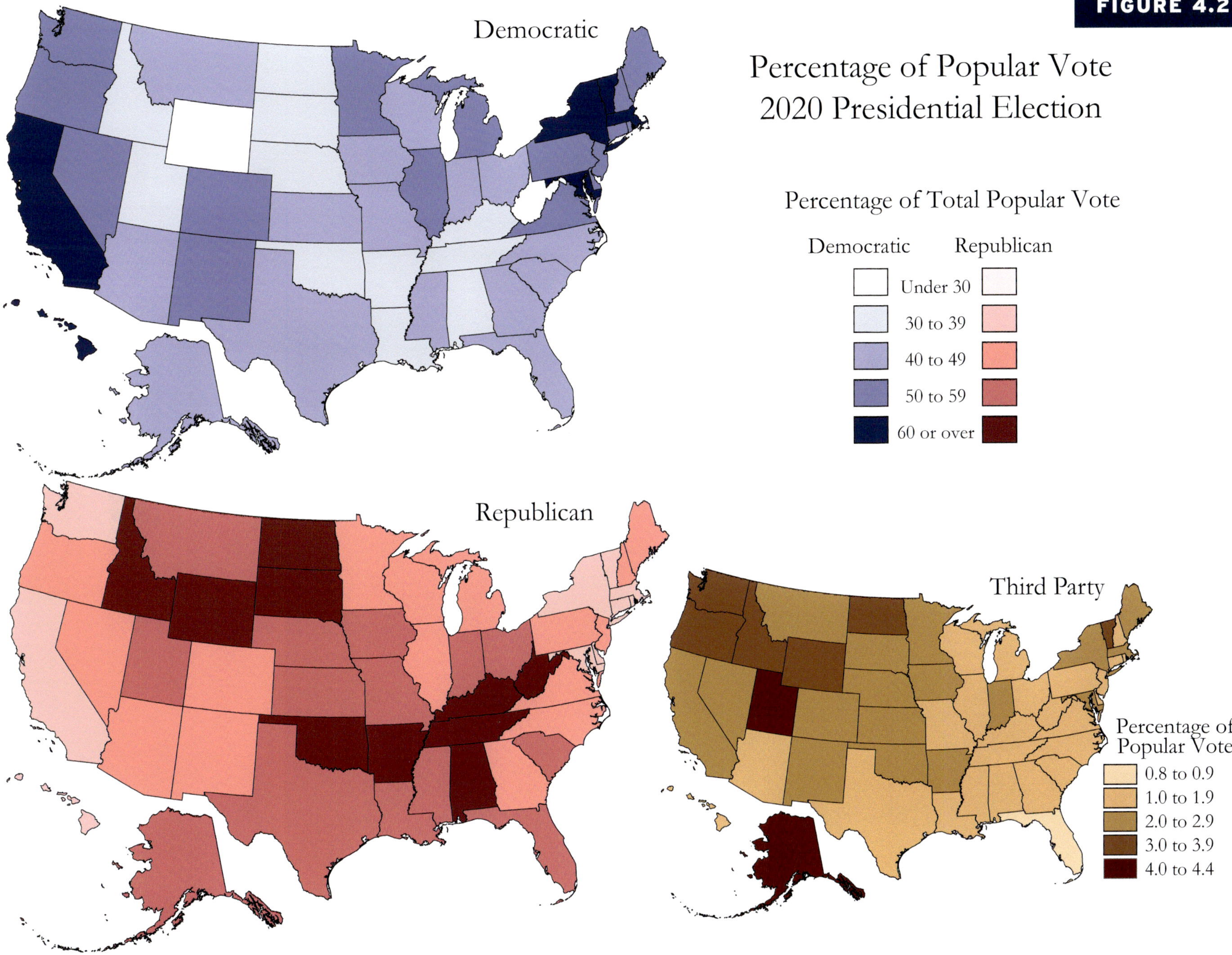

FIGURE 4.3

Presidential Election Popular Vote 2020

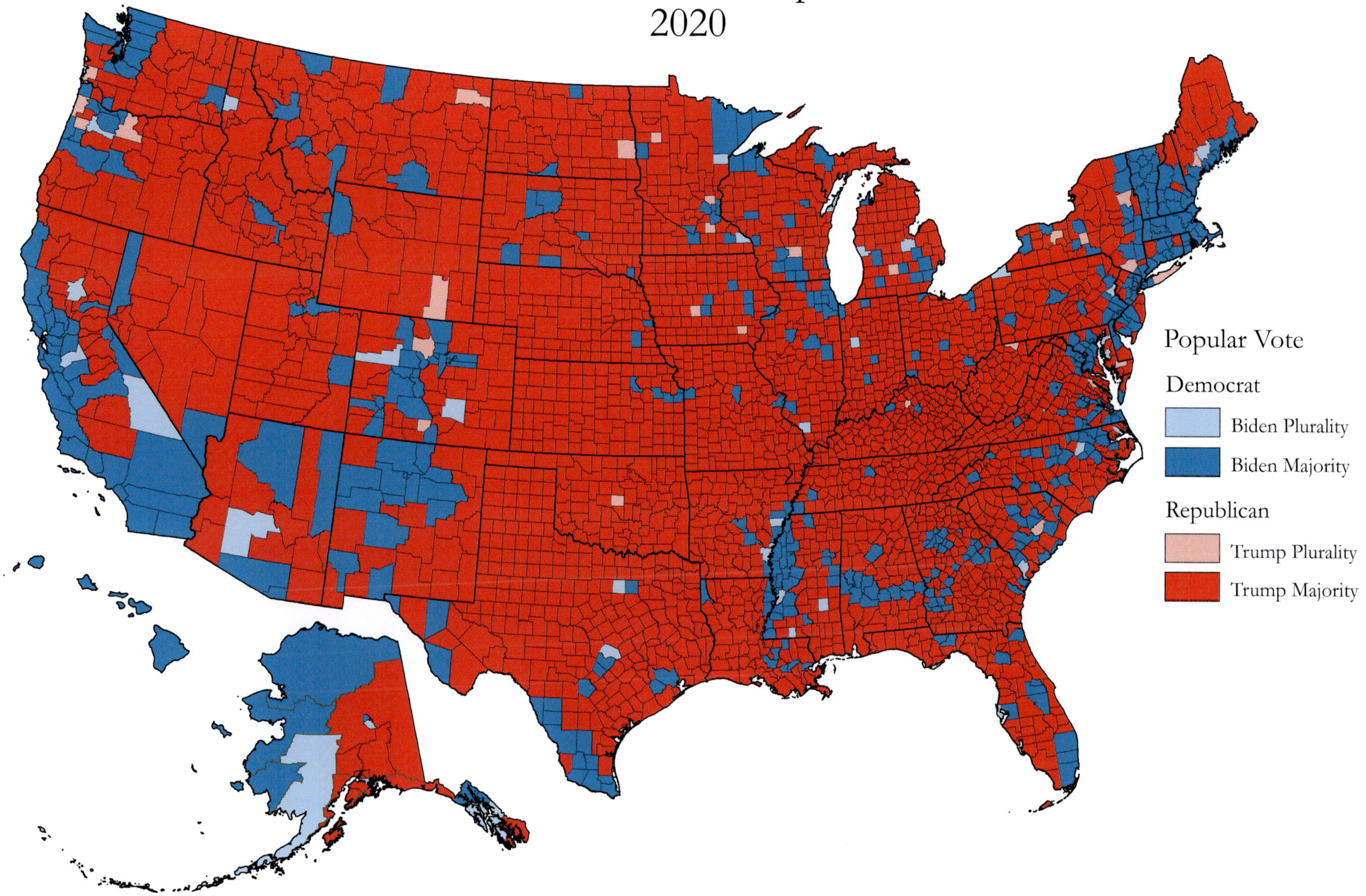

most non-urban counties. He carried Detroit, Philadelphia, and Milwaukee in Michigan, Pennsylvania, and Wisconsin respectively, but unlike 2016 his margins in these cities were larger and were augmented by increased support in suburban counties. Every county bordering Cook County (Chicago) Biden won. He won two counties—Washtenaw and Oakland—bordering Wayne County home to Detroit. Elsewhere, urban-rural differences are evident particularly in Texas, in which Trump won most counties but lost Harris County (Houston), Dallas County (Dallas), Bexar County (San Antonio), and Travis County (Austin) to Biden. Similarly, Biden won Cleveland, Columbus, and Cincinnati but lost most other counties in Ohio and lost the state. He won Kansas City and St. Louis but lost most other counties in Missouri. In summary, a common theme is the dominance of Biden in urban counties and in states with the largest urban metropolitan areas. From New York in the East, Illinois in the Midwest, and California in the West this was Biden's base. But there were notable exceptions in Texas, Florida, and Ohio. Perhaps the switch of Georgia fueled by Metropolitan Atlanta foreshadows change in other Sunbelt States.

Even more insights can be gained by looking at levels of intensity of vote for Trump and Biden, as shown in figures 4.4 and 4.5 respectively. On a percentage basis, Trump won his largest margins in rural and generally sparsely populated counties in the Great Plains. For example, he won supermajorities of the popular vote in counties of the Great Plains and Rocky Mountains. From the Canadian border to the plains of Texas he obtained incredibly high vote percentages. In Wallace County, Kansas, Trump won 93 percent of the vote. In Carter County, Montana, he won 90 percent and 92 percent in Harding County, South Dakota. Trump also exceeded 80 percent of the vote in many rural Appalachian counties with 88 percent in Clay County, Kentucky, and 85 percent in Mingo County, West Virginia.

Biden, who won the metropolitan vote nationwide, did especially well in large central-city places. He won 92 percent in Washington, DC, 87 percent in New York County, New York (Manhattan), 85 percent in San Francisco County, California, and 83 percent in Boston, Massachusetts. All these places are small in area and very densely populated. However, his margins in some urban counties with larger land areas and containing suburbs were almost as large, with 81 percent in Philadelphia County, Pennsylvania, 74 percent in Cook County, Illinois (Chicago), 71 percent in Los Angeles County, California, and 75 percent in King County, Washington (Seattle). Outside of metropolitan areas, his largest percentages were concentrated in rural areas with large minority populations with 85 percent in Claiborne County, Mississippi, 63 percent in East Carroll Parish, Louisiana, along the Mississippi River, and 88 percent on the Oglala Lakota (Sioux) reservation in South Dakota.

These trends are evident also from examination of Biden's margin of victory at the county level, as illustrated in figures 4.6 and 4.7. Biden's margins of victory were highest in the urban states along the Northeast Corridor from Massachusetts to Virginia. Biden also won big in Illinois and along the West Coast, anchored by California, Washington, and Oregon.

Biden's margins in the swing states of Georgia, Wisconsin, and Arizona tended to be lower than was the case in the West Coast and Northeast. However, he won counties in urban areas of these states by large enough margins to offset Trump's support in rural areas. Note that Trump's margins of victory tended to be lower in more urbanized counties in these states such as counties containing Knoxville (Tennessee), Wichita (Kansas), and Cheyenne (Wyoming). But because these counties have larger populations, Trump's absolute margin of victory tended to be greater in these counties, adding to his lead in the states in which they are located.

Third-party candidates did not have as much appeal in 2020 as they did in 2016. In 2016 the third parties won approximately 5 percent of the vote nationally (figures 4.8, 4.9, 4.10). In 2016 the Libertarian Party won 3.27 percent of the popular vote with highs of 9 percent in New Mexico. They exceeded 5 percent of the popular vote in Alaska, Colorado, Maine, Montana, North Dakota, Oklahoma, South Dakota, and Wyoming. In 2020 they won less than 2 percent of the vote nationally. The Libertarian Party did not have as high-profile candidates in 2020. In 2016 they had two former governors in Gary Johnson (New Mexico) and William Weld (Massachusetts) on the ticket. In 2020 the Libertarians ran Dr. Jo Jorgensen for president and Jeremy "Spike" Cohen for vice president. They earned 1.18 percent of the national vote. They received more than 2 percent of the popular vote in eight states: North and South Dakota, Utah, Montana, Alaska, Kansas, Nebraska, and Wyoming. The West has long been the most popular region for Libertarian candidates relatively speaking, but given Trump's margins in those states they were basically a weak protest vote or a principled Libertarian vote. But in three states, Georgia, Arizona, and Wisconsin, their vote could have made a difference. Biden's margins in those three states were 12,670, 10,457, and 20, 682 respectively.

FIGURE 4.4

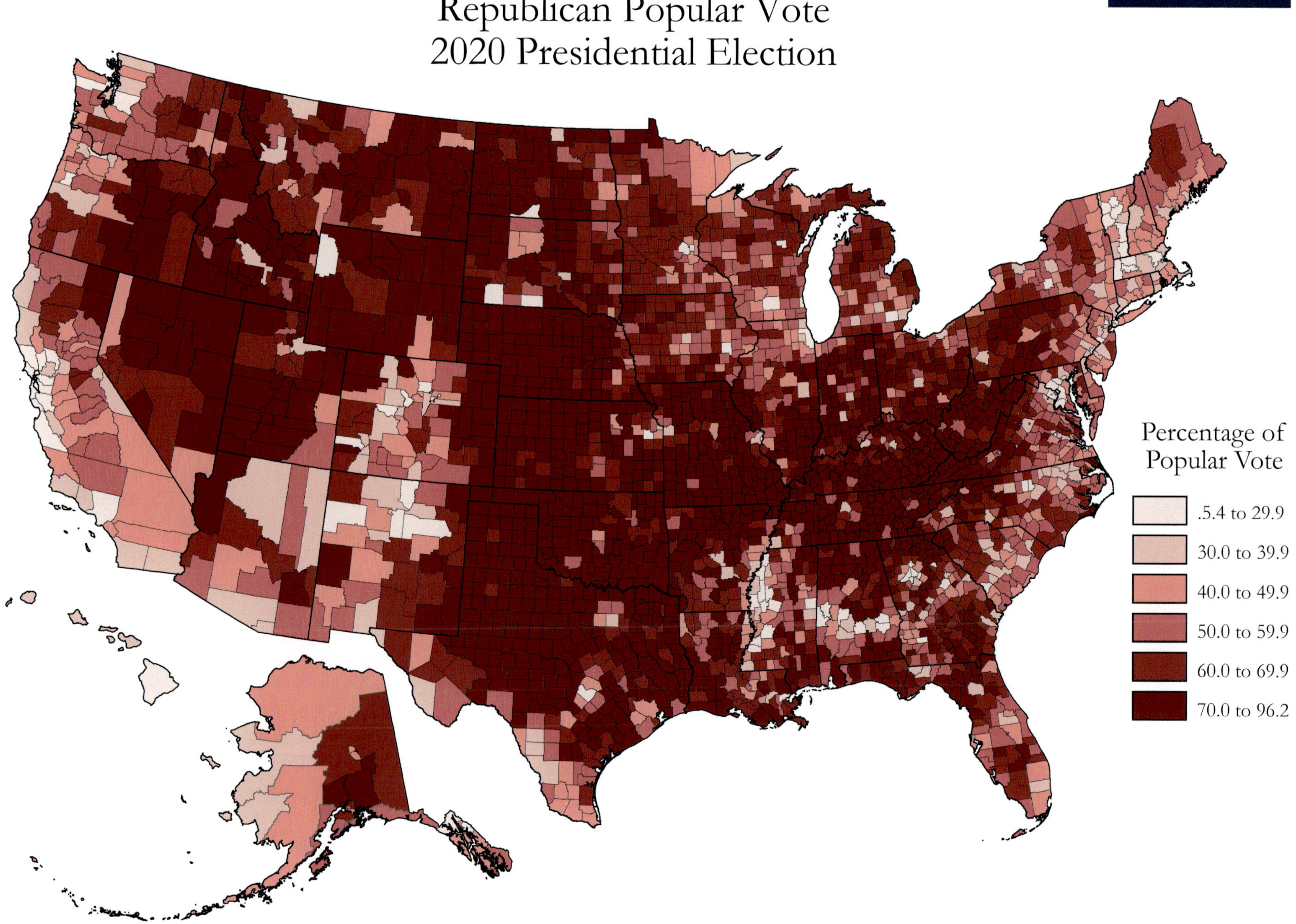
Republican Popular Vote
2020 Presidential Election
Percentage of
Popular Vote
.5.4 to 29.9
30.0 to 39.9
40.0 to 49.9
50.0 to 59.9
60.0 to 69.9
70.0 to 96.2

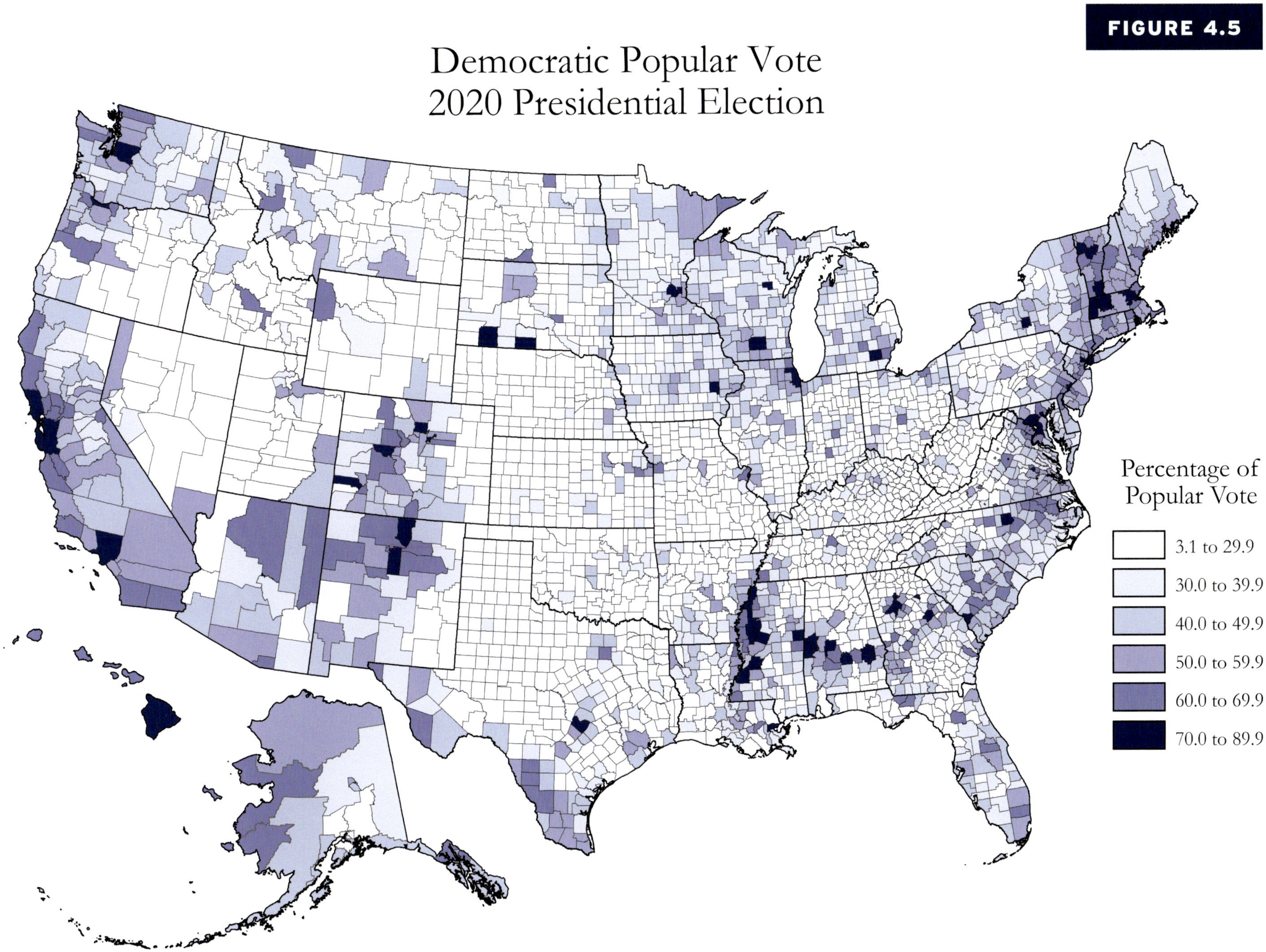
FIGURE 4.5
Democratic Popular Vote
2020 Presidential Election
Percentage of Popular Vote
3.1 to 29.9
30.0 to 39.9
40.0 to 49.9
50.0 to 59.9
60.0 to 69.9
70.0 to 89.9

FIGURE 4.6

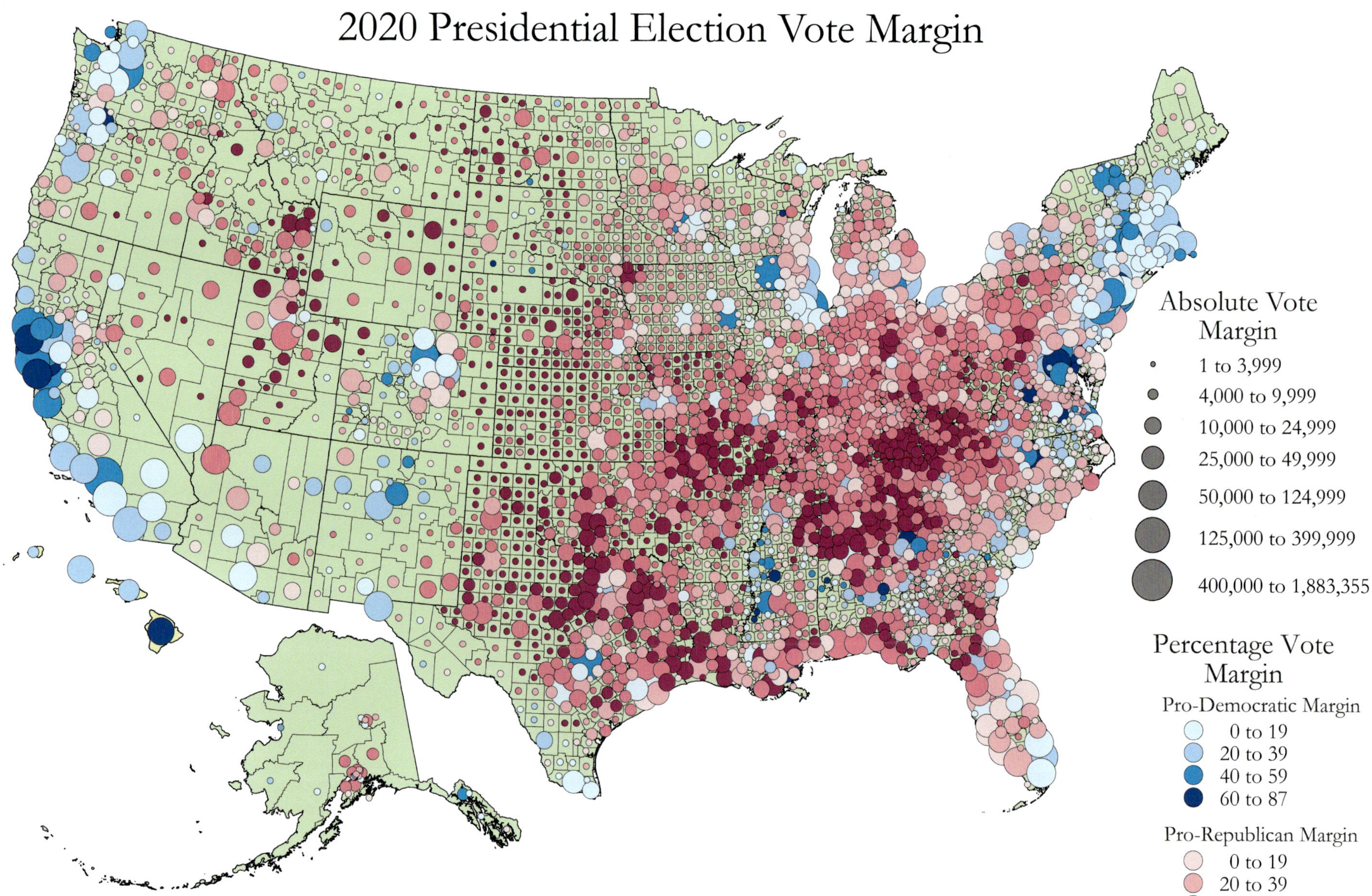
2020 Presidential Election Vote Margin
Absolute Vote Margin
1 to 3,999
4,000 to 9,999
10,000 to 24,999
25,000 to 49,999
50,000 to 124,999
125,000 to 399,999
400,000 to 1,883,355
Percentage Vote Margin
Pro-Democratic Margin
0 to 19
20 to 39
40 to 59
60 to 87
Pro-Republican Margin
0 to 19
20 to 39
40 to 59
60 to 94

FIGURE 4.7

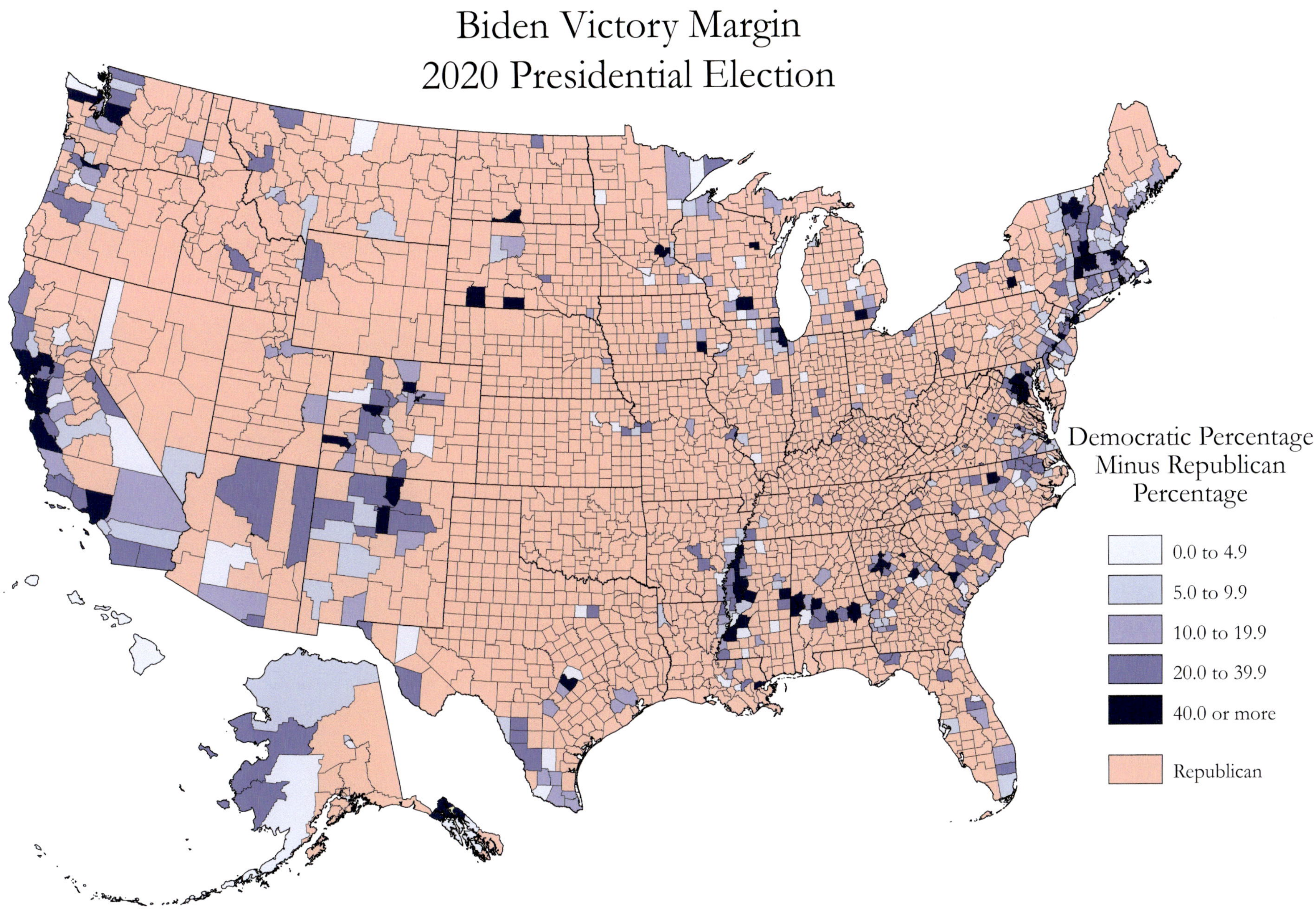
Biden Victory Margin
2020 Presidential Election
Democratic Percentage
Minus Republican
Percentage
0.0 to 4.9
5.0 to 9.9
10.0 to 19.9
20.0 to 39.9
40.0 or more
Republican

Jorgensen received 62,138, 51,465, and 38,491 in those states respectively. Prominent Libertarian-leaning Republicans like Kentucky Senator Rand Paul have found a home in the Republican Party and some Republicans associated with the Lincoln Project vowed not to vote for Donald Trump. A weak protest or principled Libertarian vote may have been the difference in those three states. At the county level, counties of the Plains, the Rocky Mountains, and electoral districts of Alaska are where Libertarian views are more popular. His highest total was in Alaska's Electoral District 2 where he received 4.7 percent of the total popular vote. In all of Alaska's forty electoral districts, Jorgensen received better than 1.67 percent of the vote. Of note is Sioux County, North Dakota, which saw the Libertarian Party win 4.0 percent of the popular vote. That was the second-highest level of the Libertarian Party in the country below the state level. Sioux County lies entirely within the Standing Rock Indiana Reservation. Jorgensen received 48 votes out of a total of 1,186 cast.

In 2016 the Green Party led by Dr. Jill Stein won 1.06 percent of the popular vote. In 2020 the Green Party vote plummeted to 0.26 percent of the vote. Howie Hawkins was the Green Party nominee. He was also the nominee of the Socialist Party USA, one of the descendants of the Socialist Party of America, as is the more well-known and larger Democratic Socialists of America. The Socialist Party of American reached its peak in the 1912 presidential election with Eugene Debs leading the ticket at 6 percent of the national popular vote. The Green Party was on the ballot in thirty-six states plus the District of Columbia and only topped 1 percent of the vote in Maine. In no state did the Greens perform better than the Libertarians when both were on the ballot. After Maine, only in Alaska, Hawaii, Maryland, Massachusetts, Illinois, District of Columbia, and Oregon did the Green Party hover above or around 0.50 percent of the vote. In counties in Maine, the West Coast, and New Mexico, and in Alaska electoral districts, Hawkins managed to do better than his national average. In only ten counties of the Lower 48 did Hawkins manage more than 1 percent of the popular vote, with eight of those counties in Maine. He did better in Alaska's electoral districts. He topped 1 percent of the total vote in six districts with a high of 2.1 percent in Electoral District 38. Hawaii was more fertile ground for the Greens; in Hawaii County, the biggest county of Hawaii, Hawkins received a little over 1 percent of the vote.

As noted, Trump won the 2016 presidential election by winning the electoral votes of Florida, Iowa, Michigan, Ohio, Pennsylvania, and Wisconsin, all of which had given their electoral votes to Barack Obama in 2012. Biden's ability to win back Pennsylvania, Michigan, and Wisconsin, and take away Georgia and Arizona, for a swing of seventy-three electoral votes, was vital to his victory. Even though Trump won most of the counties, it was the urban and suburban counties that saw an increase in Democratic votes from 2016 that was key to Biden's victory. Of note is that in many counties Trump increased his share of the vote from 2016 also (figures 4.11 and 4.12). Figure 4.11 maps the change in certified Republican votes counted from the 2016 to 2020 election. The change is expressed as a percentage. This map represents where the greatest percentage increase of counted votes for Trump happened from 2016 to 2020. Figure 4.12 shows the same phenomenon for Biden. It maps the change in the certified Democratic votes counted from 2016 to 2020. This map represents where the greatest percentage increase of counted votes from Clinton to Biden happened from 2016 to 2020.

An analysis and comparison of the two maps provides keen insights. First the West showed the greatest increase in votes for both parties. Washington, Oregon, California, Arizona, Nevada, Idaho, Montana, New Mexico, Utah, Colorado, and Wyoming saw increases at the county level of greater than 30 percent. Of note is that Democratic votes saw more counties with an increase over 30 percent than Republican. California, Oregon, and Washington, Montana, Idaho, Wyoming, Utah, and New Mexico were not considered battleground states, but it does demonstrate greater participation and interest. Arizona, Nevada, and Colorado were considered states more in play. Utah saw an increase in Democratic votes counted of 30 percent or more in every county. Trump's vote increased also. The 2016 election featured the candidacy of Mormon Evan McMullin, who ran a third-party campaign as a principled conservative alternative to Donald Trump. Some residue of his campaign can be found perhaps in the increased votes for Biden compared to Clinton, as some Utah voters still could not pull the lever for Donald Trump. In the western battleground states of Arizona, Nevada, and Colorado, there was greater enthusiasm for Biden versus Clinton, and Trump was not able to match the increased enthusiasm.

Texas and Florida present interesting cases. Southern Texas and the Rio Grande River Valley is home to a traditionally strong Democratic vote with high percentages of Mexican American voters. Trump showed greater increases in counted votes in those areas than Biden and Biden saw a decrease or very little increase from 2016. Though the Urban Triangle of Texas from Houston to

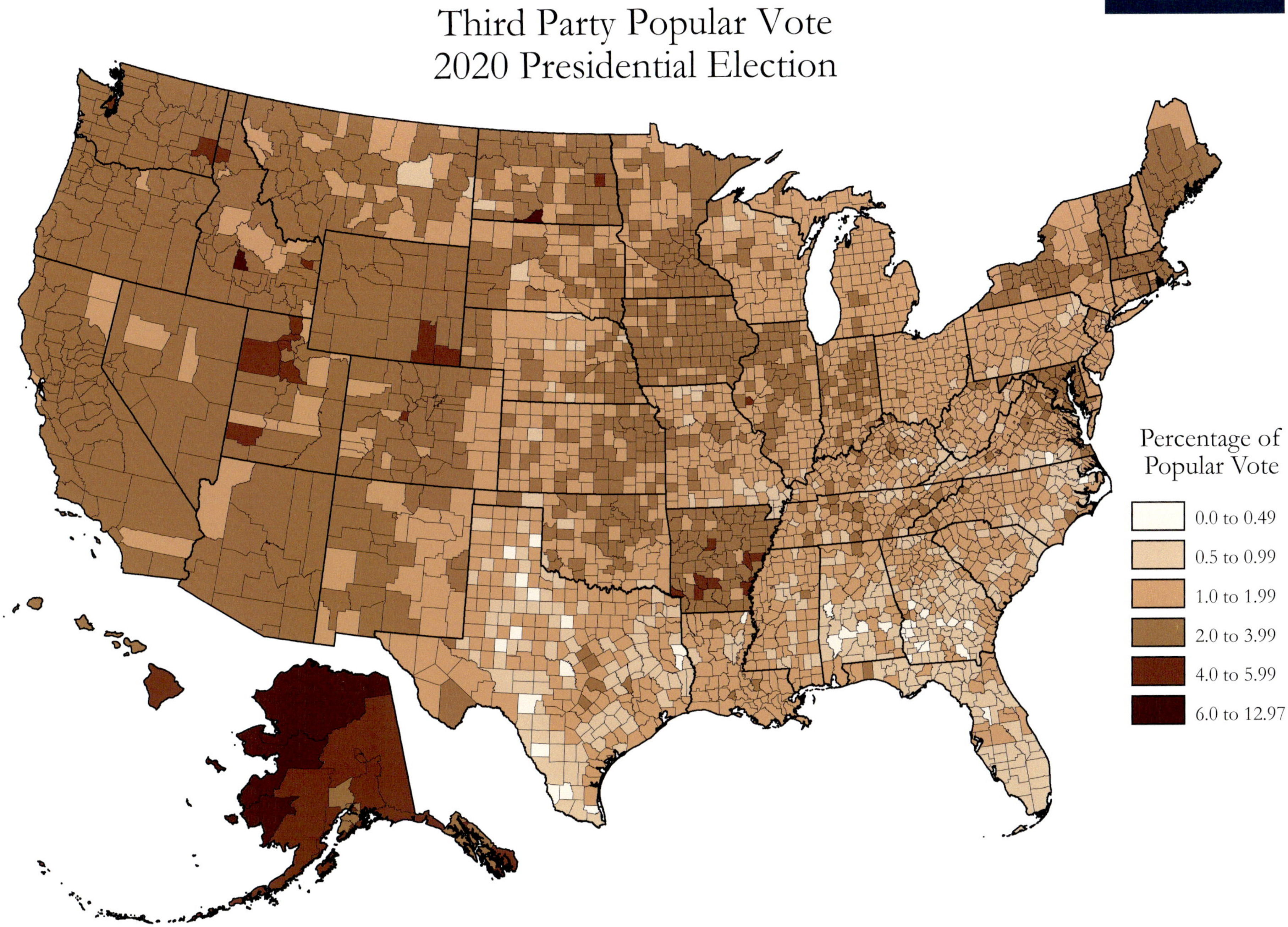
Third Party Popular Vote
2020 Presidential Election
Percentage of Popular Vote
0.0 to 0.49
0.5 to 0.99
1.0 to 1.99
2.0 to 3.99
4.0 to 5.99
6.0 to 12.97

FIGURE 4.9

Jo Jorgensen, Libertarian Party Popular Vote
2020 Presidential Election

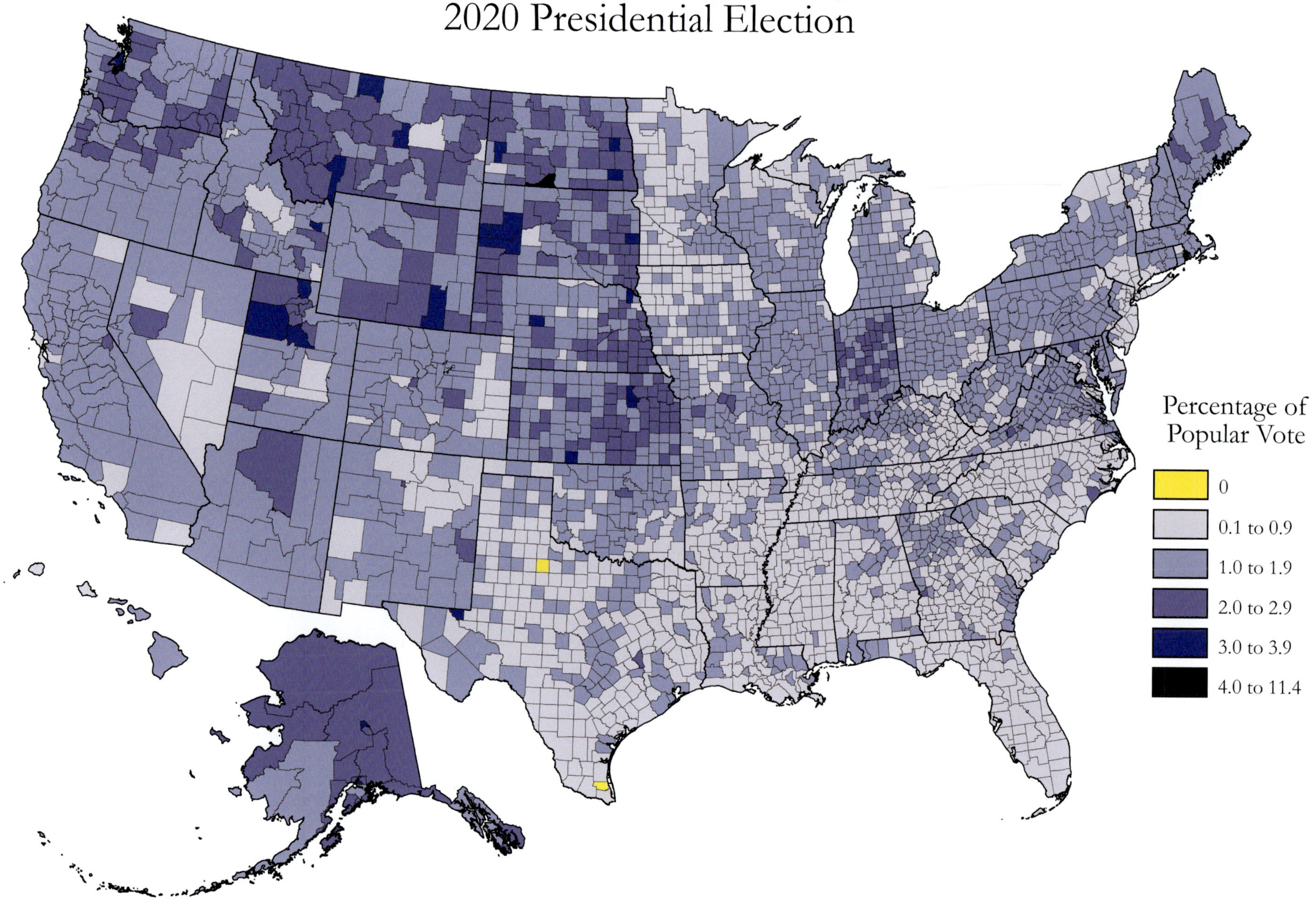

FIGURE 4.10

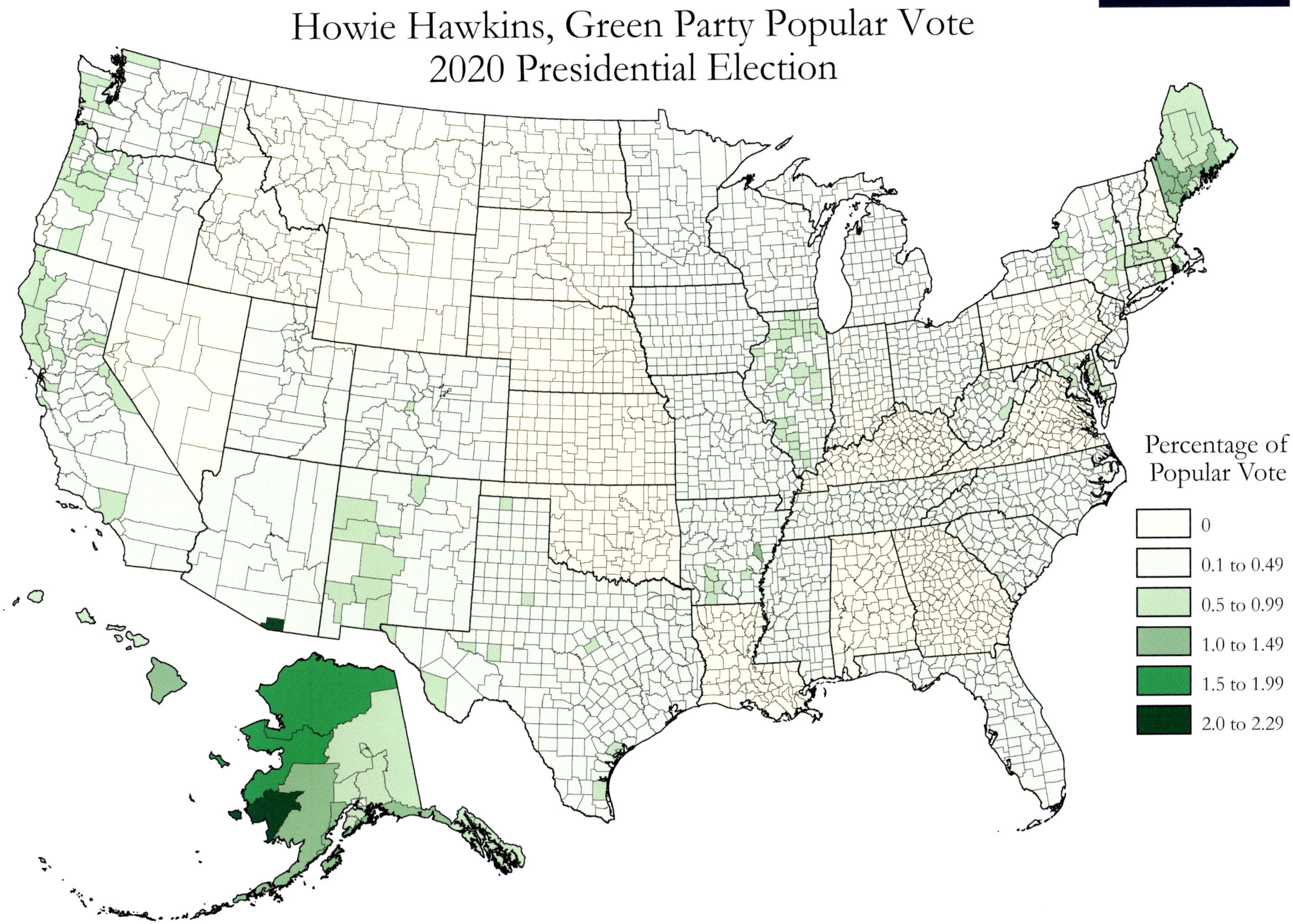
Howie Hawkins, Green Party Popular Vote
2020 Presidential Election
Percentage of Popular Vote
0
0.1 to 0.49
0.5 to 0.99
1.0 to 1.49
1.5 to 1.99
2.0 to 2.29

FIGURE 4.11

Change in Certified Republican Votes Counted 2016 to 2020 Presidential Elections

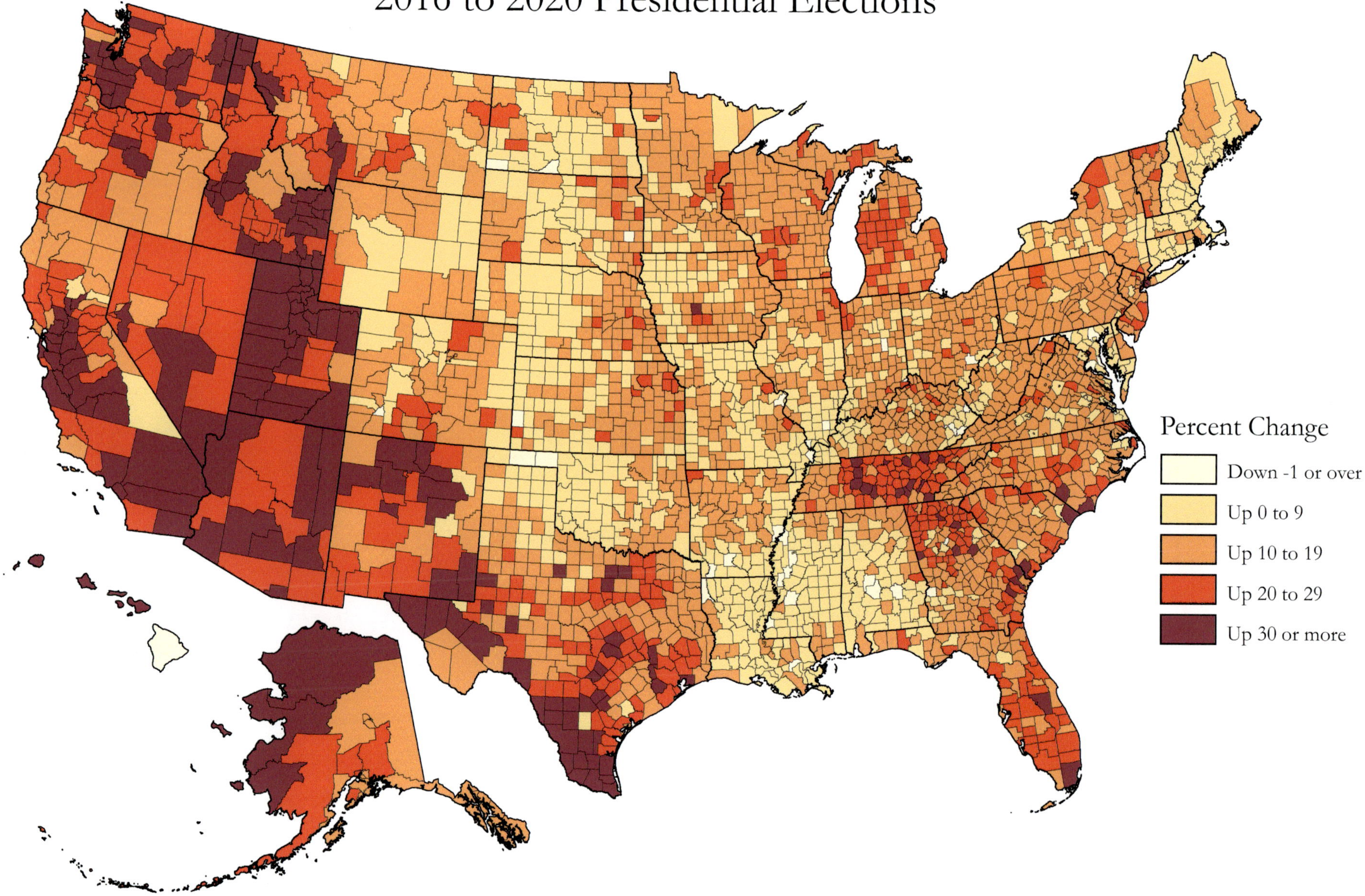

FIGURE 4.12

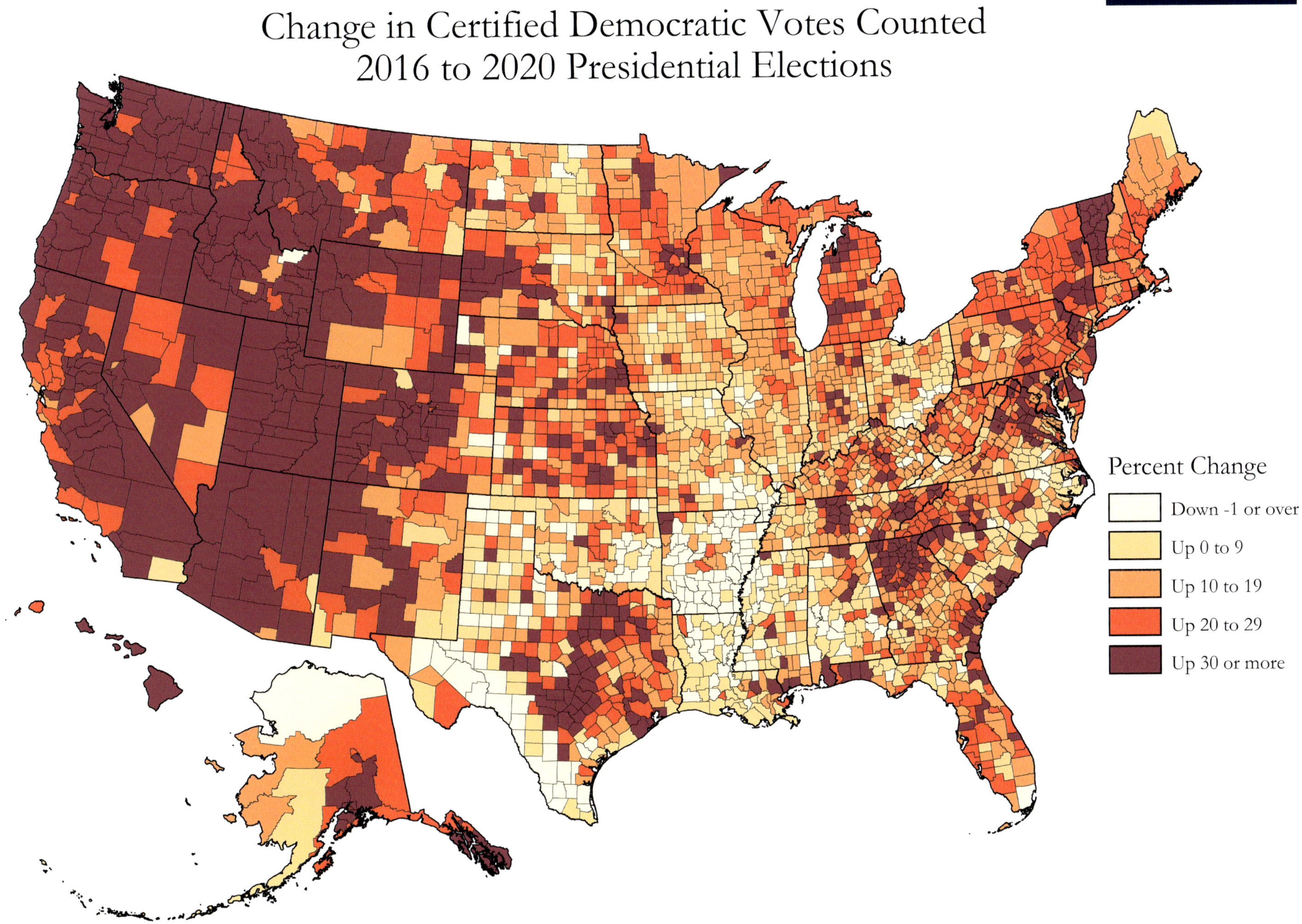
Change in Certified Democratic Votes Counted
2016 to 2020 Presidential Elections
Percent Change
Down -1 or over
Up 0 to 9
Up 10 to 19
Up 20 to 29
Up 30 or more

FIGURE 4.13

Persistence and Change
2016 and 2020 Presidential Elections

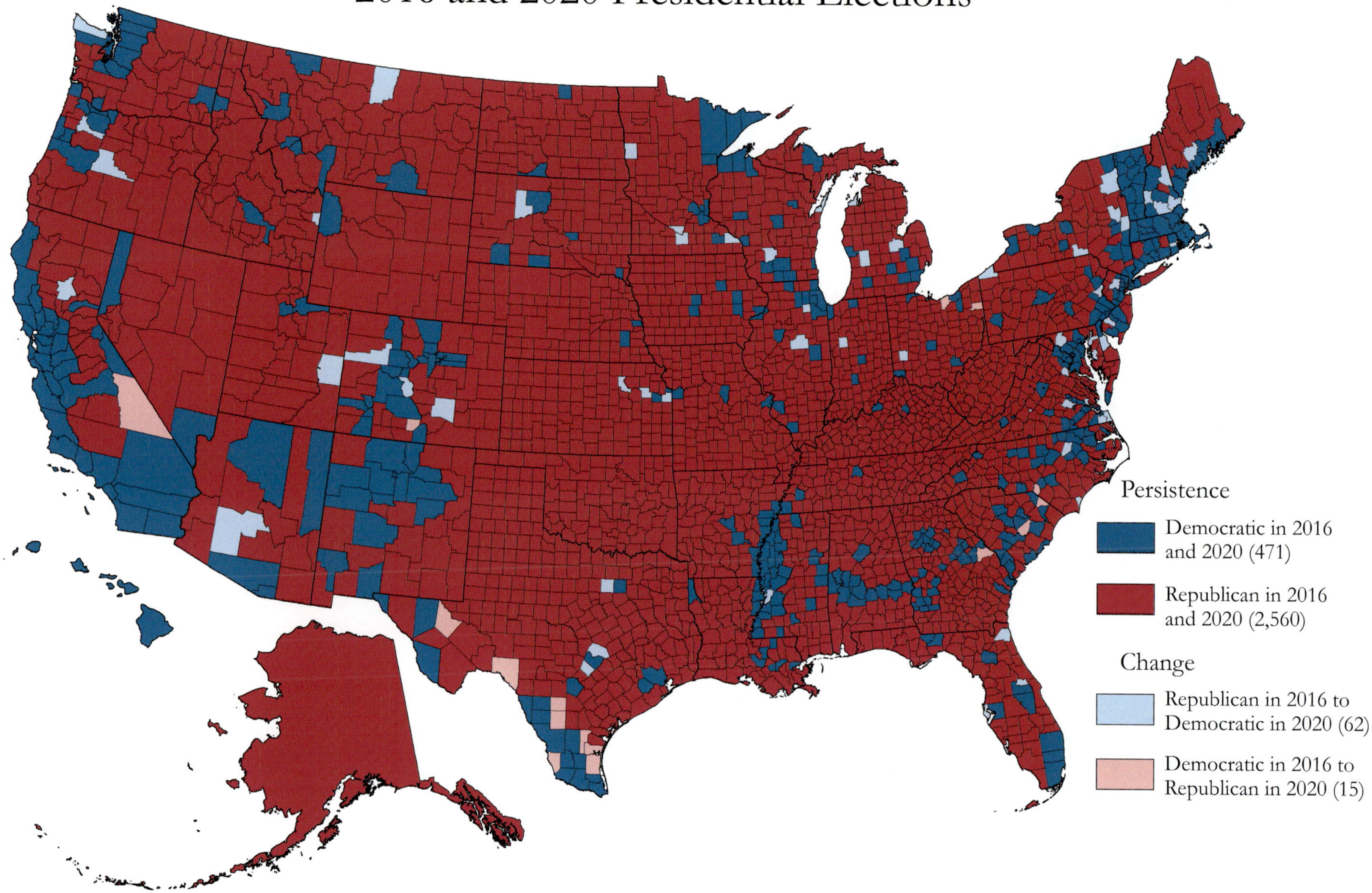

San Antonio to Dallas–Fort Worth with Austin in the middle saw large increases in Democratic votes. Florida remained in the Trump column; his certified votes increased in southern Florida and Biden's remained flat, especially in Miami–Dade County where Cuban American voters hold considerable power.

In other parts of the country there is a general pattern of increased votes for both parties, but the Democratic vote increase in the Midwest, South, and Northeast is more robust, especially in the Atlanta region, middle Tennessee, Pennsylvania, the Northeast, and New England. Vermont saw an increase of note. Perhaps any hard feelings from the Clinton-Sanders battle of 2016 were forgiven. Michigan, Wisconsin, and the Twin Cities region also saw a more robust increase of Democratic votes compared to Republican. The overall pattern is greater Democratic increases in urban and suburban areas in most of the battleground regions and increased enthusiasm for Democrats in traditional areas of strength on the West Coast, Northeast, and Upper Midwest. Trump also saw increases in certified votes counted, but not at the same level as Biden.

Persistence and change have long been hallmarks of presidential elections. Future chapters and articles will highlight persistence and change. This was no major realignment, but a map of persistence and change (figure 4.13) offers some insights into what has and has not changed. While the absolute change in counties flipping from red to blue was not that great, it occurred in counties with more voters and hence more electoral clout given the Electoral College. About three thousand counties in the United States supported the same party in 2016 and 2020. There wasn't much change. Only a small number of counties changed their support. Light blue counties are counties where the county switched from Republican to Democrat from 2016 to 2020 and pink counties are counties that switched from Democrat to Republican from 2016 to 2020. Of note are counties in South Texas in the Rio Grande Valley that are traditionally Democratic and heavily populated by Hispanics. Additionally, some pink counties in the Southeast bordering majority-Black counties switched. Although numbering only fifteen, these counties warrant further study. Light blue counties are scattered throughout the country, but they number sixty-two and include some important counties. First are some New England counties bordering counties that remained Democratic. Counties in the Midwest and West tend to be counties in extended suburban or exurban areas. Of note is the switch of Kent County, Michigan. Kent County is home to Grand Rapids, and is a longtime Republican county. Kent County has slightly more than six hundred thousand people and it was a swing county in a swing state. The swing of Kent County can be considered the most iconic in a swing state in a swing election.

Examination of maps of state- and county-level votes for Trump, Biden, and third-party candidates in 2020 provides insights into the outcome of the election and the possible motivations of voters within places. The results illustrate the continuing high degree of political and societal polarization within the county, the deepening divide between urban and rural, and the degree to which outcomes are correlated to patterns of race and ethnic identity.

VOTER PARTICIPATION IN THE 2020 ELECTION

CHRIS MAIER

With the votes counted in the 2020 presidential election, we can attempt to answer questions about voter participation and subsequent geographic patterns that emerged from the election. Figures 4.14 and 4.15 show levels of voter turnout in the 2020 election on a statewide and county-by-county basis. Both figures show voter turnout at the state and county level as a percentage of the overall voting-age population who voted in the 2020 presidential election. According to the US Elections Project, roughly 66 percent of eligible voters cast ballots in the 2020 presidential election (McDonald 2020). Voter turnout was up significantly from 2016, when roughly 60 percent of eligible voters participated in the election (Maier 2018). There are many possible explanations for the significant increase in voter participation in the 2020 presidential election, which include the COVID-19 pandemic and subsequent lockdowns, the Black Lives Matter movement, access to absentee or universal mail-in balloting, and President Donald Trump, whose candidacy motivated both Republicans and Democrats to get to the polls.

Figures 4.14 and 4.15 show that turnout rates varied significantly across the country. Historically, northern states, especially in the Upper Midwest, produce higher voter participation rates, which tend to decline as one moves from north to south across the country (Shelley 2014). The 2020 presidential election continued this pattern with a few exceptions. Washington, Oregon, and Montana, in the Pacific Northwest, Iowa, Michigan, Minnesota, and Wisconsin, in the Upper Midwest, and Maine, New Hampshire, and Vermont, in New England saw voter participation rates above 70 percent. Figure 4.14 shows the increased voter participation seen across the United States. In 2016, only Maine had a voter participation rate above 70 percent. In 2020, twelve states had a voter participation rate above 70 percent. Even states with the lowest voter participation rates saw significant increases in voter turnout when compared to the 2016 presidential election. In 2016, there were nine states with voter participation rates below 50 percent. In 2020, no state had a voter participation rate below 50 percent.

Except for Colorado, a state with a relatively high voter participation rate in 2016 and one that automatically sends out mail-in ballots, and North Carolina, a competitive swing in the 2020 election, voter participation generally declines as one moves south. Like the 2016 election, the lowest participation rates were generally found in Appalachia and across the southern part of the United States. Using figure 4.15, there are four states, Arkansas, Oklahoma, Texas, and West Virginia, with a participation rate below 55 percent. Using figure 4.15 one can see a generally lower-than-average turnout in a large swath of counties in Appalachia, stretching from West Virginia south through Tennessee and Arkansas, eventually reaching the southern plains states of Oklahoma and Texas.

Using figure 4.14, one can see the significant growth in voter participation rates that took place in Arizona, Hawaii, and Utah between the 2016 and 2020 presidential elections. The three states saw at least a 30 percent increase in certified votes counted between the 2016 and 2020. In Hawaii and Utah, the increase in voter participation could be attributed to ballot access, especially during the COVID-19 pandemic, as both states automatically mail ballots to all voters (Swasey 2020). Arizona's significant increase in voter participation may have a couple of explanations. First, Arizona's increase in voter participation could be attributed to ballot access as well. While Arizona does not automatically mail a ballot to all registered voters, they have a permanent early voting list, which provides all enrollees access to a mail-in ballot. Second, Arizona quickly became one of the most competitive and consequential states in 2020 election cycle, as it was home to a competitive Senate race as well, which may have driven additional voters to the polls. In the end, the increased voter turnout may have been what turned the state blue for the first time since 1996 and gave Joe Biden a much-needed victory that helped propel him to the White House.

REFERENCES

Maier, C. 2018. Voter Participation. In R. H. Watrel, R. Weichelt, F. M. Davidson, J. Heppen, E. H.

Voter Turnout

FIGURE 4.14

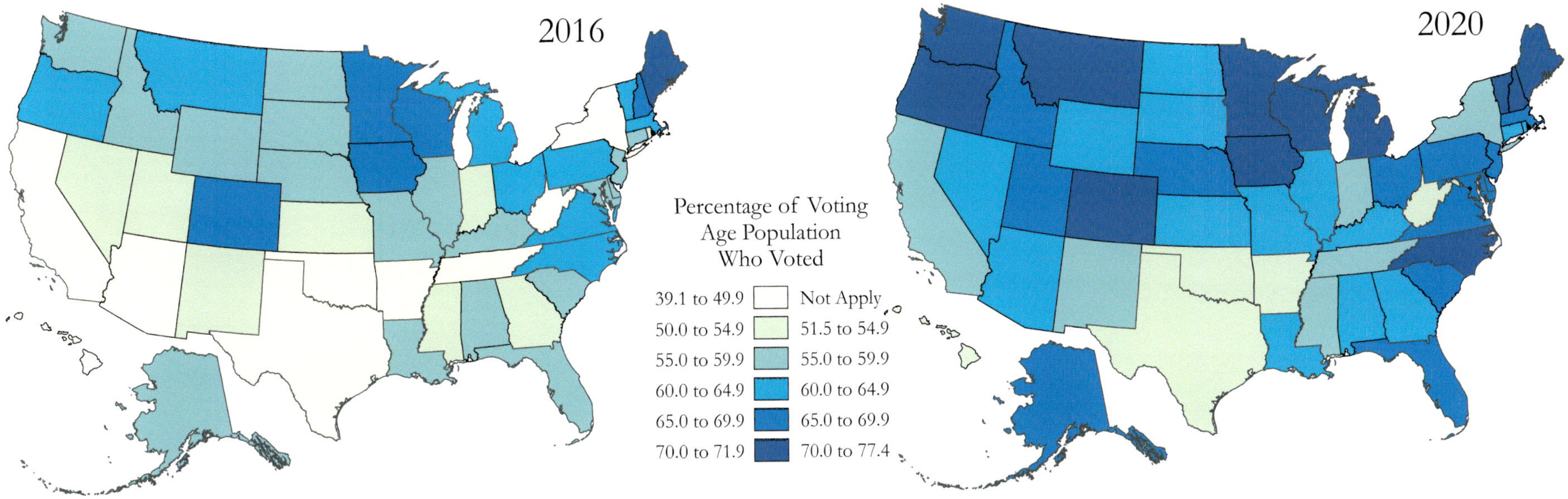

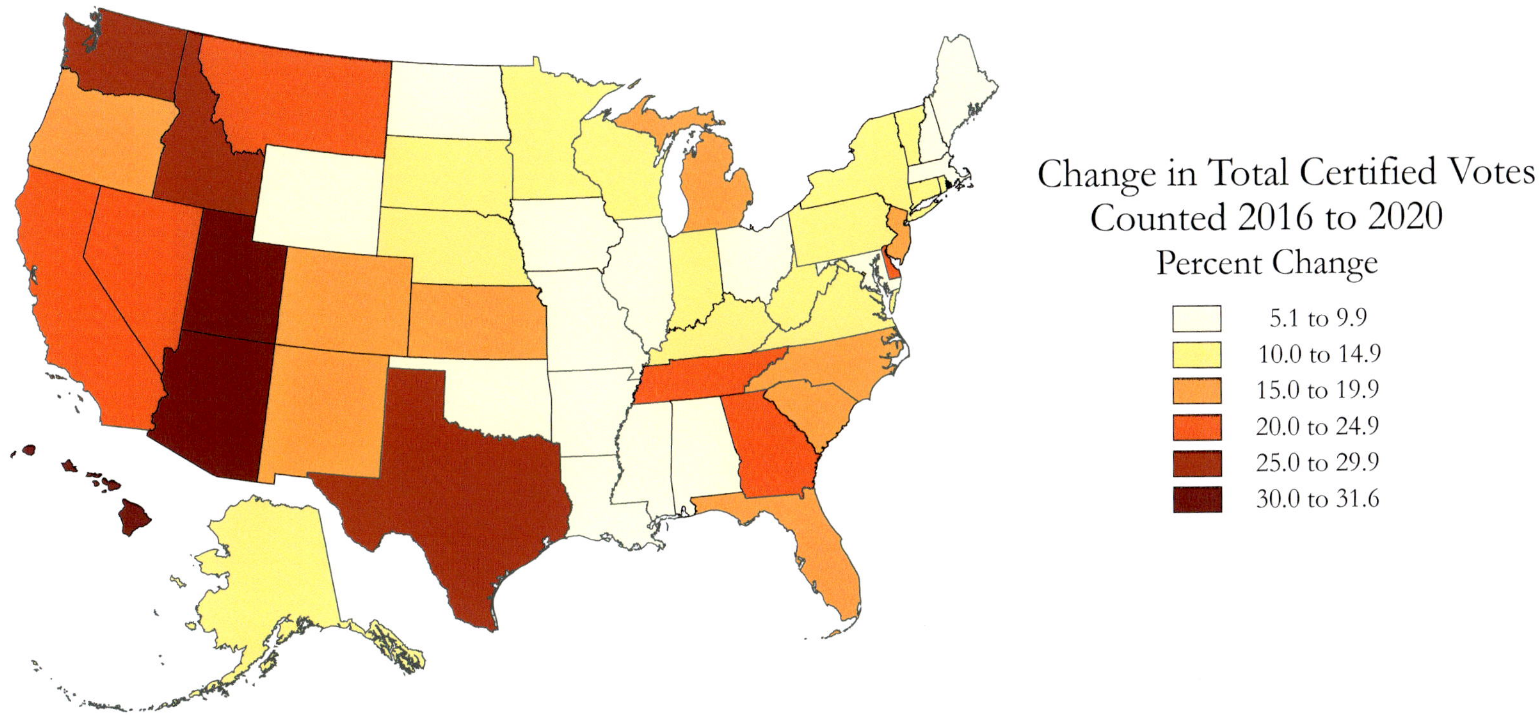

FIGURE 4.15

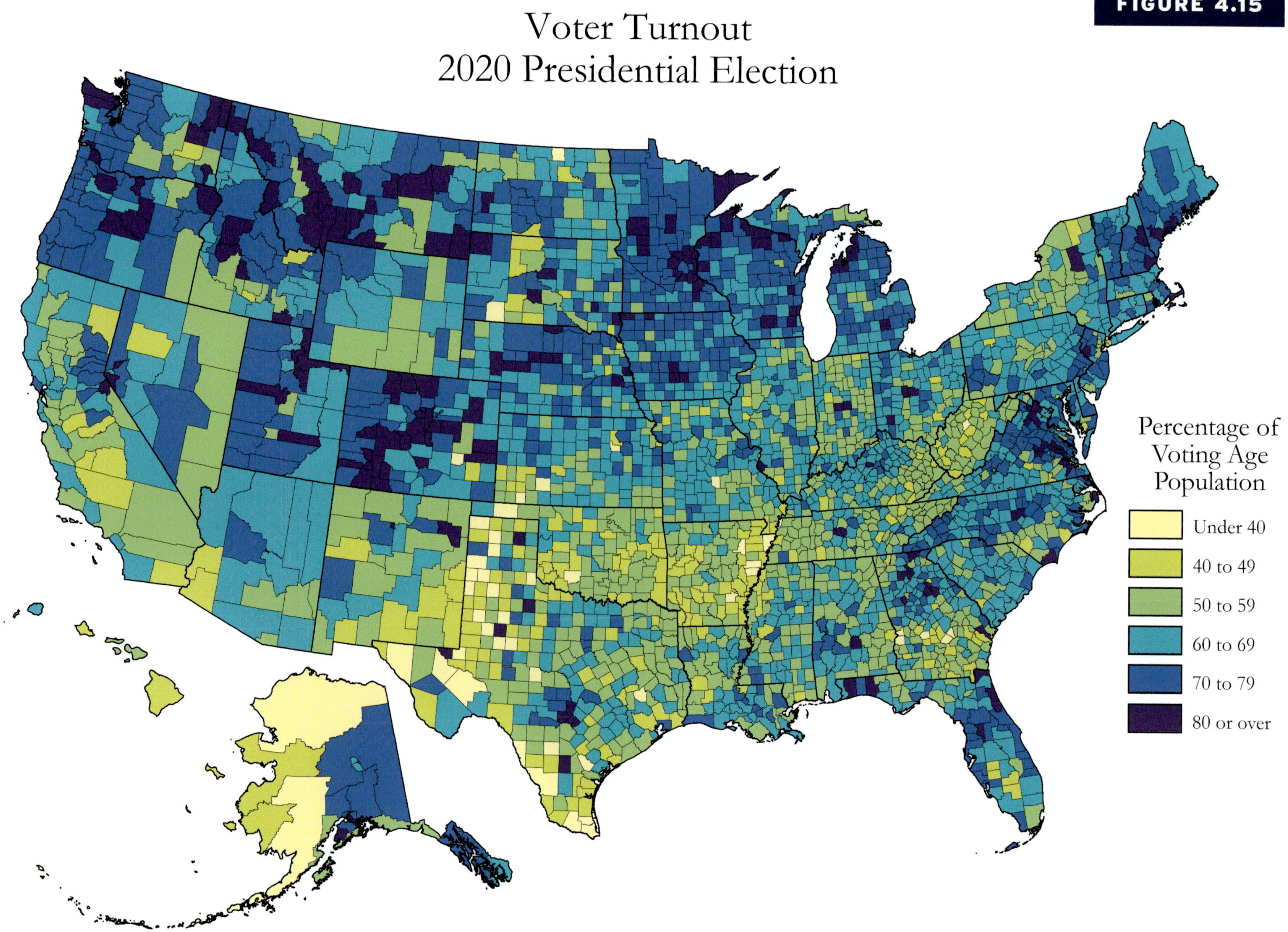
Voter Turnout
2020 Presidential Election
Percentage of Voting Age Population
Under 40
40 to 49
50 to 59
60 to 69
70 to 79
80 or over

Fouberg, J. Clark Archer, R. Morrill, F. M. Shelley, and K. C. Martis, eds., *Atlas of the 2016 Elections*, 76–78. Lanham, MD: Rowman & Littlefield.

McDonald, M. P. 2020. 2020 November General Elections Turnout Rates. US Elections Project, University of Florida. http://www.electproject.org/2020g (last accessed June 23, 2021).

Shelley, F. 2014. Results of the 2012 Presidential Election at the State and County Levels. In J. Clark Archer, R. H. Watrel, F. Davidson, E. H. Fouberg, K. C. Martis, R. L. Morrill, F. M. Shelley, and G. R. Webster, eds., *Atlas of the 2012 Elections*, 77–98. Lanham, MD: Rowman & Littlefield.

Swasey, B. 2020. Map: Mail-In Voting Rules by State—and the Deadlines You Need. *NPR*. https://www.npr.org/2020/09/14/909338758/map-mail-in-voting-rules-by-state (last accessed June 22, 2021).

THE 2020 PRESIDENTIAL ELECTION IN HISTORICAL PERSPECTIVE

FRED M. SHELLEY, J. CLARK ARCHER, AND ROBERT H. WATREL

HISTORICAL CHANGES IN VOTER PARTICIPATION

The numbers of voters who have participated in US presidential elections has expanded greatly since the mid-nineteenth century, as additional states joined or rejoined the Union after the US Civil War and the practice of selecting each state's presidential electors by popular vote rather than by appointment became universal. The growth of the US population, and the expansion of the franchise to women and to ethnic minority groups, also contributed to the overall but sometimes intermittent increase in voter participation. As shown in figure 4.16, the total number of votes cast in each presidential election grew from about 6.5 million votes in 1872 to nearly 158.5 million votes in 2020. The increase was very slow in the late nineteenth century when "grandfather clauses," poll taxes, and outright physical intimidation limited Black voter participation in the former Confederacy. Woman's suffrage was adopted nationally for the 1920 election, and the Voting Rights Act of 1965 and its successors led to significant increases. But a notable feature of the long-term trend is quite apparent fluctuations in numbers of votes cast from election to election during the past half century from about 1980 onwards. A more than 16 percent increase in total votes cast during the last two election cycles is especially prominent on the graph, representing an increase from 136.5 million total votes in 2016 to nearly 158.6 million total votes in 2020. On a partisan basis, the number of Republican votes grew from just under 63 million in 2016 to 74.2 million in 2020, and the number of Democratic votes grew from 65.8 million in 2016 to nearly 81.3 million in 2020. In other words, the 2020 election drew an unprecedented total number of Americans to the polls.

Voter turnout, which compares actual participation to potential participation, has fluctuated considerably since 1872. Figure 4.17 depicts voter turnout as a percentage of gender-eligible and age-eligible population for each presidential election from 1872 to 2020. Citizenship status is not accounted for in the statistics used for the graph, in part because only very approximate voting-age citizen population estimates are available for most past elections, and in part because several states actually allowed non-citizens to vote in federal elections up to about the time of World War I. The vertical axis on the graph is based on male voting-age population up to 1916, and total voting-age population from 1920 to the present.

The end of Reconstruction is apparent on the graph since voter turnout jumped from 1872, when Federal forces were still stationed in several southern states to 1876, by which time Union occupation had ended. National turnout remained at very high levels near or above 80 percent until 1896, when Populist-Democratic candidate William Jennings Bryan of Nebraska was defeated by Republican candidate William McKinley of Ohio. Thereafter, voter participation plummeted during the next quarter century, especially in the South (Shelley et al. 1996). The circumstances are relevant to the present, albeit with intriguing party role reversals. Several Black politicians were elected to the US Congress from several former Confederate states during the last half of the nineteenth century, in most cases under fusion Republican-Populist Party labels. In the West, several fusion Democratic-Populist candidates were elected to local, state, and national offices during this same period. Bryan's defeat in 1896 led Republicans to write off the South and to remove civil rights planks from later Republican Party national platforms. This led to a half century of one-party Democratic dominance in the "Solid South," reinforced by a wave of Jim Crow anti–voting rights laws that drove Black election participation rates down to single digits in many former Confederate States. The Voting Rights Act of 1965 started the former Democratic Solid South careening off in the opposite direction toward the Grand Old Party of Lincoln.

The nadir in voter turnout came in 1920 when women could vote in federal elections for the first time, and the denominator for calculating turnout thus newly summed over voting-age adults of both sexes. Turnout rates grew during the next decade, as adult women became more adept at exercising their newly won franchise rights. There was again a sharp dip in 1940 and 1944 when the Great Depression and World War II led to lower voting rates, followed by a rise to a temporary plateau over

FIGURE 4.16

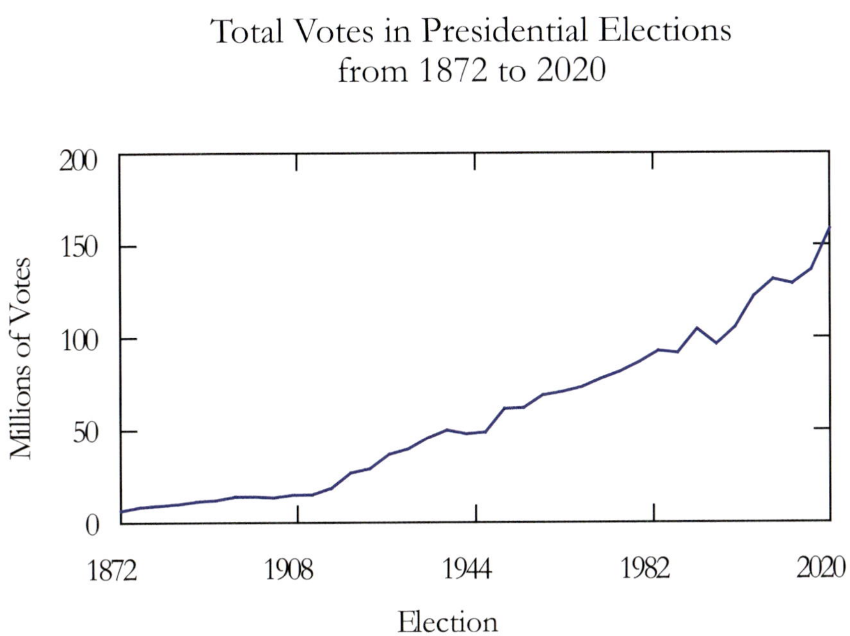

FIGURE 4.17

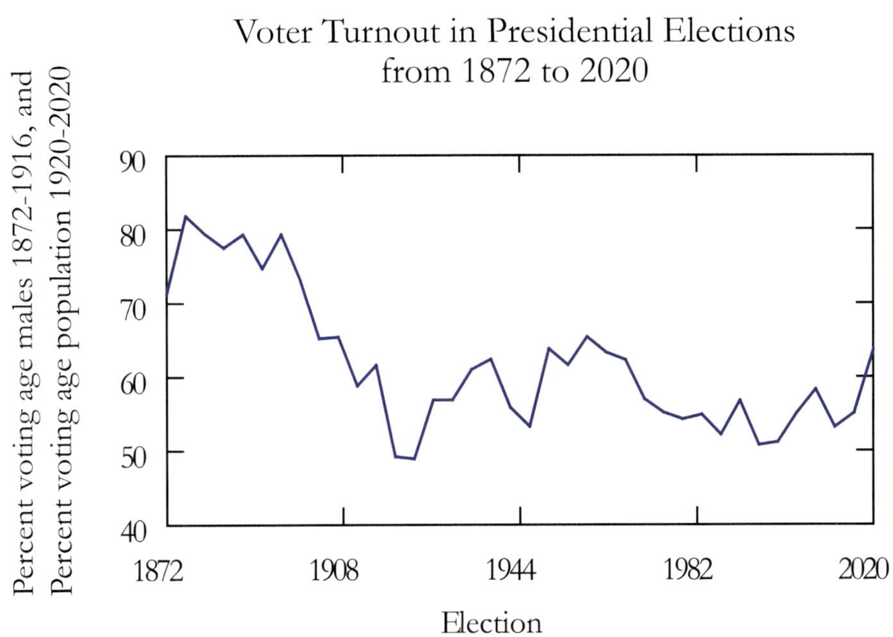

60 percent during the early post–World War II period. Voter turnout declined by several percentage points following ratification of the Twenty-Sixth Amendment in 1971, which lowered the voting age to eighteen in federal elections. Younger voters continue to cast ballots at lower rates than older voters. The US Census estimated that voter turnout in the 2020 election was only 48.0 percent for persons ages eighteen to twenty-four and rose systematically by age grouping to as high as 73.0 percent for persons ages sixty-five to seventy-four (US Census 2021). The overall voter turnout rate reported by the Census for the 2020 election was 61.3 percent of voting-age population. However, the Census estimate of the "Number Reported Voted" in 2020 was 154,628,000, which is considerably below the 158,461,909 figure of total votes cast in November, 2020, used in this study. This value was calculated by summing numbers of "certified" or "final" presidential votes reported on the official websites of each of the fifty states and the District of Columbia. When the "actual" national total of 158,461,909 votes is divided by the Census estimate of 252,274,000 "Total Population of Voting Age" on November 3, 2020, a turnout percentage of 62.8 percent is obtained; this is the percentage shown on figure 4.17 for the 2020 election. As an aside, our total vote value does not include "spoiled ballots," "provisional ballots," "undervotes," "overvotes," or other categories of ballots that were not counted in the "certified" election results reported by states. However, no matter which value is used, it is apparent that the 2020 US presidential election reached a turnout level not previously exceeded since the 1964 and 1968 elections a half century ago.

COMPARING THE 2020 ELECTION WITH PAST ELECTIONS

How did the geographic distribution of popular votes in the 2020 presidential election compare with distributions of votes in previous elections? At the state level, forty-five of the fifty states gave their electoral votes to the same party in both elections. However, Biden won five states that Trump had won in 2016—Arizona, Georgia, Michigan, Pennsylvania, and Wisconsin. Capturing these states enabled him to secure enough electoral votes to win the election in the Electoral College.

The margins of victory for the two candidates in 2020 also paralleled the margins of victory for Trump and for Hillary Clinton in 2016. Trump won more than 65 percent of the popular vote in Oklahoma, West Virginia, and Wyoming in both elections. The Democratic nominees, Clinton in 2016 and Biden in 2020, both won California, Hawaii, Maryland, and Massachusetts with more than 60 percent of the popular vote. Furthermore, the five states won by Biden in 2020, but lost by Trump in 2016, were among the six states in which the Democrats received the smallest margins of victory in 2020.

How did the 2020 election compare with the 2016 election, and with earlier elections, at the county level? Again, the geographic distributions of votes for the two elections were quite similar, as indicated in the county-level maps of the election results found earlier in this chapter. In both cases, the Democrats won large majorities in central-city counties, including those containing New York, Los Angeles, Chicago, Boston, and San Francisco. They also won large majorities in rural counties whose populations are dominated by Blacks, Latinos, and Native Americans. On the other hand, in both cases Trump won two-thirds or more of the popular vote in most primarily white rural counties stretching from Appalachia and the Deep South westward through the Great Plains and Rocky Mountain states. As discussed elsewhere in this atlas, the 2020 election was characterized by sharp divergence between urban and rural places.

In some states, however, Trump's large margins in rural counties were not sufficient to offset Democratic majorities in more populated counties in their states. For example, Biden won only two of Nevada's sixteen counties, but these two counties contain the state's three largest cities—Las Vegas, Henderson, and Reno. Similarly, Biden carried only fourteen of Wisconsin's seventy-two counties. However, Biden won more than 75 percent of the vote in Dane County, including the city of Madison, and nearly 70 percent of the popular vote in Milwaukee County, which contains the city of Milwaukee. His combined margin of more than 364,000 popular votes in these two counties alone allowed him to win Wisconsin by 20,608 popular votes. Excluding Alaska and Hawaii, Trump carried 2,575 counties, or 82.9 percent of all counties within the contiguous United States.

HISTORICAL CHANGES IN VOTE DISTRIBUTION: COMPARING 2020 WITH 1908 AND OTHER EARLIER ELECTIONS

Historically, the county-level patterns of votes in presidential elections in the late nineteenth and early twentieth centuries were quite different from those in the early twenty-first century. At that time, the northern tier of states including New England and the Upper Midwest was solidly Republican, while the "Solid South" was dominated by the Democrats. The primary electoral battlegrounds were the Middle Atlantic and Great Lakes states in between, notably New York, New Jersey, Pennsylvania, Ohio, Indiana, and Illinois. Generally, the party that was more successful in many or all these states won the election.

This pattern can be illustrated by the state and Electoral College distributions of votes from the election of 1908, in which Republican William Howard Taft defeated Democrat William Jennings Bryan. Taft won 51.6 percent of the national popular vote, as compared to 43.0 percent for Bryan. (The remaining popular votes were cast for third-party candidates.) Taft won the election by a margin of 321 to 162 in the Electoral College. Taft carried all six of the battleground states, which together cast 150 or 31 percent of the total votes in the Electoral College. By contrast, these six states cast a total of 112 or 21 percent of the total votes in the Electoral College in 2020, with Trump carrying Ohio and Indiana and Biden carrying the other four. In 1908 the two most closely contested states were Maryland and Missouri, both of which Taft carried by a margin of less than 1 percent of the popular vote.

This earlier electoral geography is also clear from the map of county-level results in 1908 shown in figure 4.22. Taft carried every county in New England, winning many of them by 65 percent or more of the vote. He also won every county in Michigan, North Dakota, Washington, and Oregon and almost every county in New York outside of New York City. On the other hand, Bryan won every county in Louisiana, Mississippi, and South Carolina and all but one county in Florida. However, Taft carried many counties in Appalachia, including eastern Tennessee and western North Carolina. Many residents of this region had opposed secession before the Civil War and became loyal to the Republican Party after the war ended and the southern states were readmitted to the Union, and Republican strength in this region continues today. In looking at the 1908 map, note that New Mexico, Arizona, Alaska, and Hawaii were still territories in 1908 and had yet to be admitted to the Union. Bryan carried Nevada by a narrow margin, although county-level results were not reported for some counties.

The county-level patterns of the 1908 and 2020 elections can be compared numerically using the Pearson product-moment correlation coefficient, which is a numerically standardized measure that ranges from -1.0 to +1.0. Values near -1.0 show inverse relationships of high values for one variable and low values for another variable; values near 0.0 depict no systematic relationship; and values near +1.0 indicate strong positive relationships of similar values for both variables. After "pairwise exclusion of missing data" to avoid confounding numerical results due to counties in states not yet participating in presidential elections in 1908 from the analysis, a correlation of +0.017 was calculated for the elections of 1908 and 2020 using percent Democratic vote as the measure in both elections and counties as observation units (N = 3,007). In other words, there is essentially no systematic

FIGURE 4.18

Correlation of 2020 Percent Democratic Popular Presidential Vote by County with Percent Democratic Popular Presidential Vote by County in Previous Elections from 1872 to 2016

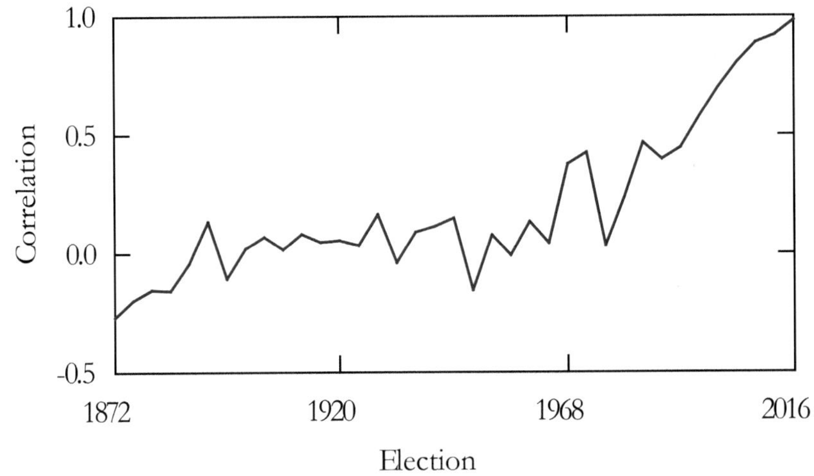

relationship between the county-level distributions of popular support for the Democratic candidacies of William Jennings Bryan versus Taft in 1908 and for Joe Biden versus Trump in 2020. In comparison, the county-level correlation between percent popular Democratic support for Hillary Clinton versus Trump in 2016 and for Joe Biden versus Trump in 2020 was calculated to be +0.982 (N = 3,079 with the state of Alaska included as a "super county" in the calculations). In other words, maps of percent Democratic support at county level in 2016 and in 2020 are essentially identical to one another, even though the numerical tallies of votes and the Electoral College results that were determined by these vote tallies produced a Republican presidential victory in 2016 and a Democratic presidential victory in 2020.

A gaming metaphor helps to achieve insights into the implications of these correlations. The electoral deck was so completely reshuffled between the 1908 and 2020 elections that voters' collective responses to the issues and candidates presented in the 1908 election were found to be statistically unrelated to voters' collective responses to the issues and candidates presented to them in the 2020 election at a county level of observation across the United States. Although the major-party labels were the same in both elections, all other electorally relevant considerations were thoroughly jumbled across the political landscape of America, as some issues and candidates faded from memory while others rose to prominence over the passage of time between these elections. It is tempting to interpret such changes in generational terms, since on average old voters were replaced by new voters approximately three times between 1908 and 2020. But long-term generational changes had minimal influence on voters' collective responses to the issues and candidates presented to them in the elections of 2016 and 2020. Indeed, the Republican presidential and vice presidential candidates were the same in both elections, and on the Democratic side the presidential candidates in both elections had served previously in the Obama administration as either secretary of state or as vice president. So the electoral deck from which the cards were drawn in the 2020 election was essentially the same as the electoral deck from which the cards had been drawn four years earlier in the 2016 election.

The American political landscape is always in motion, as new issues, candidates, and voters enter the scene; but rates of change can vary as some alterations happen abruptly while others emerge more slowly. Using the 2020 election as the basis for comparison, figure 4.18 graphs the Pearson product-moment correlations between the county-level pattern of percent Democratic vote in the 2020 election with the county-level pattern of percent Democratic vote in each earlier election from 1872 to 2016. One obvious message in the graph is that no election before the late twentieth century exhibited a geographical pattern at county level like that of the 2020 election. Another obvious message is that the negligible congruence between the 2020 election and the 1908 election discussed above illustrates a pattern that in fact predominated from the late nineteenth century until the mid-twentieth century. And a somewhat less obvious message is that early post–Civil War elections exhibited geographical patterns that were somewhat obverse to the geographical pattern of the 2020 election, as indicated by negative correlations between the elections of 1872, 1876, 1880, and 1884 and the election of 2020. In other words, county-level patterns of Democratic voting in these elections were a mirror image of patterns of Democratic voting a century and a half later in the 2020 election.

THE T-MODE FACTOR ANALYSIS PROCESS

Clearly, the county-level geography of popular votes in presidential elections has changed dramatically, as illustrated by comparing the maps of the 1908 and 2020 results and by examining correlations between the 2020 election and earlier elections since the Civil War. How and why did these changes occur? Stronger statistical analysis methods are needed to try to untangle the currents and countercurrents of American electoral geography over long periods of time. As was done in previous studies by the authors, the identification of long-run changes in votes for the two major parties at the county level can be undertaken using T-mode factor analysis (Archer, Shelley, et al. 1988; Archer and Taylor 1981; Archer, Watrel, et al. 2014; Shelley et al. 1996; Watrel et al. 2018). The technique is a means by which a large, complicated matrix of correlation coefficients is reduced to a small number of separate factors, each of which represents a period in which the distributions of popular vote percentages by county are similar to one another. The T-mode factor analysis technique allows distinctive long-run "normal vote" patterns of varying duration to be separately identified, and it helps to uncover insights into how to answer questions about when and why changes occurred. For more details about factor analysis, please refer to chapter 11.

T-Mode Oblimin Factor Pattern Loading Profiles: Percent Democratic Vote by County, 1872 to 2020 Presidential Elections

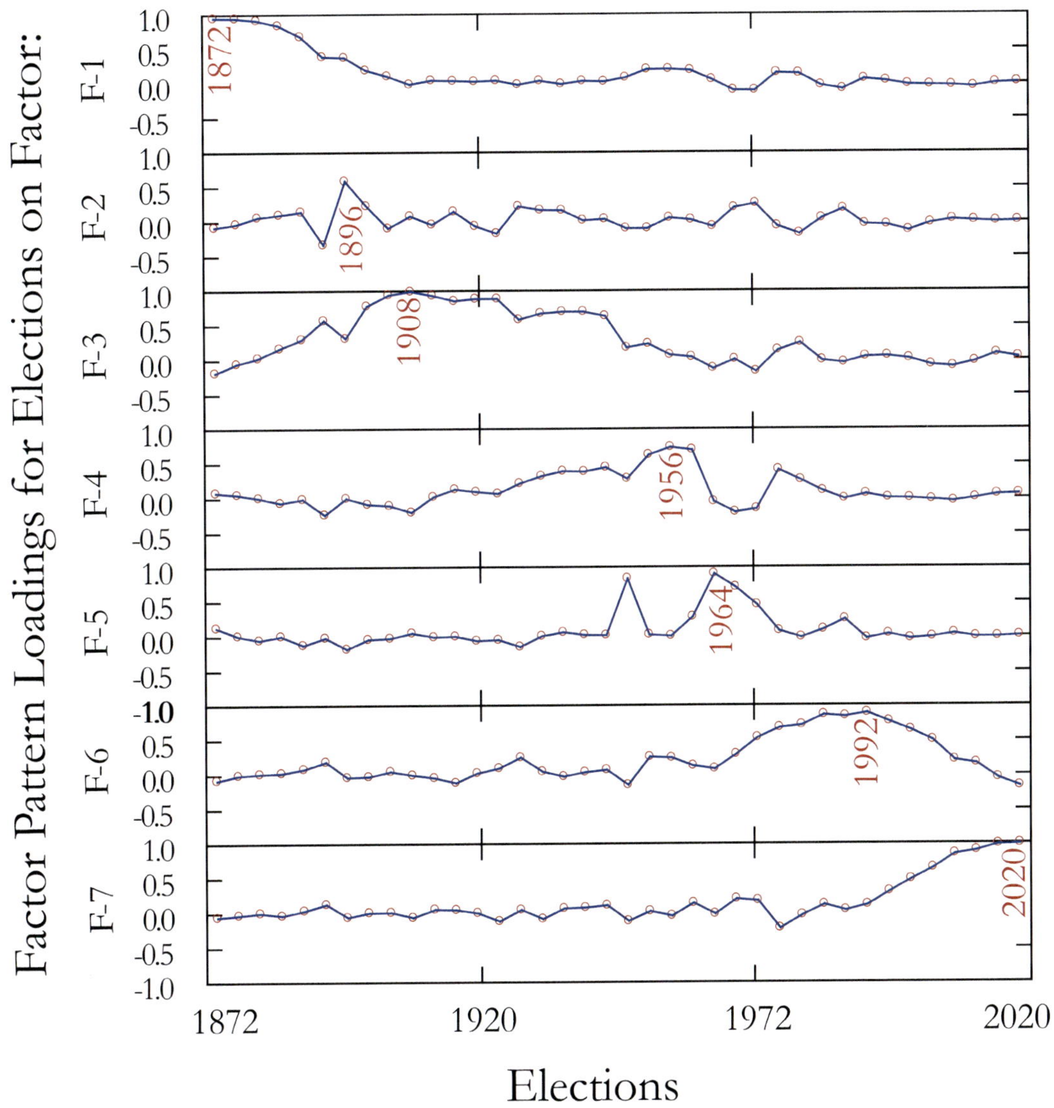

Highest loading election for each factor labeled in red

FIGURE 4.20

Presidential Election Popular Vote 1872

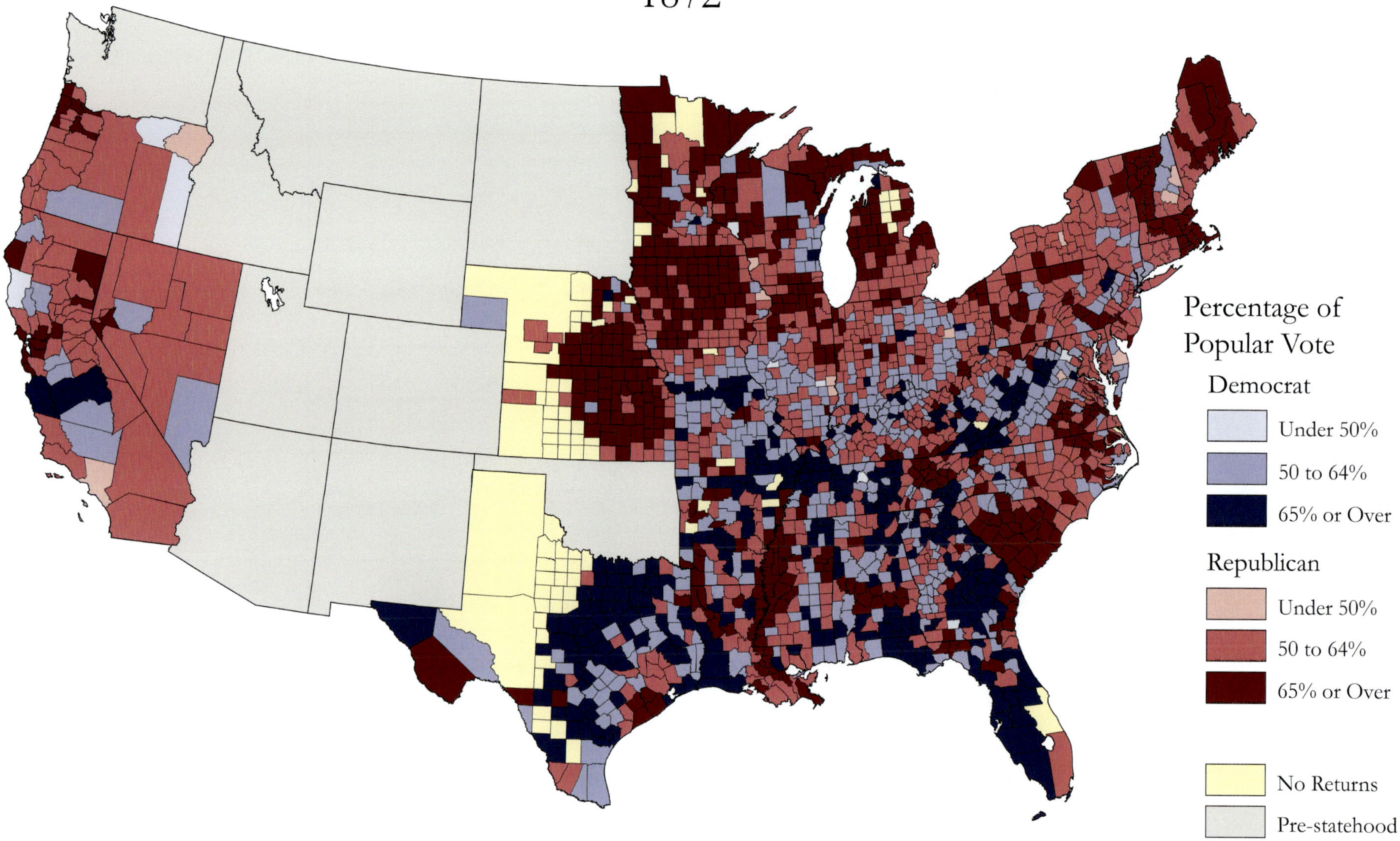

FIGURE 4.21

Presidential Election Popular Vote
1896

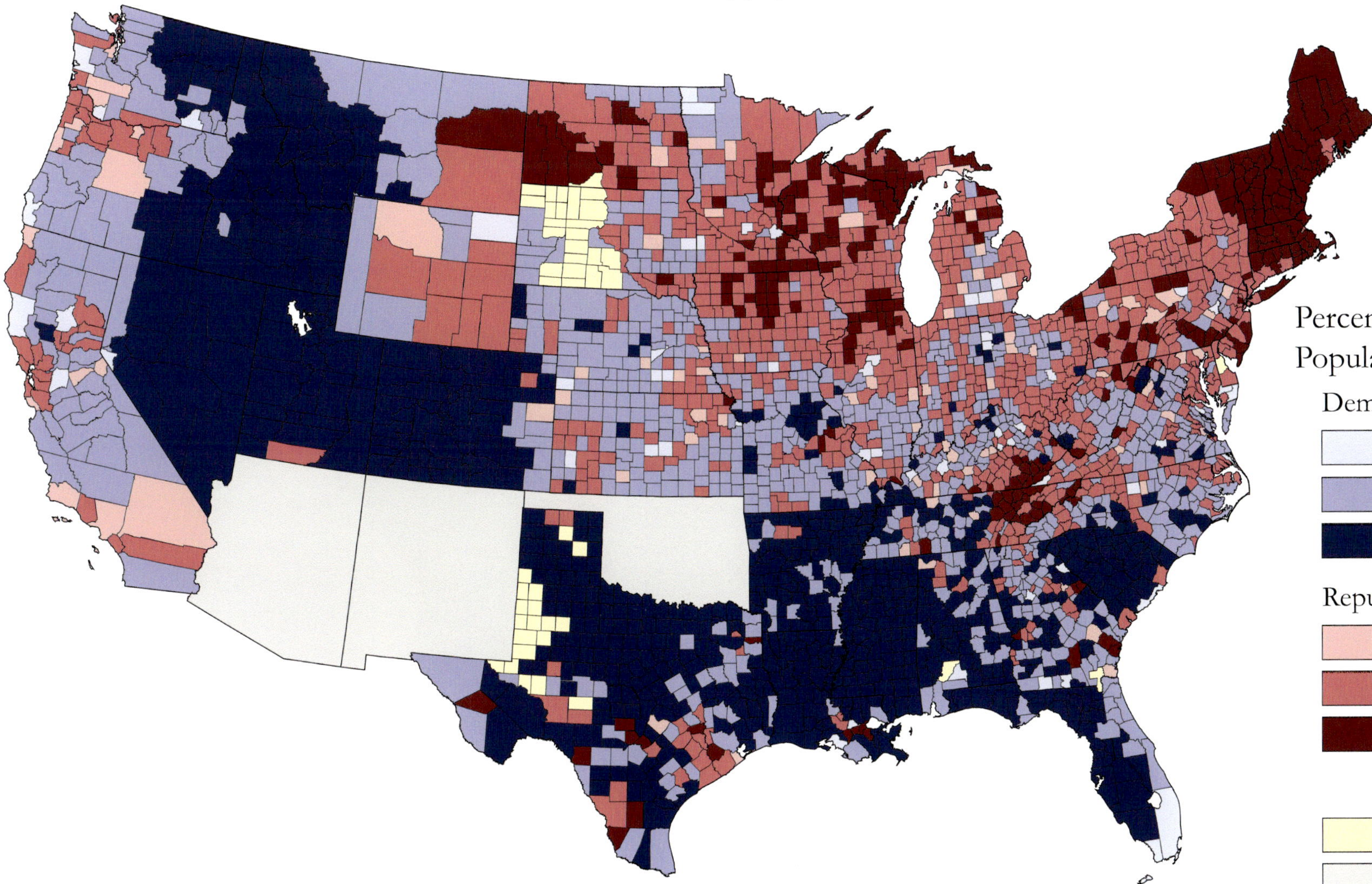

Percentage of Popular Vote

Democrat

- Under 50%
- 50 to 64%
- 65% or Over

Republican

- Under 50%
- 50 to 64%
- 65% or Over

- No Returns
- Pre-statehood

FIGURE 4.22

Presidential Election Popular Vote 1908

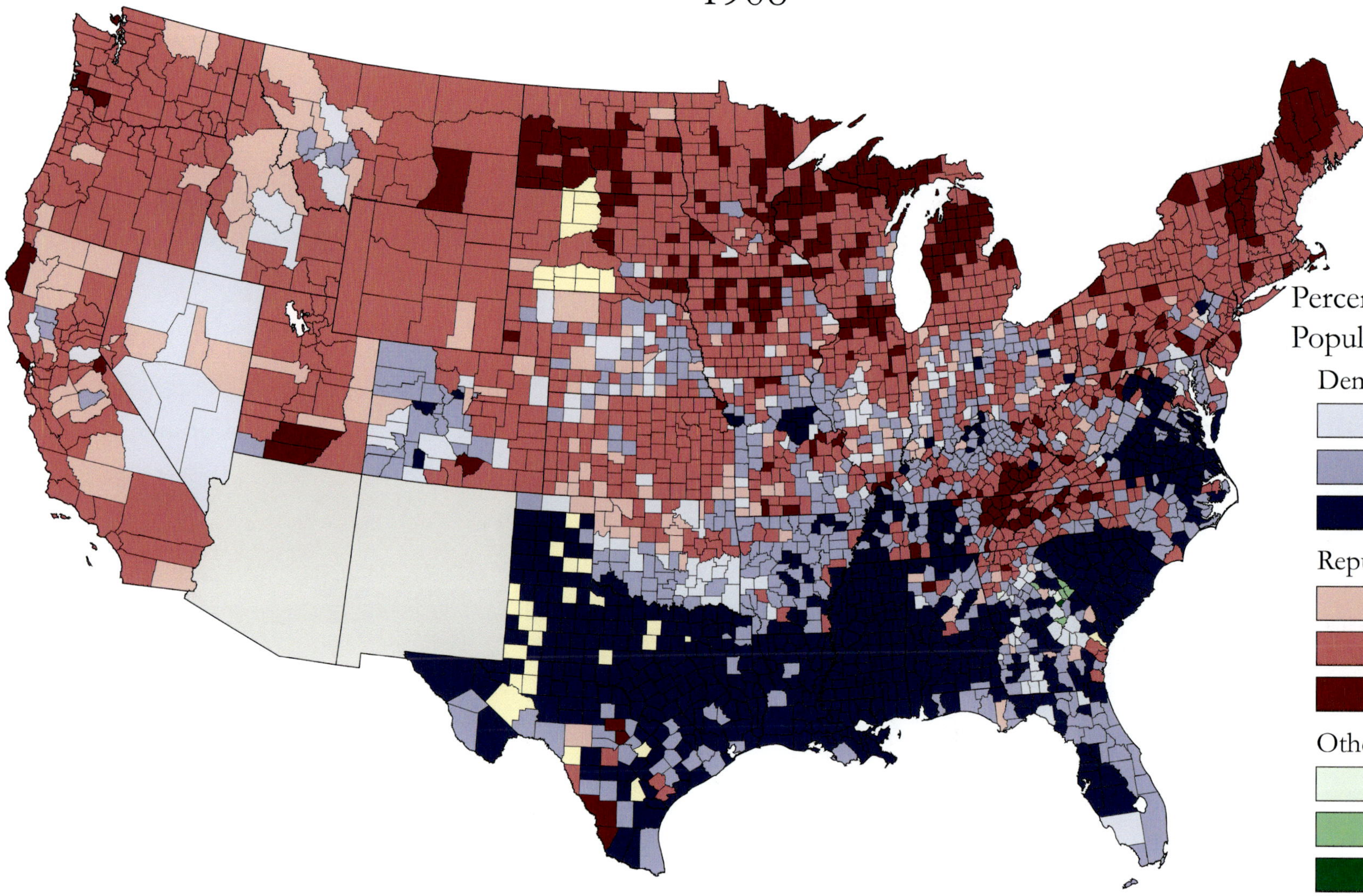

Percentage of Popular Vote

Democrat
- Under 50%
- 50 to 64%
- 65% or Over

Republican
- Under 50%
- 50 to 64%
- 65% or Over

Other
- Under 50%
- 50 to 64%
- 65% or Over

- No Returns
- Pre-statehood

FIGURE 4.23

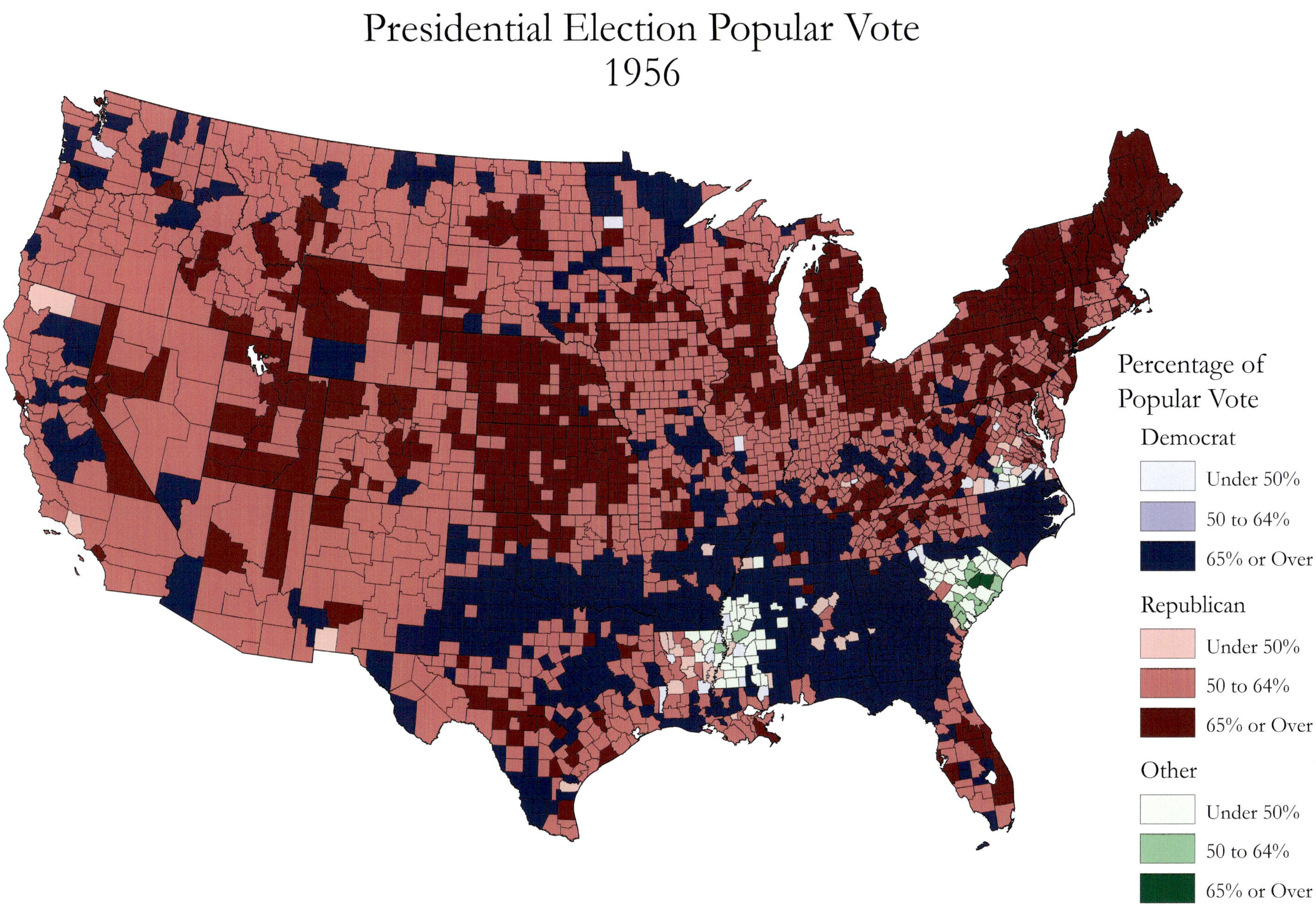
Presidential Election Popular Vote
1956
Percentage of
Popular Vote
Democrat
Under 50%
50 to 64%
65% or Over
Republican
Under 50%
50 to 64%
65% or Over
Other
Under 50%
50 to 64%
65% or Over

FIGURE 4.24

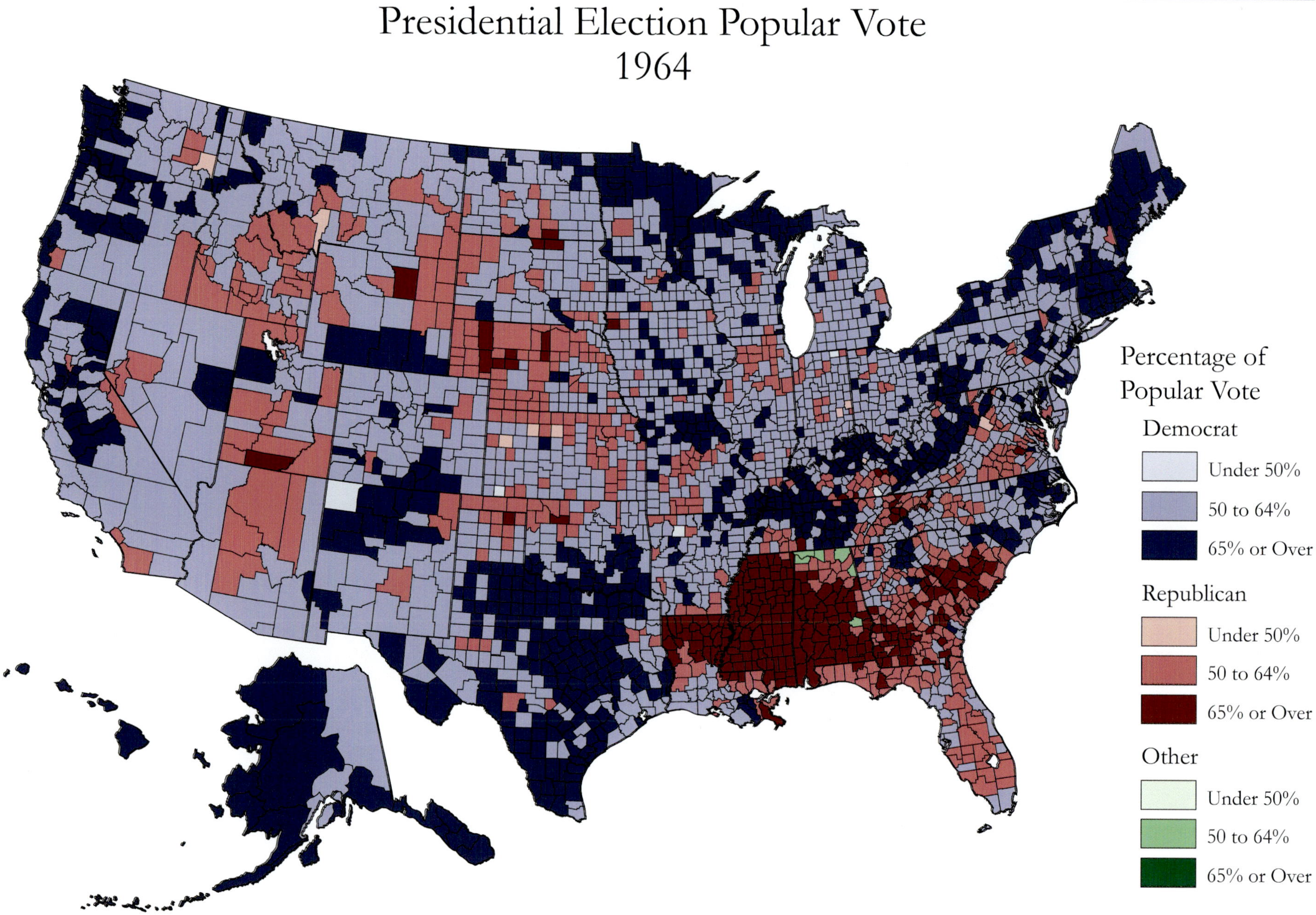
Presidential Election Popular Vote
1964
Percentage of Popular Vote
Democrat
Under 50%
50 to 64%
65% or Over
Republican
Under 50%
50 to 64%
65% or Over
Other
Under 50%
50 to 64%
65% or Over

FIGURE 4.25

Presidential Election Popular Vote 1992

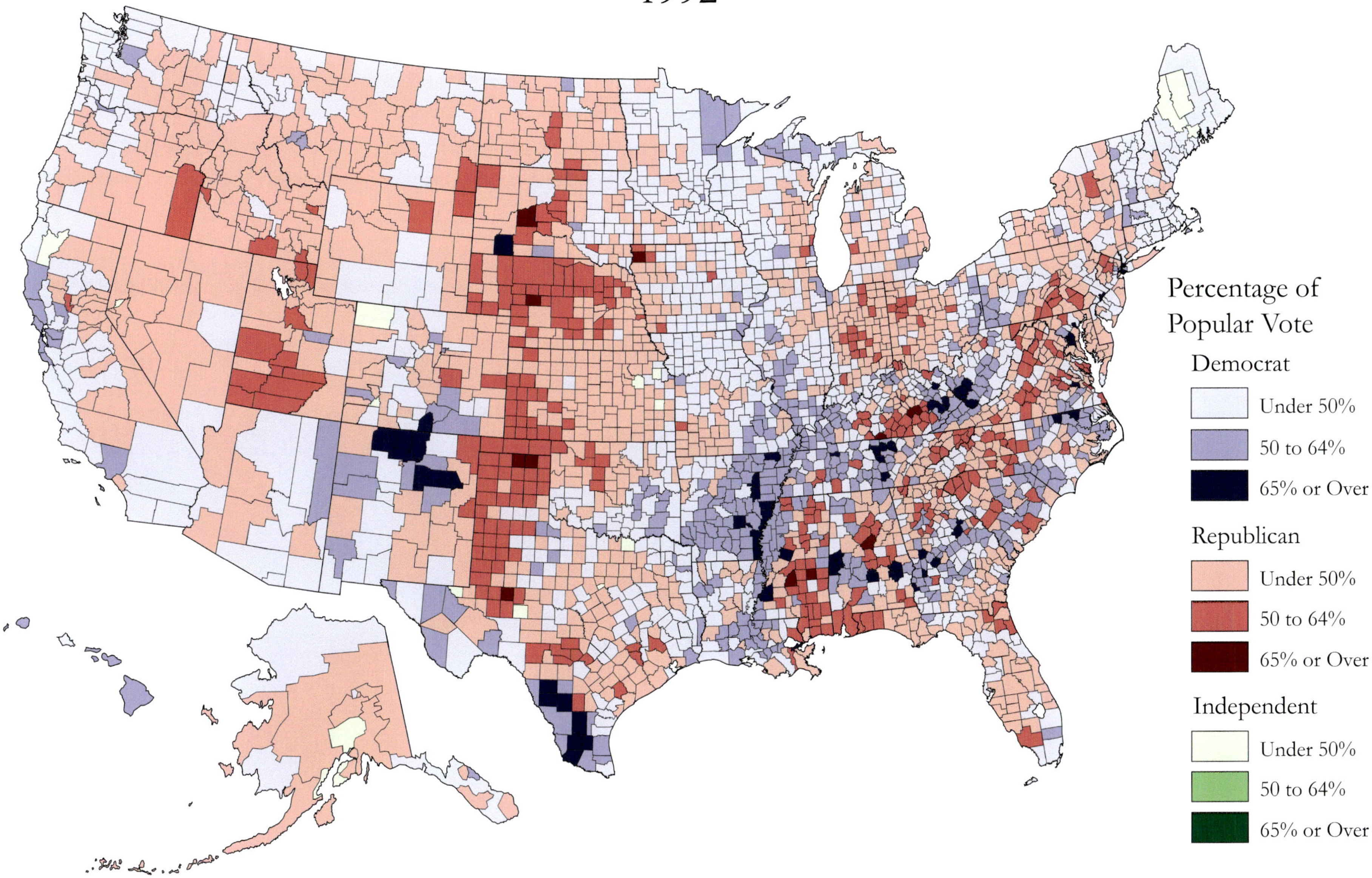

FIGURE 4.26

Presidential Election Popular Vote 2020

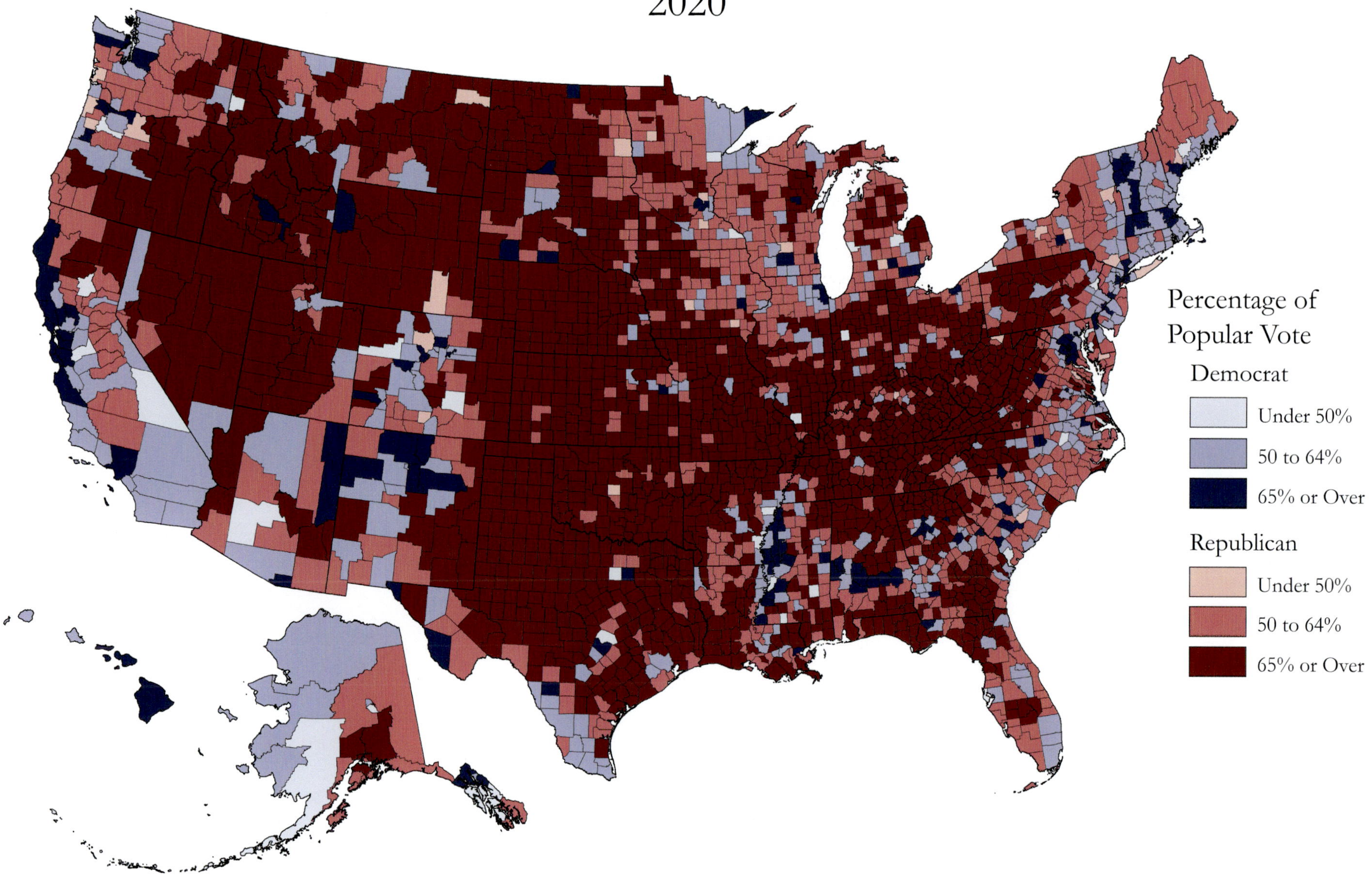

RESULTS OF THE T-MODE ANALYSIS

Figure 4.19 is a graph of factor pattern loadings for each of the twenty-eight elections from 1872 to 2020 on each of the seven retained oblimin rotated factors. The factors are numbered and arranged vertically on the graph in the chronological order of the election with the highest factor pattern loading on each of the factors. The direct oblimin rotation factor pattern loadings represented are somewhat similar to standardized regression coefficients that show the importance of a factor in accounting for the variance of an election. The potential range is close to -1.0 to +1.0, which are the values shown on the vertical axis for each factor. The horizontal axis for the entire graph depicts the time of each election, from 1872 to 2020. Each election is depicted by a small red circle along the horizontal plot of factor pattern loadings for each factor. The election with the highest factor pattern loading on a particular factor is labeled in red on the graph. And a map is shown below for each of these highest loading elections (figures 4.20 to 4.26). The subgraph for each factor and the map of the election with the highest loading on that factor should be examined jointly. Larger values on each subgraph show which elections are best represented by that factor, and the corresponding map shows how the geographical pattern of electoral support varied across the United States in the election that is most representative of that factor. Notice that just as in the case of the graph of the correlations between the 2020 election and earlier elections shown and described above, most of the values on each subgraph for a factor tend to fluctuate around 0.00, which reemphasizes that the "election deck" tends to be reshuffled between the "normal vote" epochs represented on the graph of factor pattern loadings.

The election with the largest factor pattern loading on Factor 1 is the election of 1872, in which former Union general and incumbent Republican President Ulysses S. Grant won reelection by a national popular vote of 55.6 percent and 286 electoral votes (Archer et al. 2006). Liberal Republican-Democrat challenger Horace Greeley received 43.8 percent of the national popular vote but died before the remaining sixty-three electoral votes were split symbolically among four others. This was a very controversial election in which the Republican-controlled Congress refused to accept Greeley electoral votes from Arkansas, Louisiana, and Georgia. Many interior western states were still territories and not yet participating in presidential elections. And as can be seen on the map, Grant won strong election support in several areas in former Confederate states, since newly freed Black voters were strongly pro-Republican in their voting and former Confederate soldiers were denied the franchise. Support for the Grand Old Party of Lincoln was especially prominent in New England and the Midwest. This Reconstruction pattern continued to be influential in presidential elections until the late 1880s.

The highest positive factor loading on Factor 2 is for the election of 1896, and the highest negative factor loading on Factor 2 is for the election of 1892 only four years earlier. These two "deviating" elections reflected a widespread electoral reshuffling that happened during the transition from the fading Reconstruction pattern to a newly rising sectional normal vote pattern. Although this brief deviating sequence coincided with and was partly caused by a national economic downturn, the most important political challenges of the time involved the end of Reconstruction and the resurgence of Democratic voting strength in the South, and the adolescent rise of many former territories to statehood in the West. These produced tumultuous tensions both within and between the two major parties. The 1892 election was a three-way race between incumbent Republican Benjamin Harrison, who lost with 43.0 percent of the national popular vote and 145 electoral votes, victorious Democratic challenger Grover Cleveland, who received a plurality of 46.1 percent of the popular vote and 277 electoral votes, and Populist Party candidate James B. Weaver, who received 8.5 percent of the popular vote and 22 electoral votes. In 1896 Grover Cleveland, who had won earlier in 1884 but lost in 1888, was deemed ineligible to run for a third term. The nomination of William Jennings Bryan for president on a national fusion ticket by both the Populist Party and the Democratic Party in 1896 both reflected and precipitated political tensions, especially in the South, where Populist and Republican candidates had run on fusion tickets for legislative and state and local administrative offices in 1892 and 1894. Republican candidate William McKinley received a slim 51.0 percent majority of the popular vote and won with 271 electoral votes, versus Bryan's 46.7 percent of the popular vote and 176 electoral votes. These two elections have factor loadings with opposite algebraic signs on Factor 2 because Populist Weaver carried several states and many counties in the West in 1892 but Democrat-Populist Bryan returned most to the Democratic column in 1896, causing these elections to have somewhat inverse geographical patterns nationally. As pointed out above, the Republican Party then abandoned the South for a half century after the election of 1896.

The election with the highest factor pattern loading on Factor 3 is the election of 1908. This was Bryan's third and last try for the presidency after Republicans William McKinley and Theodore Roosevelt had won in 1896, 1900, and 1904. Very popular "Teddy" Roosevelt had succeeded to the presidency after McKinley's death in 1901 and declined to run for a third term in 1908, although he would run again as a splinter "Progressive" or "Bull Moose" Republican candidate in 1912. The 1908 election was won by Roosevelt's secretary of war, William Howard Taft, with 51.6 percent of the popular vote and 321 electoral votes, to Bryan's 43.0 percent of the popular vote and 162 electoral votes. Ongoing tensions within the American body politic were reflected in the 2.6 percent of the popular vote that was cast for Socialist Eugene V. Debs. As is apparent on the subgraph of factor pattern loadings for Factor 3, the 1908 election is the most representative of an election pattern that had begun to replace the earlier Reconstruction pattern shortly before the deviating elections of the 1890s and that then held strong prominence for nearly a half century from 1900 to 1944. As illustrated on the map of the 1908 election (figure 4.22), this pattern can quite reasonably be called the "sectional normal vote" pattern, which divided the United States into three durable sections: North, South, and West. Democratic voters dominated in the "Solid South," while Republican voters usually predominated in the North and the West. This strongly sectional division even predominated across Democrat Franklin D. Roosevelt's victories from 1932 to 1944, including the "New Deal" landslide election of 1936, which FDR won with 60.8 percent of the popular vote and 523 out of 531 total electoral votes, with only 8 electoral votes for Republican Alfred Landon. New Yorker Roosevelt's poorest showings were in "upstate" New York and New England, where traditional Yankee Republican sentiments remained strong. Democratic percentages were 70 percent or more of the popular vote in most counties in the South in 1936.

Factors 4 and 5 point to a complex transition period with contrasting geographical patterns following the breakup of the sectional normal vote. The highest factor pattern loading on Factor 4 is for the election of 1956, although the elections of 1952 and 1960 also have quite high loadings on this factor. Two of these elections were won by World War II hero General Dwight D. Eisenhower, who ran as a Republican candidate although he was reportedly also courted by the Democratic Party for their nomination in 1952. Eisenhower received 55.1 percent of the popular vote and 442 electoral votes in 1952, and 57.4 percent of the popular vote and 457 electoral votes in 1956. Democratic challenger Adlai Stevenson received 44.4 percent of the popular vote and 89 electoral votes in 1952, and 42.0 percent of the popular vote and 73 electoral votes in 1956. The geographical patterns of these elections were somewhat similar to those of the sectional normal vote period, except that Eisenhower carried many counties and several states in the Rim South for the first time since early in the twentieth century (figure 4.23).

The election with the highest loading on Factor 5 is the election of 1964, but the election of 1948 also has a very high loading on this factor. These two elections so thoroughly reshuffled the electoral deck that the once Democratic Solid South soon vanished from the map. In 1948 the Democratic platform contained strong civil rights language for the first time, using wording similar to provisions that had been in Republican platforms before the 1900 election. Delegates from several southern states walked out of the Democratic National Convention, and States' Rights Democrat ("Dixiecrat") Strom Thurmond replaced national Democratic nominee Harry Truman on official presidential ballots in several southern states. Truman received a 49.6 percent plurality of the national popular vote and won the presidency with 303 electoral votes, versus Republican Thomas Dewey's 45.1 percent of the popular vote and 189 electoral votes, and Thurmond's 2.4 percent of the popular vote and 39 electoral votes. The wall around the Solid South was crumbling, and the election of 1964 then brought it crashing down (figure 4.24).

Democrat Lindon B. Johnson had succeeded to the presidency after the assassination of President John Kennedy in 1963. Johnson's "Great Society" programs, including the 1964 Civil Rights Act, proved popular with many voters. An internal contest in the Republican Party between more liberal Governor Nelson Rockefeller of New York and conservative Senator Barry Goldwater of Arizona yielded the narrow nomination of Goldwater for president. This led to the stunning outcome that several Deep South states cast Republican electoral votes for the very first time in history in the 1964 election. And for the first time since before the Civil War the Democrats swept the North. For example, Johnson won all sixty-three counties in New York, and he also won all of Vermont's fourteen counties. Thus, Johnson became the first Democrat to win Vermont in that state's history. (Of course, Vermont is one of the most Democratic and liberal states in the country today.) Nationally, Johnson received 61.0 percent of the popular vote and 486 electoral votes, compared with 38.3 percent of the popular vote and 52 electoral votes for Goldwater.

Factor 6 depicts the first election pattern to emerge after the thorough reshuffling from 1948 and 1964. As shown on the subgraph for Factor 6, this new pattern was strong during the late twentieth century from 1972 to 2000, with the highest loading associated with the election of 1992. Five of these eight elections were won by Republican candidates (1972, 1980, 1984, 1988, and 2000), and three were won by Democratic candidates (1976, 1992, and 1996). Interestingly, all of these elections were won by candidates who were nominated from a state in the South or West: Democrats Carter and Clinton from the South, and Republicans Nixon and Regan from California, and Bush Sr. and Jr. from the Southwest. The center of population in the United States crossed the Mississippi River for the first time in 1980, and all of these winners except Carter were nominated from west of the Mississippi River, so this might well be called the "sunbelt normal vote" pattern. The 1992 election was a three-way contest between incumbent Republican George Bush Sr., Democratic challenger Bill Clinton, and Independent Ross Perot. Clinton won with 43.0 percent of the popular vote and 370 electoral votes, versus 37.4 percent of the popular vote and 168 electoral votes for Bush, and 18.9 percent of the popular vote and no electoral votes for Perot. The outcome for Perot conveyed the message once again that the electoral deck is stacked against third-party candidates. Bush did well in the Southeast and Interior West, and Clinton did well in the North, border South, and Pacific West (figure 4.25).

The subgraph for Factor 7 shows that the contemporary pattern first became most prominent in 2004, although factor pattern loadings of similar magnitude for the 2000 and 2004 elections on both Factors 6 and 7 indicate a gradational rather than a dramatic reshuffling of the electoral deck around the turn of the century. Not surprisingly, given the very high correlation between the elections of 2016 and 2020, these most recent elections have the very highest loadings on Factor 7. These high factor loadings manifest the degree to which the distributions of popular votes at the county level in these elections are very similar to one another. Each of the elections from 2004 to 2020 has a factor pattern loading of 0.64 or greater on Factor 7, which illustrates the increasingly high degree of polarization between urban and rural areas over the past two decades, with urban centers strongly Democratic and many rural areas strongly Republican. The 2020 map shown in figure 4.26 typifies this factor. It should be noted that while the Republicans won a large majority of counties across the United States in each of the five elections since 2004, the Democrats won more popular votes than did the Republicans in four of these five elections, with 2004 being the only exception.

In summary, the T-mode factor analysis of the presidential elections from 1872 through 2020 yielded seven factors, each of which represents a particular geographical distribution of voting percentages. Those elections before World War II are characterized by the sectional normal vote, with strong Republican support in the North and strong Democratic support in the South. This was followed by transitional patterns associated with the elections between 1948 and 1968, and then two subsequent patterns from 1972 to 2000 and then from 2004 to 2020. Since the massive electoral reshufflings in the mid-twentieth century ended the long-term sectional normal vote, the South has gravitated to the Republicans while the Democrats have made significant inroads into the North. The contemporary period, especially during and after the late 1990s, is associated with not only Democratic strength in the North and Republican strength in the South, but a strong divergence between metropolitan and non-metropolitan areas.

REFERENCES

Archer, J. Clark, and Peter J. Taylor. 1981. *Section and Party.* New York: Wiley.

Archer, J. Clark, and Fred M. Shelley. 1986. *American Electoral Mosaics.* Washington, DC: Association of American Geographers.

Archer, J. Clark, Fred M. Shelley, Peter J. Taylor, and Ellen R. White. 1988. The Geography of U.S. Presidential Elections. *Scientific American* 259(1):18–25.

Archer, J. Clark, Stephen J. Lavin, Kenneth C. Martis, Fred M. Shelley. 2006. *Historical Atlas of the U.S. Presidential Elections, 1788–2004.* Washington, DC: Congressional Quarterly Press.

Archer, J. Clark, Robert H. Watrel, Fiona Davidson, Erin H. Fouberg, Kenneth C. Martis, Richard L. Morrill, Fred M. Shelley, and Gerald R. Webster. 2014. *Atlas of the 2012 Elections.* Lanham, MD: Rowman & Littlefield.

Cattell, Raymond B. 1952. *Factor Analysis: An Introduction for the Psychologist and Social Scientist.* New York: Harper and Brothers.

Rummel, Rudolph J. 1970. *Applied Factor Analysis.* Evanston, IL: Northwestern University Press.

Shelley, Fred M., J. Clark Archer, Fiona M. Davidson, and Stanley D. Brunn. 1996. *Political Geography of the United States.* New York: Guilford.

US Bureau of the Census. 2021. Voting and Registration in the Election of November 2020, Table 1. Reported Voting and Registration by Sex and Single Year of Age, November 2020. April 29, www2.census.gov/programs-surveys/cps/tables/p20/585/table01.xlsx (last accessed May 4, 2021).

Watrel, Robert H., Ryan Weichelt, Fiona M. Davidson, John Heppen, Erin H, Fouberg, J. Clark Archer, Richard L. Morrill, Fred M. Shelley, and Kenneth C. Martis. 2018. *Atlas of the 2016 Elections.* Lanham, MD: Rowman & Littlefield.

SPATIAL ANALYSIS OF THE 2020 PRESIDENTIAL ELECTION

JOHN HEPPEN

The 2020 presidential election presented evidence, despite the reversal from 2016, that electoral and especially spatial patterns have remained consistent since 2000 at the state level, where the vast majority of electoral votes are awarded to the winner of the popular vote in each state. Spatial analysis of presidential election results presents a case for a consistent level of spatial autocorrelation since the Bush-Gore election in 2000. Moran's *I* is a global measure of spatial autocorrelation (Moran 1948, 1950). Moran's *I* analysis at the state level presents a case for a consistent level of spatial autocorrelation since 2000. Spatial autocorrelation is when similar values of a particular variable cluster and correlate with each other. In electoral geography it can be used to provide a measure of or degree to which similar or dissimilar voting levels are clustered across levels of analysis (state, counties, precincts). As Moran's *I* approaches 1, it indicates higher levels of spatial autocorrelation; as it approaches –1, it approaches a checkerboard pattern; 0 indicates no spatial autocorrelation (complete randomness). The Democratic vote for president from 1892 to 2020 at the state level was tested for spatial autocorrelation. Eighteen ninety-two was the first election where the United States was contiguous from the Atlantic to the Pacific. Hawaii and Alaska are not included in the analysis because they are not contiguous, but the District of Columbia is. The Democratic vote was paired with a spatial-weights matrix, which recorded each state's neighbors to calculate the global level of spatial autocorrelation. Figure 4.27 presents Moran's *I* from 1892 to 2020.

The highest level of spatial autocorrelation was during the Solid South era of one-party dominance of the Democrats in the American South. White southern voters in this era of segregation and voting-rights violations felt that maintaining one-party unity was vital in maintaining white political, social, and economic power. But factionalism in southern states between conservative and populist wings existed. The Democratic vote for president was highly clustered in the South from 1892 to 1944. The period from 1948 to 1988 was an era of immense political and social change in the South and the rest of the country from the civil rights movement, the turbulent 1960s, recessions of the 1970s and the Reagan Revolution. There was volatility, but with a pattern of lowering spatial autocorrelation through time in a saw-toothed manner. The South stopped being the realm of the Democratic Party and transitioned from two-party competition to arguably being the realm of the Republican Party in current times. The Bill Clinton election in 1992 started a period of increased spatial autocorrelation as Red and Blue America started to form and solidified in the infamous 2000 election. Moran's *I* has measured near 0.4 since 2000, which is statistically significant and is manifested with Democrats being strongest in the Northeast, Pacific Coast, and Great Lakes and Republicans generally being strongest in the South, Great Plains, and Intermontane West. The current era has the most consistent level of spatial autocorrelation during the study period, providing spatial-temporal evidence in addition to political evidence of the era of Red and Blue America. In summary, at the state level the country has had two decades of spatial polarization from the George W. Bush to Trump elections.

The county level, with more than three thousand units of analysis, offers a more nuanced view of Red and Blue America using local measures of spatial association. Local Moran *I*'s is a local indicator of spatial association or LISA (Anselin 1995, 2020a, 2020b). Each county has a measure of how it contributes to the global level of spatial autocorrelation, allowing mapping of counties in our case. The local Moran's *I* tested to see if the percentage of the vote in a county was spatially and statistically significant in relation to its neighbors. The same methods were used in calculating the local Moran as in calculating Moran's *I* at the state level, but the unit of analysis was refined to the county. Local Moran's *I* classifies counties to one of five categories. The first are counties that are not significant. Counties that are High-High are counties of high vote percentage bordered by counties of similar, high values. These are counties of high levels of support clustered together. Low-Low counties are counties of lower vote percentages bordered by similar counties. These are counties of low support clustered together. Those two cases represented clustering of similar values. High-Low counties are counties where a higher vote is contiguous with

FIGURE 4.27

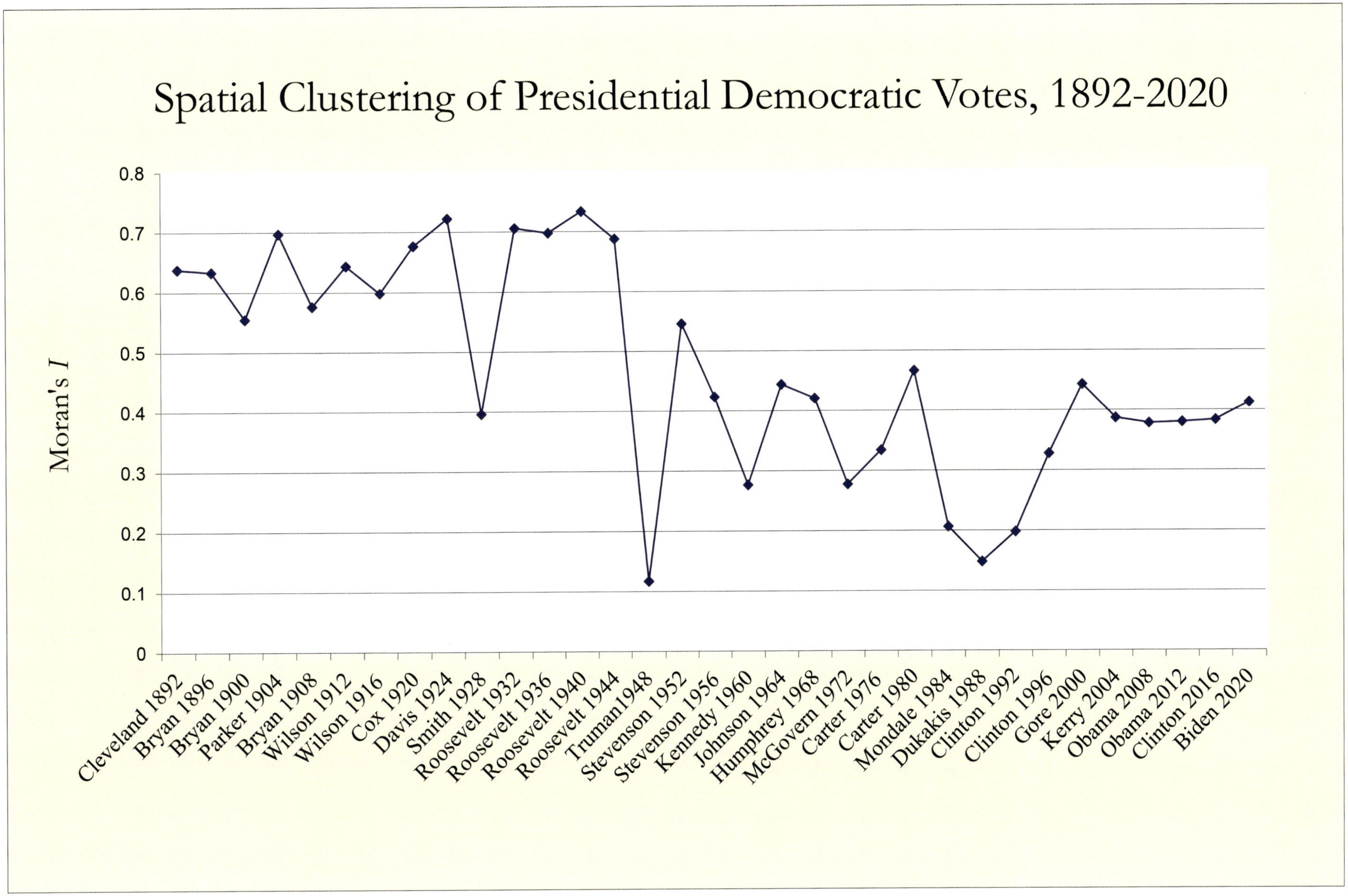
Spatial Clustering of Presidential Democratic Votes, 1892-2020
Moran's I
0.8
0.7
0.6
0.5
0.4
0.3
0.2
0.1
0
Cleveland 1892
Bryan 1896
Bryan 1900
Parker 1904
Bryan 1908
Wilson 1912
Wilson 1916
Cox 1920
Davis 1924
Smith 1928
Roosevelt 1932
Roosevelt 1936
Roosevelt 1940
Roosevelt 1944
Truman1948
Stevenson 1952
Stevenson 1956
Kennedy 1960
Johnson 1964
Humphrey 1968
McGovern 1972
Carter 1976
Carter 1980
Mondale 1984
Dukakis 1988
Clinton 1992
Clinton 1996
Gore 2000
Kerry 2004
Obama 2008
Obama 2012
Clinton 2016
Biden 2020

FIGURE 4.28

Biden Voting Clusters, 2020

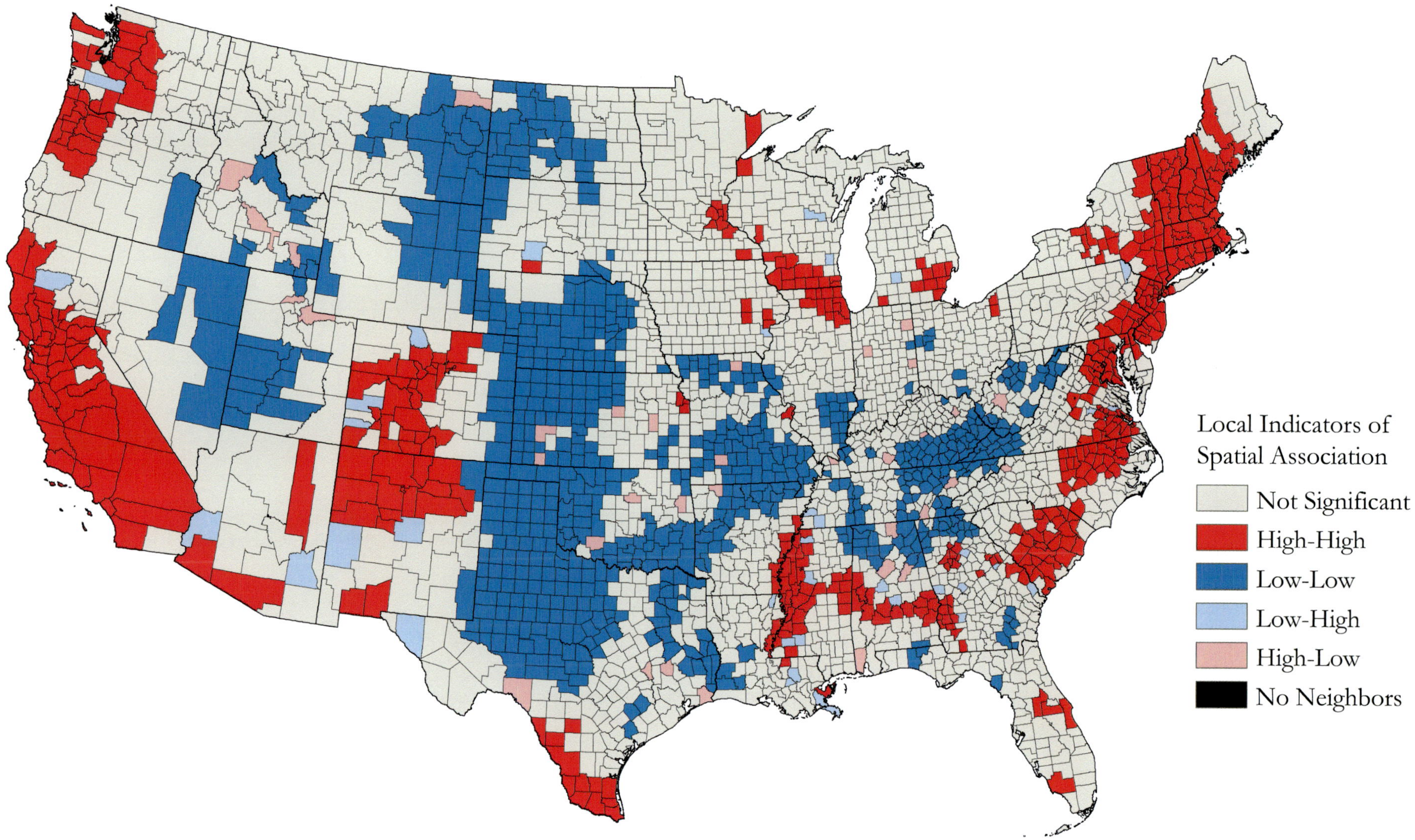

FIGURE 4.29

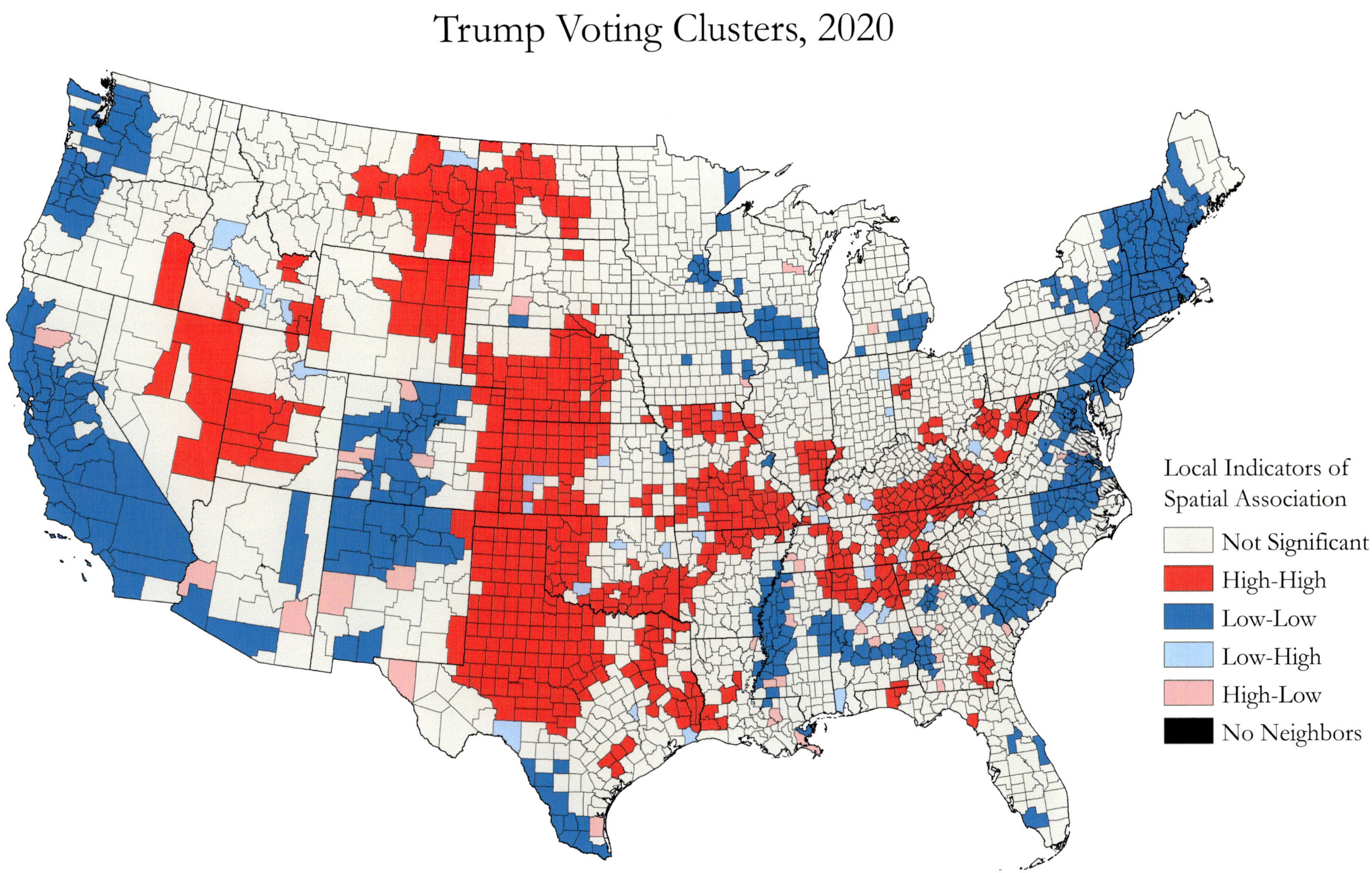
Trump Voting Clusters, 2020
Local Indicators of Spatial Association
Not Significant
High-High
Low-Low
Low-High
High-Low
No Neighbors

neighbors' counties of low support. Low-High counties are counties of low votes with neighbors of high votes. Those last two categories represent outliers and are often only one county in size. These counties in our example are often counties of American Indian reservations, or counties with a markedly different racial or economic makeup from their neighbors.

Figure 4.28 presents counties of various levels of support for Biden and figure 4.29 shows the level of support for Trump. Counties of High-High support for Biden appear in New England and the urban Northeast from Boston to northern Virginia, the Pacific Coast, the high-back population counties of the South, southern Texas, and northern New Mexico and Colorado. Comparing figures 4.28 and 4.29 shows levels of support for Trump that are counties of Low-Low Biden support. Trump's High-High counties are clustered in the Great Plains, Appalachia, the Ozark Counties of Missouri and Arkansas, Wyoming, Montana, Utah, and portions of Nevada. Biden had high levels of support in the largest urban counties in the country on the East and West Coast and in the Chicago and Detroit areas. Trump, on the other hand, had huge swaths of High-High counties in states and regions with low populations as in the Great Plains and Appalachia and in states that he safely carried in 2016 and 2020. The Electoral College base for the Democrats is based on the urban counties of the Northeast, Great Lakes, and California. Dominating New York, Los Angeles, Chicago, San Francisco, Boston, Seattle, and Portland provided an electoral vote of 134 as the states of those cities were not contested by Trump. Increasing the urban vote in Detroit and Philadelphia created High-High counties in those states and brought back thirty-six electoral votes to Biden. High-High counties for Biden in major urban areas means space and time can be used more productively.

Trump having High-High counties dominated by rural counties in states with fewer electoral votes save Texas is problematic for any future Republican strategy. Texas, Florida, and Ohio were the bedrock of Trump's strategy. But the lack of any High-High counties in major urban areas of those states should be a warning. The lack of voters in densely populated places means organizing voters and get-out-the-vote activities face spatial difficulties. Organizing across rural and less-populated counties faces challenges of time and space and unlike volunteers and money, time stretched across more and more space will run out sooner.

REFERENCES

Anselin, L. 1995. Local Indicators of Spatial Association—LISA. *Geographical Analysis* 27:93–115.

———. 2020a. Local Spatial Autocorrelation (1) LISA and Local Moran. http://geodacenter.github.io/workbook/6a_local_auto/lab6a.html (last accessed June 7, 2021).

———. 2020b. GeoDa: An Introduction to Spatial Data Analysis. https://geodacenter.github.io/ (last accessed June 7, 2021).

Moran, P. A. P. 1948. The Interpretation of Statistical Maps. *Journal of the Royal Statistical Society Series B*: 245–51.

———. 1950. Notes on Continuous Stochastic Phenomena. *Biometrika*, 37:17–23.

METROPOLITAN CONCENTRATION OF THE DEMOCRATIC VOTE, 2020

JASON COMBS AND RICHARD L. MORRILL

The polarized rural-urban pattern witnessed in the last few elections continued in 2020 and large metropolitan areas once again proved decisive. Stretching back to the 1950s and 1960s, the now familiar rural-urban split was even more pronounced in 2020 (Kanik and Scott 2020). The divide is not simply between states or regions either, but also within states (Damore, Lang, and Danielsen 2021). Republican Donald Trump ultimately carried most counties across the country—metropolitan, micropolitan, and rural—but Democrat challenger Joe Biden won key metropolitan areas, which translated into an Electoral College victory.

Figure 4.30 demonstrates the spatial breakdown of support across the country for major-party candidates. Incumbent Donald Trump won most of the metropolitan counties, but his challenger—Democrat Joe Biden—won metropolitan places in key locations (figure 4.30). Of the 1,247 metropolitan counties in the United States, Trump won 878 counties (70.4 percent) while Biden won just 369 (29.6 percent). Winning counties is one thing; considering the number of votes captured is another.

Voter turnout in 2020 was quite high and metropolitan areas led the way (Lee 2020; Rainwater 2020). Metropolitan votes tallied 144,287,495, which equals 91.1 percent of all ballots cast in 2020 (158,394,605). Of the metropolitan votes, Biden captured 53.3 percent (76,929,166) compared to Trump's 44.8 percent (64,667,983), but just sharing the total metropolitan vote tends to hide several fascinating trends in that category.

Consider, for example, the ten, twenty-five, fifty, and one hundred largest metropolitan counties in the United States. Of the top ten metropolitan counties by population—which includes cities such as Anaheim, Brooklyn, Chicago, Dallas, Houston, Los Angeles, Miami, Phoenix, Riverside, and San Diego—Biden won all ten of the most populated counties by an astounding 4,772,063 votes—10,940,574 (63.9 percent) to 6,168,511 (36.1 percent).

Of the largest twenty-five metropolitan counties, which includes Clark County, Nevada (Las Vegas), San Bernardino County, California (Victorville), Tarrant County, Texas (Fort Worth), Wayne County, Michigan (Detroit), Alameda County, California (Oakland), Middlesex County, New Jersey (Edison), Philadelphia County, Pennsylvania (Philadelphia), and Palm Beach County, Florida (West Palm Beach), Biden won all twenty-five by 9,250,837 votes—19,404,606 (65.6 percent) to 10,153,769 (34.4 percent). Extending that to the fifty most populated counties, a similar pattern is found. Biden won forty-eight of the fifty largest metropolitan areas, capturing 28,537,064 votes (63.7 percent) compared to Trump with 15,462,919 (34.5 percent).

Of the hundred largest metropolitan areas, Biden again found tremendous support across the country, which equaled a boost regarding the Electoral College (Still and Mellnik n.d.) (figure 4.31). In this category, Biden captured 40,845,438 votes (62.8 percent) compared to Trump with 23,006,923 (34.5 percent) for an advantage of 17,838,515 votes. Biden ultimately carried ninety of the hundred most populated metropolitan counties ranging from Los Angeles County, California, with 4,264,365 votes (Biden captured 3,028,885) to Dane County, Wisconsin, which registered 344,728 votes (Biden won 260,121) in 2020. Of the hundred most populated counties, Trump carried just ten with one in Colorado (El Paso), three in Florida (Brevard, Lee, and Polk), one in Michigan (Macomb), two in New Jersey (Monmouth and Ocean), one in New York (Suffolk), and two in Texas (Collin and Denton). Trump won these ten metropolitan counties by 437,001 votes—2,362,574 (55.1 percent) to 1,925,573 (44.9 percent).

Next consider both the top Democrat counties by percent and the number of votes cast. The top ten Democrat counties by percent include the District of Columbia (92.15 percent), Prince George's (Baltimore, 89.26 percent), New York (Manhattan, 86.74 percent), San Francisco (San Francisco, 85.26 percent), Bronx (New York, 83.44 percent), DeKalb (Atlanta, 83.12 percent), Philadelphia (Philadelphia, 81.44), Suffolk (eastern Long Island, 80.64 percent), Alameda (Oakland, 80.21 percent), and Denver (Denver, 79.55 percent). It was, however, not just a matter of winning by a significant percentage. Biden captured 4,146,592 votes (83.8 percent) in these ten

2020 Presidential Vote Leader
Metropolitan Counties

FIGURE 4.30

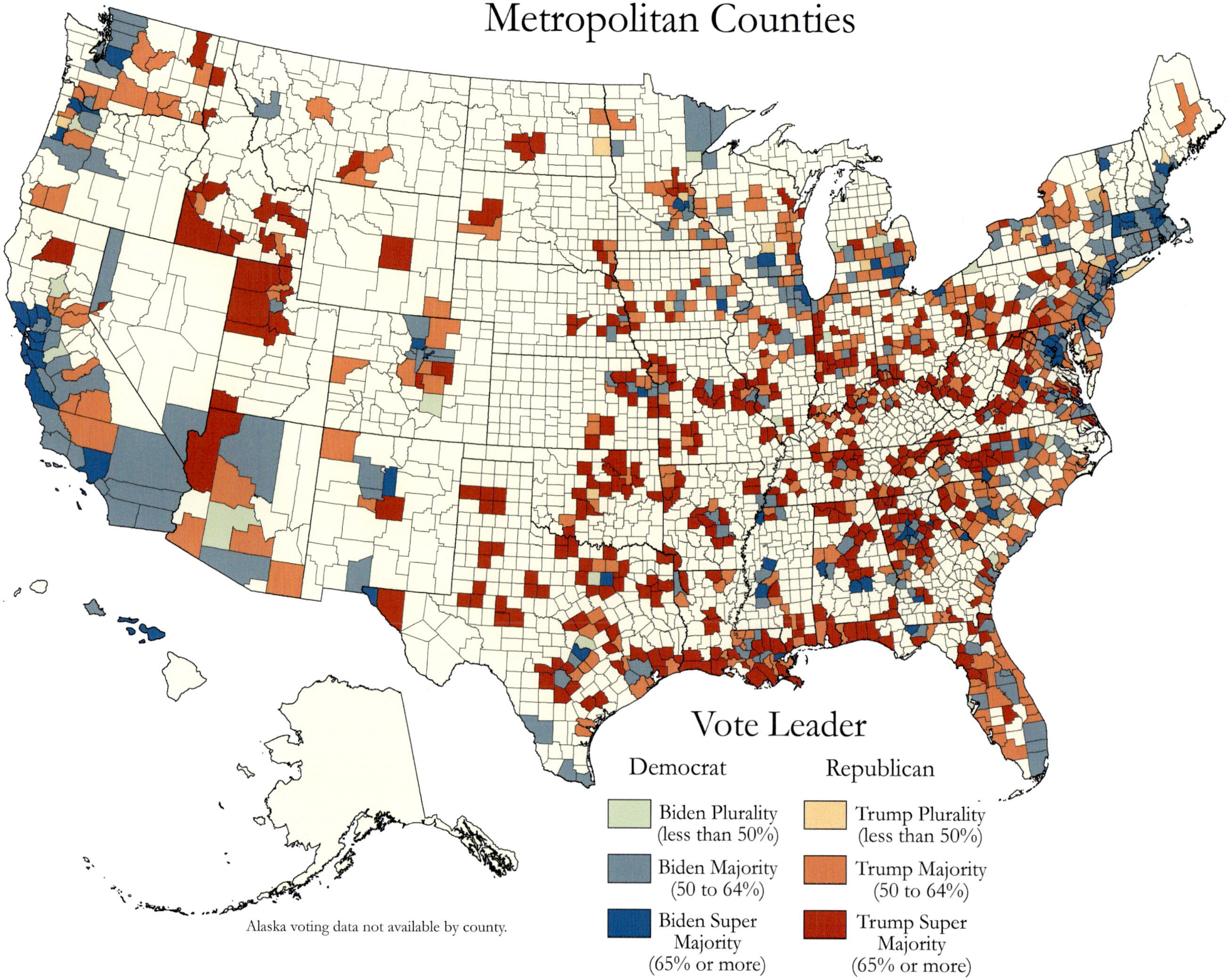

FIGURE 4.31

2020 Popular Presidential Vote in Metropolitan Areas With Populations of 1 Million or More

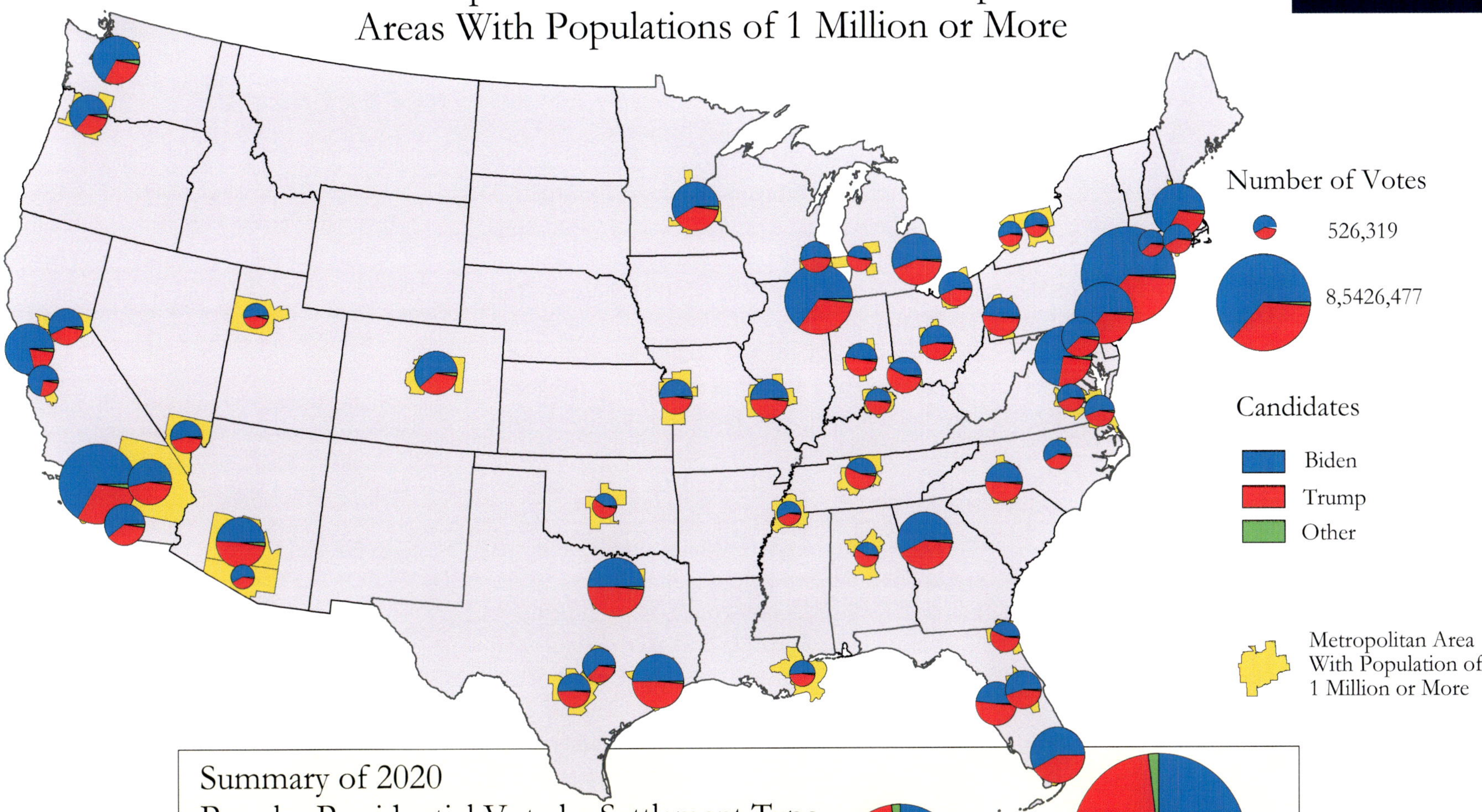

Summary of 2020 Popular Presidential Vote by Settlement Type

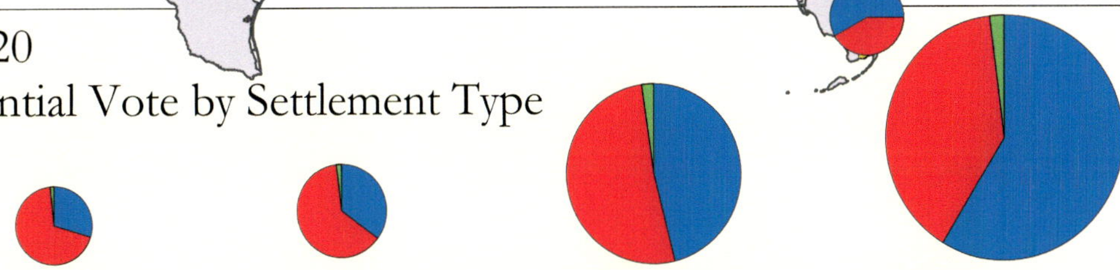

	Non-CBSA	Micropolitan	Metropolitan Under 1 Million	Metropolitan 1 Million or More
Biden	2,695,641	4,549,214	22,104,966	51,912,893
Trump	6,170,546	8,129,536	24,825,372	35,070,644
Other	148,743	245,561	957,194	1,587,441
Total	**9,014,930**	**12,924,311**	**47,887,532**	**88,570,978**

counties in comparison to Trump's 722,671 votes (14.6 percent).

In comparison, consider the top ten Republican counties by percent and the number of votes involved in the 2020 contest. Trump carried the following counties with at least 83 percent of the vote in each: Blount (Alabama, 89.57), Cullman (Alabama, 88.12), Campbell (Wyoming, 86.76), Hardin (Texas, 86.33), Brown (Texas, 85.78), Van Zandt (Texas, 85.56), DeKalb (Alabama, 84.37), Livingston (Louisiana, 84.13), Murray (Georgia, 84.11), and Marshall (Alabama, 83.70). Trump won these ten by a wide margin—86.9 percent to 13.1 percent—yet victories in these ten counties provided only a 224,245-vote margin (264,170 to 39,925) for the Republican.

Next consider cases where a single metropolitan region constitutes more than half of the state's population. In 2020, Biden carried Arizona, Colorado, Georgia, Illinois, Minnesota, Nevada, New York, Oregon, and Washington. Each of these states contains a single metropolitan area that contains most of the state's total population, and by winning these key metropolitan areas Biden was able to capture each state's electoral votes. Combined, these states contain 111 Electoral College votes; two of these (Arizona and Georgia flipped from Republican in 2016 to Democrat in 2020) hold twenty-seven Electoral College votes.

It should also be noted that in many cases Trump increased his vote totals from 2016, even in several key metropolitan places—but lost (Kanik and Scott 2020; Still and Mellnik n.d.). Biden won many metropolitan areas, due in part to increasing turnout totals. In 2016, Trump carried Tarrant County, Texas (Fort Worth), for example, with a 52.2 percent to 43.5 percent margin. In 2020, the county witnessed a 24 percent turnout jump and Biden won Tarrant County by a slim margin—49.3 percent to 49.1 percent.

Even in states that voted Republican, Democrat Joe Biden made significant inroads primarily in metropolitan areas. Texas, for instance, is an excellent example. Of the state's largest metropolitan areas—Austin, Dallas–Fort Worth, Houston, and San Antonio—Democrat Hillary Clinton carried only Austin in 2016 (Lee 2020). In 2020, those four areas held a higher percentage of the state's votes and challenger Joe Biden won all four, which ultimately tightened the race in Texas—Trump carried Texas by 9.2 percent in 2016 and only 5.6 percent in 2020 (Damore, Lang, and Danielsen 2021). If that trend continues and Democrats capture Texas, a Republican path to victory in the Electoral College becomes much more difficult.

Voting patterns dating back to the 1950s and 1960s continued in 2020. As the United States becomes more urbanized and rural counties continue their decline in the share of population, philosophical and political differences will most likely not only continue but escalate. Moreover, reapportionment following the 2020 census is set to place even greater emphasis on this divide. Several southern and western states (Arizona, Colorado, Georgia, and Texas, for example) are assumed to capture additional House seats and more Electoral College votes. As metropolitan areas in many states continue to expand and control a larger percentage of the state's share of votes, the focus on large metropolitan areas will most likely intensify.

REFERENCES

Damore, D., R. Lang, and K. Danielsen. 2021. In 2020, the Largest Metro Areas Made the Difference for Democrats. *The Brookings Institution*. February 4, https://www.brookings.edu (last accessed March 30, 2021).

Florida, Richard. 2020. How Metro Areas Voted in the 2020 Election. *CityLab*. December 4, https://www.bloomberg.com/news/features/2020-12-04/how-metro-areas-voted-in-the-2020-election (last accessed March 30, 2021).

Kanik, A., and P. Scott. 2020. The Urban-Rural Divide Only Deepened in the 2020 US Election. *City Monitor*. November 11, https://citymonitor.ai (last accessed April 4, 2021).

Lee, E. 2020. Fact Check: Joe Biden Outperformed Hillary Clinton in Most Major Metro Areas. *USA Today*. November 20, https://www.usatoday.com (last accessed March 20, 2021).

Rainwater, Brooks. 2020. How Urban Areas Played a Pivotal Role in the Presidential Election. *U.S. News & World Report*. December 4, https://www.usnews.com/news/cities/articles/2020-12-04/how-cities-made-the-difference-in-the-2020-presidential-election (last accessed June 22, 2021).

Still, A., and T. Mellnik. n.d. Democrats Gained in Nearly Every Battleground State, Running up Margins in Cities. *Washington Post*. https://www.washingtonpost.com (last accessed March 30, 2021).

THE RURAL VOTE

CHRISTOPHER R. LAINGEN

In 2020, there were 1,961 non-metropolitan counties, the term used to define "rural" in this summary, in the United States as defined by the Office of Management and Budget (US Department of Agriculture, Economic Research Service 2021). These include all counties except those not categorized in "metropolitan" classes: counties in metro areas fewer than 250,000, counties in metro areas of 250,000 to one million, and counties in metro areas greater than one million. These counties contained just over 14 percent of all 2020 voters.

Every rural county in Nevada, Oklahoma, Kansas, Nebraska, Iowa, Missouri, Illinois, Indiana, Kentucky, West Virginia, Pennsylvania, Florida, and Massachusetts had a majority vote for Donald Trump (figure 4.32). Rural counties that voted for Joe Biden are loosely scattered around the western states, although clustering does occur in western Colorado and northern New Mexico, as well as in the Black Belt of the southeastern United States and in rural New Hampshire and Vermont.

In sum, 90 percent of the 1,961 rural counties voted Republican. Donald Trump received 64.9 percent of all rural votes, Joe Biden received 33.2 percent, and others received 1.8 percent. Twenty-seven counties had greater than 90 percent of votes cast for Donald Trump. Roberts County, Texas, topped the list at 96.2 percent with Borden County, Texas, in second place at 95.4 percent. The greatest Democratic voting percentages for rural counties were 88.4 percent in Oglala-Lakota, South Dakota, and 85.6 percent in Hawaii County, Hawaii.

Leading up to the 2020 election, many Republicans pondered whether President Trump's share of the rural vote would wane. Midterm election results in 2018 revealed a shift in voting preferences (from red to blue) nationally, but specifically in the battleground states of Michigan, Pennsylvania, Wisconsin, and Iowa. However, these trends did not translate into the 2020 presidential election, though many rural counties' D-R gap did indeed narrow. Ultimately Joe Biden's victory in the 2020 election was due "mostly to voters in large metropolitan suburbs, especially in important battleground states," not a 2018 midterm redux (Frey 2020).

Indeed, the agricultural industry was hit hard between the 2016 and 2020 elections: financial woes related to Chinese tariffs, regional weather-related disasters, and COVID-19 (Siegler 2020). But despite these hurdles, most farmers did not blame President Trump and 85 percent planned to vote for him a second time in 2020 (Bunge 2019). Offsets between Chinese trade deals that created new avenues for exports and subsequent tariffs that may have cost farmers billions in lost revenue were overshadowed by farmers' overwhelming approval of President Trump's stance against crippling environmental regulations—regulations that Joe Biden would quickly reinstate in the so-called fight against climate change.

The 2020 presidential election results showed a clear and continuing polarization between rural and urban people—a polarization that we can see playing out across broad swaths of cultural, socioeconomic, environmental, and political debates: republican counties/regions become more republican and democrat counties/regions become more democrat (McGee 2020). In terms of explaining this widening chasm geographically, there is no better summation than what Bryant Moy of the University of Maryland said when interviewed by Washington University in St. Louis (Savat 2020): "Life experiences shape our perceptions of the world. On the other hand, we tend to overlook how the social environment—outside of race, gender and income—plays a role in our partisan identity. And that is the main takeaway from our research (Gimpel et al. 2020). The environment around us—the distance we live away from a metropolitan area and population density—shapes what we think about the political world and the partisan labels we adopt."

As geographers we already know this: "where" matters. Where shapes us: where we were born, where we grew up, where we went to school, where we got a job, and where we raise our family. For some of us during our lives our "where" changes, but pieces of those experiences are forever a part of us. The rural "where" for someone who grew up in an urban area is just as foreign a place as is an urban "where" for someone who grew up in a rural area, and aligning those perceptions and worldviews,

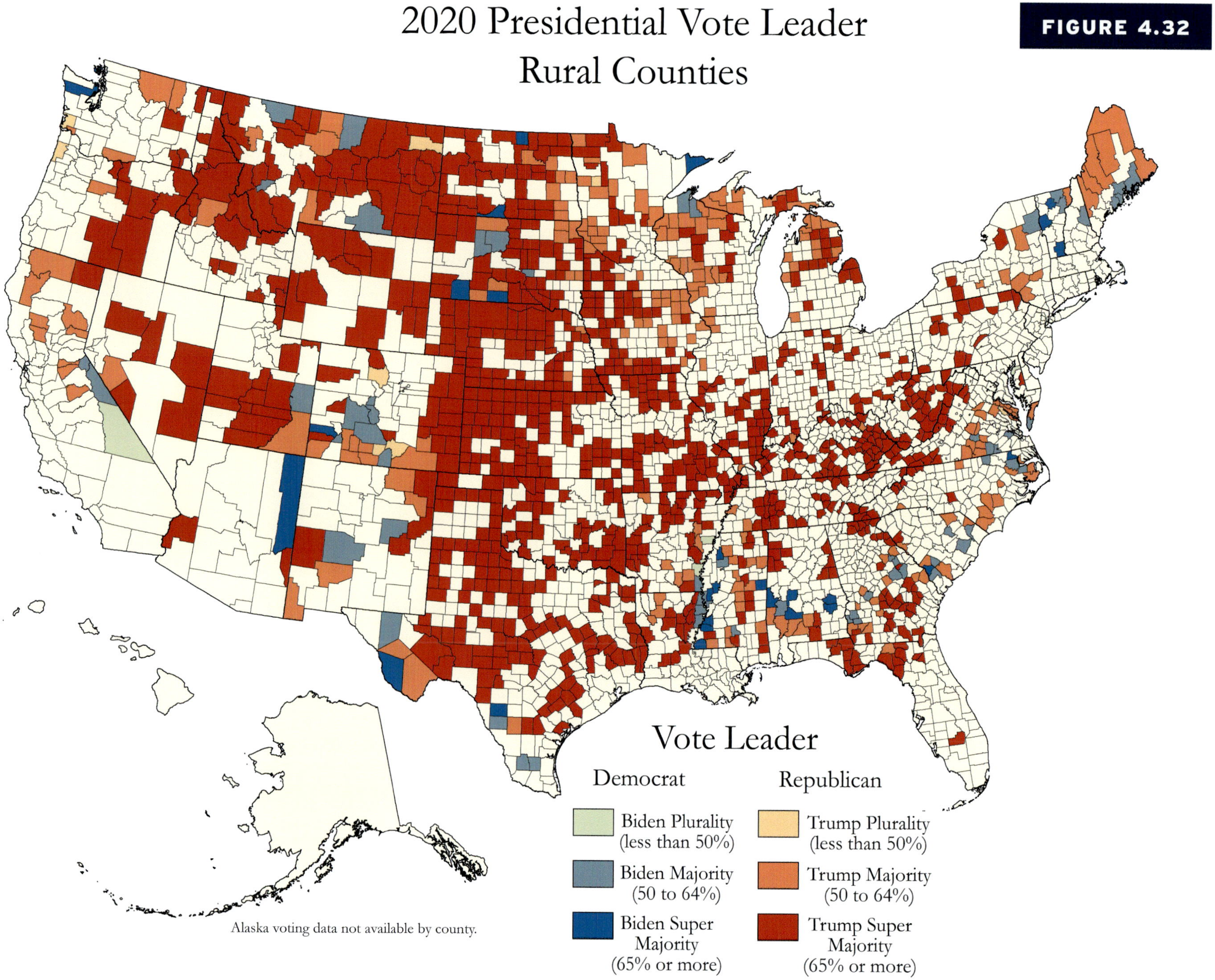
FIGURE 4.32
2020 Presidential Vote Leader
Rural Counties
Vote Leader
Democrat
Republican
Biden Plurality (less than 50%)
Biden Majority (50 to 64%)
Biden Super Majority (65% or more)
Trump Plurality (less than 50%)
Trump Majority (50 to 64%)
Trump Super Majority (65% or more)
Alaska voting data not available by county.

correct or not, is difficult. And as our population continues to become more clustered in urban areas (greater than 80 percent of the population resides on 2–3 percent of the country's land area) the partisan divides will likely only grow wider. Neither "where" is the wrong choice; each is as important to the future success of our country as the other.

REFERENCES

Bunge, J. 2019. Farmers Stick with Trump, Despite Trade-War Pain. *Wall Street Journal.* October 19, https://www.wsj.com/articles/farmers-election-trump-biden-trade-war-11602693377 (last accessed June 8, 2021).

Frey, W. H. 2020. Biden's Victory Came from the Suburbs. *Brookings Institution.* https://www.brookings.edu/research/bidens-victory-came-from-the-suburbs/ (last accessed June 8, 2021).

Gimpel, J. G., N. Lovin, B. Moy, and A. Reeves. 2020. The Urban-Rural Gulf in American Political Behavior. *Political Behavior* 42:1343–68. https://doi.org/10.1007/s11109-020-09601-w (last accessed June 8, 2021).

McGee, G. 2020. What 2020 Election Results Tell Us About America's Growing Urban-Rural Divide. *UVAToday.* https://news.virginia.edu/content/what-2020-election-results-tell-us-about-americas-growing-urban-rural-divide (last accessed June 8, 2021).

Savat, S. 2020. The Divide Between Us: Urban-Rural Political Differences Rooted in Geography. *The Source.* February 18, https://source.wustl.edu/2020/02/the-divide-between-us-urban-rural-political-differences-rooted-in-geography/ (last accessed June 8, 2021).

Siegler, K. 2020. Biden's Win Shows Rural-Urban Divide Has Grown Since 2016. *NPR.* November 18, https://www.npr.org/2020/11/18/934631994/bidens-win-shows-rural-urban-divide-has-grown-since-2016 (last accessed June 8, 2021).

US Department of Agriculture, Economic Research Service. 2021. *Rural-Urban Continuum Codes* https://www.ers.usda.gov/data-products/rural-urban-continuum-codes.aspx (last accessed June 8, 2021).

MICROPOLITAN COUNTY VOTING

JASON COMBS

The micropolitan county vote pattern in 2020 was like recent elections; however, the percentage of votes from this category declined as metropolitan counties captured a larger share of total votes. In 2012, the micropolitan vote equaled 9.9 percent of the total turnout and in 2016 that number dipped to 8.6 percent. In 2020, the micropolitan percentage further declined to 8.2 percent. In comparison, the metropolitan counties increased their share from 83.7 percent in 2012, 85.4 percent in 2016, to 91.1 percent in 2020. In contrast, the percentage of votes from rural areas remained somewhat constant between 2012 (6.4 percent) and 2016 (6.0 percent), yet tumbled to 0.76 percent in 2020. It should be noted that the declines in the percentages for micropolitan and rural areas in 2020 does not necessarily point to depopulation or low turnout rates. The presidential election in 2020 witnessed a turnout rate of 66.3 percent, the highest in more than a century, with 91.1 percent of all votes coming from metropolitan counties across the country (DeSilver 2021; Montanaro 2020; Schaul, Rabinowitz, and Mellnik 2020).

It is interesting to note that in all three categories (metropolitan, micropolitan, and rural) Trump carried most counties. At the micropolitan level, Trump won 568 of the 658 counties (8,119,879 votes or 62.9 percent) in this category, with Biden capturing just 90 counties (4,539,655 votes or 35.2 percent). Trump in 2016 fared better than the two previous Republican nominees (McCain in 2008 and Romney in 2012) in the micropolitan category, which helped him flip six key states—Florida, Iowa, Michigan, Ohio, Pennsylvania, and Wisconsin. Trump did well in the micropolitan category again in 2020, yet Biden captured a handful of micropolitan counties in key swing states, including Michigan, Pennsylvania, and Wisconsin, which proved decisive (Lee 2020).

Trump in 2020 captured 62.9 percent of all micropolitan votes, which equaled a tremendous advantage at the county level (figure 4.33). Trump won 568 micropolitan counties (86.3 percent) in 2020; in comparison Biden carried 90 counties, just 13.7 percent of the total. Trump enjoyed success across the country and found significant micropolitan support in the Great Plains (North Dakota all the way south to Texas), Midwest (Ohio west to Missouri), Mountain West (Nevada and Utah, for instance), and throughout the South (Alabama, Florida, and Tennessee, for example). Moreover, it should be noted that Trump carried five micropolitan counties by at least 90 percent of the vote (Roberts County, Texas, McPherson County, Nebraska, Loving County, Texas, Logan County, Nebraska, and Ellis County, Oklahoma) and another seventy-six counties with at least 80 percent of the vote. Extending that further, Trump carried an additional 213 counties with at least 70 percent of the vote.

Biden did not have widespread support at the micropolitan level, but did enjoy success along the West Coast (California and Washington), in Colorado and south to New Mexico, in the lower Mississippi River Delta (Arkansas and Mississippi), further east in North Carolina, and in the far northeastern United States in New Hampshire and Vermont. In contrast to Trump, Biden carried only two micropolitan counties with at least 80 percent of the vote—Hawaii County (Hilo) at 85.6 percent and Menominee, Wisconsin (Shawano) at 82.0 percent—and just eight more counties with at least 70 percent of the vote: Dukes County (Vineyard Haven, Massachusetts) 77.5 percent, Taos County (Taos, New Mexico) 76.4 percent, Pitkin County (Glenwood Springs, Colorado) 75.2 percent, Hancock County (Milledgeville, Georgia) 71.7 percent, Washington County (Barre, Vermont) 71.4, Coahoma County (Clarksdale, Mississippi) 70.8 percent, Leflore County (Greenwood, Mississippi) 70.2 percent, and Sunflower County (Indianola, Mississippi) 70.0 percent.

For the entire micropolitan category, Trump won 62.9 of the counties and carried 568 of the 658 counties by a wide margin—3,580,224 votes. As astounding as those numbers are, consider those totals in comparison to the largest metropolitan counties. Biden won the top ten largest metropolitan counties by 4,772,063 votes, more than a one million advantage over Trump's margin of victory in 568 micropolitan counties. In 2020, despite an overwhelming advantage in both micropolitan and rural counties, Trump was not able to overcome Biden's dominance in America's largest metropolitan centers.

2020 Presidential Vote Leader
Micropolitan Counties

FIGURE 4.33

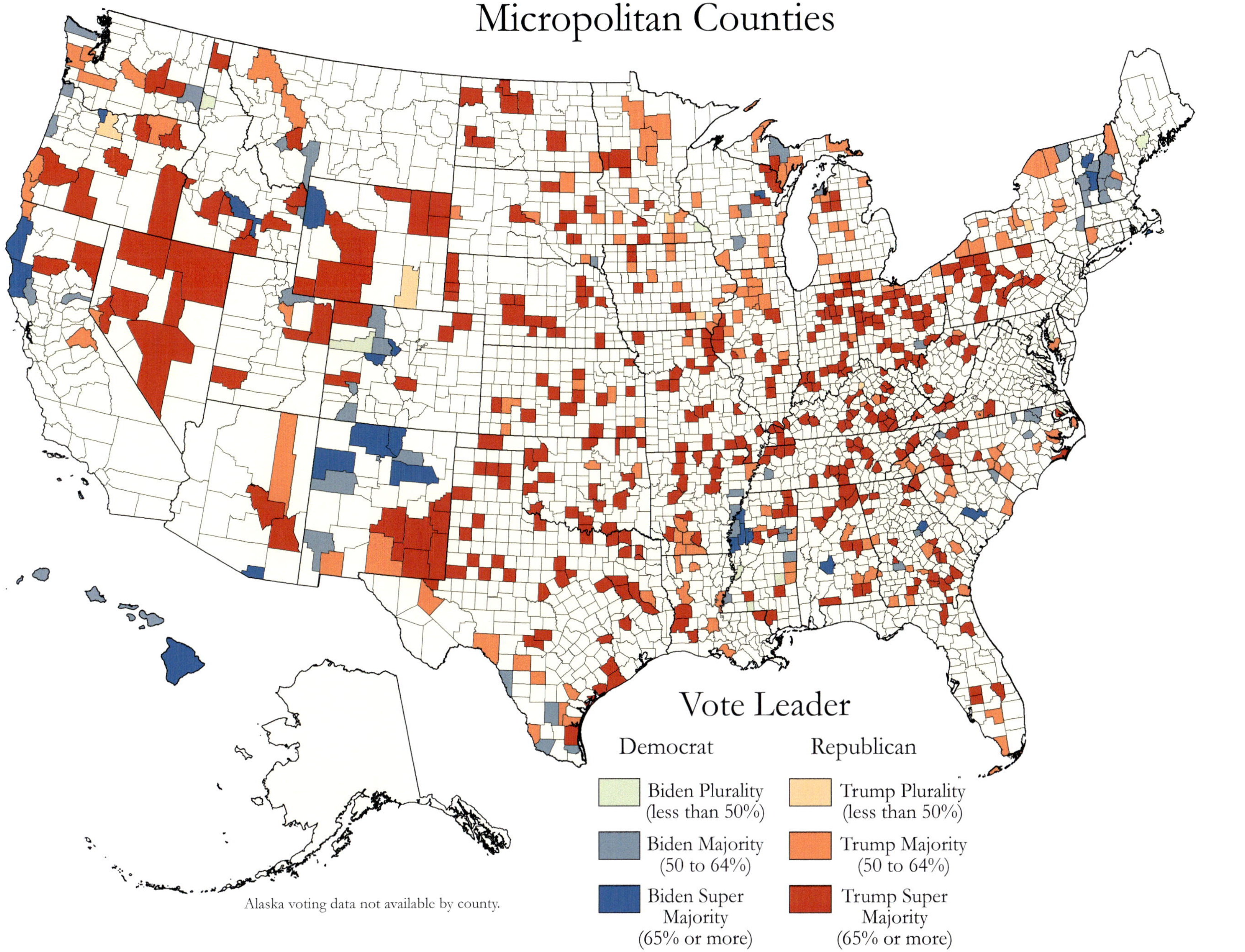

REFERENCES

DeSilver, D. 2021. News in the Numbers. *Pew Research Center*. January 28, https://www.pewresearchcenter.org (last accessed March 20, 2021).

Lee, E. 2020. Fact Check: Joe Biden Outperformed Hillary Clinton in Most Major Metro Areas. *USA Today*. November 20, https://www.usatoday.com (last accessed March 20, 2021).

Montanaro, D. 2020. President-Elect Joe Biden Hits 80 Million Votes in Year of Record Turnout. *National Public Radio*. November 25, https://www.npr.org (last accessed April 4, 2021).

Schaul, K., K. Rabinowitz, and T. Mellnik. 2020. 2020 Turnout is the Highest in Over a Century. *Washington Post*. December 28, https://www.washingtonpost.com (last accessed April 21, 2021).

"FLIPPED" COUNTIES IN THE 2020 ELECTION

JASON COMBS

As in the last few presidential elections where "flipped" counties were essential, those counties once again played a pivotal role in 2020. In 2016, Republican Donald Trump flipped 221 counties that Democrat Barack Obama had won in 2012. During the same election, Democrat Hillary Clinton carried seventeen counties that Republican Mitt Romney won in 2012, for a total of 238 flipped counties. The 2020 election had a total of seventy-seven flipped counties, with sixty-two that Republican Donald Trump had carried in 2016 switching to Democrat Joe Biden, while Trump captured just fifteen counties that Democrat Hillary Clinton won in 2016. Even though fewer counties were "flipped" in 2020, with election results so tight they were essential.

Of the fifteen counties that Trump flipped in 2020, only four were metropolitan counties; eight were micropolitan, and three were rural. The low numbers are not a huge surprise for an incumbent president, especially for someone who won 91.3 percent of the rural counties along with 86.8 percent of the micropolitan counties in 2016 (Crump 2018). Trump's flipped counties in 2020 were located in just six states—Colorado (one county), Georgia (one county), North Carolina (one county), Ohio (two counties), South Carolina (two counties), and Texas (eight counties; see figure 4.34). Overall, Trump carried these fifteen flipped counties by a slim margin—200,865 votes (51.5 percent) to 189,112 votes (48.5 percent). In comparison, Democrat Joe Biden flipped sixty-two counties in 2020; forty-three were metropolitan, ten micropolitan, and nine rural. Biden's flipped counties were sprinkled throughout the country, ranging from Maine, New Hampshire, and New York in the northeast to California, Oregon, and Washington on the West Coast (figure 4.34).

Several clusters of flipped counties reflect the country's current rural-urban divide and are noteworthy considering the thin margins witnessed across the nation. Three states—Florida, Kansas, and Texas—that typically lean Republican, for example, had several counties that flipped Democrat. Make no mistake, Trump still won those three states, yet vote totals reveal slim margins and an overall trend. In Florida, Trump won 51.2 percent of the votes, garnering 5,668,731 votes compared to 47.9 percent or 5,297,045 votes for Biden. Biden, however, flipped three large metropolitan areas in Florida—Jacksonville, Tampa-St. Petersburg-Clearwater, and Orlando-Kissimmee-Sanford—and carried those three by 27,322 votes. A similar situation played out in both Kansas and Texas. Trump carried Kansas by a wide margin (56.2 percent to 41.6 percent), but Biden flipped three metropolitan areas and carried those three by 32,355 votes—Kansas City, Kansas, Manhattan, Kansas, and Topeka, Kansas. Texas was even tighter. Trump carried the state with 52.1 percent of the votes compared to Biden's 46.5 percent. Biden, however, was able to flip three metropolitan Texas counties—two in the Austin metropolitan area and another in the Dallas–Fort Worth metropolitan area—by the tiny margin of 17,736 (614,886 to 597,150).

While flipped counties in places like Florida, Kansas and Texas did not upset the overall electoral count, others did and proved decisive. In 2016, Trump won the presidency in large part to outcomes in three key states—Michigan, Pennsylvania, and Wisconsin—by a combined 77,774 votes that Barack Obama won in 2012 (Michigan—Trump +10,704, Pennsylvania—Trump +44,292, and Wisconsin—Trump + 22,748). In 2020, Biden flipped those three states—and their combined forty-six Electoral College votes—and a handful of key counties played an integral part (Alberta 2020). In Michigan, just north of Detroit is Macomb County. Democrats won Macomb County by nearly nine points in 2008 and by 4 percent in 2012, and then Trump flipped the county in 2016 and carried it by 12 percent (Trump won Michigan 47.6 percent to 47.3 percent over Clinton in 2016). Trump won Macomb County again in 2020, yet the margin of victory dipped to 8 percent and Trump ultimately lost the state by 154,188 votes (2,804,040 to 2,649,852).

Lackawanna County, Pennsylvania, is home to Scranton, which has historically been a Democrat stronghold. In 2012, Obama won Lackawanna County by twenty-seven points, yet four years later Clinton carried Lackawanna by just three points. Lackawanna was deemed essential by both campaigns in 2020 and the result was decisive for

FIGURE 4.34

"Flipped" Counties between the 2016 and 2020 Presidential Elections

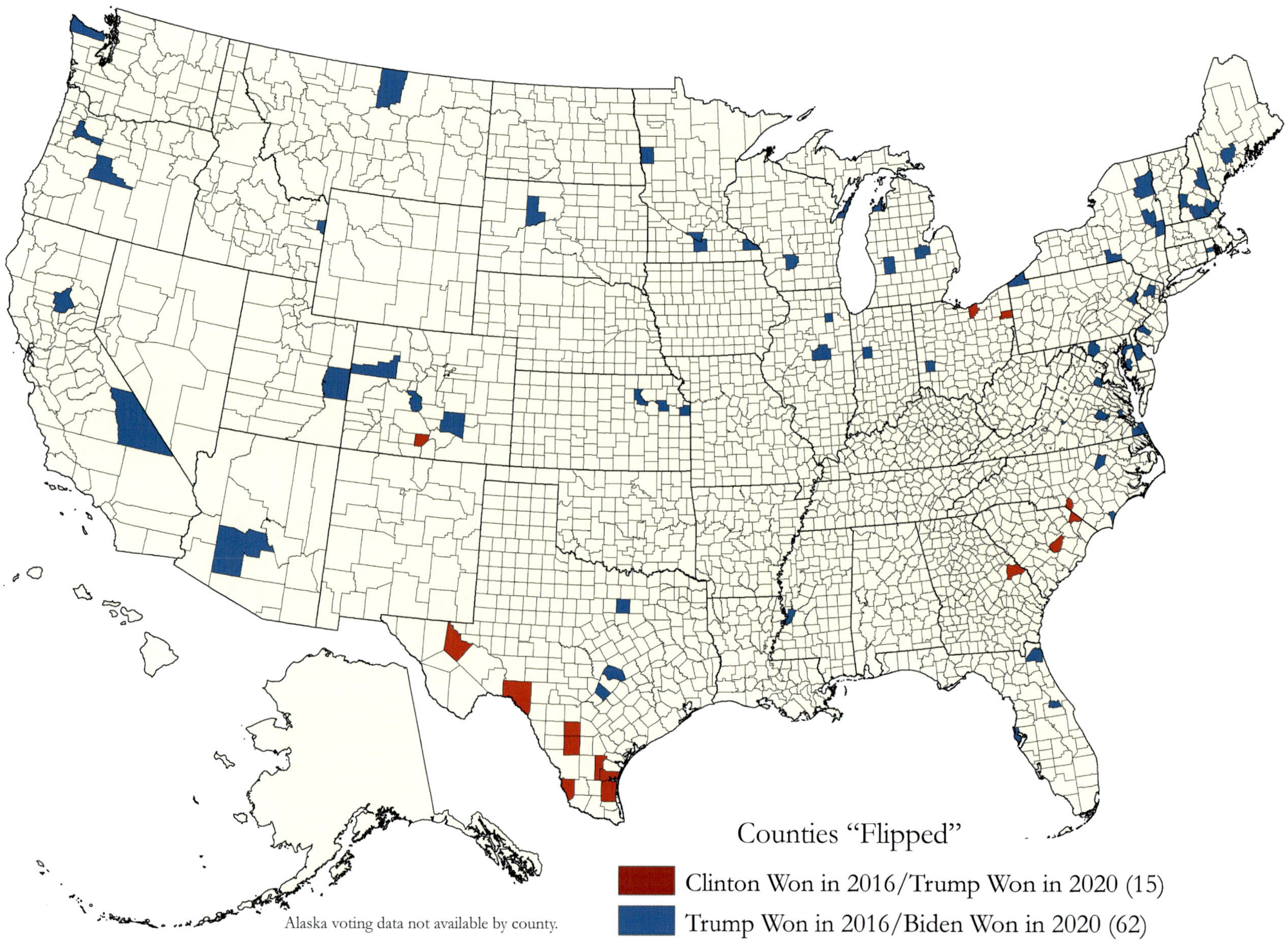

Biden as he won the county by eight points and ultimately carried Pennsylvania by 80,555 votes (3,458,229 to 3,377,674) (Alberta 2020).

Finally, in Wisconsin, Milwaukee's suburbs represent one of the most conservative areas in the country. Three counties—Waukesha, Ozaukee, and Washington—the so-called WOW counties—have voted Republican in every election since 1968 (Alberta 2020). Trump carried these three by overwhelming margins in 2016 and won handily in 2020, but in each the margin of victory was less—Trump captured those three counties with 65.4 percent of the votes in 2016 (227,706 of 347,926) and that percent dipped to 61.8 percent in 2020 (253,798 of 410,871). That number, along with other examples, might seem trivial or inconsequential. Understand, however, that Biden was able to flip Wisconsin in 2020, carrying the entire state by just 20,682 votes and capturing its ten Electoral College votes.

Campaign spending and campaign stops in key battleground states were tied to slim vote margins, which in many cases resulted in flipped counties and states. Regarding spending in 2020, through mid-October some $1.5 billion had been spent on presidential advertisements; in 2016 the amount during the same time span totaled $496 million (Corasaniti, Cai, and Lu 2020). Additionally, more than $1 billion was spent by early October in 2020 in just thirteen states—Biden spent roughly $600 million, while Trump and his campaign invested just over $400 million (Montanaro 2020). The bulk of the money was utilized in six states—Arizona, Florida, Michigan, Pennsylvania, North Carolina, and Wisconsin—and clearly it worked. Biden flipped thirteen counties in these six states while Trump flipped only one, and ultimately flipped and won four of the six states (Montanaro 2020).

"Flipped" counties and states were central not only to campaign spending, but to campaign events as well. Consider campaign events by major-party nominees. In 2020, from late August through early November twelve states received 96 percent of all general-election events by major-party presidential and vice presidential candidates (National Popular Vote n.d.). Of all events held in late 2020, Pennsylvania—which flipped from Trump in 2016 to Biden in 2020—led the way with forty-seven campaign events, followed by Florida with thirty-one (National Popular Vote n.d.).

Ultimately, "flipped" counties played a critical role in 2020. Biden held a distinct advantage over Trump (sixty-two to fifteen flipped counties) with many "flips" found in battleground states that equaled a decisive Electoral College victory. Battleground counties and states will most likely play a pivotal role in upcoming presidential elections and again receive the bulk of attention from major-party campaigns.

REFERENCES

Alberta, T. 2020. Three Reasons Biden Flipped the Midwest. *Politico*. https://www.politico.com/news/magazine/2020/11/04/three-reasons-biden-flipped-the-midwest-434114 (last accessed March 18, 2021).

Corasaniti, N., W. Cai, and D. Lu. 2020. Flush with Cash, Biden Eclipses Trump in War for the Airwaves. *New York Times*. October 17.

Crump, J. 2018. The 2016 Presidential Vote: The Rural Vote. In Robert Watrel, Ryan Weichelt, Fiona Davidson, John Heppen, Erin Fouberg, J. Clark Archer, Richard Morrill, Fred Shelley, and Kenneth Martis, eds., *Atlas of the 2016 Elections*, 97–98. Lanham, MD: Rowman & Littlefield.

Montanaro, D. 2020. Presidential Campaign TV Ad Spending Crosses $1 Billion Mark in Key States. *NPR*. October 13, https://www.npr.org/2020/10/13/923427969/presidential-campaign-tv-ad-spending-crosses-1-billion-mark-in-key-states (last accessed March 20, 2021).

National Popular Vote. n.d. Map of General-Election Campaign Events and TV Ad Spending by 2020 Presidential Candidates. https://www.nationalpopularvote.com/map-general-election-campaign-events-and-tv-ad-spending-2020-presidential-candidates (last accessed March 12, 2021).

ELECTION OUTCOMES: ABSOLUTE AND RELATIVE

RICHARD L. MORRILL AND KIMBERLY JOHNSON MAIER

Discussed here are three bivariate maps (figures 4.35, 4.36, 4.37) that display both absolute volume of votes, represented by the size of the circle, and percentage of the vote, represented by the color of the circle. Two additional maps (figures 4.38 and 4.39) show the percent change for the Democratic candidate in 2016 and 2020 and the margin of victory for the Democratic candidate in the 2020 presidential election. This is particularly useful in the 2020 election, because of the radically different geography of their areas of strength, Biden in the metropolitan cores, Trump across much of the less dense America.

Examining the first two maps, for Biden (figure 4.35) and Trump (figure 4.36), we see there is a size difference in absolute votes county by county. Looking carefully, we see that on the Biden map the dark blue circles are in large metropolitan core counties, which he dominated. When comparing to the 2016 election, Biden did far better in counties in the Northeast, West, and parts of the South. Additionally, these circles are much larger than the pink circles for Trump for the same counties. Conversely, the much larger number of dark pink circles for Trump dominate smaller counties across the South, North-South borderlands, and the Great Plains states. When compared to the 2016 election, Trump did better in some places such as the South. For the same regions, we can see Biden's vote was smaller, indicated by the light blue circles in these counties. The same is true for the dark pink Trump majority counties, larger than the same counties on the Biden map, indicating a minority vote.

The map for other/third-party candidates (figure 4.37) reveals higher shares for third-party candidates in much of the West and New England. Furthermore, we can see larger, dark brown circles in Mormon areas (i.e., Utah) and in the mountain states (i.e., Colorado). Unlike in 2016, third parties played a minimal role.

The margin of victory percent map for 2020 shows support for Biden in the Northeast, metropolitan cores, and along the West Coast. Furthermore, we see that Biden made substantial gains in counties in Michigan, Wisconsin, Georgia, Pennsylvania, and Arizona as noted by the dark blue circles. Like 2016, we see that Trump has maintained and gained some support in the Great Plains, greater Appalachia, and the rural South as indicated by the small and medium-size dark pink circles in figure 4.36. Some areas do have a closer vote and less dramatic margins—for example, Tarrant in Texas, Inyo in California, and Ontario and Warren in New York.

Finally, the last map shows change in the percent for the Democrat, Clinton in 2016 and Biden in 2020 (figure 4.39). We see an increase in Democratic support in parts of the South, Northeast, and West. In traditionally Democratic-leaning areas like Michigan, Wisconsin, and Pennsylvania that Clinton lost in 2016, Biden was able to regain Democratic support in 2020. Biden also made considerable gains in Democratic support in states such as Georgia. Additionally, there were some notable decreases in Democratic support, namely in parts of southern and northern California, the borderlands of Texas, and southern Florida, namely Miami-Dade, where we see a 10.26 percent decrease.

FIGURE 4.35

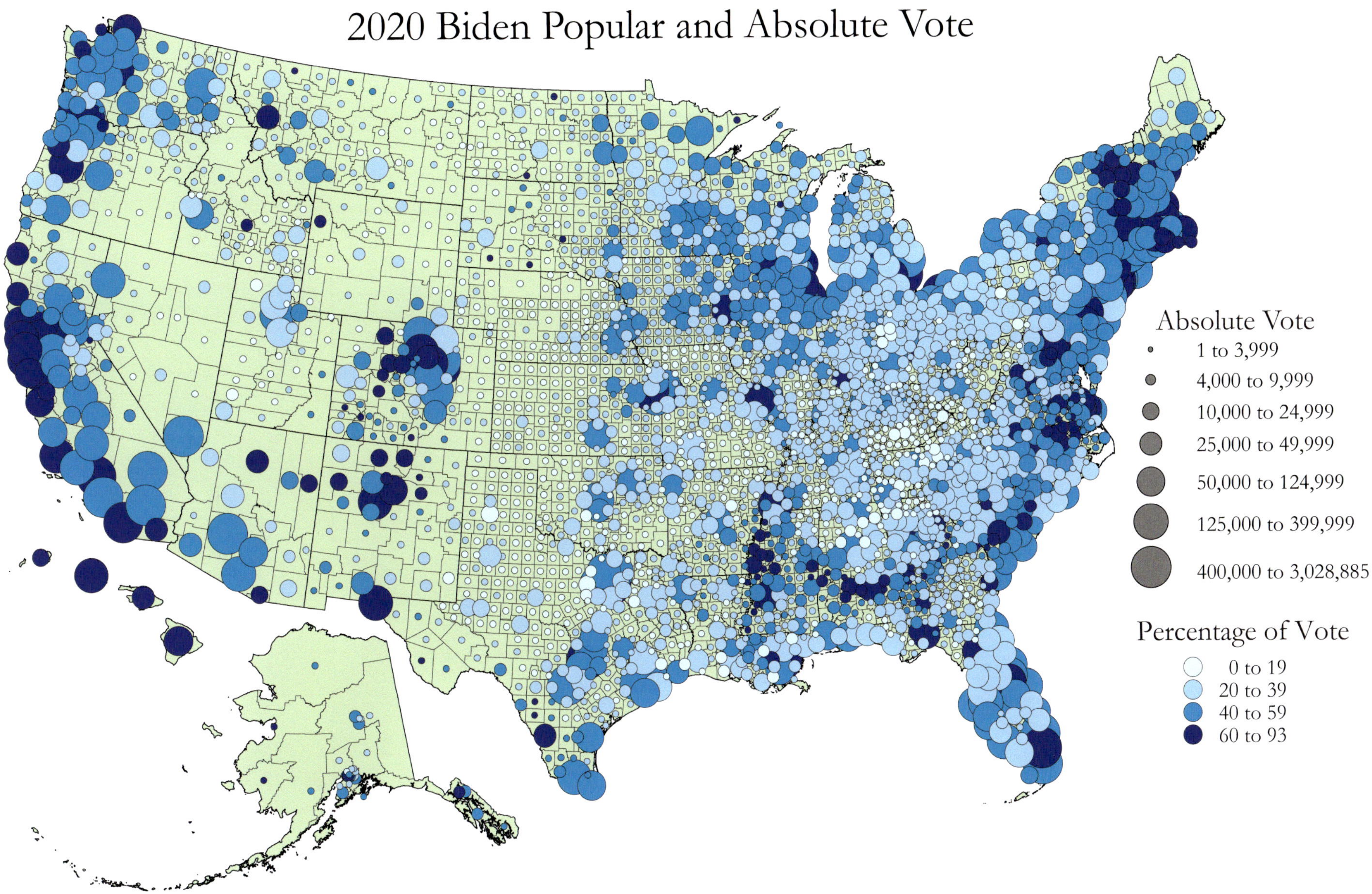
2020 Biden Popular and Absolute Vote
Absolute Vote
1 to 3,999
4,000 to 9,999
10,000 to 24,999
25,000 to 49,999
50,000 to 124,999
125,000 to 399,999
400,000 to 3,028,885
Percentage of Vote
0 to 19
20 to 39
40 to 59
60 to 93

FIGURE 4.36

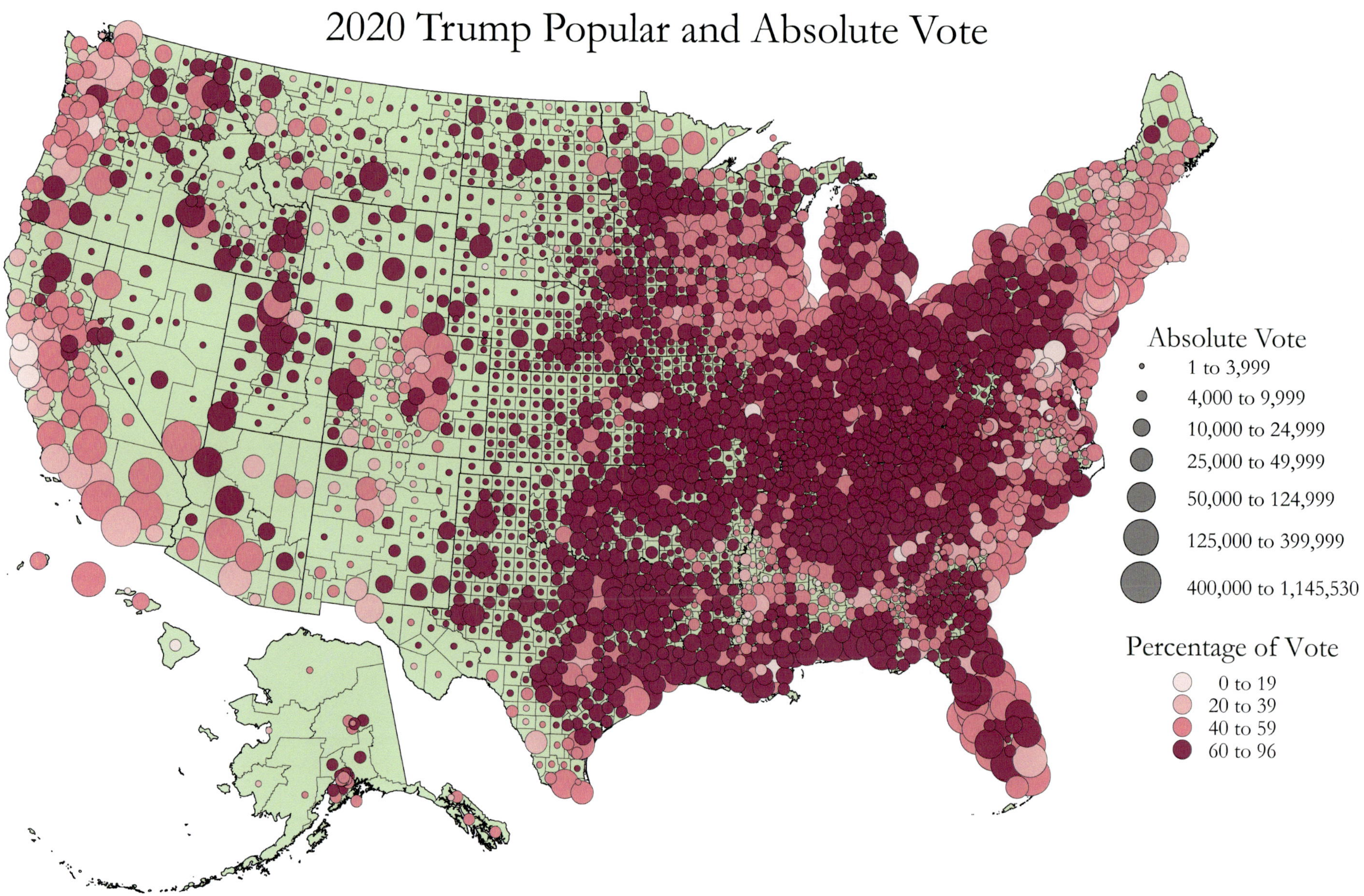
2020 Trump Popular and Absolute Vote
Absolute Vote
1 to 3,999
4,000 to 9,999
10,000 to 24,999
25,000 to 49,999
50,000 to 124,999
125,000 to 399,999
400,000 to 1,145,530
Percentage of Vote
0 to 19
20 to 39
40 to 59
60 to 96

FIGURE 4.37

Other/Third Party Popular and Absolute Vote

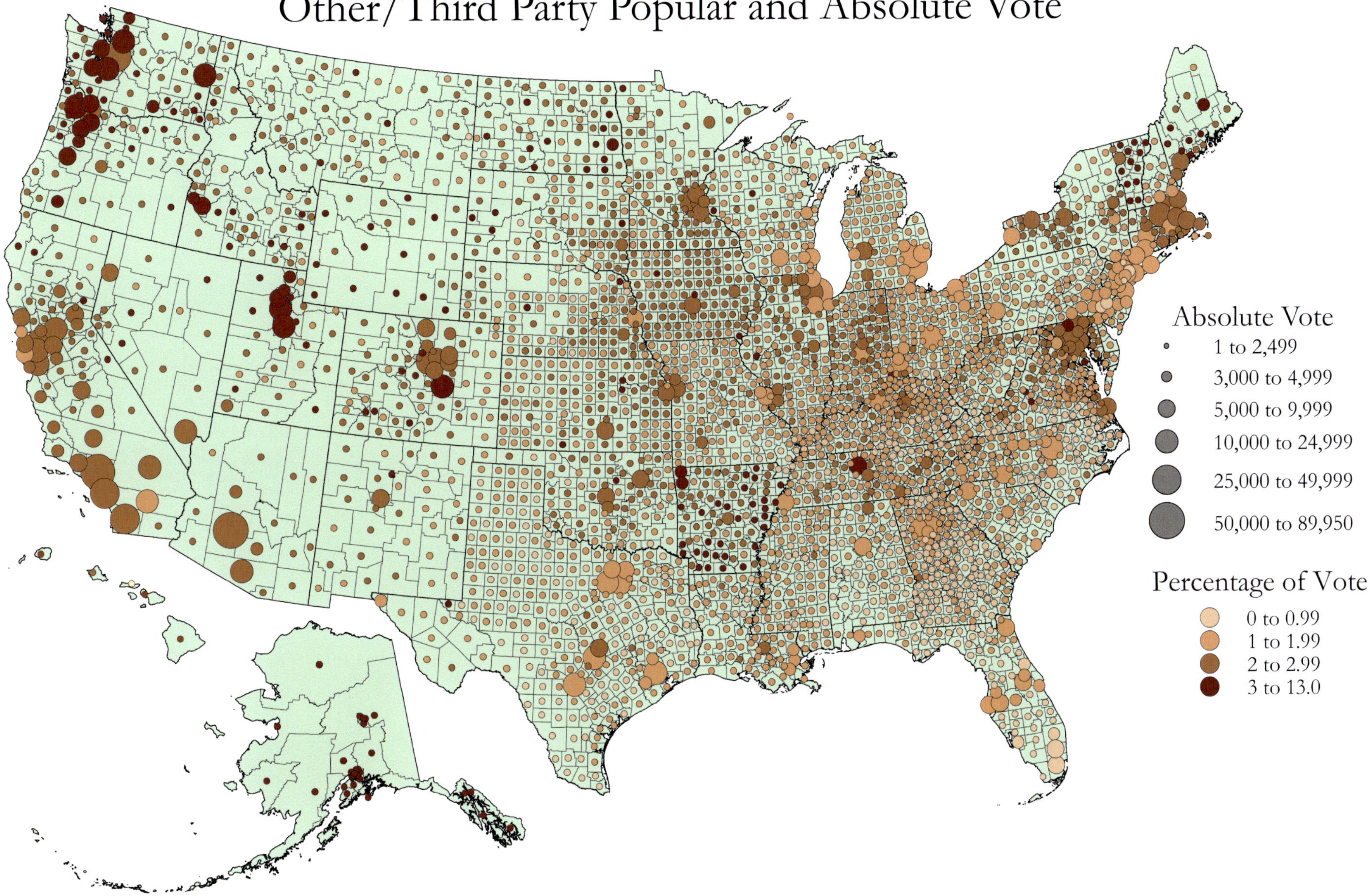

FIGURE 4.38

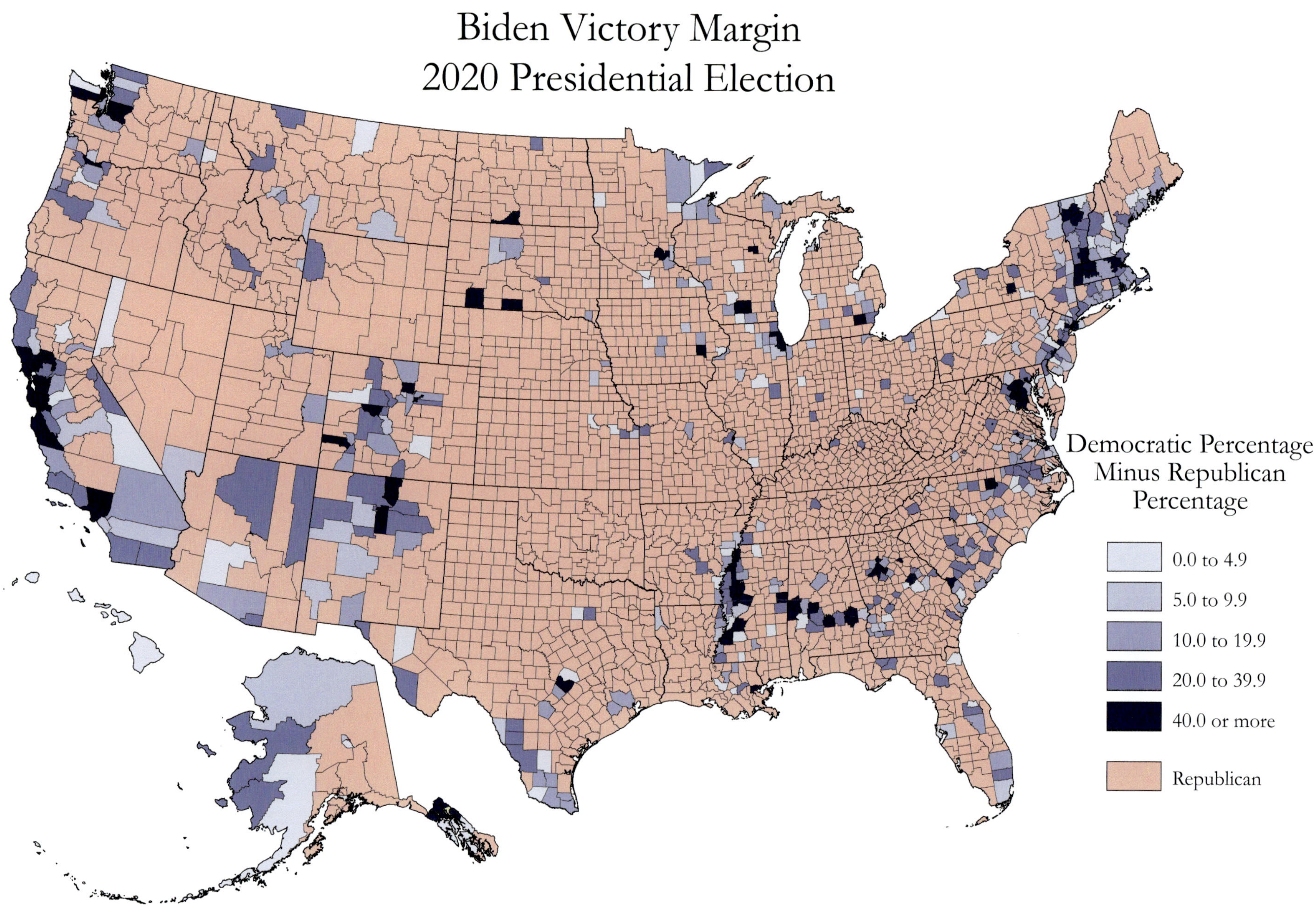
Biden Victory Margin
2020 Presidential Election
Democratic Percentage
Minus Republican
Percentage
0.0 to 4.9
5.0 to 9.9
10.0 to 19.9
20.0 to 39.9
40.0 or more
Republican

FIGURE 4.39

Change in the Democratic Vote from the 2016 Election to the 2020 Election

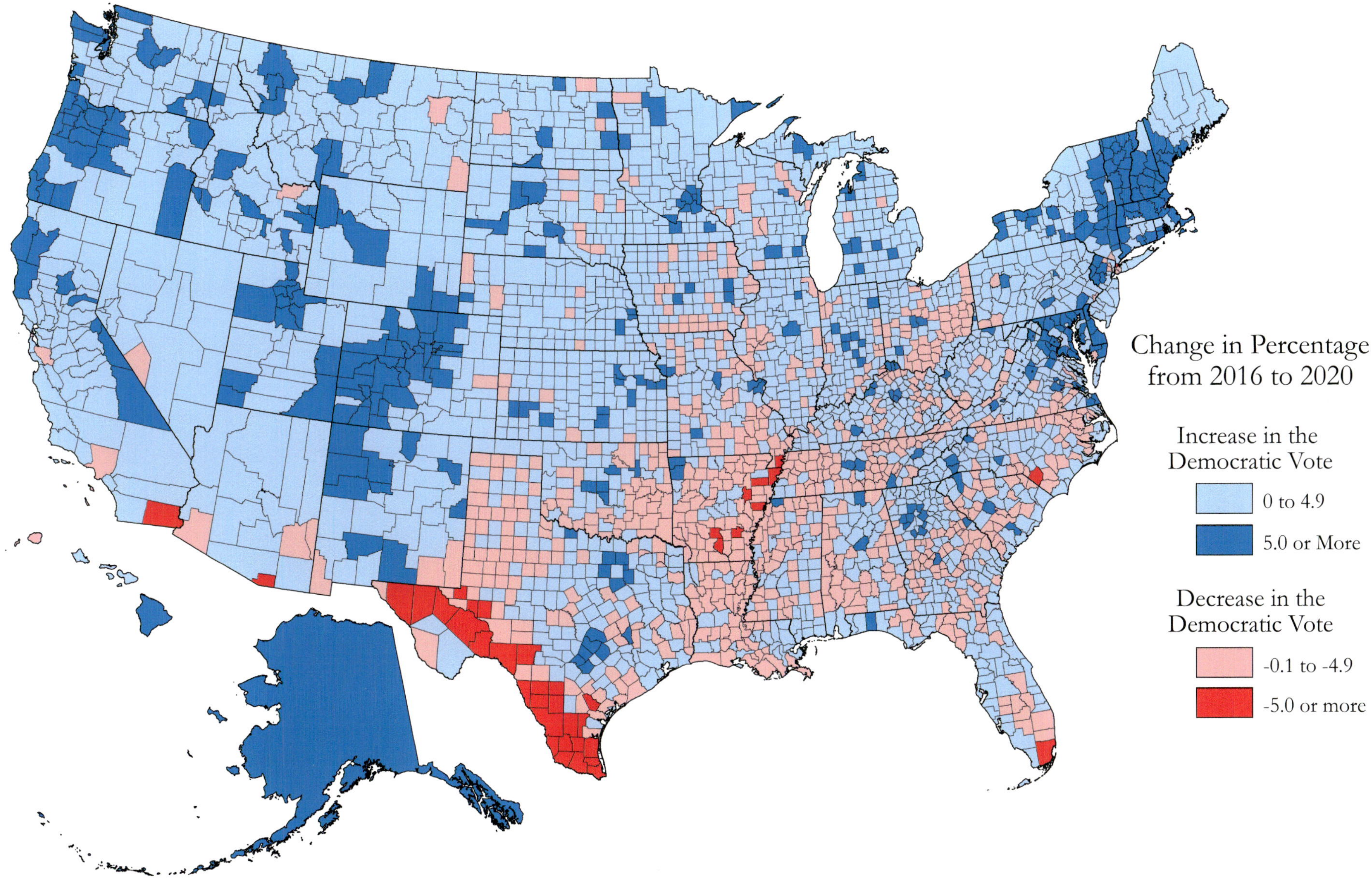

POSTELECTION VOTER SUPPRESSION LAWS: THE GEORGIA EXAMPLE

RICHARD ENGSTROM

Another outcome of the presidential election in 2021 has been a spike in vote suppression laws and legislative proposals. Vote suppression refers to the use of election policies and practices that impede the ability of a group of voters to cast their votes. It has been practiced by the Republican Party for over a decade (Engstrom 2018). Groups targeted by Republicans are primarily African Americans and Latinos, who tend to vote largely for Democratic candidates. The recent adoption of many vote suppression laws by Republican-controlled states is now widely referred to as Jim Crow 2.0, in remembrance of the laws adopted in the past in the American South to keep African Americans from voting.

The first state to adopt a new omnibus vote suppression law following the 2020 election was Georgia. The vote suppression features of the law are identified in the complaints filed in lawsuits that maintain that the law violates Section 2 of the federal Voting Rights Act (VRA), which prohibits election arrangements that deny voters in minority groups protected by that act an opportunity to elect candidates of their choice equal to that of other voters, and also several provisions of the US Constitution, such as the Equal Protection Clause of the Fourteenth Amendment. African Americans, the largest minority group in Georgia, is one of the groups protected by the VRA, and the new law is said to target its voters with "uncanny accuracy" (Waldman 2021).

The complaint in the case brought by the New Georgia Project, Black Lives Matter Fund, and Rise, Inc. (hereinafter NGP Complaint 2021) clearly identifies in its statement of facts the importance of mail-in absentee ballots to the mobilization of African American voters in the 2020 general election. Data maintained by the state revealed that "Nearly 30 percent of Black voters cast their ballots by mail . . . compared to only 24 percent of white voters." The statement also noted that voters "relied heavily" on drop boxes as a way to submit their mail ballots. These boxes were located outside on government property and were available to voters on a 24/7 basis with video surveillance for protection. Placing their ballot in a drop box assured voters that their ballots were received by election officials, and it also avoided potential problems with the mail. In majority nonwhite Fulton County, over half of the absentee ballots were submitted via drop boxes. The statement also noted that in order to make voting in Fulton County "easy and efficient," the county also provided voters across the county with mobile voting units, buses that contained eight to ten voting stations (NGP Complaint 2021, 12).

The complaint filed by the Georgia State Conference of the NAACP, League of Women Voters of Georgia, Common Cause, and other organizations (hereinafter NAACP Complaint 2021) identifies in its statement of facts a way that African Americans are mobilized to vote during the early voting period. It is a traditional practice known as "Souls to the Polls," a Sunday GOTV event organized by predominately African American churches in which people are provided with rides to polling places after worship services. The practice significantly boosts African American turnout. The statement of facts reveals that statistics compiled by the secretary of state show that during the 2020 general election, African Americans constituted 36.5 percent of the Sunday voters, 10 percentage points higher than on other days, for which it was 26.8 percent. Indeed, the African American percentage of turnout in the state on Sundays exceeded its percentage of the voter registration by about six percentage points, while the comparable figure for whites was minus six (NAACP Complaint 2021, 23–24, 34).

The 2020 presidential election in Georgia was hotly contested and produced a high voter turnout at the polls on Election Day. The statement of facts in the NGP Complaint notes that this resulted in "long lines at many polling places—especially at polling places located in Black neighborhoods." Referenced in it is "a recent study that found the average wait time in Georgia after polls were scheduled to close was six minutes in neighborhoods that were at least 90 percent white, and almost 51 minutes in places that were at least 90 percent nonwhite" (NGP Complaint 2021, 13). Long wait times have resulted in a practice known as "line warming." This is described in the statement of facts contained in the NAACP Complaint as "well-meaning individuals and organizations who hand out water or snacks, umbrellas, or chairs to ease the burden on voters standing in line for protracted periods of time" (NAACP Complaint 2021:26).

Vote suppression laws appear facially neutral. The statements of facts by plaintiffs, however, reveal that they are not neutral given the context in which they operate. Forms of voting disproportionally relied upon by African Americans in Georgia have now been made more difficult to use, curtailed, or even eliminated by the new law. It is this feature of the law that qualifies for the appellation Jim Crow 2.0.

Prior to the new law in Georgia, voters could simply request an absentee ballot from county election officials who, upon verifying their eligibility to vote, would mail the appropriate ballot to that person. Voters would have to sign their returned ballot, which had to match a signature on file in order to be valid. This is a form of voting, as revealed above, now more prevalent among African Americans than whites in Georgia. The new law, however, makes getting an absentee ballot by mail a more complicated procedure. Now a person must include in their absentee ballot application a Georgia driver's license number or Georgia State Identification number. If they do not have one of these, they are required to attach a photocopy of another form of specified identification to both their ballot application and to the inside of their absentee ballot envelope when returning the voted ballot (NGP Complaint 2021, 17 and NAACP Complaint 2021, 21). In addition, the window for applying for a mail absentee ballot has been seriously contracted. Previously a voter could request one as early as six months prior to an election and up to four days before. The new law sets the start time at eleven weeks before the election, while the latest is increased from four days to eleven.

Under the new law, submitting ballots in drop boxes, another form of voting revealed above as relied upon heavily by African American voters, has also been seriously curtailed. Drop boxes may no longer be located outside for voters to access on a 24/7 basis. They must now be located inside of early voting locations, which restricts access to them to voters able to enter those facilities during the early voting hours. In short, as noted in the NAACP Complaint, drop boxes "are essentially useless to voters who can vote early in person or who cannot access early voting hours at all due to work or other commitments during early voting hours" (NAACP Complaint 2021, 22, NGP Complaint 2021, 18); it further notes that "Because Black adults are more likely to work multiple jobs as compared to workers of other races, eliminating before- and after-hours voting opportunities will disproportionately burden their electoral participation."

The new law also contains a ban on mobile polling places, such as those that have been used in majority non-white Fulton County, and also allows county registrars or absentee ballot clerks to decide, on their own, whether there will be any Sunday voting in their county. They can do so regardless of the size of the African American community in that county. In short, a county official can choose to preclude Souls to the Polls activities in their county, even if that county has a substantial number of African American voters.

The provision of the new law that has received the most extensive and most negative attention across the country is the prohibition on providing voters waiting in line to vote with water or coffee. This applies no matter how long they have been waiting and how long they might still have to wait, and no matter how hot and humid, or cold and rainy, the weather might be. Given that African American voters are the group most likely to confront long wait times in Georgia (which is not unique to Georgia; see Bipartisan Policy Center 2019), it is not hard to infer the type of voter that Georgia's Republican lawmakers would most like to see give up their place in line on a cold November Tuesday. These and other provisions of the law are considered vote suppression measures by voting rights attorneys and the voters they represent.

The Brennan Center for Justice has identified twenty-two new laws, through May 14, that have been adopted by states since the 2020 election and that restrict access to voting, and also sixty-one bills being processed by the legislatures of eighteen states that will do the same. These new laws and bills generally call for stricter constraints on absentee voting, restricting the use of drop boxes, curtailing early voting, making voter registration more difficult, expanding voter purges, and yes, even prohibitions on providing water to people facing long waits to vote. The number of these laws and bills that are vote suppression measures is not yet known, but there is little doubt that many are.

A round of litigation had already begun in the spring of 2021 as civil rights groups filed cases that seek to block the use of Jim Crow 2.0 laws prior to the 2022 midterm elections. The US Department of Justice in June filed suit against the new Georgia law for having not just a racially discriminatory effect but also a racially discriminatory purpose, in violation of Section 2 of the Voting Rights Act (Johnson 2021). The US Justice Department announced in June 2021 that it would double the civil rights division's voting rights enforcement staff in anticipation of the large number of new laws, like that in Georgia, that might be found to violate the voting rights of minority voters. It will not be surprising if litigation of this type will continue into 2023 and even 2024.

REFERENCES

Bipartisan Policy Center. 2019. The 2018 Voting Experience: Polling Place Lines. Washington, DC. https://bipartisanpolicy.org/download/?file=/wp-content/uploads/2019/11/The-2018-Votin-Experience.pdf (last accessed June 30, 2021).

Complaint for Declaratory and Injunctive Relief, Georgia State Conference of the NAACP et al. v. Raffensperger et al. (U.S. D.C. N.D. GA). March 28, 2021.

Complaint for Declaratory and Injunctive Relief, New Georgia Project et al. v. Raffensperger et al. (U.S. D.C. N.D. GA). March 25, 2021.

Engstrom, Richard L. 2018. Vote Fraud or Vote Suppression: The Debate Before and After the 2016 Election. In Robert H. Watrel, Ryan Weichelt, Fiona M. Davidson, John Heppen, Erin H. Fouberg, J. Clark Archer, Richard L. Morrill, Fred M. Shelley, and Kenneth C. Martis, eds., *Atlas of the 2016 Elections*, 47–49. Lanham, MD: Rowman & Littlefield.

Johnson, Kevin. 2021. Justice Department Suing Georgia over Election Law It Says Restricts Black Voters' Access. *USA Today*. June 25, https://www.usatoday.com/story/news/politics/2021/06/25/georgia-election-law-prompts-doj-sue-over-voter-restriction-claims/5344151001/ (last accessed June 30, 2021)

Waldman, Michael. 2021. Georgia's Voter Suppression Law. Brennan Center for Justice. https://www.brennancenter.org/our-work/analysis-opinion/georgias-voter-suppression-law (last accessed June 30, 2021).

5 REGIONS

NORTHEAST

WILLIAM BERENTSEN

The populous Northeast has more than 20 percent of US presidential electoral votes, and in the 2020 election Joe Biden won nearly every one of them—111 of 112. Incumbent Donald Trump won a lone electoral vote in Maine, one of its four, by carrying the geographically large but thinly populated Second Congressional District. Biden's Northeast electoral votes represented more than 40 percent of those he needed to reach the minimum 270 required to be elected president. A second prominent 2020 election outcome in the Northeast is that Biden won all electoral votes in the Mid-Atlantic region, most importantly a pivotal pickup victory in battleground Pennsylvania. Many observers viewed the state as a "must win" for Donald Trump, who won it narrowly in 2016.

These outcomes extend the decades-long electoral dominance of both the Mid-Atlantic and New England regions by Democratic presidential candidates. In New England, the election results also follow on the Democrats' decades of dominance in congressional elections. Senator Susan Collins (Maine) remains the only Republican in New England to still hold a congressional seat. The Democrats retain control of thirty others, as well as two independents (Maine, Vermont), who caucus with Democrats. Congressional representation in the Mid-Atlantic states remains more contested. Here there are forty-six House Democrats and twenty Republicans, including a 9–9 split in Pennsylvania. As in New England, there is only one Republican senator from the Mid-Atlantic, Pat Toomey (Pennsylvania).

Another notable 2020 Northeast presidential election outcome is that after the 2012 and 2016 presidential election cycles, in which GOP presidential candidates made wide geographical vote gains at the county level, Democrats gained a higher proportion of votes than Republicans in 2020 versus 2016 in all but 13 of the Northeast's 245 counties (Archer et al. 2014; Watrel et al. 2018). Although the 2020 geographic pattern of political party wins in Northeastern counties is very similar to the 2016 election, Biden won in 2020 in part by small but widespread vote gains across the Northeast. He won ("switched") just seventeen counties in the Northeast that Trump won in 2016, but Trump won no counties won by Clinton in that election. As the Change in Democratic Vote map clearly indicates (figure 5.1), there was a rising "blue tide" in the Northeast in 2020. There was an actual electoral "blue tsunami" in the District of Columbia, where Biden captured more than 92 percent of all votes in the US seat of government.

Biden's key win in battleground Pennsylvania was accomplished by the widespread "blue tide," plus some substantial wins in Philadelphia's suburban counties. Biden won all counties in Pennsylvania won by Hillary Clinton in 2016, plus

Erie County in the northwest and Northampton County in the east. Those wins, however, accounted for a small portion of the vote gain that Biden accumulated on the way to win the state. In 2020, Biden-Harris gained about 2 percent more of the total vote than Clinton had in 2016, in order to narrowly win the state by about 1 percent. Eastern Pennsylvania, particularly near Philadelphia, where Democrats have long been strong, especially helped Biden-Harris. Pennsylvania counties just east and northeast of Philadelphia and bordering Biden's home state of Delaware—Chester County and Delaware County in Pennsylvania—provided about 40 percent of the Democrats' net 124,000 vote gain vis-à-vis the GOP between the 2016 and 2020 presidential elections (Pramuk and Schoen 2020). Joe Biden thus appears to have been decisively supported by the Delaware-neighboring suburbs of populous and ethnically diverse Philadelphia, which itself voted 63 percent for Biden-Harris.

In general in the Northeast, as in many other parts of the United States, Biden did especially well in both urban and suburban areas of metropolitan counties, especially in the most populous (Damore, Lang, and Danielsen 2021; Florida 2000). This is well illustrated by Biden's many metropolitan area wins from Portland, Maine, straight through the megalopolis corridor to Washington, DC. Although Trump did win many metro areas in the Northeast, Biden won nearly all of the Northeast's largest ones (i.e., those within megalopolis) and did so with 58 to 72 percent of the vote (figure 5.1). Trump's wins were mostly in less populous metro areas north and west of megalopolis. The most notable of these was a narrow win in metro Pittsburgh. Most of Trump's successes were otherwise in thinly populated rural and small-town areas in northern New England, upstate New York, the Delmarva Peninsula, northern and western Maryland, and, especially, central and western Pennsylvania.

As noted above, in 2020 Donald Trump increased his percentage of vote compared to 2016 in just thirteen Northeastern counties. Interestingly, among these were the combined city/county of Philadelphia and also eight New York City metro counties (including Bronx County, also a New York City borough). Locally based observers believe that these results largely resulted from increased support for Trump among conservative Orthodox Jews and Latinos in New York, and residents of North Philadelphia, especially Latinos (Benshoff 2020; Geringer-Sameth and Max 2020; Kolko and Monkovic 2020; Tamari and Lai 2020; Vielkind and Tucker 2020).

As Democrats so often have done in recent presidential elections, Biden-Harris generally won states that have higher-than-US-average voter turnout rates. In 2020, nineteen of their twenty-five state wins had 2020 voter turnouts above the national average (nine of eleven in the Northeast). Perhaps crucially, in 2020, Pennsylvania had one of its highest voter turnouts since the early nineteenth century. The 2020 wins for Trump were most often in states with below-average voter turnouts (fifteen of twenty-five states; McDonald 2020). In general, the Northeast has not only relatively high turnout rates, but also relatively well educated and ethnically diverse voters—while Trump's typical voters are more often less well educated and white. Biden-Harris also gained some ground among several key voter groups that leaned more toward Trump in 2016 than in 2020, including older voters and those with military service.

The Northeast, and especially most of New England, has not been a good fit for the GOP or Trump. As this author noted after the 2016 election, "the conservative platforms and actions of the Republican Party seem likely to continue to be an unpopular option for the majority of voters in the Northeast" (Berentsen 2018). Still, Trump got 12.6 million votes in 2020, won scores of counties, and received 40 percent of the total vote. Although the Democrats clearly dominate the Northeast, Pennsylvania, New Hampshire, and northern Maine have provided electoral votes to GOP candidates for decades—although just one in November 2020. As in both the 2016 and 2020 presidential elections, the GOP's most likely key to success in the Northeast would be targeting Pennsylvania. It seems quite plausible that the GOP could win more electoral votes in future elections in the Northeast; however, gaining a majority in any near-term presidential election seems implausible.

REFERENCES

Archer, J. C., F. Davidson, E. H. Fouberg, K. C. Martis, R. L. Morrill, F. M. Shelley, R. H. Watrel, and G. R. Webster, eds. 2014. *Atlas of the 2012 Elections*. Lanham, MD: Rowman & Littlefield.

Benshoff, L. 2020. Trump's Gains in North Philly Show Latino Voters Are Not a "Gimme" for Democrats. *WHYY News*. November 20, https://whyy.org/articles/trumps-gains-in-north-philly-show-latino-voters-are-not-a-gimme-for-democrats (last accessed April 28, 2021).

Berentsen, W. H. 2018. Northeast. In R. H. Watrel, R. Weichelt, F. Davidson, J. Heppen, E. H. Fouberg, J. C. Archer, R. L. Morrill, F. M. Shelley, and K. C.

FIGURE 5.1

Northeast Region Presidential Elections

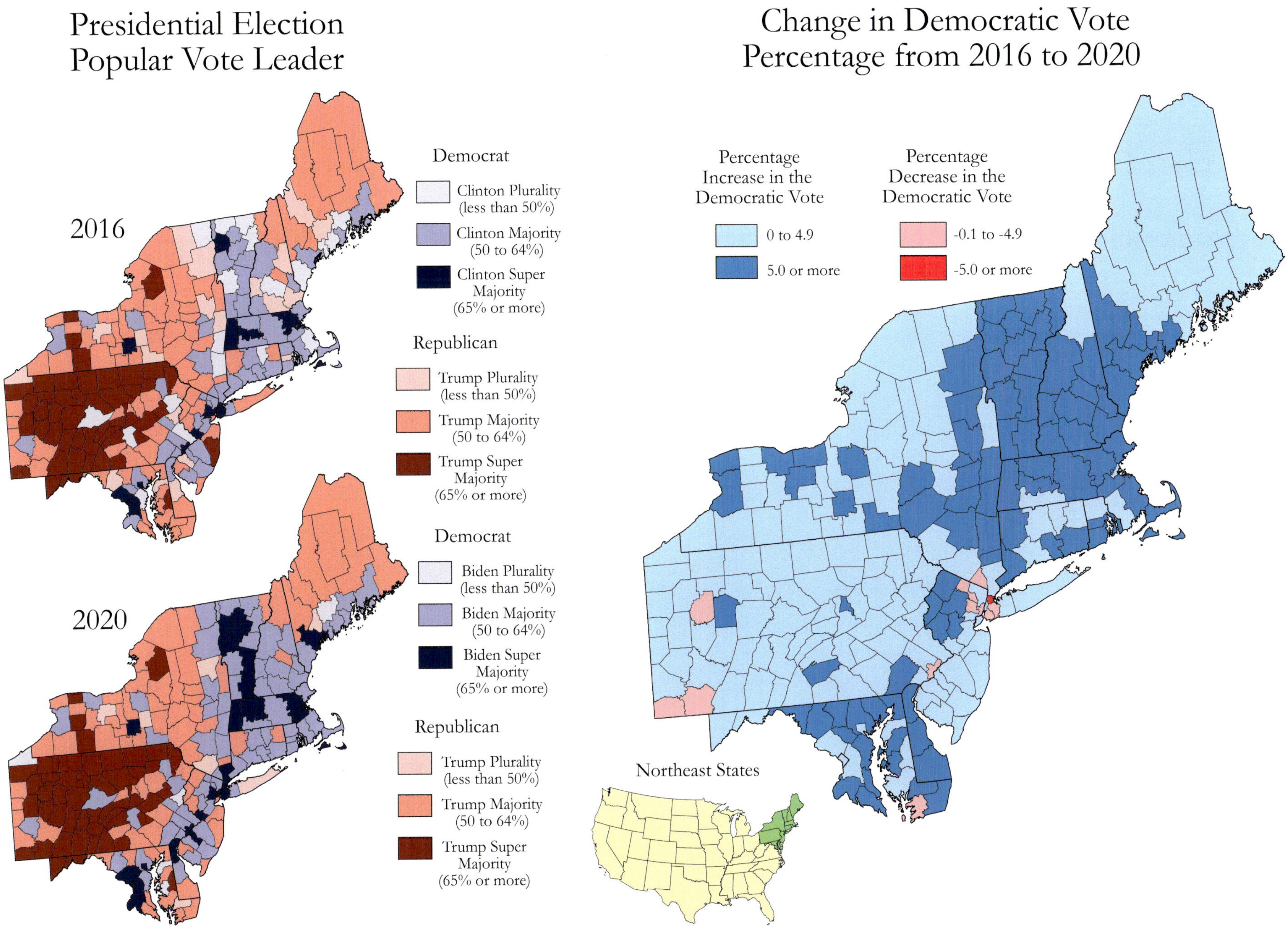

Martis, eds., *Atlas of the 2016 Elections*. Lanham, MD: Rowman & Littlefield.

Damore, D. F., R. E. Lang, and K. Danielsen. 2021. In 2020, the Largest Metro Areas Made the Difference for Democrats. Brookings. February 4, https://www.brookings.edu/research/in-2020-the-largest-metro-areas-made-the-difference-for-democrats (last accessed April 28, 2021).

Florida, R. 2020. How Metro Areas Voted in the 2020 Election. Bloomberg CityLab. December 4, https://www.bloomberg.com/news/features/2020-12-04/how-metro-areas-voted-in-the-2020-election (last accessed April 28, 2021).

Geringer-Sameth, E., and B. Max. 2020. Breaking Down Final 2020 Vote Tallies across New York City. *Gotham Gazette*. December 4, https://www.gothamgazette.com/city/9961-breaking-down-2020-vote-new-york-city-biden-trump (last accessed April 28, 2021).

Kolko, J., and T. Monkovic. 2020. The Places That Had the Biggest Swings Toward and Against Trump. *New York Times*. December 7, https://www.nytimes.com/2020/12/07/upshot/trump-election-vote-shift.html (updated January 15, 2021).

McDonald, M. P. 2020. 2020 November General Election Turnout Rates. United States Elections Project. December 7, http://www.electproject.org/2020g (last accessed April 28, 2021).

Pramuk, J., and J. W. Schoen. 2020. This Map Shows How Biden Won Back Pennsylvania in the 2020 Election. *CNBC*. November 9, https://www.cnbc.com/2020/11/09/this-map-shows-how-biden-won-back-pennsylvania-in-the-2020-election.html (last accessed April 28, 2021).

Tamari, J., and J. Lai. 2020. With the Vote Count Now Over, Here's How Pennsylvania Broke for Joe Biden. *Philadelphia Inquirer*. November 29.

Vielkind, J., and E. Tucker. 2020. More New Yorkers Voted for Donald Trump in 2020 than in 2016: Biggest Gains Are Notched in Majority-Latino South Bronx. *Wall Street Journal*. November 12.

Watrel, R. H., R. Weichelt, F. Davidson, J. Heppen, E. H. Fouberg, J. C. Archer, R. L. Morrill, F. M. Shelley, and K. C. Martis, eds. 2018. *Atlas of the 2016 Elections*. Lanham, MD: Rowman & Littlefield.

APPALACHIA

KENNETH C. MARTIS

The Appalachian region stretches southwestward from southern New York State to northern Alabama and Georgia and northeastern Mississippi. Appalachia can be defined in three ways. First, it is a physiographic region containing the rolling hills of the Piedmont, mountain peaks of the Ridge and Valley, and dendritic drainage pattern and complex incised topography of the Plateau. Appalachia is also a long-recognized cultural region characterized by a high percentage rural white population; large areas with low rankings in educational, social, and economic indicators; high reliance on agriculture and extractive industries such as coal, timber, and natural gas; distinctive music and food heritage; and high adherence to fundamentalist Protestant and non-affiliated Christian churches. Appalachia is also a rare phenomenon in American geography because it is a precisely defined region by the US Congress. In 1965 the Appalachian Regional Commission (ARC) was created to promote economic development in this lagging region. In order to funnel the assistance, a map was created to define the Appalachian region by county. Today the expanded ARC area encompasses 420 counties in 13 states, the entire state of West Virginia, and the Appalachian portion of Alabama, Georgia, Kentucky, Maryland, Mississippi, New York, North Carolina, Ohio, Pennsylvania, South Carolina, Tennessee, and Virginia. The ARC boundary lines are used in the electoral analysis for this region.

HISTORICAL POLITICAL PARTY SUPPORT

Since it includes parts of northern states like New York and Pennsylvania, and parts of southern states like Georgia and Alabama, Appalachia has a complex political party voting history. During the Civil War, large areas of the upcountry South remained loyal to the Union and thereafter to the Republican Party. Throughout the Great Depression years, the Democrats gained support because of New Deal programs, such as the Tennessee Valley Authority, and the unionization efforts of the United Mine Workers. This support was sustained by Democrat President Lyndon Johnson's "war on poverty" programs, much of it aimed at this region (including the creation of the ARC). As late as the 1996 presidential election, Democrat Bill Clinton received support throughout much of Appalachia.

The turning point in contemporary Appalachia voting history came in the 2000 presidential contest between Republican George W. Bush and Democrat Al Gore (Webster and Morrill 2015). Gore had a history of promoting the idea of fossil fuel–induced global warming and gun control, two issues that conflicted with the economic base and cultural traditions of the region. These issues, plus a plethora of socially conservative "family values" concerns, which the Republican Party championed, facilitated the rapid decline of the Democrats in Appalachia. The Democrat Obama administration's perceived "war on coal" all but solidified the shift to the Republicans.

West Virginia is the only state wholly within the Appalachian Regional Commission boundary. This state is a textbook example of voter realignment. Table 5.1 shows the Democrat presidential vote percentage and national Democrat support ranking of West Virginia from the late twentieth century to 2020. West Virginia was reliably one of the top Democratic states since the Great Depression. As the table numbers illustrate, beginning in 2000, there was a rapid and sharp departure from

TABLE 5.1. West Virginia Presidential Vote, 1988–2020

Election	Percent Democrat Vote	Rank in States in Democrat Support
1988	52.2	6
1992	48.4*	5
1996	51.5	13
2000	45.6	25
2004	43.2	31
2008	42.5	37
2012	35.5	46
2016	26.5	49
2020	29.7	49

* Democrat candidate Clinton won the electoral vote in a three-way race by a 13 percent margin.

the Democrats. In 2020 Joe Biden received only 29.7 percent of the vote in West Virginia, trailing only Wyoming as the most Republican state in the Union.

THE 2020 ELECTION

The Biden campaign portrayed his candidacy as "working class Joe" (Lerer 2020). This was met with skepticism throughout Appalachia—even more so after seemingly conflicting statements about hydraulic fracking and natural gas extraction (Rizzo 2020). The gas-rich Marcellus Shale formation is extensive in the northern part of the region and an expanding element of employment and economic growth. This is particularly so in the critical swing states of Pennsylvania and Ohio. The Trump campaign jumped on Biden's confusing statements, falsely claiming he wanted to ban all fracking. All this, plus the long-term antipathy of the Democratic Party against fossil fuels, especially coal and now natural gas, led to another devastating defeat for the party throughout the region.

In 2016 and 2020 Donald Trump portrayed himself as a friend of coal and natural gas fracking. Many of the campaign signs in the region read "Trump Digs Coal." In 2016 he promised to roll back many of the Obama admiration's clean energy policies. Appalachian voters, especially in hard-hit coal counties, saw the Republican candidate evoking hope out of despair. In addition, Trump's other "Make America Great Again" populist policies also resonated with the rural, mostly white, socially conservative population. Despite the Trump promises, coal employment declined and continued to diminish as a percentage of US energy production during his administration (Lipton 2020). In spite of this, virtually all coal county voters agreed it would have been much worse without him.

The Popular Vote Leader maps in figure 5.2 show the Republican dominance in the Appalachian region. The 2016 map illustrates Donald Trump's massive victory over Hillary Clinton, carrying most counties with a super majority of 65 percent or more. The 2020 map shows virtually the same results, an overwhelming sea of red. Although candidate Biden won by more than seven million votes nationwide, he lost the vast majority of Appalachian counties by wide margins. Amazingly, of the four-hundred-plus Appalachian counties, Biden won only eight not on the fringe of the region: Allegany (Pittsburgh), Lackawanna (Scranton [Biden's birthplace]), and Centre (home of Penn State University) in Pennsylvania; Athens (home of Ohio University) in Ohio; Buncombe (Asheville) and Watuaga (home of Appalachian State University) in North Carolina; Jefferson (Birmingham) in Alabama; and Oktibbeha (home of Mississippi State University) and Clay (majority African American) in Mississippi. In historically Democrat West Virginia, President Trump not only won every county, but by a super majority in forty-nine of the fifty-five.

The Democrats were soundly defeated in Appalachia in 2020. Nevertheless, the Change in Democratic Vote Percentage from 2016 to 2020 map in figure 5.2 reveals a swing toward the Democrats in a majority of counties. This swing was especially important in three critical battleground states, Pennsylvania, Ohio, and Georgia. In fact, it could be asserted the Appalachian portion of these states were central to who won the electoral vote.

In Pennsylvania all but three of the Appalachian counties moved toward the Democrats. Joe Biden won the state 50.01 percent to 48.84 percent in one of the most controversial contests of 2020. The Trump campaign contested the election in court for weeks after the election. In addition, the Pennsylvania results were challenged by a majority of Republicans in the House of Representatives in the final Electoral College certification. The other state to swing to the Democrats was Georgia. Biden won Georgia by the slimmest of margins, 49.51 percent to 49.25 percent. The vast majority of Appalachian Georgia swung to the Democrats, especially the counties surrounding Atlanta. In Ohio the story was the opposite. The majority of the Appalachian counties bucked the regional and national trend by giving even more support to President Trump. This backing helped him to solidify a comfortable win in this historically battleground state.

In spite of some 2020 swings to the Democrats, the Appalachian region is solidly in the Republican camp for the foreseeable future. This is in keeping with the national trend of rural white America. There are four swing states in the region, Pennsylvania, Ohio, North Carolina, and now Georgia. The mountainous portions of these states have played a role, and will continue to play a role, in national and statewide elections (Weigel 2020).

REFERENCES

Lerer, Lisa. 2020. Joe from Scranton Didn't Win Back the Working Class. *New York Times*. December 5.

Lipton, Eric. 2020. "The Coal Industry Is Back," Trump Proclaimed. It Wasn't. *New York Times*. October 5.

FIGURE 5.2

Appalachia Region Presidential Elections

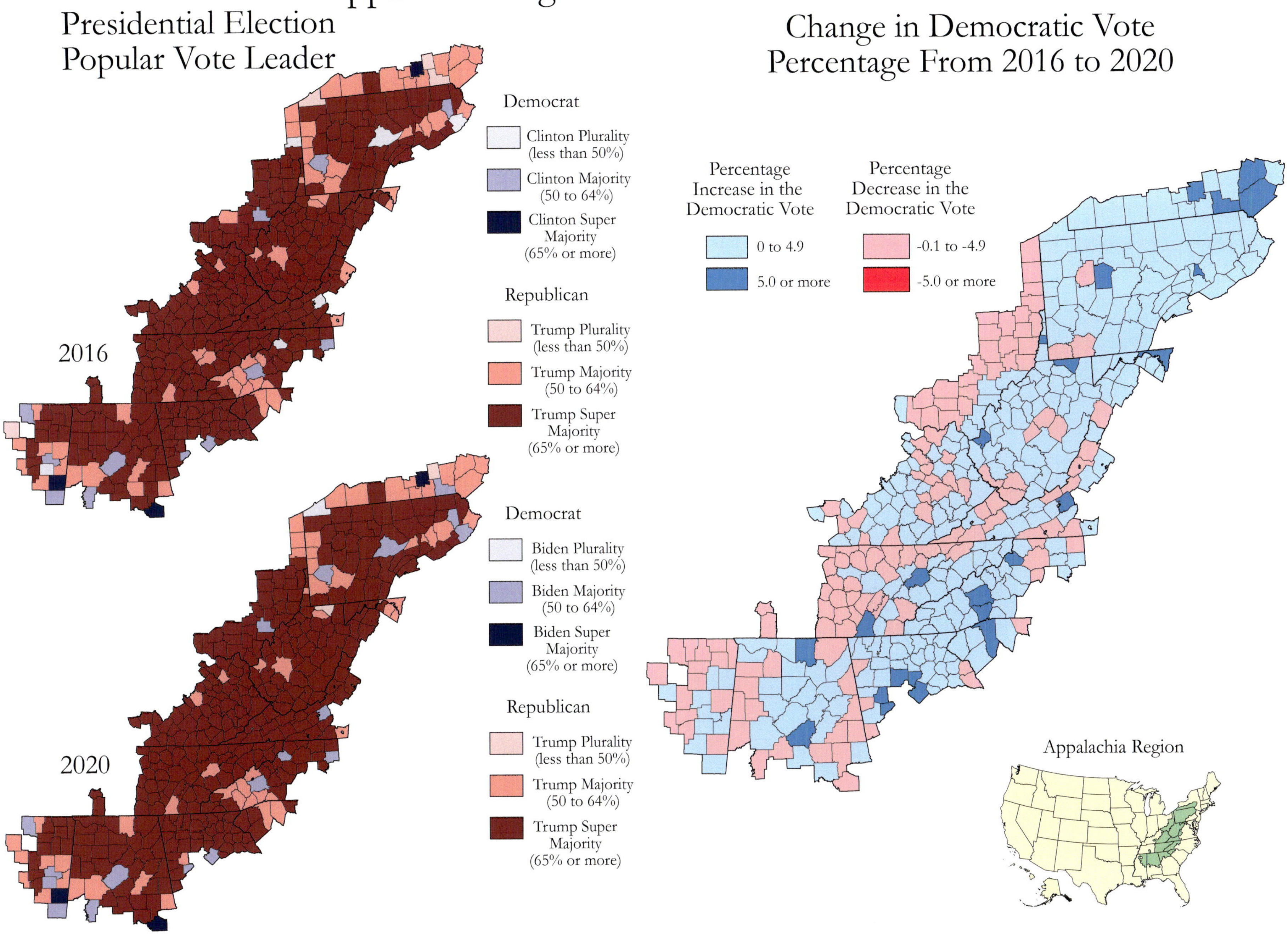

Rizzo, Salvador. 2020. Fact-Checking the Biden Fracking Fracas. *Washington Post*. March 19.

Webster, Gerald R., and Richard L. Morrill. 2015. Regional Realignments: Appalachia and the Upper South, in Spatial and Political Realignment of the U.S. Electorate, 1988–2012. *Political Geography* 48: 93–107.

Weigel, David. 2020. The Six Political States of North Carolina. *Washington Post*. August 23.

SOUTHEAST

JONATHAN I. LEIB

While the highlight of the 2020 presidential election was the change in power from Republican Donald Trump to Democrat Joe Biden, the 2020 presidential election in the twelve-state Southeast region was marked by both continuity and change. This section examines the continuity and change in the Southeast at three different scales: regional, state, and local.

In terms of continuity in the Southeast, the two-party presidential vote percentages regionwide were similar in 2016 and 2020. In 2020, Donald Trump won 54.1 percent of the two-party vote regionwide, compared to 54.3 percent in 2016. The number of states won by Trump in the region was also similar. Whereas Donald Trump carried eleven of the Southeast's twelve states in the 2016 election, he won ten in 2020 (Georgia being the one state that flipped from Trump to Biden).

At the same time, there was also considerable change regionwide. With the COVID-19 pandemic raging, it might have been assumed that fewer people would vote in 2020 than before; just the opposite occurred. Turnout increased dramatically across the region. Indeed, turnout increased so significantly that Democrat Joe Biden received more votes in 2020 while losing the region by 8.2 percentage points than Donald Trump did in winning the region in 2016 by 8.6 percentage points.

At the state scale, eleven of the twelve Southeastern states saw a two-party percentage point vote increase for Joe Biden in 2020 over that of Hillary Clinton in 2016. However, most of the increases were minor and only four states had an increase of more than 2 percentage points: Georgia (+2.8), Kentucky (+2.5), Virginia (+2.4), and West Virginia (+2.4). In terms of Kentucky and West Virginia, while Biden's two-party percentage increases were among the highest in the region, that increase made little practical difference in the overall outcome in these two states; the Democratic nominee increased their share of the vote to only 36.15 percent and 29.70 percent, respectively.

However, the increases in the Democratic percentage were important in Virginia, North Carolina, and Georgia. In Virginia, Joe Biden received 54.11 percent of the vote, by far the highest in the region. Indeed, Virginia is continuing its trend of moving from a toss-up state in national elections to a lean Democratic state. This trend is also evident in state elections, where Democrats (as of June 2021) hold the three statewide elected offices (governor, lieutenant governor, and attorney general), both US Senate seats, and, in 2019, won control of both chambers of the state's general assembly for the first time since the early 1990s. Virginia holds statewide elections in off years, so the next test of Democratic control of state government will be in November 2021.

In North Carolina, Donald Trump defeated Joe Biden by only 1.34 percentage points, or a margin of about 74,000 votes out of 5.4 million cast. The 2020 results continue a trend of elections over the past decade as the state has become very competitive. As was the case in 2016, statewide elections across the board in North Carolina were closely contested in 2020. Of the eighteen statewide contested elections in North Carolina in 2020, fourteen were won by a margin of fewer than two hundred thousand votes out of more than five million cast.

The most noteworthy state in the region in the 2020 presidential election was Georgia, which was won by Joe Biden and which saw the Democrats' largest two-party percentage point increase in the region (+2.8). In the *Atlas of the 2016 Elections*, voter and demographic shifts in Georgia were enumerated that could increase its potential as a toss-up state in future presidential elections. While Hillary Clinton lost Georgia to Donald Trump in 2016 by 5 percentage points, she did increase the Democratic share of the vote in the state compared to Barack Obama's 2012 total (Leib 2018). Indeed, Georgia's 2016 result stood out in a region where most states saw increases in their Republican percentage of the vote. A continuation of the Democratic trend in 2020, as well as a tremendous decade-long registration and get-out-the-vote effort, led most notably by Stacey Abrams (Epstein and Herndon 2021), led Biden to a narrow twelve-thousand-vote win out of nearly five million cast (49.51 percent to 49.25 percent). Biden became the first Democrat to win Georgia since Bill Clinton in 1992, and the first non-southern-born Democratic to win the state since John F. Kennedy in 1960. Indeed, whereas Obama lost Georgia in

2012 by 8 percentage points, Biden was able to win it by 0.26 points in 2020.

While eleven of the twelve states in the region saw an increase in their Democratic share of the two-party vote when compared to 2016, the one exception was Florida. Long a toss-up state, Florida shifted slightly more Republican in 2020, with Donald Trump increasing his two-party percentage of the vote by 1.1 points over 2016, winning the state by 3.36 percentage points, a margin of victory larger than in any presidential election since 2004. Given that it is the most populous state in the region, Trump's success in Florida offset most of the Democratic popular vote gains in the rest of the region.

At the local scale, several recent trends contributed to the 2020 election changes. At the county level, as shown in figure 5.3, a number of the counties with the largest increases in percentage of the Democratic vote were found in fast-growing suburban areas. While Joe Biden did not win all of these counties, the shift toward the Democratic nominee is notable. Among the suburban areas shifting the most toward the Democrats are the northern Virginia suburbs of Washington, DC, suburbs in the Richmond and Hampton Roads region of Virginia, suburbs of South Carolina's three largest metropolitan areas (Charleston, Columbia, and Greenville-Spartanburg), and the suburbs of Atlanta, Nashville, Memphis, and the northern Kentucky suburbs of Cincinnati. This Southeast trend follows the nationwide trend of the suburbs voting more Democratic in recent years. Of these metropolitan areas, in terms of the overall election outcome, the swing toward the Democrats in the suburbs of northern Virginia and Atlanta are the most important. At the same time, there is one main outlier in this group, the Miami–Fort Lauderdale–West Palm Beach metropolitan area in Florida, which swung toward the Republicans.

In northern Virginia, the tremendous population growth of the Washington, DC, suburbs has had a major impact on statewide elections, and has shifted the state from Republican toward the Democrats. In 2020, all of northern Virginia's counties voted more Democratic than in 2016, with the most populous counties shifting 5 percentage points or more to the Democrats. Today, northern Virginia accounts for approximately 40 percent of the state's population, compared to 25 percent in 1980. Much of the increase in the region has been the result of migration, both international and interregional, changing the partisan dynamics of the region. Thus, winning northern Virginia has been key to recent Democratic victories statewide.

The most important regional shift in the 2020 elections was in the suburbs surrounding Atlanta. Atlanta's inner ring suburban counties all saw a greater than 5 percentage point swing to the Democrats, with the outer suburbs shifting toward the Democrats as well. Indeed, the northern counties of Cobb and Gwinnett, long a base of Republican support, and won most recently by Mitt Romney in 2012, cast 56 percent and 58 percent of their vote respectively for Joe Biden. The importance of metro Atlanta to Georgia elections cannot be understated. One of the fastest-growing metropolitan areas in the United States, metro Atlanta's population of approximately six million has almost doubled since 1990. A considerable amount of this population increase has resulted from interregional migration, and Atlanta has become a center for international migration as well. In the 2010s, Atlanta was only second to the New York City metro area in terms of interregional migration, with an important component of this interregional migration being the in migration, in both the city and suburbs, of African Americans. At this point, approximately two million African Americans live in metro Atlanta (one-third of Atlanta's population), which has led Atlanta to surpass Chicago as the second-largest metropolitan area in terms of African American population.

The combination of Atlanta's growth (the metro area makes up more than half the state's population) and its growing diversity has made it a key area for Democrats. Indeed, while an oversimplification of the regionalization of Georgia, the growth of Democratic strength in Atlanta and its suburbs, which in this case helped lead to Biden's victory, highlights the "Two Georgias" thesis first proposed in the 1980s (King and Avery 1988), namely that Georgia can be divided into the Atlanta metro area and the rest of the state.

While Democrats gained ground in most metro areas of the South, the major exception was the southeastern Florida conurbation of Miami–Fort Lauderdale–West Palm Beach. While the three counties in the region (Miami-Dade, Broward, and Palm Beach) each cast a majority of votes for Joe Biden (and make up one-quarter of the counties in the state that voted for Biden), they shifted more Republican in 2020 than 2016. Indeed, Miami-Dade, the largest county in population, had the greatest shift to the Republicans in 2020 from 2016 with Republicans increasing their percentage of the two-party vote by 11.4 points. While Biden still defeated Trump in Miami-Dade, Trump was able to win two hundred thousand more votes in 2020 than in 2016, narrowing the Democrats' margin of victory in the county from 30.2 points in 2016 to only 7.4 points in 2020. Given the large margins that Democrats need to

FIGURE 5.3

Southeast Region Presidential Elections

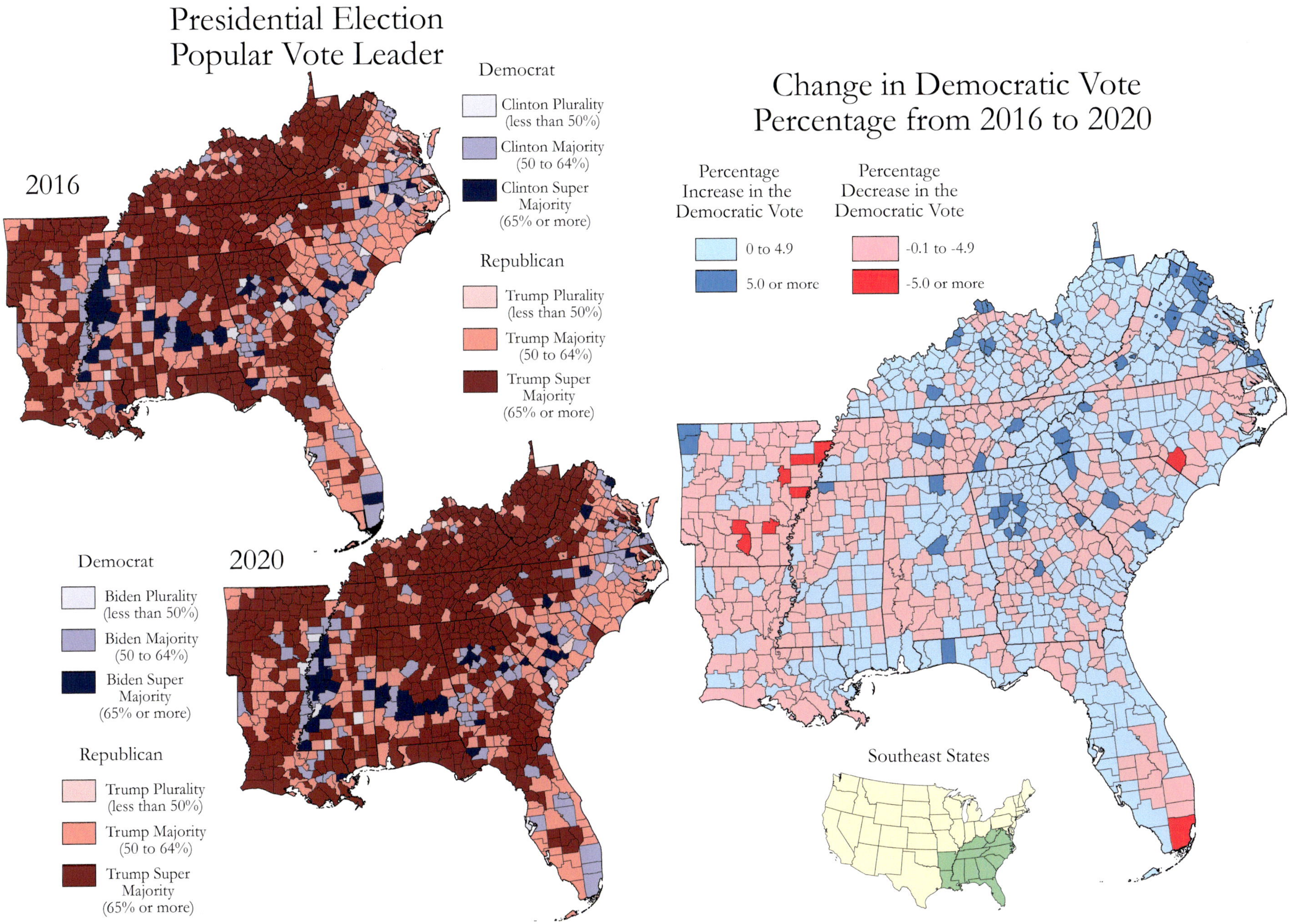

win in southeastern Florida to offset Republican votes in the rest of the state, the swing in Miami-Dade and the rest of southeastern Florida was too much for Democrats to overcome.

According to reports following the election, Trump's success in southeastern Florida was partly a result of putting more time and effort into campaigning among the region's large Latino population. The Trump campaign was successful at micro-targeting messages to the diversity of ethnic and national groups that make up that population (Boryga, Valdez, and Ariza, 2020). The Biden campaign, on the other hand, was slow to respond to the Trump campaign's efforts.

Thus, the 2020 election in the Southeast was marked by continuity and change, and a continuation of recent demographic and partisan trends. At the same time, while demographics and partisan trends tell part of the story of the 2020 election, concerted efforts to register new voters, get-out-the-vote drives, and effective campaigns still matter in the Southeast.

REFERENCES

Boryga, Andrew, Yvonne H. Valdez, and Mario Ariza. 2020. How Trump Won Big with Latinos in Florida—and Then Some. *South Florida Sun Sentinel.* November 4, https://www.sun-sentinel.com/news/politics/elections/fl-ne-latinos-miami-voting-20201104-eyydu3ltkbfqrkklsg2jgtvy2i-story.html.

Epstein, Reid J., and Astead W. Herndon. 2021. The 10-Year Stacey Abrams Project to Flip Georgia Has Come to Fruition. *New York Times.* January 6, https://www.nytimes.com/2021/01/06/us/politics/stacey-abrams-georgia.html.

King, B. Frank, and David Avery. 1988. "A Tale of Two Georgias." *Economic Review—Federal Reserve Bank of Atlanta* 73 (January/February): 24–35.

Leib, Jonathan. 2018. Southeast. In Robert H. Watrel, Ryan Weichelt, Fiona Davidson, John Heppen, Erin H. Fouberg, J. Clark Archer, Richard L. Morrill, Fred M. Shelley, and Kenneth C. Martis, eds., *Atlas of the 2016 Elections*, 125–27. Lanham, MD: Rowman & Littlefield.

MIDWEST

JOHN HEPPEN

The Midwest in 2020 was the most important battleground region in the country. Ohio, Michigan, Wisconsin, Iowa, and Minnesota were targeted based on the 2016 results. In 2016 Ohio, Michigan, Wisconsin, and Iowa flipped Republican after voting Democratic in 2012, and Minnesota was closer than expected. Trump's Midwest victories were attributed to his strength among white, less educated/working-class voters (Smith and Hanley 2021). In 2020 the electoral votes of the Midwest helped push the outcome in Joe Biden's favor. Both candidates placed importance on the region early. President Trump visited Green Bay, Wisconsin on, June 25 for a televised town hall meeting. The Democrats originally scheduled their convention in Milwaukee but moved it online due to COVID-19. Trump was eight points behind Biden in Wisconsin based on a poll released on June 24 (Glauber 2020). In late October, Biden in one Michigan poll had an eight-point lead (Peters and Paz 2020). Both candidates made multiple visits to Michigan, Wisconsin, and Minnesota late in the campaign (figures 3.5, 3.6, and 3.7). The good news for Biden was that Michigan, Minnesota, and Wisconsin all elected or reelected Democratic governors in 2018, with Michigan and Wisconsin being pickups for the Democrats. Ohio, Iowa, and Indiana had Republican governors and Missouri elected a Republican governor on November 3. Michigan, Minnesota, and Wisconsin have same-day voter registration, adding to the drama on Election Day and reinforcing the importance of turnout even in a COVID-19 election. The Midwestern states had high rates of absentee and mail-in voting due to the pandemic, which later fueled charges by Republicans of a rigged vote and stolen election. Early voting was important; early and absentee voting accounted for 59 percent of all votes in Wisconsin (Wisconsin Elections Commission 2021). In Michigan, 57 percent of all votes were cast by absentee ballot (Gibbons 2020).

The COVID-19 crisis hit Michigan particularly hard as President Trump openly feuded with Democrat Governor Gretchen Whitmer on her measures to slow the spread of the virus. Michigan also had a competitive US Senate campaign between incumbent Democrat Gary Peters and upstart Republican John James. In late May and early June 2020, Minnesota was rocked by the George Floyd murder as civil disturbances and street protests against policing in the midst of the pandemic added uncertainty to the vote. In addition, the Black Lives Matter (BLM) movement resounded in Wisconsin with the shooting of Jacob Blake on August 23 in Kenosha and later the shooting of BLM protestors by Kyle Rittenhouse.

Michigan and Wisconsin proved vital to Biden's electoral vote victory. Biden's taking of the twenty-six electoral votes from those two states swung on increased voter turnout in Detroit, Milwaukee, and Madison; those urban areas are keys to Democratic victories. Additionally, suburban counties surrounding Milwaukee and Detroit shifted their vote to varying degrees in favor of Biden. Trump had hoped to continue using his populist appeal and rhetoric to keep Wisconsin (Weichelt 2021).

The number of votes increased for both parties from 2016 to 2020. The electoral and geographic key to Biden's gains were increased votes in Milwaukee, Dane (Madison), and Wayne (Detroit) counties. Whereas Hillary Clinton underperformed in 2016, Biden increased his vote by more than 70,000 compared to Clinton in Wayne County as Clinton lost the state in 2016 by about 11,000 votes. Statewide, Biden won Michigan by more than 150,000 votes. The margin in Wisconsin was not as large, but Biden improved in Milwaukee and Dane counties. Kenosha County, the site of the Blake and Rittenhouse incidents, saw Trump's vote decrease by about 2,000 votes from 2016, but Biden couldn't top Clinton's total of 42,641 and in fact underperformed at 42,193 votes. But Biden improved by about 40,000 votes in Dane County and Trump upped his vote by about 8,000 there as well. Milwaukee County proved to be decisive. Biden upped Clinton by about 29,000 votes while Trump improved by about 8,000. Suburban counties also had an increase of votes for both candidates as Trump increased from approximately 142,000 votes to about 159,000 votes in Waukesha County, which is Milwaukee's largest suburban county and the third largest in the state. Biden improved on Clinton's vote from about 79,000 to about 103,000 in Waukesha County. This lends some credence to the narrative that enough suburban voters tired of Trump and allowed Biden to eke out a victory in Wisconsin by a scant 20,685 votes (0.63 percent).

Suburban voters in Michigan boosted Biden. Biden won the state by 2.8 million to Trump's 2.65 million. Biden won suburban Oakland County north of Detroit and increased his vote by more than 101,000 versus Clinton and Trump declined by about 36,000 votes. Combined with Wayne County's increase of 70,000 by Biden, two counties made all the difference in Michigan. Much ballyhooed Macomb County to the east of Oakland, and home of the Reagan Democrats and working-class whites, remained in Trump's column, although with about 18,000 fewer votes in 2020 compared to 2016. It can be argued that

FIGURE 5.4

Midwest Region Presidential Elections

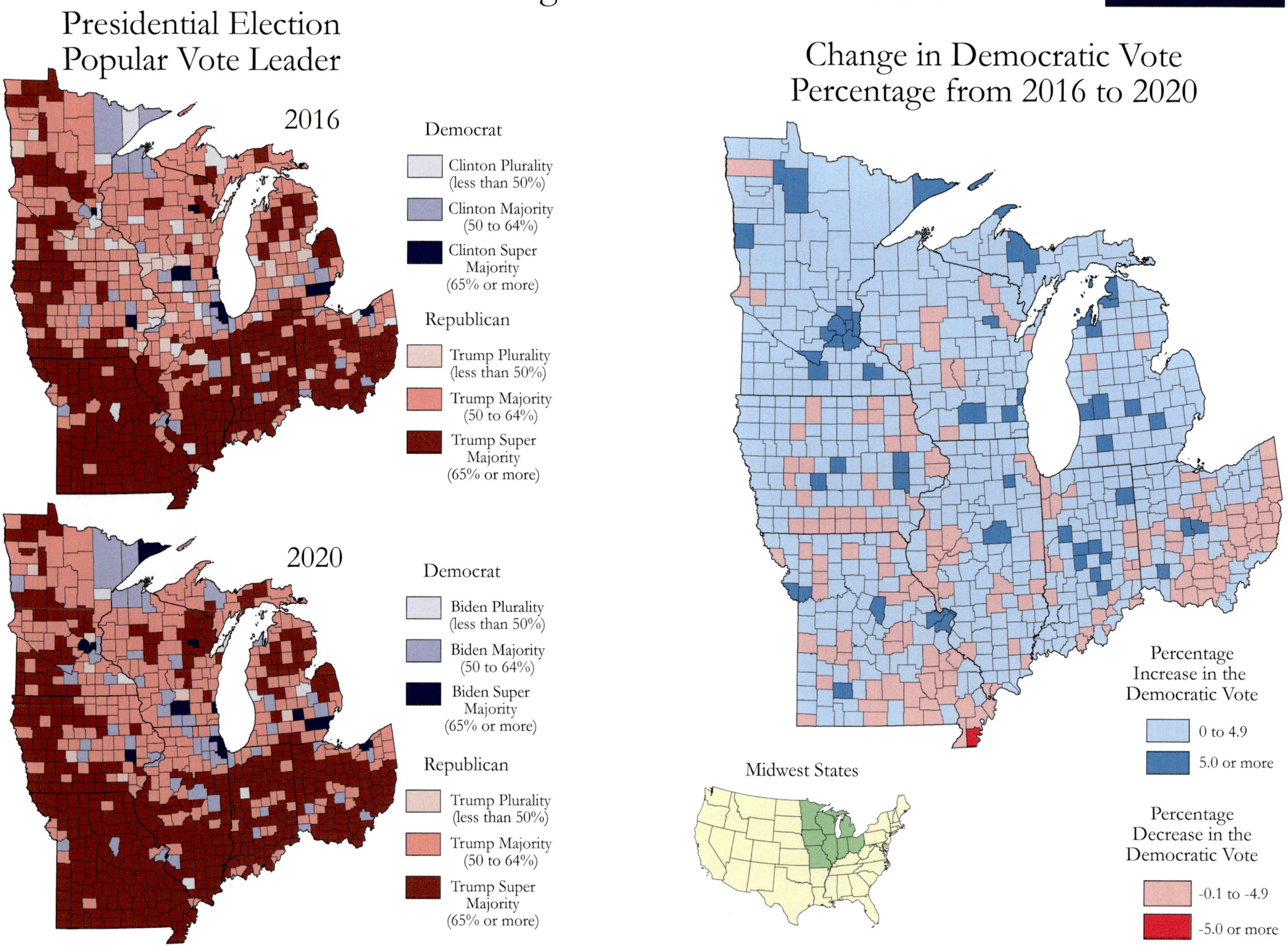

Michigan's Oakland County was the most important suburban county in the Midwest.

Illinois, Indiana, Iowa, Minnesota, Missouri, and Ohio did not flip their electoral votes; all except Illinois and Minnesota saw their electoral votes go in the Trump column. Minnesota was closer than expected in 2016, but Biden outperformed Trump for its ten electoral votes. The killing of George Floyd on May 25 and the subsequent protests, which apexed from May 26 to May 29 amid COVID-19 restrictions, introduced a wild card in the Gopher State as Clinton won Minnesota by only about 40,000 votes in a near upset. Biden won by more than 230,000 votes while both he and Trump increased their vote totals. Biden increased by more than 300,000 from Clinton, and Trump on his 2016 total by more than 150,000 votes. Of note is Biden taking Minnesota's suburban Washington and Dakota counties, where he improved upon Clinton's performance. Biden in fact increased by greater than 5 percent throughout the counties of the Twin Cities region (figure 5.4).

Illinois voters have supported Democrats for president since 1992 and show no signs of changing. Biden's margin increased to 57.24 percent from Clinton's 55.24, and he won the state by about one million votes based on his margin in Cook County, home to Chicago and near suburbs. He won Cook by more than one million votes and won every suburban county sharing a border with Cook except for McHenry, which he lost by about 2 percentage points. The reliability of Illinois forms the base for the Democratic strategy in the Midwest.

Indiana, Iowa, Ohio, and Missouri also saw increased turnout and votes for both candidates compared to 2016, but Trump scooped up all their electoral votes. Ohio was a state that seemed winnable for Biden; Obama carried it twice with him on the ticket. Biden won the urban counties of Cuyahoga (Cleveland), Summit (Akron), Franklin (Columbus), Lucas (Toledo), Hamilton (Cincinnati), and Montgomery (Dayton). But Biden was not able to win a single suburban or collar county to those central city counties, unlike Michigan, where he won suburban Oakland and Washtenaw counties. Biden in fact lost suburban Lorain County next to Cleveland, which had supported Clinton. Biden's trouble in Ohio was that despite outperforming Clinton by more than two hundred thousand, votes Trump improved upon 2016 by about three hundred thousand votes. Ohio, despite having several major metropolitan areas, behaved more like Indiana than Michigan.

Despite winning Indiana in 2008, Democrats have not been close in recent elections and their only reliable wins are Marion (Indianapolis) and Lake counties (Gary), and the college counties of Tippecanoe (Purdue), Monroe (Indiana University), and St. Joseph (Notre Dame). Missouri paints a similar picture as Jackson (Kansas City), Boone (University of Missouri), St. Louis County, and St. Louis City were the only counties captured by Biden in an uncontested state that Trump won by 15.39 percentage points. Iowa was not close; Trump won it by 8.20 points. The urban-university pattern was repeated in Iowa as Biden won the counties home to its bigger cities and major universities: Des Moines (Polk), Cedar Rapids and Waterloo (Linn and Black Hawk), Davenport (Scott), and the University of Iowa (Johnson), and Iowa State (Story). Although Biden did increase his total vote by more than one hundred thousand from 2016, Trump improved by about ninety-seven thousand votes.

In summary, increased turnout for both candidates favored Biden; he recaptured Michigan and Wisconsin as part of his blue wall in the North based on winning favor with more suburban voters and by increasing his vote total over Clinton in the urban giants of Detroit and Milwaukee. He staved off any Trump hopes for taking Minnesota by upping the ante in Minneapolis–St. Paul and its surrounding suburbs.

REFERENCES

Gibbons, L. 2020. One Big Winner in Michigan's 2020 Election Cycle: No-Reason Absentee Voting. *Mlive*. November 11, https://www.mlive.com/politics/2020/11/one-big-winner-in-michigans-2020-election-cycle-no-reason-absentee-voting.html (last accessed June 7, 2021).

Glauber, B. 2020. Marquette Poll: Joe Biden Leads Donald Trump by 49% to 41% in Wisconsin. *Milwaukee Journal-Sentinel*. June 26, https://www.jsonline.com/story/news/politics/elections/2020/06/24/marquette-poll-joe-biden-leads-donald-trump-49-41-wisconsin/3250329001/ (last accessed June 7, 2021).

Peters, J. W., and I. Grullon Paz. 2020. Biden Holds 8-Point Lead in Michigan as He Tries to Reconstruct the "Blue Wall." *New York Times*. November 3, https://www.nytimes.com/2020/10/28/us/politics/michigan-poll-trump-biden.html (last accessed June 7, 2021).

Smith, D. N., and E. Hanley. 2021. The Heart of Whiteness: Patterns of Race, Class, and Prejudice in the Divided Midwest. In B. Warf, ed., *Political Landscapes of Donald Trump*, 111–28. New York: Routledge.

Weichelt, R. 2021. The 2016 Presidential Election and Trump's Populist Rhetoric. In B. Warf, ed., *Political Landscapes of Donald Trump*, 92–110. New York: Routledge.

Wisconsin Elections Commission. 2021. https://elections.wi.gov/publications/statistics/absentee (last accessed June 7, 2021).

GREAT PLAINS

ROBERT H. WATREL

The 2020 election was another landslide win for the Republican Party at all levels of government in the Great Plains region. These states have long been supporters of Republican presidential candidates and the last time a state gave all its electoral votes to a Democratic candidate was in 1976 (Hurt 2011). However, as in 2008, Nebraska cast one of its electoral votes for a Democratic candidate. In 1992, Nebraska changed its winner-take-all method to the congressional district method of distributing electoral votes. In the Second District, which includes the city of Omaha, Joe Biden received a majority of the popular vote (51.95 percent) and its Electoral College vote.

At the state level, Great Plains voters have never been friendly to Democratic candidates, and 2020 was no exception. Republican candidate Donald Trump's vote returns for the six Great Plains states continued a pattern that Republican presidential candidates have enjoyed for several decades. No state gave less than 52 percent (Texas) of the popular vote to Trump and three states gave returns greater than 60 percent (North Dakota, Oklahoma, and South Dakota). Three of the six Great Plains states had a margin of victory greater than 20 percent, while Texas had the lowest margin of victory at 5.58 percent. Oklahoma led the largest popular vote with 65.37 percent for Trump, while North Dakota had the largest margin of victory for Trump at 33.35 percent. However, these margins of victory were less for Trump in all six states than in the 2016 election. Democratic candidate Joe Biden's largest popular vote from any Great Plains state was Texas with 46.48 percent. All the other states gave less than 41 percent of the popular vote, with North Dakota polling the lowest at 31.76 percent. Third-party and write-in candidates did not fare as well as those in the 2016 election, with five states polling less than 2.6 percent. North Dakota polled the highest at 3.1 percent for third-party and write-in candidates.

The wave of Republican support is even more impressive when observed at the county level. Of the 648 counties that comprise the Great Plains states, 609 counties returned 50 percent or more for Republican incumbent Donald Trump, with 26 counties polling over 90 percent. Granted, many of these counties have small populations, but it shows the popularity of the Republican Party in these mostly rural areas. If you include the additional counties where Trump received a plurality, he carried a total of 611 counties. On the other hand, Joe Biden received a majority in 35 counties and a plurality in two for a total of 37 counties.

Counties where voters gave Joe Biden a majority or plurality are counties that have long been bastions of support for Democratic candidates over the last several election cycles (figure 5.5). These Democratic-majority counties roughly fall into four categories: counties with large American Indian populations, counties with large Hispanic populations, several metropolitan counties mainly in Texas, and several counties with colleges.

In the northern Great Plains, almost all the Democratic-majority counties have large American Indian populations, which have historically been strong supporters of the Democratic Party (Wishart 2004). In North Dakota, Rolette County (Turtle Mountain Reservation) and Sioux County (Standing Rock Reservation) gave Democrats super majorities. This pattern also holds true for South Dakota. Counties with large American Indian populations, such as Dewey (Cheyenne River Reservation), Oglala Lakota (Pine Ridge Reservation), Todd (Rosebud Reservation), and Buffalo (Crow Creek Reservation) gave a majority or super majority to Joe Biden.

Another area of traditional Democratic support is located in the southern Great Plains, close to or along the US-Mexico border. These are counties with historically large Hispanic populations. Some of these counties have given strong support to the Democratic Party for many decades. In fact, Brooks and Jim Hogg counties have never given a Republican presidential candidate a majority of their votes. All these counties have Hispanic populations over 75 percent (Barcus and Simmons 2013). However, when comparing the 2016 and 2020 Popular Vote Leader maps and the Change in Democratic Vote map (figure 5.5), some border counties in southern Texas did flip and gave a majority or plurality to Donald Trump and had

FIGURE 5.5

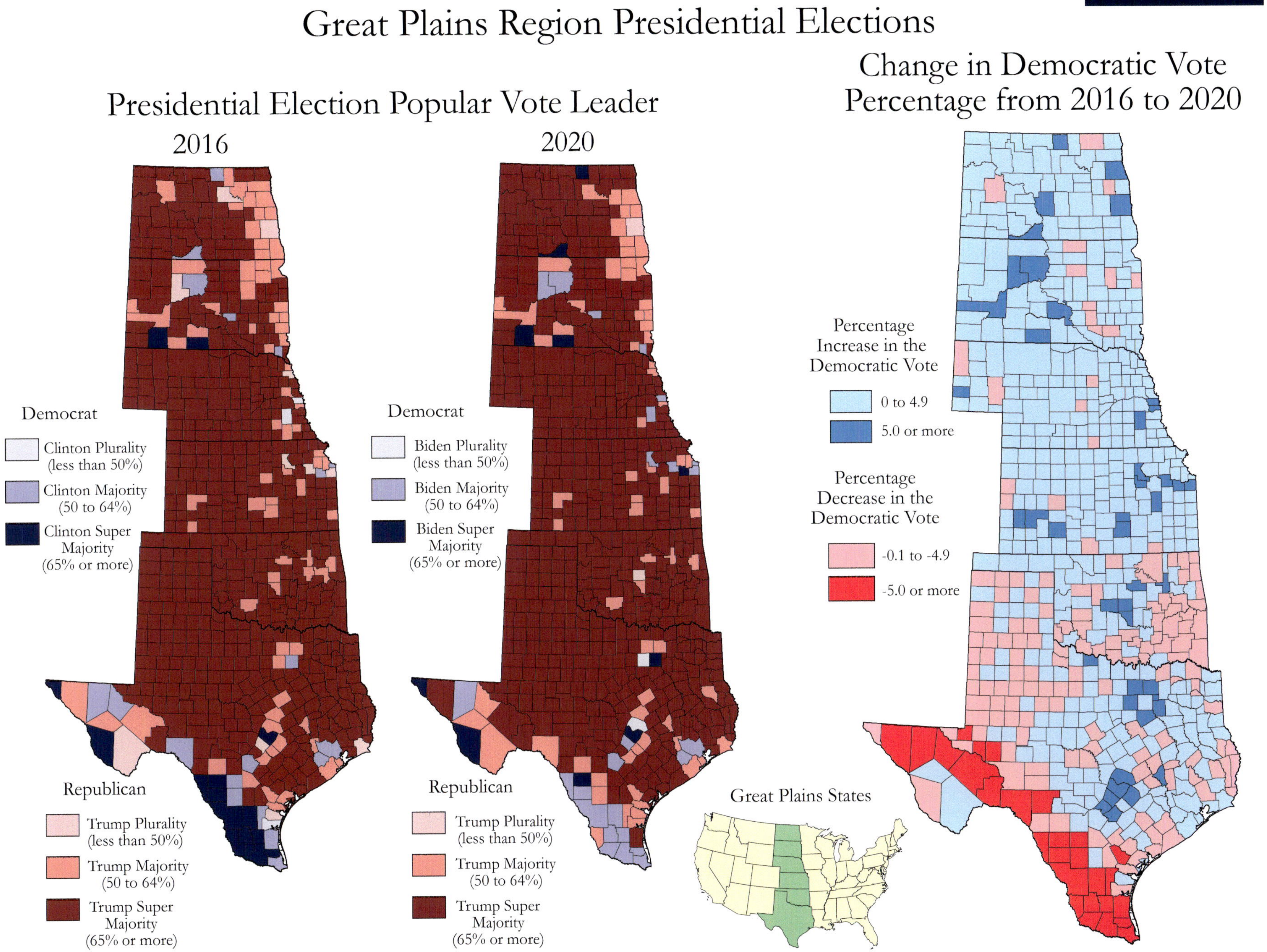
Great Plains Region Presidential Elections
Presidential Election Popular Vote Leader
2016
2020
Democrat
Clinton Plurality (less than 50%)
Clinton Majority (50 to 64%)
Clinton Super Majority (65% or more)
Republican
Trump Plurality (less than 50%)
Trump Majority (50 to 64%)
Trump Super Majority (65% or more)
Democrat
Biden Plurality (less than 50%)
Biden Majority (50 to 64%)
Biden Super Majority (65% or more)
Republican
Trump Plurality (less than 50%)
Trump Majority (50 to 64%)
Trump Super Majority (65% or more)
Change in Democratic Vote Percentage from 2016 to 2020
Percentage Increase in the Democratic Vote
0 to 4.9
5.0 or more
Percentage Decrease in the Democratic Vote
-0.1 to -4.9
-5.0 or more
Great Plains States

a swing of over 5 percent or more away from the Democratic Party. For example, Zapata County voted 65.7 percent for Hilary Clinton in 2016, but voted 52.5 percent for Donald Trump in 2020. Whether this is a onetime phenomenon remains to be seen.

Larger urban counties, where Biden received majorities or pluralities, include Douglas County, Nebraska (Omaha); Wyandotte County, Kansas (Kansas City); and Dallas County (Dallas), Harris County and Fort Bend (Houston), and Bexar County (San Antonio) in Texas. The remainder of counties that gave Biden majorities or pluralities have universities: Clay County, South Dakota (University of South Dakota); Lancaster County, Nebraska (University of Nebraska–Lincoln); Douglas County, Kansas (University of Kansas), and Riley County, Kansas (Kansas State University); and Travis County, Texas (University of Texas–Austin).

When comparing the vote returns between the 2016 and 2020 elections, a shift in the electorate does not look that apparent (figure 5.5). However, when looking over changes in the percentage of the Democratic votes from the 2016 to 2020 elections by county, some patterns emerge. It must be noted that there was an increase in voter turnout for both candidates in the 2020 election, and this must be kept in mind, but two main patterns occur. The first is the overall increase in Democratic vote from 2016 to 2020 as indicated by the dark- and light-blue-shaded counties. This included almost all the central and northern Great Plains and large swaths of the southern Great Plains. The second noticeable change is along the southern border of Texas. Here most counties had a decrease of 5 percent or more in Democratic support from 2016 to 2020, where many of these counties have large Hispanic populations.

Not only has the Republican Party dominated presidential politics in Great Plains states, but also over the last ten to fifteen years the Republicans have increased their hold over state governments and delegations to the US Congress, so much so that several of these states can be considered a one-party state. For example, currently, in five of the six Great Plains states Republicans hold all state-level executive offices (e.g., governor, lieutenant governor, attorney general, etc.). Only Kansas has an elected Democrat governor, lieutenant governor, and state treasurer, which makes their elected executive offices an even 50 percent split. In three states, North Dakota, South Dakota, and Oklahoma, Republicans have supermajorities—more than 66 percent representation—in both houses of the legislature. South Dakota leads the way with a 91 percent Republican majority in the state house and 88 percent majority in the state senate. Nebraska, which has a nonpartisan unicameral legislature, is dominated by conservative legislators and can essentially be thrown into the supermajority category. In Texas, the Republicans have majorities, with under 60 percent representation, in both legislative bodies.

This trend in Republican popularity is also evident in US congressional delegations. As of the 2020 election, there is one Democratic member (Kansas) in the US House of Representatives in the five Great Plains states north of Texas. Although this does not equate to very many House seats—for example North and South Dakota have only one at-large member, Nebraska three, Kansas four, and Oklahoma five—many of these seats were competitive for Democratic candidates just ten years ago. Texas is the only Great Plains state that has more than one Democratic member in its US House of Representatives delegation with thirteen members out of a total of thirty-six for the state. In the US Senate, as of the 2020 election, there are no Democratic senators out of twelve seats from the six Great Plains states. There were five senatorial elections during the 2020 election and Republican candidates reclaimed all five seats.

The prospect for future Democratic presidential candidates in the Great Plains states looks bleak. There are a few metropolitan counties and pockets of counties with some ethno-cultural diversity (e.g., American Indians, Hispanics, and African Americans) that support the Democratic Party. Most of the remaining counties are rural, lightly populated, and predominantly white with conservative political values.

The Great Plains states of North Dakota, South Dakota, Nebraska, Kansas, and Oklahoma will probably not cast their electoral votes for a Democratic presidential candidate anytime soon (with the possible exception of an electoral vote from Nebraska). However, the state of Texas offers a glimmer of hope for the Democrats. The margin of victory of Republican presidential candidates over Democratic candidates decreased from 15.8 percent in 2012, to 9 percent in 2016, and to 5.6 percent in 2020. This should make the Republican Party nervous of the future in Texas. However, as noted above, the Democratic Party may well be nervous too because of the relative shift to the Republican Party in some counties in southern Texas with sizable Hispanic populations. The tipping point of Texas to the Democratic Party in presidential elections may be nearing, but may not happen in another four years.

REFERENCES

Barcus, Holly R., and Simmons, Laura. 2013. Ethnic Restructuring in Rural America: Migration and the Changing Faces of Rural Communities in the Great Plains. *Professional Geographer* 65(1):130–52.

Hurt, R. Douglas. 2011. *The Big Empty: The Great Plains in the Twentieth Century.* Tucson: University of Arizona Press.

Wishart, David J. 2004. *Encyclopedia of the Great Plains.* Lincoln: University of Nebraska Press.

MOUNTAIN WEST

TONY ROBINSON

The ocean of blue that dominates the Change in Democratic Vote map (figure 5.6) suggests a dramatic decline in fortune for the Republican Party in the Mountain West between 2016 and 2020. Digging into the numbers behind the western blue wave clarifies the grim news for the GOP. Compared to 2016, the Democrats increased their share of the 2020 vote in 267 western counties. Republicans received a higher percentage of the 2020 vote in just twelve counties across eight western states. The Democrats won three of the four most populated states (Arizona, Colorado, and Nevada) as well as New Mexico (the fifth most populated). The GOP won the three least populated states (Idaho, Montana, and Wyoming) plus Utah (the third most populated). About three times as many people live in the four states captured by Democrats than in the four smaller states captured by Republicans, 18.2 million to 6.6 million. That population difference translated into a thirty-one to sixteen Electoral College vote advantage for the Democrats in the 2020 Mountain West. Ominously for the GOP, the state of Arizona joined regular Democratic strongholds of Colorado, New Mexico, and Nevada in voting blue, the first time Arizona voted Democrat for president since 1996 and only the second time Arizona has voted blue since 1952. The last time the Democratic party won four states in the Mountain West was 1964, in Lyndon Johnson's landslide victory over Barry Goldwater.

The 2020 election was not an anomaly. The last several presidential elections have shown an unmistakable trend toward Democratic strength in the West. For the fourth presidential election in a row, Democrats won at least three of the five most populated states in the West (they won four of the top five in 2020). In the ten preceding presidential elections 1968–2004, the Democrats repeated this feat on just two occasions, Bill Clinton's 1992 and 1996 victories. Correspondingly, in the ten presidential elections between 1968 and 2004, Democrats won just 10 percent of all possible electoral votes in the Mountain West. In the last four presidential elections, Democrats have won 48 percent of all Mountain West electoral votes.

The fact that Biden won four of the most heavily populated states in the West is hidden in the Popular Vote Leader map, which suggests Republican dominance across the West. The maps of the 2016 and 2020 election winners reveal an enduring sweep of red and pink GOP counties, showing that Republicans won a substantial majority of all western counties in both 2016 and 2020. Moreover, the deep red of many of these counties—showing areas of Trump super-majorities—helps explain the sizable victories Trump notched in several states. Trump won Wyoming by 43.39 percent, Idaho by 30.77 percent, Utah by 20.49 percent, and Montana by 16.37 percent. However, Wyoming, Montana, and Idaho are the least populated states in the region, so racking up large-scale victories across dozens of counties in those states doesn't translate into many electoral votes. The Democratic pattern is the inverse: although Joe Biden won only 22 percent of the counties in the West, those counties account for about 50 percent of all votes cast. This is why the Democrats won both Nevada and Arizona in 2020, although the 2020 Popular Vote Leader map shows that only seven among the twenty-nine counties in those two states voted Democrat.

The Democratic Party is strong in heavily populated counties, and this strength grew between 2016 and 2020. In the six states with multiple counties where the Democrats grew their share of the vote by 5 percent or more, almost all of these were the most populous in the state. All five of Colorado's and Utah's most heavily populated counties posted Democratic gains of 5 percent or more in 2020 compared to 2016. Four of Idaho's five most populous counties posted Democratic gains of 5 percent or more. Three of the top five counties in Montana and New Mexico and two of the top five counties in Wyoming posted Democratic gains of 5 percent or higher. The Republican Party didn't post gains in a single one of the forty most populated counties across the eight-state region. Moreover, only a single county in the entire region posted Republican gains of more than 5 percent of the vote share, Arizona's Santa Cruz County, which Biden still won by more than 65 percent. All told, the Democrats experienced a 5 percent growth in vote share in seventy-nine western counties.

Mountain West Region Presidential Elections

FIGURE 5.6

Presidential Election Popular Vote Leader

Change in Democratic Vote Percentage from 2016 to 2020

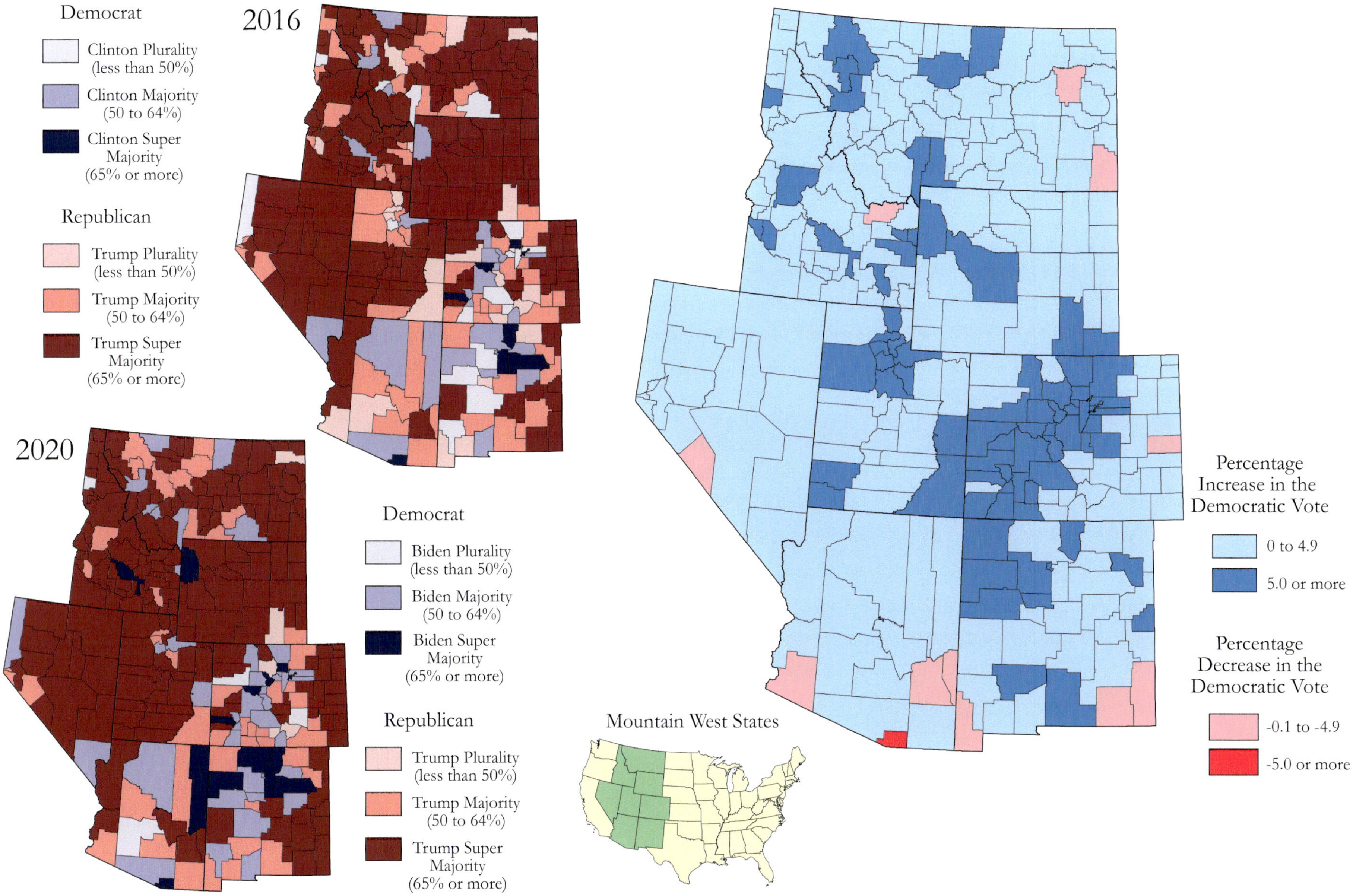

Many of the heavily populated, increasingly Democrat-leaning counties in the Mountain West have common characteristics: rapidly growing, ethnically diverse, urban, and suburban. Meanwhile, the population in rural, socially conservative counties (the swaths of deep red in these maps) is shrinking as a percentage of the West. Census data shows that between 2010 and 2020, the Mountain West was the fastest-growing region in the country, with an average growth rate of 12 percent, twice the national rate. Most of this growth has been in urban areas like the Colorado Front Range, Nevada's Las Vegas and Reno, or Arizona's Phoenix area.

As the West becomes more urban, the Democrats are finding improved prospects. The rapidly growing urbanized areas of the Mountain West tend to be more ethnically diverse, have younger voters, and feature a higher percentage of well-educated professionals, which are more liberal demographics than the more elderly and white voters of the resource-extraction and agriculture economies of the rural West. Correspondingly, National Election Pool data shows that urban areas are increasingly voting Democratic (a +3 Democrat advantage in 2010 grew to +13 in 2020). Rural areas counter this trend by increasingly voting Republican (a +18-point GOP rural advantage in 2000 grew to a +30 advantage in 2020). The problem for the GOP is that more and more Mountain West residents live in urbanized areas. Three of the four western states won by Joe Biden (Arizona, Colorado, and Nevada) are rated among the top fifteen "most urbanized" states in the United States (defined as percentage of people living in urban/suburban areas). In contrast, three of the four won by Donald Trump are among the fifteen least urbanized (Idaho, Montana, and Wyoming). Montana and Wyoming rank forty-ninth and fiftieth on the urbanization scale. Highly urbanized Utah bucks this pattern due to the influence of conservative Mormon voters in Salt Lake and Provo, while rural New Mexico counters the trend due to its unusually high concentration of highly Democratic Latinx voters.

All of this grim news for the Republicans is tempered by two GOP-favoring dynamics. First, regardless of undeniable trends toward the Democratic Party across the Mountain West, the overall region remains majority Republican. The deep red maps of 2016 and 2020 election winners show continuing Republican majorities in most counties across the West. In fact, Donald Trump received about 200,000 more votes than Joe Biden in the region, 6,054,491 versus 5,847,175. The winner-take-all aspect of the Electoral College tilted the region's electoral vote to Biden (thirty-one to sixteen) because of his narrow wins in Arizona and Nevada, while Donald Trump racked up blowout wins in Idaho, Montana, Utah, and Wyoming. The elections in Arizona and Nevada were so close that a flip of just 44,053 votes in the two states would have resulted in a GOP Electoral College advantage of thirty-three to fourteen in the Mountain West. Biden won Arizona by only 0.30 percent and Nevada by 2.39 percent of the vote, indicating that two of the four western states that Democrats won in 2020 will be competitive for the Republicans going forward. On the other hand, none of the four states won by Trump are competitive for the Democrats. Trump won Idaho, Utah, and Wyoming by double digits—and none of these states have voted Democratic since 1964.

A second arguably favorable dynamic for the GOP is that the only high-population counties that shifted toward the GOP in the 2020 election have something in common: they are all border counties in Arizona and New Mexico (these are among the red counties near the southern border of the Change in Democratic Vote map). This pattern suggests that the GOP may continue to have meaningful electoral traction in one of Donald Trump's signature issues: border security and immigration control. The demographics of the West are changing, and not in the GOP's favor. But in the largest state in the region and in the state most narrowly won by the Democrats (Arizona), it appears that border-county concerns over immigration may help keep that state in play for the GOP in future election cycles.

PACIFIC REGION

JOHN AGNEW AND JUAN HERRERA

In 2016, Republican Donald J. Trump performed much better in relative terms across the Pacific Region than the Republican candidates in the previous two presidential elections. This occurred despite Trump having a mixed personal history of party registration and proclaiming many policies well away from mainstream Republican ones on trade and national security. He still lost heavily to Hillary Clinton in all of the states save Alaska. Trump did, however, establish large margins in the most rural areas with declining shares of the region's overall population. He picked up some votes among traditionally Democratic constituencies such as timber workers in towns that had not voted Republican since the 1920s. Beyond that, Clinton's huge margin of victory in California, the most populous state, with a historic high 63.48 percent of the vote, accounts for why she won the national popular vote in 2016 even as Trump was winning in the Electoral College because of his very narrow victories in Michigan, Pennsylvania, and Wisconsin.

The 2020 election occurred in the midst of a once-in-a-lifetime pandemic and national mobilizations against police brutality and anti-Black racism. This led to both increased focus on the election from people trapped at home and intense social media frenzy stirred up particularly by Trump supporters regarding the legitimacy of the election. The election resulted in the highest voter turnouts because of fewer restrictions on voting by means of increased mail-in balloting and greater overall electoral enthusiasm because it was defined in almost existential terms. Many cities in the region erupted in intense protests and uprisings in the aftermath of the murder of George Floyd by a Minneapolis policeman in May 2020. This constituted an unprecedented leftward leap in public opinion on race, criminal justice, and the Black Lives Matter (BLM) movement (Cohn and Quealy 2020). In Seattle and Portland, in particular, standoffs occurred between Black Lives Matter protestors and police. In addition, there were direct encounters between the BLM protestors and far-right racist and misogynist groups such as the Proud Boys. Many of the latter apparently came from outlying areas of Oregon and Washington State rather than from the cities in question. Largely peaceful protests over the Floyd murder also broke out elsewhere across the Bay Area and in southern California during early June 2020.

At the state level in 2020, the story was much the same: higher turnout on both sides. According to survey data, particularly in California, Trump was consistently unpopular (DiCamillo 2021). County-level returns tell a more complex story (figure 5.7). They show enormous variation across the region that challenges the stereotype of the liberal "left coast" (HoSang 2010). California as a whole, as the state of the region with easily the largest number of Electoral College votes, went heavily for Joe Biden in 2020, as it had for Barack Obama in 2008 and 2012, and Hillary Clinton in 2016. However, at the county level, the state is divided in political complexion between its interior and coastal counties. Donald Trump captured all of the same counties he had in 2016. Much of this interior-to-coastal contrast reflects rural-urban differences in ethnic composition and religious traditions. The largely Anglo electorates in the agribusiness central valley and rural-mountainous counties overwhelmingly breaking for Republican candidates in contrast to the urban-metropolitan counties in southern California and the Bay Area with more diverse electorates going for Democrats. This is now a well-established pattern in California and in Oregon and Washington as well. In the latter two states it is metropolitan Portland and Seattle-Tacoma, on the one hand, and much of the rest of the state on the other. If the rural-urban divide leads to Democrats winning across the Pacific Region (with the Alaska exception), that is what leads them to be at a disadvantage nationwide with the pro-rural state bias of the US Senate and Electoral College (Rodden 2019).

The geographical polarization throughout the Pacific Region remained much the same in 2020 as in 2016. There were limited swings to Biden in rural areas, even as he was winning nationally, and strong support for Biden in the coastal counties, including historically Republican Orange County south of Los Angeles. This can be ascribed to the rightward drift of the Republican Party, narrowing its appeal in the region to very specific cultural-geographical constituencies (evangelical Christians, Mormons, farmers, small town conservatives) and

FIGURE 5.7

Pacific Region Presidential Elections

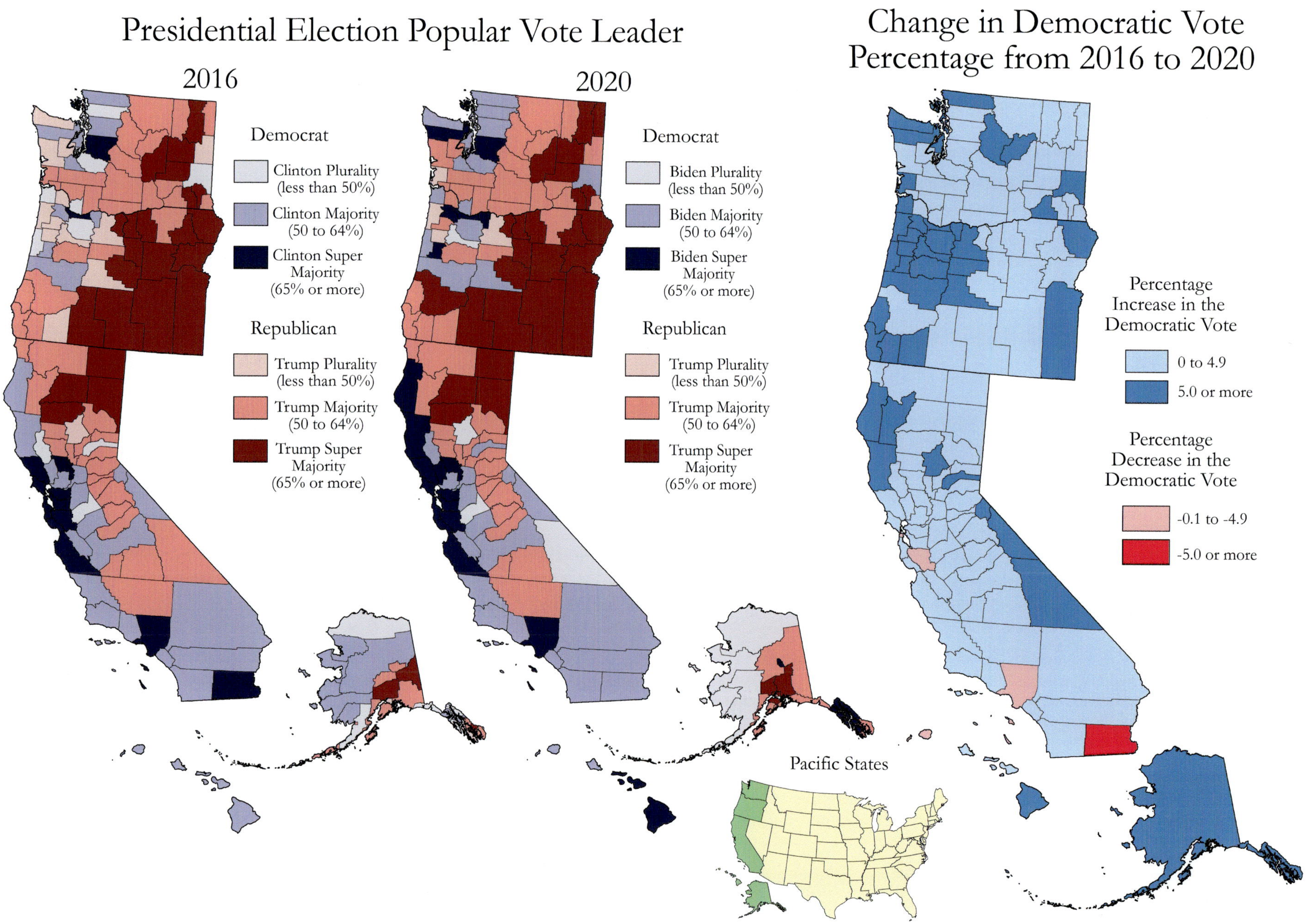

producing candidates in state and congressional elections who are so extreme, particularly on cultural issues, that they do not appeal to moderate voters. The strident southern-style white Christian nationalism of the Republican right does not sell well across the region as a whole.

Nevertheless, the 2020 election did include some novelties. California is an increasingly Latino state, even if population numbers have only slowly translated into greater electoral participation (Pastor 2019). Counties with large urban areas are now often majority Latino. Though still only at 24 percent of the eligible voting-age population, Latinos are estimated to be 38 percent of the total population of California. Over the next twenty years, the former number will undoubtedly increase at a faster rate than the latter. As is well known, the Latino vote in California, and other Pacific Region states, tends, partly because of Republican positions on immigration and economic issues, to vote increasingly Democratic. This was an important factor in 2016, given Trump's campaign was based from the beginning on antipathy, not only to undocumented immigrants, but all people of Mexican descent, a large share of the Latino population in California with citizenship. Trump's bizarre and unsubstantiated claim that he lost the national vote in 2016 because of "illegal" voters in California fits into this picture. But there were signs in 2020 that some Latinos voted for Trump. Some of this could be attributed to the fact that immigration was a much more muted theme of Trump's 2020 campaign than it had been in 2016. More importantly, however, as elsewhere in states with large Latino populations, the Latino vote in California is less monolithic than hitherto alleged. Trump picked up votes more from Latino men than women and in suburban and exurban areas rather than in central cities. Heavily Latino Imperial County in the southeast corner of the state swung heavily to Trump even as he lost there. All this suggests that, in California at least, there is increasing political differentiation among Latino voters.

The rural-urban divide is even more extreme in Oregon than in California and Washington State. This is particularly so between the sparsely populated southeast on the one hand and Portland and the population centers of the Willamette Valley on the other. In 2020, Trump did as well in southeast Oregon as he had in 2016. Even though it still went for Biden in 2020, as it did for Clinton in 2016, Oregon remains more competitive in presidential elections than California and Washington. This perhaps reflects the fact that it is generally more rural than the other two with only one metropolitan area of any size. However, Biden did increase the Democratic percentage in every Oregon county in 2020. Washington State tells much the same story; here also the basic west-east pattern is entrenched. The Seattle-Tacoma metropolitan area exerts such a strong overall effect on the statewide outcome that even with increased local turnout for Trump in 2020, Biden still managed to win the state relatively easily.

Hawaii and Alaska represent opposite ends of the political spectrum as far as the 2020 and other recent presidential elections are concerned. Ever since statehood they have been dominated respectively by the Democratic (Hawaii) and Republican (Alaska) parties in just about all types of elections. In Hawaii Clinton in 2016 and Biden in 2020 failed to match the success of Obama, born in Hawaii, yet both still acquired a substantial majority of the vote. Hawaii remains a difficult state for Republican candidates. With most of the population of native Hawaiian and Asian heritage, Hawaii has a completely different ethnic profile from the other states in the region. In Alaska, Trump kept the Republican tradition alive. His affinity for fossil fuels and the long-standing local animosity to the federal government worked on his behalf. Alaska remains the anomaly in the Pacific Region as the only state where Republicans are the ticket to beat.

REFERENCES

Cohn, N., and Quealy, K. 2020. How US Public Opinion Has Moved on Black Lives Matter. *New York Times*. June 10.

DiCamillo, M. 2021. Majorities of California Voters Hold Strong, Negative Views of Donald Trump and Support His Conviction in the Senate. Berkeley: Institute of Governmental Studies. February 8.

HoSang, D. M. 2010. *Racial Propositions: Ballot Initiatives and the Making of Postwar California*. Oakland: University of California Press.

Pastor, M. 2019. *State of Resistance: What California's Dizzying Descent and Remarkable Resurgence Mean for America's Future*. New York: New Press.

Rodden, J. 2019. *Why Cities Lose: The Deep Roots of the Urban-Rural Political Divide*. New York: Basic Books.

6 DEMOGRAPHICS AND IDENTITY

POPULATION SUBGROUPS

FIONA M. DAVIDSON AND RICHARD L. MORRILL

County-level analysis of demographic data and voting patterns provides a wealth of information about who supported the two primary political parties during the 2020 presidential election. This chapter of the atlas investigates a wide variety of variables, from population change to income, ethnicity, gender, religious adherence, and employment to uncover the relationships between these characteristics and voting behavior. Occasionally, national-level trends are mentioned, with data taken from *New York Times* exit polls and the Associated Press pollcasts that were issued at the time of the election, with the understanding that these are only approximate predictors of behavior and may vary from actual voting behavior. The maps themselves employ certified county-level election results and demographic data taken from the 2014–2018 Census Community Survey. The chapter begins with a survey of population size and change, age, racial/ethnic identification, age, education, employment, and income variables at the county level. Other sections of the chapter go into more detail about specific groups—Hispanic voters, Black voters, and women voters—and includes two segments on the influence of religion in voting patterns in the 2020 presidential election.

POPULATION SIZE

Counties in the United States demonstrate an enormous variety in the size of their population, from the 10.03 million inhabitants of California's Los Angeles County to the tiny population of Loving County, Texas, with 169 inhabitants in 2020. Mirroring the nationwide pattern of high population versus low population counties, these two counties had radically different electoral outcomes. The population of Los Angeles voted 71 percent Democratic and 27 percent Republican, while only four people in Loving County voted for Biden (6.1 percent) against sixty votes for Trump (90.9 percent). Overall, the smaller counties, which cover most of the physical area of the United States, predominantly voted Republican, particularly in the Great Plains, Appalachia, and parts of the Mountain West (figure 6.1). The voters of these counties supported Trump over Biden 67–33 percent, but the population total for these counties was only 3.9 million. Of the 1024 counties with populations under 15,000, 949 (90 percent) voted Republican; however, there were several areas of the country in which smaller counties trended Democratic: the Black Belt across the Southeast, counties along the southern border, American Indian reservation areas, and some environmental amenity areas in the Rockies. It is the small counties (under 15,000) that experienced large shifts to the Republican party in 2016, and

FIGURE 6.1

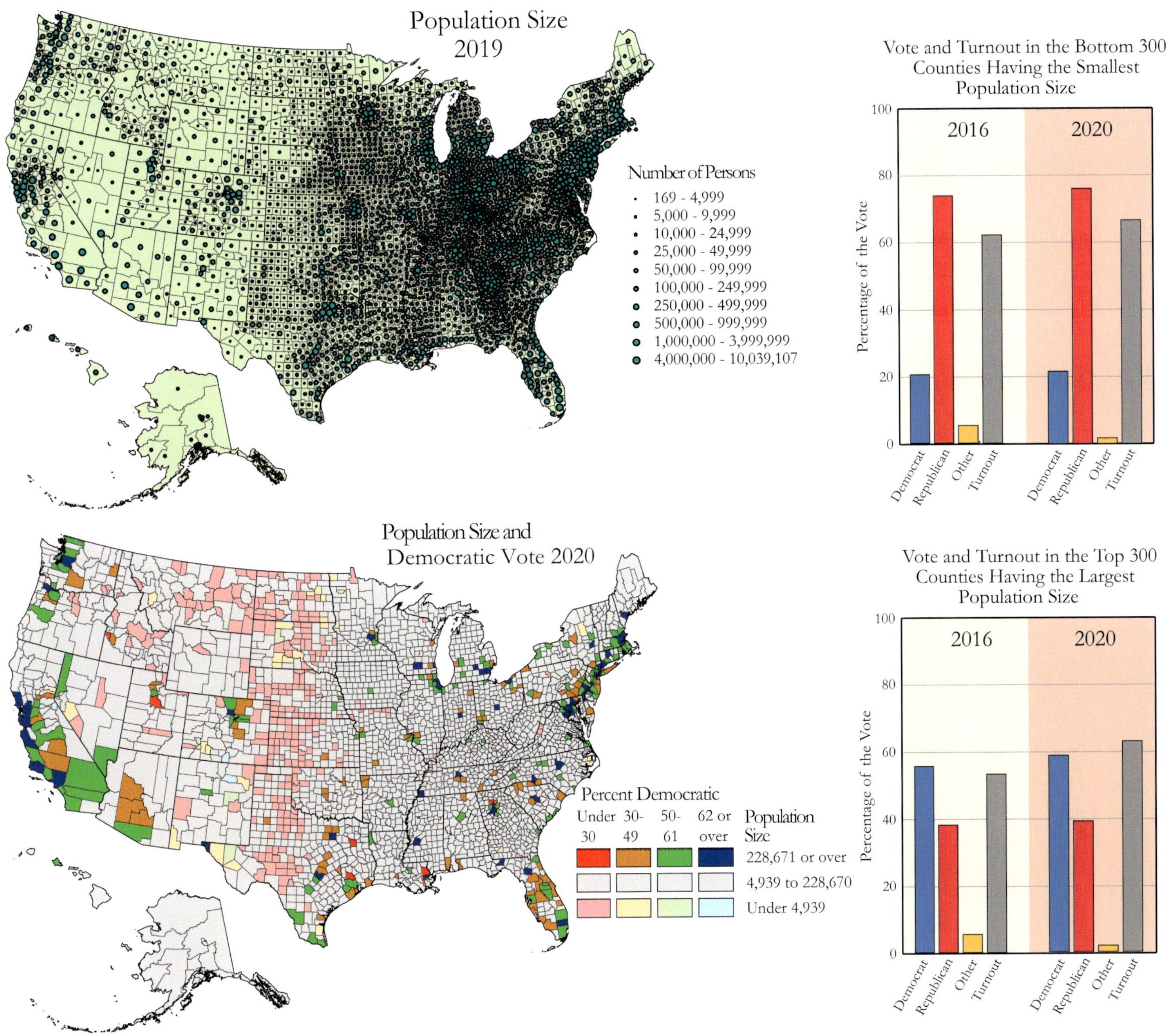
Population Size 2019
Number of Persons
169 - 4,999
5,000 - 9,999
10,000 - 24,999
25,000 - 49,999
50,000 - 99,999
100,000 - 249,999
250,000 - 499,999
500,000 - 999,999
1,000,000 - 3,999,999
4,000,000 - 10,039,107
Vote and Turnout in the Bottom 300 Counties Having the Smallest Population Size
2016
2020
Percentage of the Vote
100
80
60
40
20
0
Democrat
Republican
Other
Turnout
Democrat
Republican
Other
Turnout
Population Size and Democratic Vote 2020
Percent Democratic
Under 30
30-49
50-61
62 or over
Population Size
228,671 or over
4,939 to 228,670
Under 4,939
Vote and Turnout in the Top 300 Counties Having the Largest Population Size
2016
2020
Percentage of the Vote
100
80
60
40
20
0
Democrat
Republican
Other
Turnout
Democrat
Republican
Other
Turnout

in 2020 a smaller shift continued. In the three hundred counties with the lowest populations, the share of the vote for Republicans increased from 74.14 in 2016 to 76.64 in 2020. However, increased turnout in these counties (+5 percent) and the absence of third-party candidates meant that the vote share of the Democrats also increased in these counties (20.54 percent to 21.68 percent), albeit by a smaller margin.

In contrast, the largest counties (those with populations over two hundred thousand) provided most of the support for the Democrats. From the Pacific Coast to the Northeastern megalopolis corridor and throughout the larger metropolitan counties in the center of the country, counties with more than two hundred thousand residents voted overwhelmingly for Biden and, despite the small number of these counties, they housed more than ninety million voters. From Los Angeles County, California (71 percent Biden) to Cook County, Illinois (74.4 percent), DeKalb County, Georgia (83.1 percent) to metropolitan areas such as Travis County, Texas (71.1 percent), in otherwise red states Democrats not only dominated these counties, but slightly increased their share of the vote over 2016. In 2016 the three hundred counties with the largest populations voted Democratic by a margin of 56 percent to 38 percent, and while both Biden and Trump benefited from the absence of strong third-party candidates in 2020, the Democratic share of the vote increased by 2.88 percent while the Republican vote share was up by only 0.78 percent.

While there is no map showing the relationship of population density to vote, it is clear from the metropolitan concentration of the Democratic vote that areas of high population density largely voted for Biden, while lower density suburban and rural areas tended to vote Republican. In the three hundred largest counties, turnout increased dramatically, and the Democratic share of the vote increased over 2016, to 60 percent.

POPULATION CHANGE

There is both a regional and a density component to the voting patterns of counties that gained and lost significant populations over the last decade (figure 6.2). For counties that experienced at least an 8 percent increase in population, most of which were either in Texas and the West, or in metropolitan areas in the South, Democrats won handily in the counties that were in blue states, California, Oregon, Washington, and Colorado, and in the Atlanta metro area counties in Georgia. In the interior coastal areas (central California, eastern Washington), Republicans fared a little better in fast-growing counties, reflecting the coastal/interior political split as well as in suburban and rural areas of Colorado and Georgia, reflecting the urban/suburban political divide. In states that voted for Trump, the same general pattern is evident; rapidly growing counties voted for Trump if they were suburban, for example the counties that surround Dallas (Tarrant County), or Austin (Travis County), or the rapidly growing suburban counties around Nashville, Tennessee. In rural areas such as southwestern Texas and the western Dakotas, the expansion of energy exploitation in the last two decades contributed to rapid population growth, but given the small initial base of the population these are still small rural counties and accordingly voted overwhelmingly for Trump. Overall, in partisan split three hundred counties with the fastest-growing populations, shifted in favor of the Democrats, going from 47/47 percent in 2016 to a slight Democratic advantage 51/47 in 2020.

Counties that lost population, particularly those in the Great Plains, Appalachia, and the Midwest, largely voted Republican, while counties with populations of color that lost population, in New Mexico, the western Texas border, the Mississippi delta, and southern Black Belt, were much more likely to support Biden. Sixty-one percent of the voters in the three hundred counties that experienced the greatest population decline voted Republican, a slight increase over the 59 percent that voted Republican in 2016, while the Democratic vote declined by 0.08 percent.

MIGRATION

Net Migration

Migration in and out of counties reflects both local and state-level population changes and bears an unsurprising similarity to the population change map (figure 6.2). The map of net migration (figure 6.3) does indicate that high in-migration counties were much more likely to vote Republican than Democratic, and more likely to vote Republican than high-growth counties, indicating that shifts from urban to suburban and rural counties within states contributed to high Republican percentages in suburban Texas, Georgia, and Florida, and exurban areas of the West coast. It is important to note, however, that many of the high in-migration, high-Republican-vote counties had small initial populations and contributed relatively little to the overall Republican vote total. Counties that experienced high rates of out-migration, like the population

FIGURE 6.2

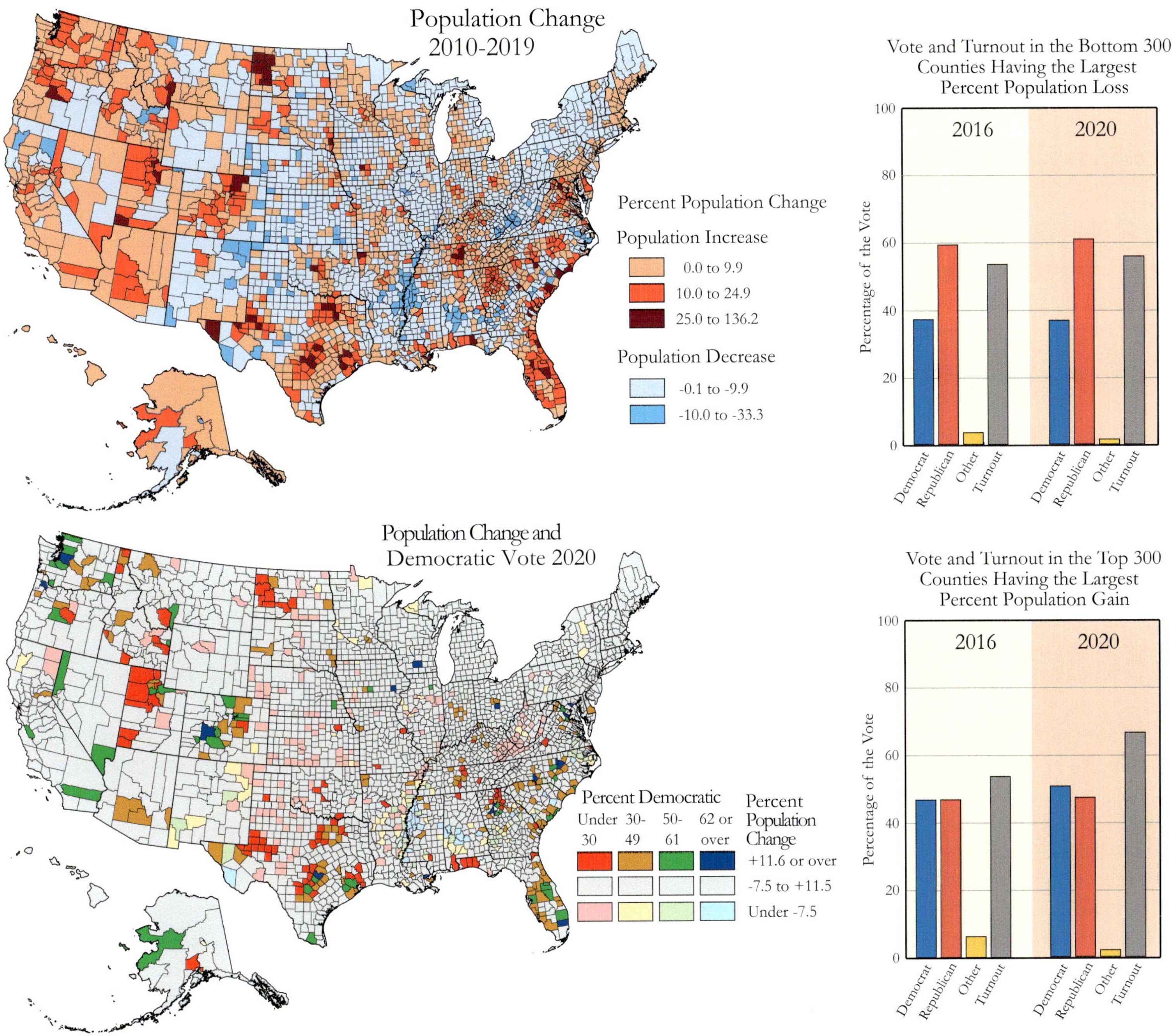

Population Change 2010-2019
Percent Population Change
Population Increase
0.0 to 9.9
10.0 to 24.9
25.0 to 136.2
Population Decrease
-0.1 to -9.9
-10.0 to -33.3
Vote and Turnout in the Bottom 300 Counties Having the Largest Percent Population Loss
2016
2020
Percentage of the Vote
0
20
40
60
80
100
Democrat
Republican
Other
Turnout
Population Change and Democratic Vote 2020
Percent Democratic
Under 30
30-49
50-61
62 or over
Percent Population Change
+11.6 or over
-7.5 to +11.5
Under -7.5
Vote and Turnout in the Top 300 Counties Having the Largest Percent Population Gain
2016
2020
Percentage of the Vote
0
20
40
60
80
100
Democrat
Republican
Other
Turnout

loss counties of figure 6.2, largely in Appalachia, the Great Plains, and the rural South including the Mississippi Delta, are more evenly split and the partisan leaning of these counties is largely racially defined, with rural counties that have high percentages of Black, Hispanic, and American Indian populations voting for Biden, while counties with predominantly white populations voted overwhelmingly Republican. Overall, the counties with the greatest gains (the three hundred counties that gained the most) increased their support for Democrats over 2016 by more than 4 percent (36.65 to 40.89), while the counties with the largest losses increased support for Republicans by 2.7 percent.

FOREIGN-BORN POPULATION

With immigration playing a major part in the rhetoric of the election cycle, the importance of the foreign population outweighs its numerical influence. While there are counties that have as much as 53 percent of their population born outside the United States, 80 percent of all US counties have less than 17 percent foreign-born population (figure 6.4). Overall, the three hundred counties with the highest percentage of foreign-born residents increased their support of the Democratic candidate by 1.97 percent from 2016 to 2020, with 62.21 percent of the votes in these counties going to Biden. Geographically, this population is concentrated along the West Coast and southern border, in South Florida and the northeastern megalopolis, and in West Texas, with some isolated urban counties and the occasional rural Great Plains county that reflects agricultural processing industries. In counties with a high foreign-born population, the partisan leanings reflect both the political leaning of the state as a whole and the urban/rural partisan divide. Counties with high foreign-born populations in Republican states, or in rural areas, tend to favor the Republicans, whereas counties with a high foreign-born population in Democratic states (the Pacific Coast and the Northeast) or in urban areas (Atlanta, Dallas, Houston, Miami) trend Democratic. Again, these patterns reflect the rural/white/interior Republican versus urban/diverse/coastal Democratic partisan split that has emerged over the last two decades.

NON-MOVERS AND PERSONS BORN IN THEIR STATE OF RESIDENCE

Figures 6.5 and 6.6 explore the tendency for Americans to remain in place over a single year or over a lifetime. Overall, in extreme counties (lowest and highest three hundred) both non-movers and persons born in their state of residence demonstrate unsurprisingly similar patterns. Counties with high percentages of non-movers and high percentages of people born in their current state of residence both show strong support for Republicans; this support increased dramatically between 2012 and 2016 and more modestly between 2016 and 2020. Conversely, the three hundred counties with the lowest percentages of non-movers and lowest percentages of people born in their state of residence showed a marked preference for Joe Biden (56 percent Democratic vote), with the share of the Democratic vote increasing by 4.2 percent in the former, and by 2.6 percent in the latter.

The maps of people who have remained in their current residence for more than a year show clusters of non-movers in almost exclusively rural counties in the Great Plains, the rural South, and Appalachia, as well as counties dominated by Native American reservations in the Southwest. Counties with sedentary populations that voted Democratic are overwhelmingly non-white (American Indian reservations, the Mississippi Delta, and the Black Belt from Mississippi to South Carolina); those who voted Republican are much more likely to be white (Appalachia and the Great Plains outside of the South Dakota reservations). More long-term stability—voters who reside in the state in which they were born—is largely a characteristic of rural areas in central and eastern states, with clusters of long-term residents in the central south, Appalachia, and the rural Midwest as well as the northern Great Plains. As with non-movers, the voting preferences of long-term residents trends white/rural/Republican and non-white/rural/Democratic. Counties that have low rates of in-state residents are more varied, with environmental amenity areas such as rural Colorado and rural New England as well as urban cores such as the Pacific Coast cities, Miami, Atlanta, and greater Washington, DC, trending Democratic while counties in interior Western states with large out-of-state populations in rural areas are much more likely to vote Republican.

AGE

National exit polls suggest that age continued to be a reliable indicator of voting preferences in the 2020 election. The tendency for older populations to lean Republican is a long-standing political trend and in 2020 the population aged sixty-five

FIGURE 6.3

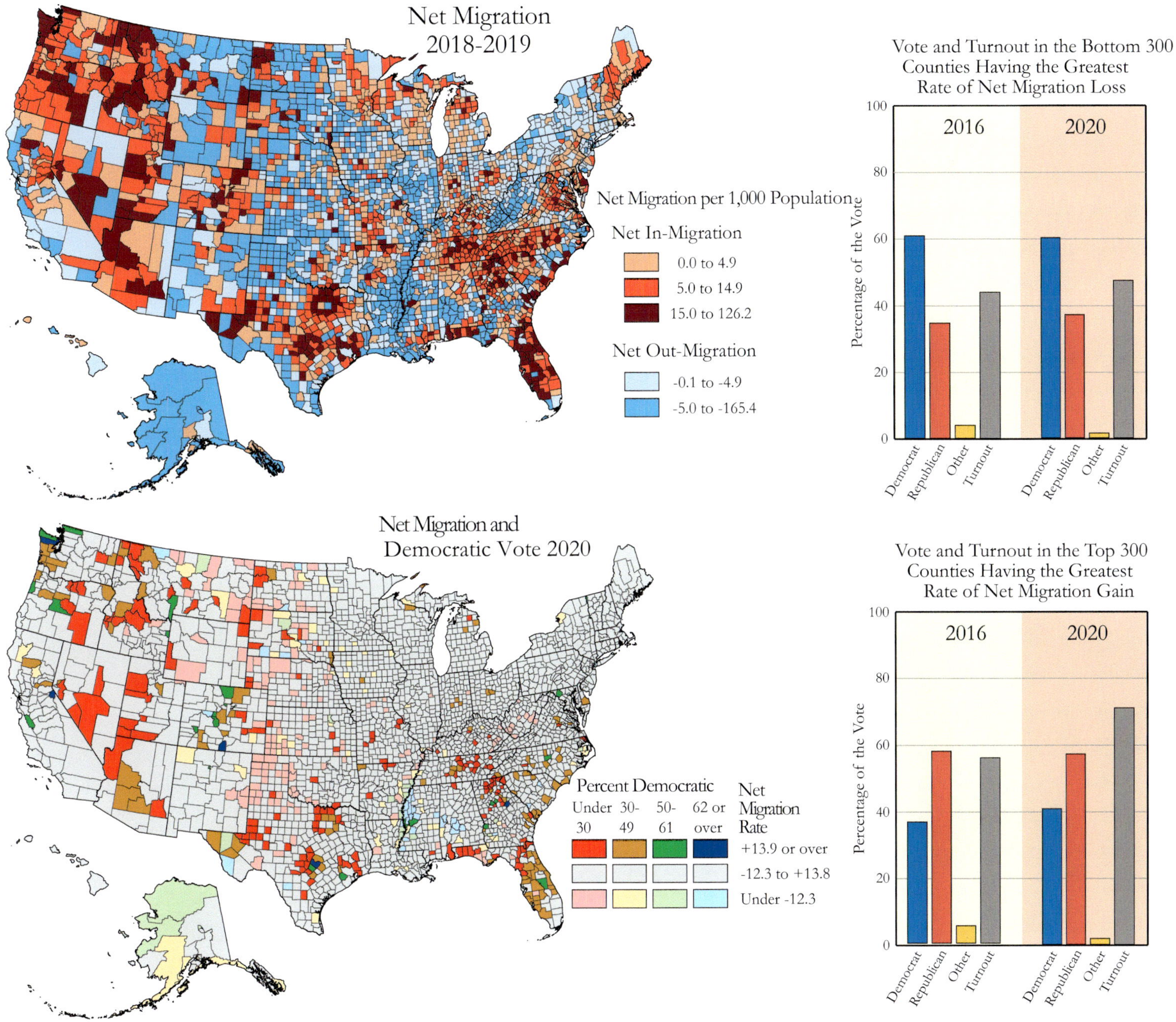
Net Migration
2018-2019
Net Migration per 1,000 Population
Net In-Migration
0.0 to 4.9
5.0 to 14.9
15.0 to 126.2
Net Out-Migration
-0.1 to -4.9
-5.0 to -165.4
Vote and Turnout in the Bottom 300 Counties Having the Greatest Rate of Net Migration Loss
2016
2020
Percentage of the Vote
100
80
60
40
20
0
Democrat
Republican
Other
Turnout
Net Migration and
Democratic Vote 2020
Percent Democratic
Under 30
30-49
50-61
62 or over
Net Migration Rate
+13.9 or over
-12.3 to +13.8
Under -12.3
Vote and Turnout in the Top 300 Counties Having the Greatest Rate of Net Migration Gain
2016
2020
Percentage of the Vote
100
80
60
40
20
0
Democrat
Republican
Other
Turnout

FIGURE 6.4

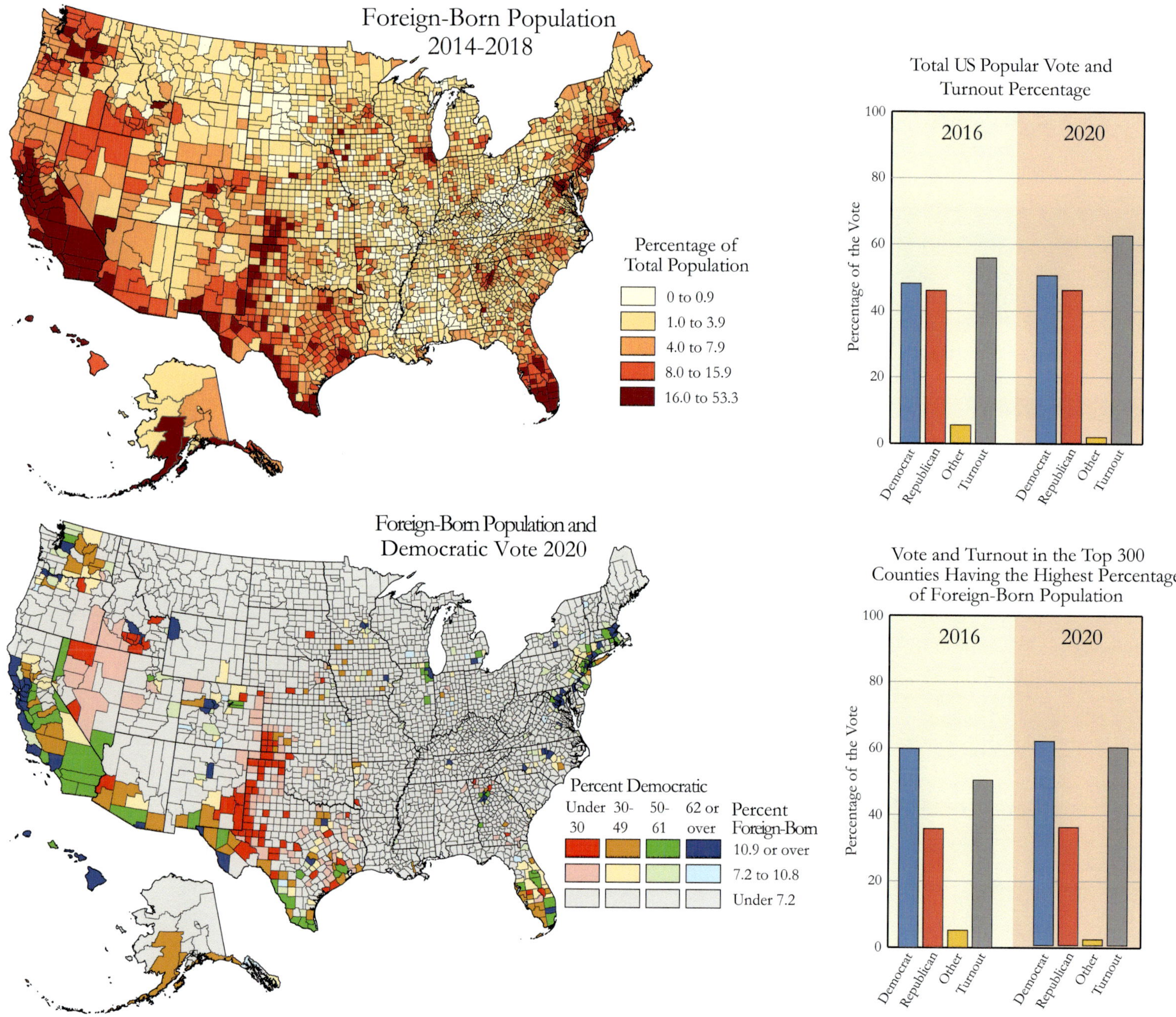
Foreign-Born Population 2014-2018
Percentage of Total Population
0 to 0.9
1.0 to 3.9
4.0 to 7.9
8.0 to 15.9
16.0 to 53.3
Total US Popular Vote and Turnout Percentage
Percentage of the Vote
0
20
40
60
80
100
2016
2020
Democrat
Republican
Other
Turnout
Foreign-Born Population and Democratic Vote 2020
Percent Democratic
Under 30
30-49
50-61
62 or over
Percent Foreign-Born
10.9 or over
7.2 to 10.8
Under 7.2
Vote and Turnout in the Top 300 Counties Having the Highest Percentage of Foreign-Born Population
Percentage of the Vote
0
20
40
60
80
100
2016
2020
Democrat
Republican
Other
Turnout

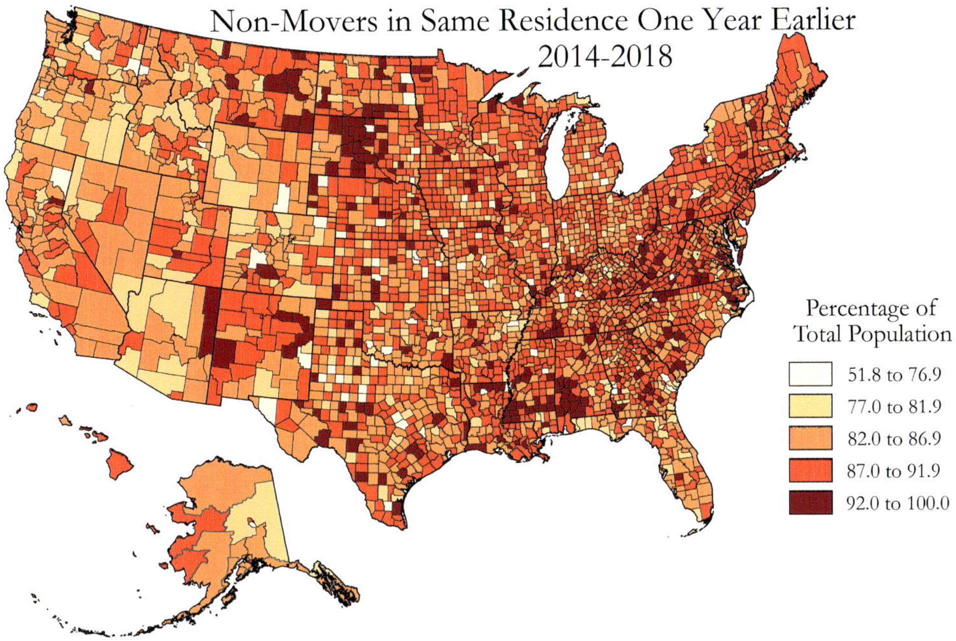
Non-Movers in Same Residence One Year Earlier
2014-2018
Percentage of
Total Population
51.8 to 76.9
77.0 to 81.9
82.0 to 86.9
87.0 to 91.9
92.0 to 100.0

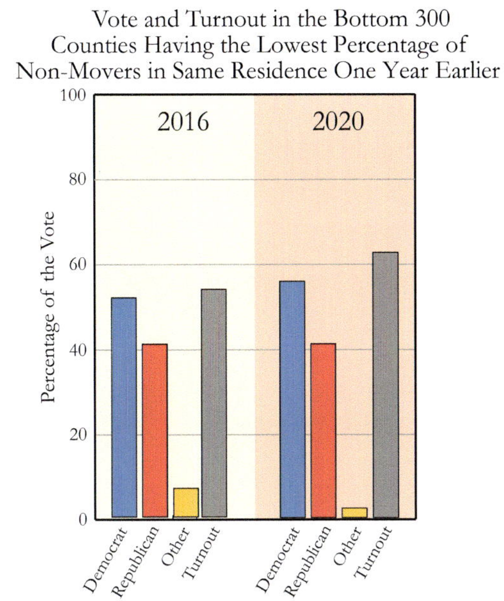
Vote and Turnout in the Bottom 300
Counties Having the Lowest Percentage of
Non-Movers in Same Residence One Year Earlier
2016
2020
Percentage of the Vote
0
20
40
60
80
100
Democrat
Republican
Other
Turnout
Democrat
Republican
Other
Turnout

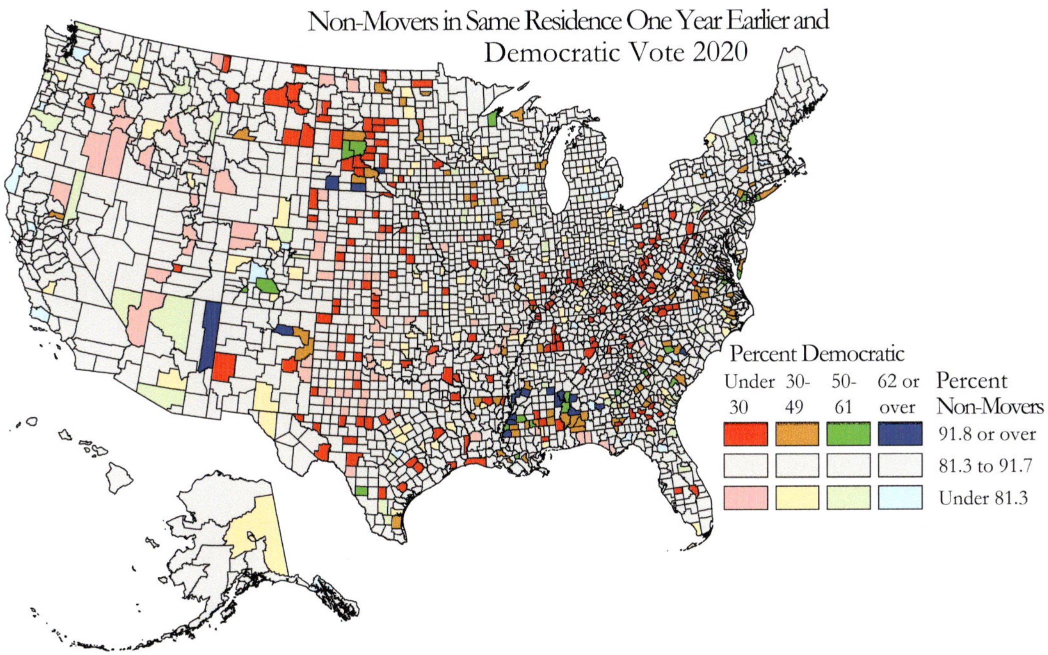
Non-Movers in Same Residence One Year Earlier and
Democratic Vote 2020
Percent Democratic
Under 30
30-49
50-61
62 or over
Percent
Non-Movers
91.8 or over
81.3 to 91.7
Under 81.3

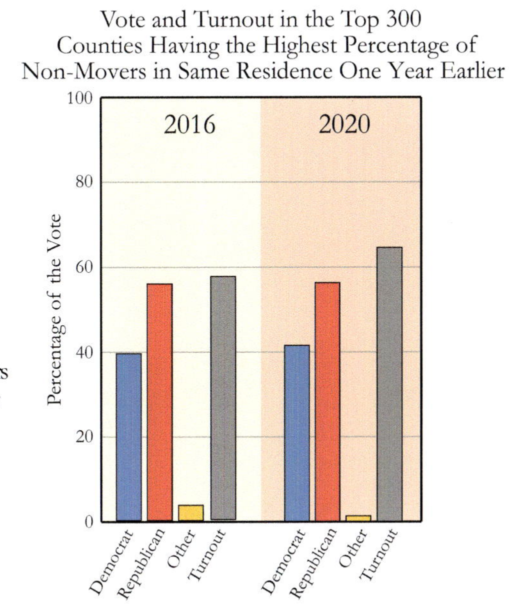
Vote and Turnout in the Top 300
Counties Having the Highest Percentage of
Non-Movers in Same Residence One Year Earlier
2016
2020
Percentage of the Vote
0
20
40
60
80
100
Democrat
Republican
Other
Turnout
Democrat
Republican
Other
Turnout

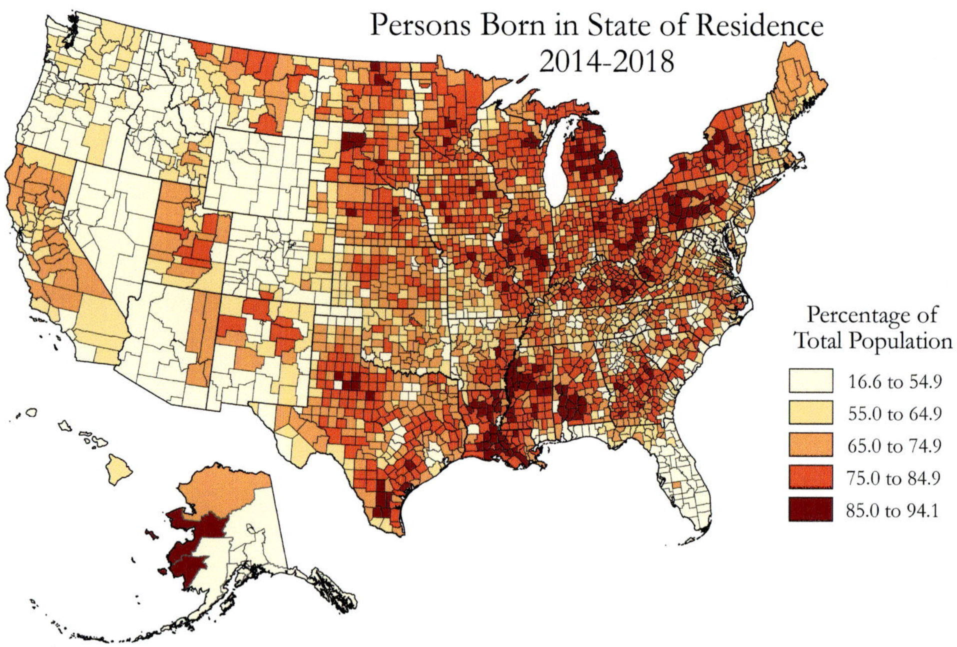
Persons Born in State of Residence 2014-2018
Percentage of Total Population
16.6 to 54.9
55.0 to 64.9
65.0 to 74.9
75.0 to 84.9
85.0 to 94.1

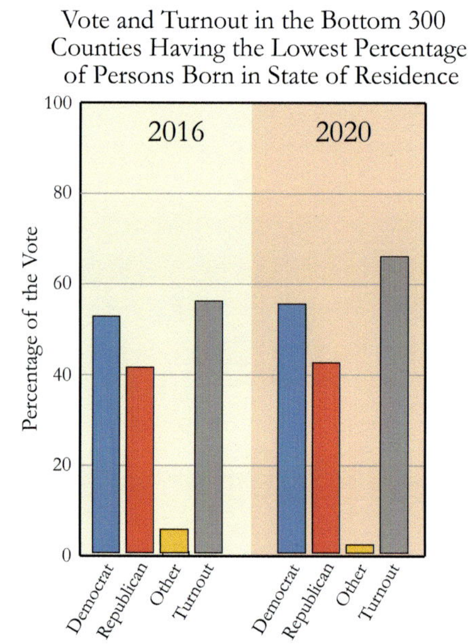
Vote and Turnout in the Bottom 300 Counties Having the Lowest Percentage of Persons Born in State of Residence
2016
2020
Percentage of the Vote
0
20
40
60
80
100
Democrat
Republican
Other
Turnout
Democrat
Republican
Other
Turnout

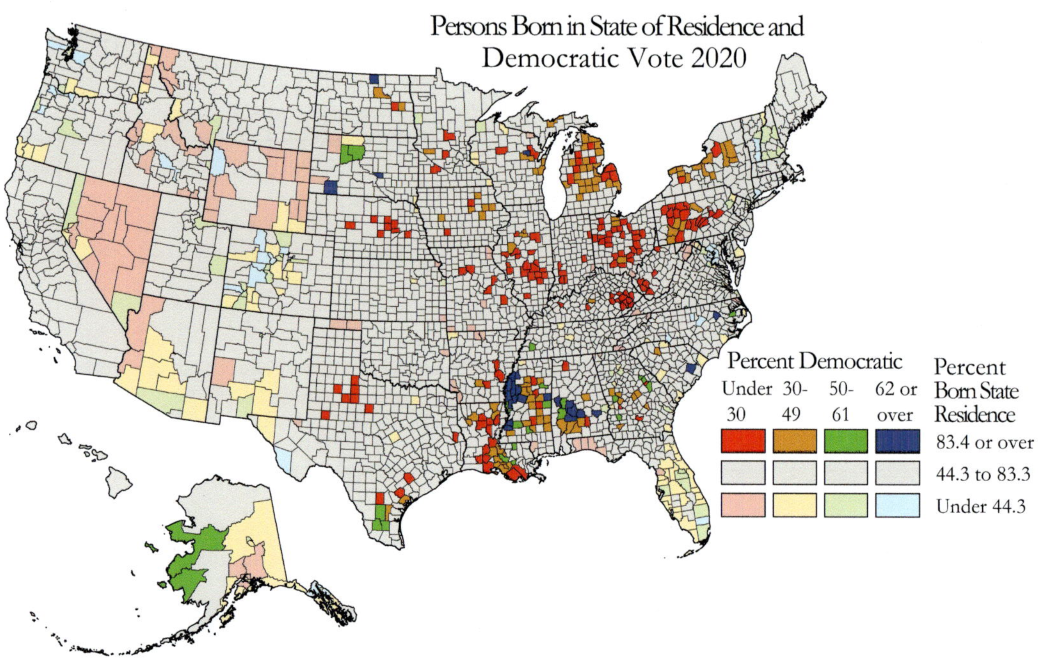
Persons Born in State of Residence and Democratic Vote 2020
Percent Democratic
Under 30
30-49
50-61
62 or over
Percent Born State Residence
83.4 or over
44.3 to 83.3
Under 44.3

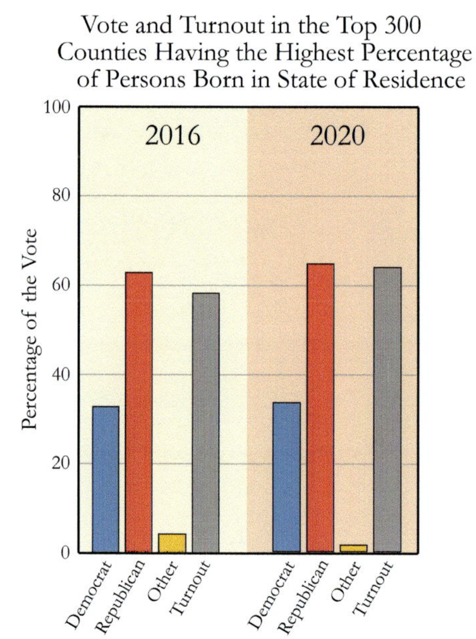
Vote and Turnout in the Top 300 Counties Having the Highest Percentage of Persons Born in State of Residence
2016
2020
Percentage of the Vote
0
20
40
60
80
100
Democrat
Republican
Other
Turnout
Democrat
Republican
Other
Turnout

and over favored Trump 52/47 while the youngest group of voters, those aged between eighteen and twenty-nine, favored Biden by 60 to 30 percent.

Population Aged Sixty-Five and Older

Across the United States, the three hundred counties that had the smallest percentage of sixty-five and over in the population were more likely to vote Republican than the general population by more than 25 percent, a figure that increased between 2012 and 2016 but narrowed slightly in the 2020 election as the Republican share of the vote remained static, but the Democratic share increased from 34.57 percent to 37.20 percent. The partisan gap for older voters was at its largest in 2012 when Mitt Romney won the older vote by 12 points; since then the margin has narrowed slightly, perhaps in response to the nontraditional nature of Trump's platform and political style. The map of US population aged sixty-five and over (figure 6.7) indicates that counties with a large elderly population are in predominantly rural regions of the Great Plains, northern Rockies, and Appalachians, with the addition of amenity areas in the South and Southwest that have attracted large numbers of out-of-state retirees (New Mexico, Arizona, and Florida). Those same counties also demonstrated overwhelming support for Trump in the 2020 election. Of the 740 counties with more than 21 percent of their population aged over sixty-five only sixty (8.1 percent) voted for Biden and those counties were scattered through rural New England, New Mexico, and parts of the West Coast.

Population Aged Twenty-Five to Forty-Four

Younger populations—specifically those aged twenty-five to forty-four—are historically much more likely to support Democratic candidates, and in 2020 the wide partisan gap of the 2016 election was sustained. In 2016 Democrats outperformed Republicans by 60 percent to 33 percent among this age group and increased their advantage by three percent by 2020, while the Republican vote also increased (by 1.5 percent) between 2016 and 2020, as third-party voters returned to the two main parties. Geographically, counties with larger percentages of younger populations are widely scattered throughout the United States, with more of those counties evident west of the Mississippi (figure 6.8). Urban counties, counties with significant minority populations, and counties with significant environmental amenity all display a high preference for Biden, while more rural counties and those that specialize in extractive and agricultural industries were more likely to vote Republican. Campbell and Teton counties in Wyoming are excellent examples of these two extremes. Both rural, low-density counties, Campbell specializes in coal and oil extraction, is 88 percent non-Hispanic white, and has voted Republican in every presidential election since 1936. In contrast, Teton County encompasses Jackson Hole, Grand Tetons National Park, and part of Yellowstone National Park and is economically tied to tourism and services. The population is less than 70 percent non-Hispanic white and has a median household income of $54,000 (although it should be noted that Teton County has one of the highest mean household incomes in the United States at $252,000 because of the large number of ultra-rich second-home owners). Teton County is the most Democratic-leaning county in the state and has voted for every Democratic president since 1992, with the exception of George Bush in 2000.

Population Aged Under Eighteen

An examination of counties with a large percentage of their populations under the age of eighteen is interesting because it illuminates the voting behavior of families with dependent children (figure 6.9). Overall, voters in counties with high percentages of minors tend to favor Republican candidates; in 2016 these counties supported Trump by 49 to 43 percent. However, by 2020 Democrats narrowed the gap by 3 percent (Republican 50.62 percent and Democratic 47.3 percent). At the county level, only 14 percent of counties with high (over 24 percent) percentages of minor children favored Biden, but many of these were large suburban counties whereas the counties with minor children that favored Trump were, once again, largely small, low-density, rural counties. The exception to this is the region centered on Utah, where the high percentage of adherents to the Church of Latter-Day Saints results in both high numbers of family units with minor children and a strong preference for Republican candidates.

RACE AND ETHNICITY

The 2020 election again revealed the fundamental importance of race and ethnicity in US voting patterns, with the shares of Black, Asian, Hispanic, and Native American voters corresponding to the vote for Biden, and the share of non-Hispanic white males even more highly predictive of the

FIGURE 6.7

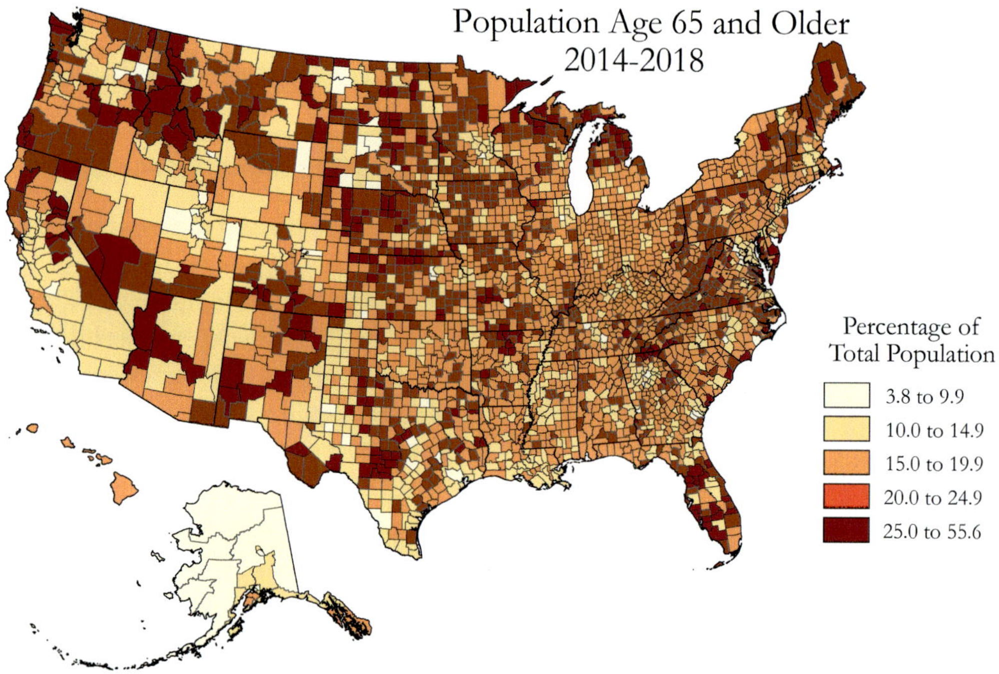
Population Age 65 and Older
2014-2018
Percentage of
Total Population
3.8 to 9.9
10.0 to 14.9
15.0 to 19.9
20.0 to 24.9
25.0 to 55.6

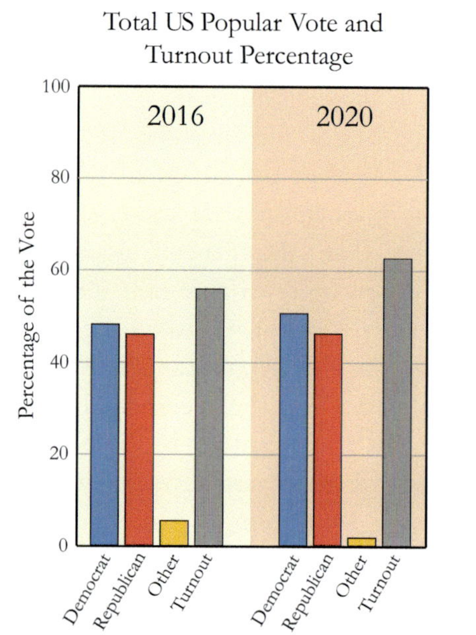
Total US Popular Vote and
Turnout Percentage
2016
2020
Percentage of the Vote
100
80
60
40
20
0
Democrat
Republican
Other
Turnout
Democrat
Republican
Other
Turnout

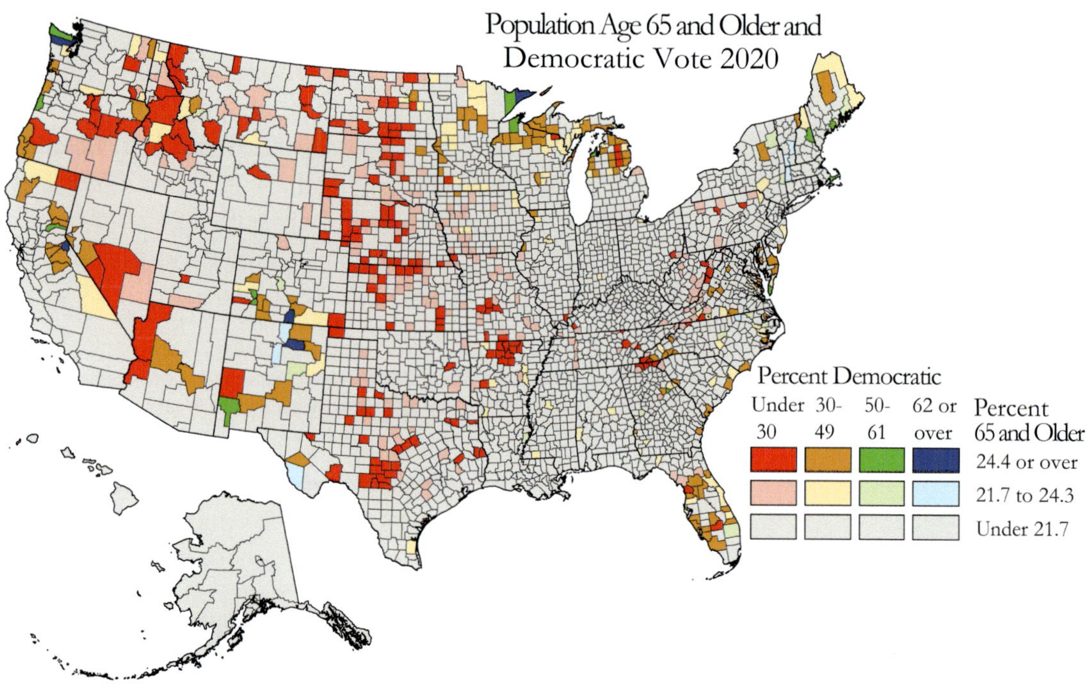
Population Age 65 and Older and
Democratic Vote 2020
Percent Democratic
Under 30
30-49
50-61
62 or over
Percent
65 and Older
24.4 or over
21.7 to 24.3
Under 21.7

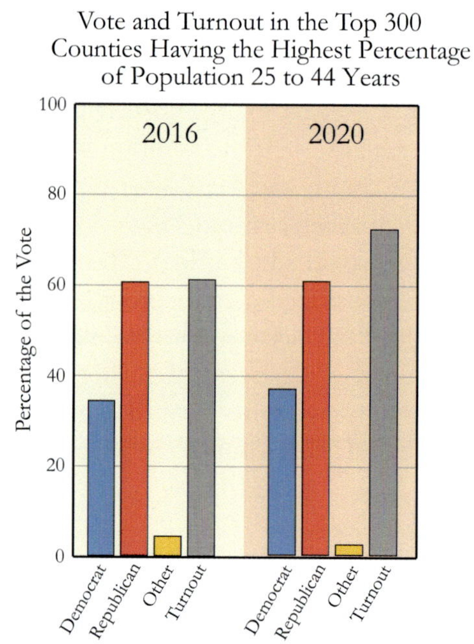
Vote and Turnout in the Top 300
Counties Having the Highest Percentage
of Population 25 to 44 Years
2016
2020
Percentage of the Vote
100
80
60
40
20
0
Democrat
Republican
Other
Turnout
Democrat
Republican
Other
Turnout

FIGURE 6.8

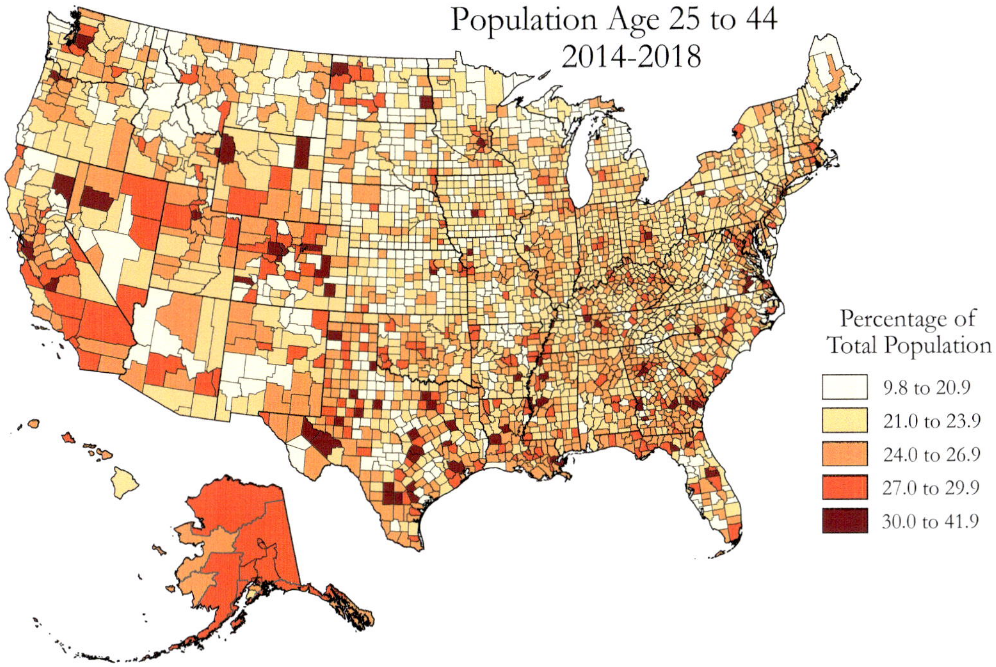

Population Age 25 to 44
2014-2018
Percentage of Total Population
9.8 to 20.9
21.0 to 23.9
24.0 to 26.9
27.0 to 29.9
30.0 to 41.9

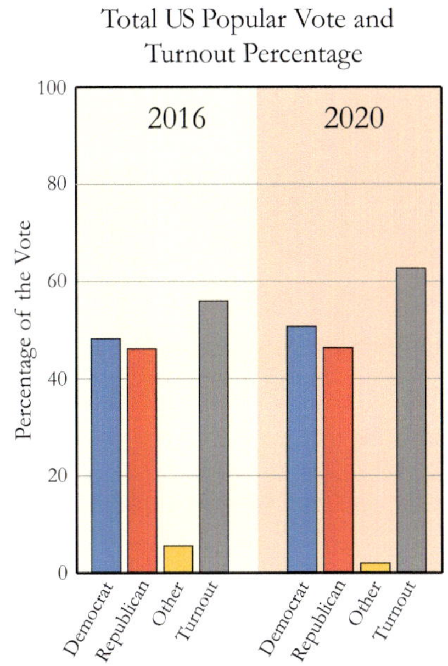

Total US Popular Vote and Turnout Percentage
2016
2020
Percentage of the Vote
0
20
40
60
80
100
Democrat
Republican
Other
Turnout

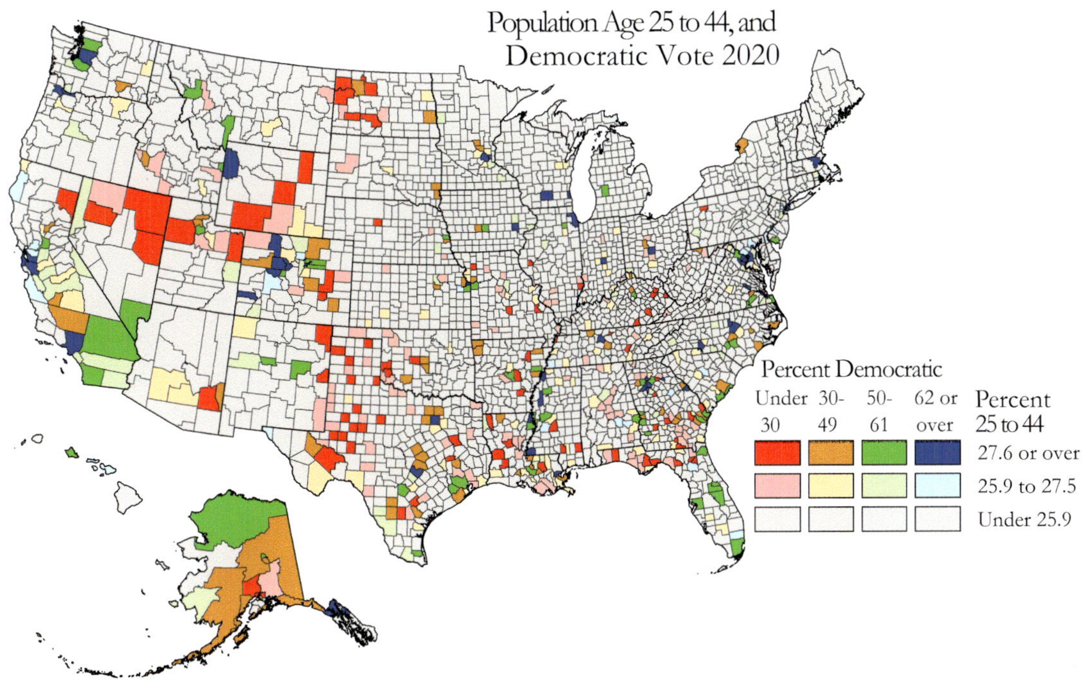

Population Age 25 to 44, and Democratic Vote 2020
Percent Democratic
Under 30
30-49
50-61
62 or over
Percent 25 to 44
27.6 or over
25.9 to 27.5
Under 25.9

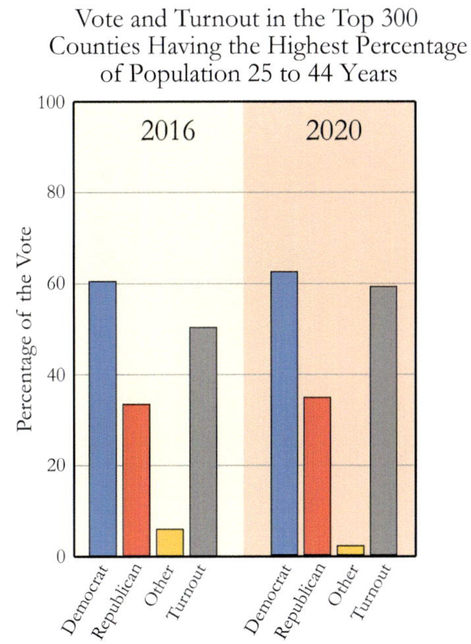

Vote and Turnout in the Top 300 Counties Having the Highest Percentage of Population 25 to 44 Years
2016
2020
Percentage of the Vote
0
20
40
60
80
100
Democrat
Republican
Other
Turnout

FIGURE 6.9

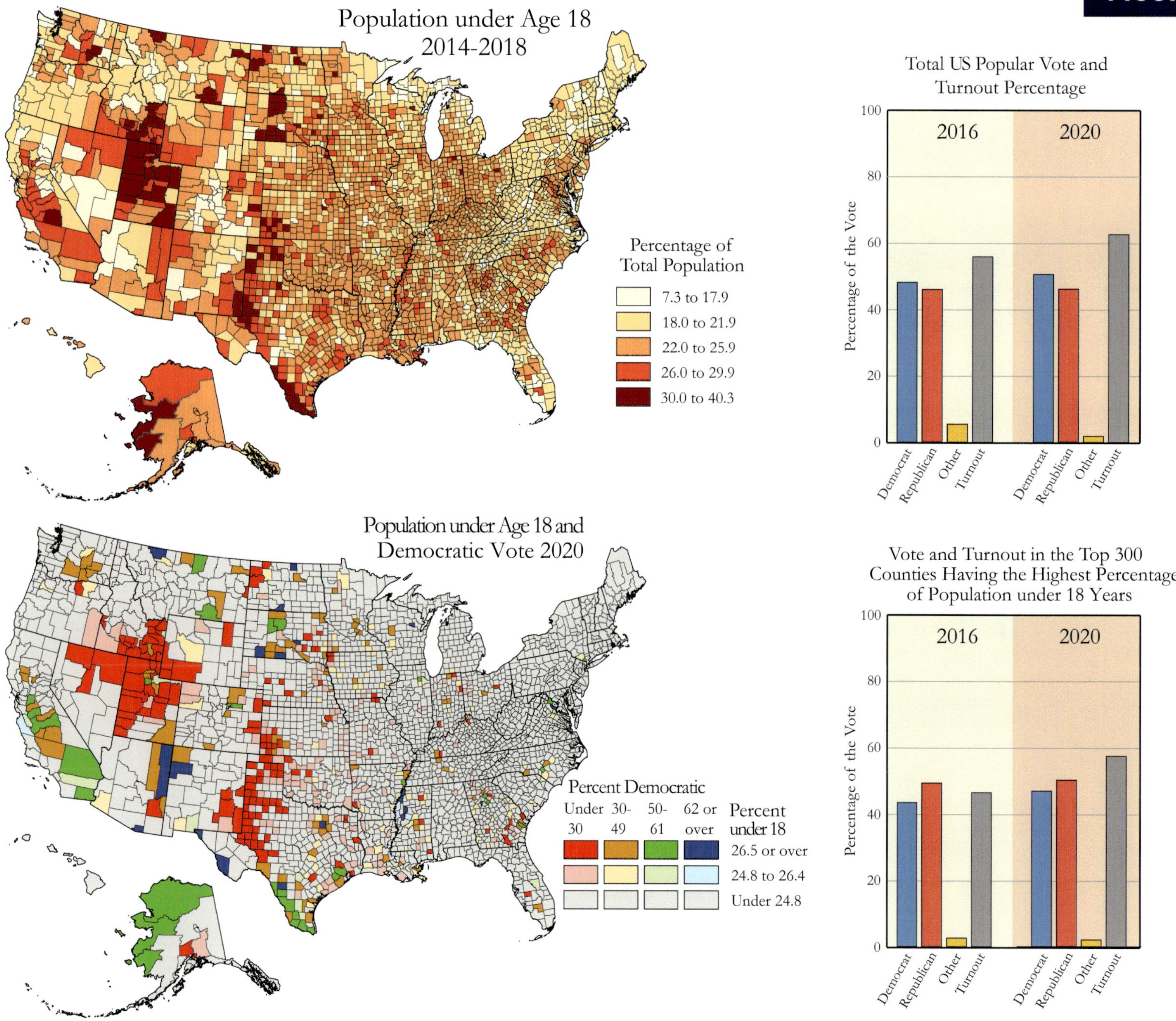

Population under Age 18
2014-2018
Percentage of
Total Population
7.3 to 17.9
18.0 to 21.9
22.0 to 25.9
26.0 to 29.9
30.0 to 40.3
Population under Age 18 and
Democratic Vote 2020
Percent Democratic
Under 30
30-49
50-61
62 or over
Percent under 18
26.5 or over
24.8 to 26.4
Under 24.8
Total US Popular Vote and
Turnout Percentage
Vote and Turnout in the Top 300
Counties Having the Highest Percentage
of Population under 18 Years
2016
2020
Percentage of the Vote
0
20
40
60
80
100
Democrat
Republican
Other
Turnout

vote for Trump—indeed, this was the single most important variable distinguishing the likelihood to vote Democratic or Republican. Both Black and Hispanic voting patterns in this election were sufficiently influential and complex that these groups will be dealt with in their own sections of the chapter.

Asian American Voters

In the United States, the population with Asian ancestry is concentrated in the coastal counties of the Pacific, and the Boston-Washington urban corridor (figure 6.10). There are a few outlying areas of Asian population in metropolitan counties in the South, Midwest, and Texas, as well as in a smattering of university communities (Raleigh-Durham, North Carolina; Gainesville, Florida; Ann Arbor and Lansing, Michigan; and Madison, Wisconsin). Counties with high (over 5 percent) populations of Asians overwhelmingly voted for Biden (66 percent to 32 percent); while there are counties in the second tier of Asian population (3.0–4.9 percent) where the votes trended Republican, these were largely in central California and suburban Texas. It should be noted that there are only three counties in the United States where the Asian population exceeds 32 percent (Maui and Kauai, Hawaii and Santa Clara, California) and only 168 where that population exceeds 5 percent, rendering any relationship between Asian population and voting trends insignificant in almost all counties.

American Indian Voters

As with the Asian population, the small overall numbers of American Indians in each county means that there are relatively few counties where the influence of American Indian voters can be separated from the more widespread rural/Republican pattern (figure 6.11). While counties that are in the top fifty of American Indian populations are marginally more likely to vote Democratic than the national average, the overall low percentages of American Indians in even those counties dilutes the effect of those Democratic votes in otherwise rural, Republican counties. An examination of the counties with more than 32 percent American Indian population shows that the population of these counties largely voted for Biden; this is particularly evident in Alaska, where all but one of the borough/census districts with more than a 32 percent American Indian population voted for Biden, despite the otherwise Republican bent of the state. Of the twenty-six counties in the United States where the American Indian population exceeds 50 percent, all but four voted Democratic, eleven of them by more than 62 percent, even in states that otherwise trended Republican (Montana and the Dakotas). In areas where the American Indian population is present but makes up less than 30 percent of the population, the rural nature of the American Indian population concentrations leads to a strong Republican vote. This is especially noticeable in eastern Oklahoma and northern Nevada, as well as in Montana and the Dakotas.

Non-Hispanic White Male Population Eighteen and Older

The relationship between adult non-Hispanic white males and voting preference is one of the clearest demographic trends in this section (figure 6.12). Counties with the lowest levels of non-Hispanic white males consistently show a 30 percent or greater preference for Democratic candidates, while those with the highest levels of non-Hispanic white males have seen an increasing tendency to prefer Republican candidates, by widening margins over the last three elections. In 2016 the top three hundred counties in this group supported Donald Trump by 62.73 percent to 31.59 percent, a margin that declined slightly in 2020. The percentage of votes for both parties increased in 2020, reflecting a shift away from third-party candidates, which favored the Democrats. However, in counties with very low percentages of non-Hispanic white males, the reverse happened, with third-party votes going primarily to the Republicans, who increased their vote share by 2.85 percent against an almost static Democratic vote (65.32 increasing to 65.38 in 2020).

From the northern Rockies through the Great Plains to Appalachia, a higher-than-average percentage of white males is strongly related to very low (below 30 percent) votes for Biden, and, with the exception of parts of Texas, low percentages of white males are related to much higher (50 percent or above) Democratic voting. Across the United States, of the 830 counties with more than 36 percent white male population, only 53 (6.4 percent) voted for Biden. Those Democratic counties are concentrated in New England, Colorado, the Upper Midwest, and the Pacific Northwest, with a few outliers in the Great Plains and Appalachia, which represent college towns (Story County, Iowa; Athens County, Ohio; Portage County, Wisconsin; or Center County, Pennsylvania). Conversely, of the 1523 counties with fewer than 26 percent white males in the population, the partisan split is more even. In the South and Texas, many of those counties voted Republican, more than half of

FIGURE 6.10

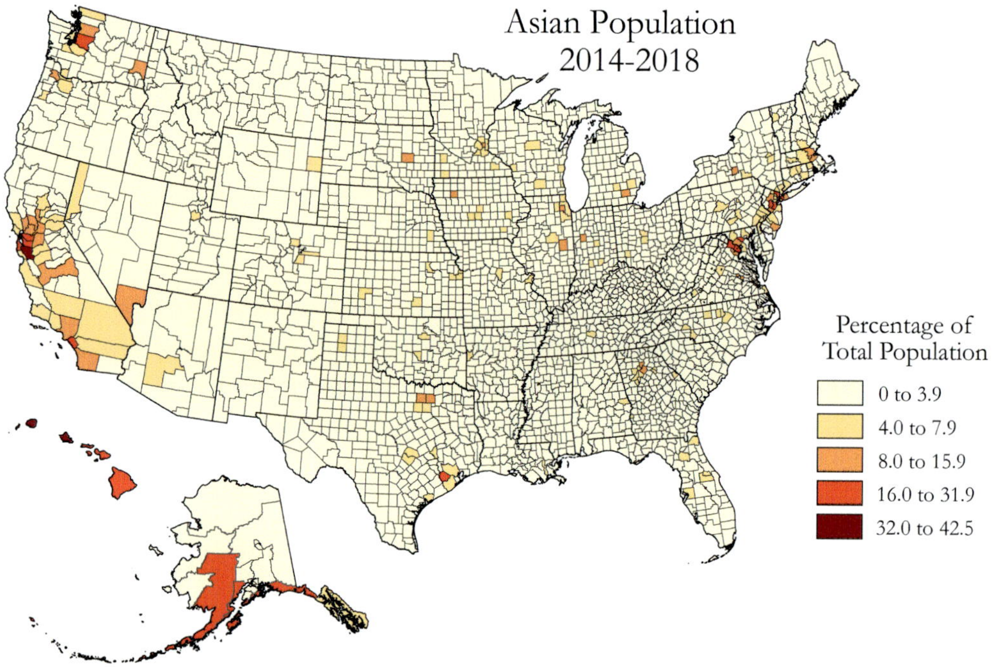

Asian Population
2014-2018
Percentage of
Total Population
0 to 3.9
4.0 to 7.9
8.0 to 15.9
16.0 to 31.9
32.0 to 42.5

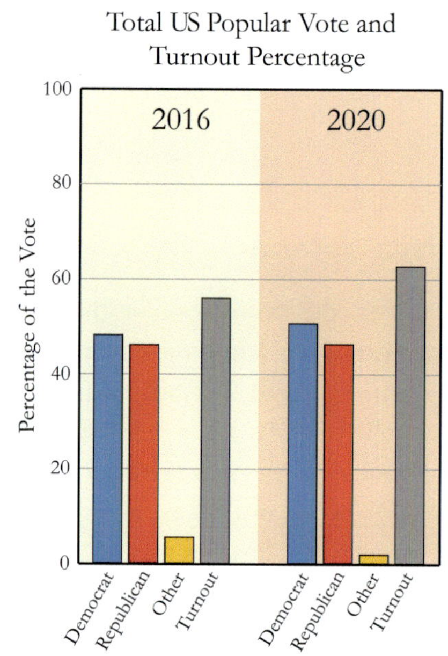

Total US Popular Vote and
Turnout Percentage
2016
2020
Percentage of the Vote
0
20
40
60
80
100
Democrat
Republican
Other
Turnout
Democrat
Republican
Other
Turnout

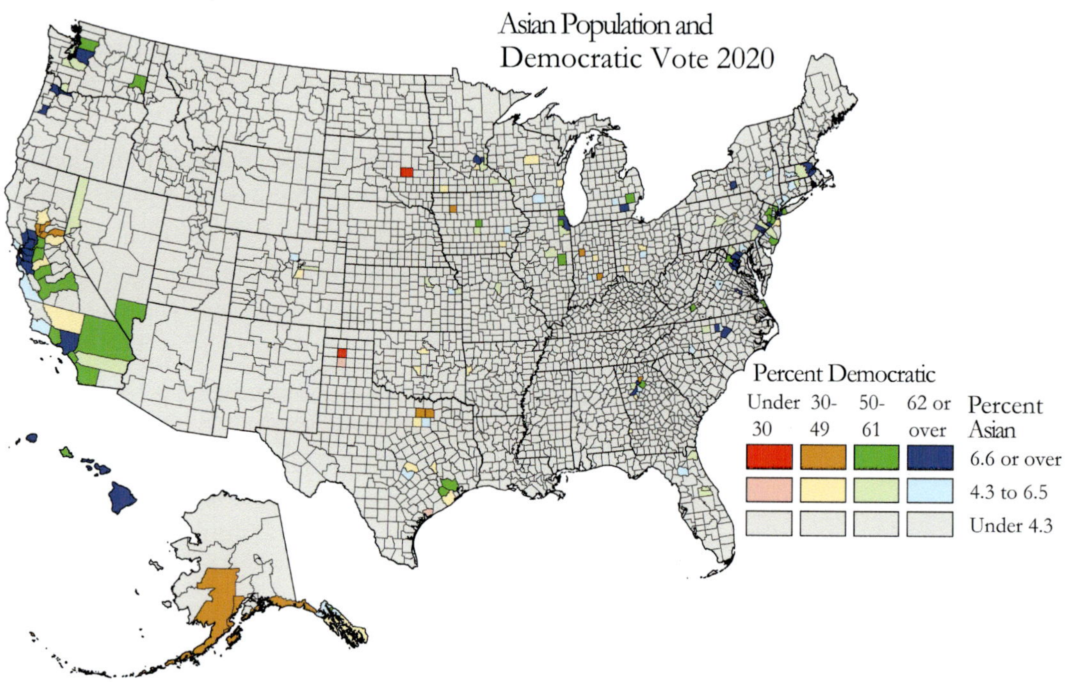

Asian Population and
Democratic Vote 2020
Percent Democratic
Under 30
30-49
50-61
62 or over
Percent Asian
6.6 or over
4.3 to 6.5
Under 4.3

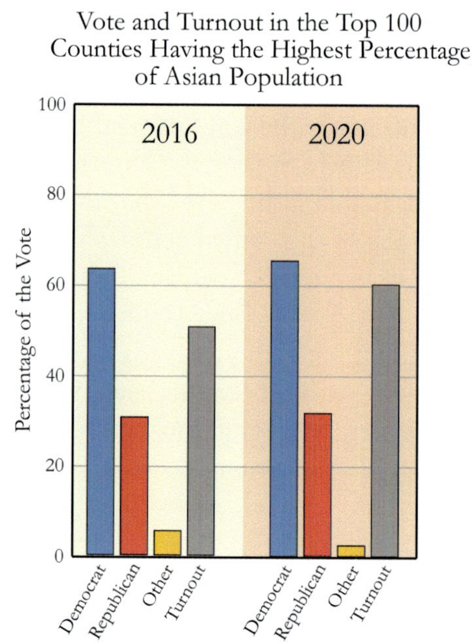

Vote and Turnout in the Top 100
Counties Having the Highest Percentage
of Asian Population
2016
2020
Percentage of the Vote
0
20
40
60
80
100
Democrat
Republican
Other
Turnout
Democrat
Republican
Other
Turnout

FIGURE 6.11

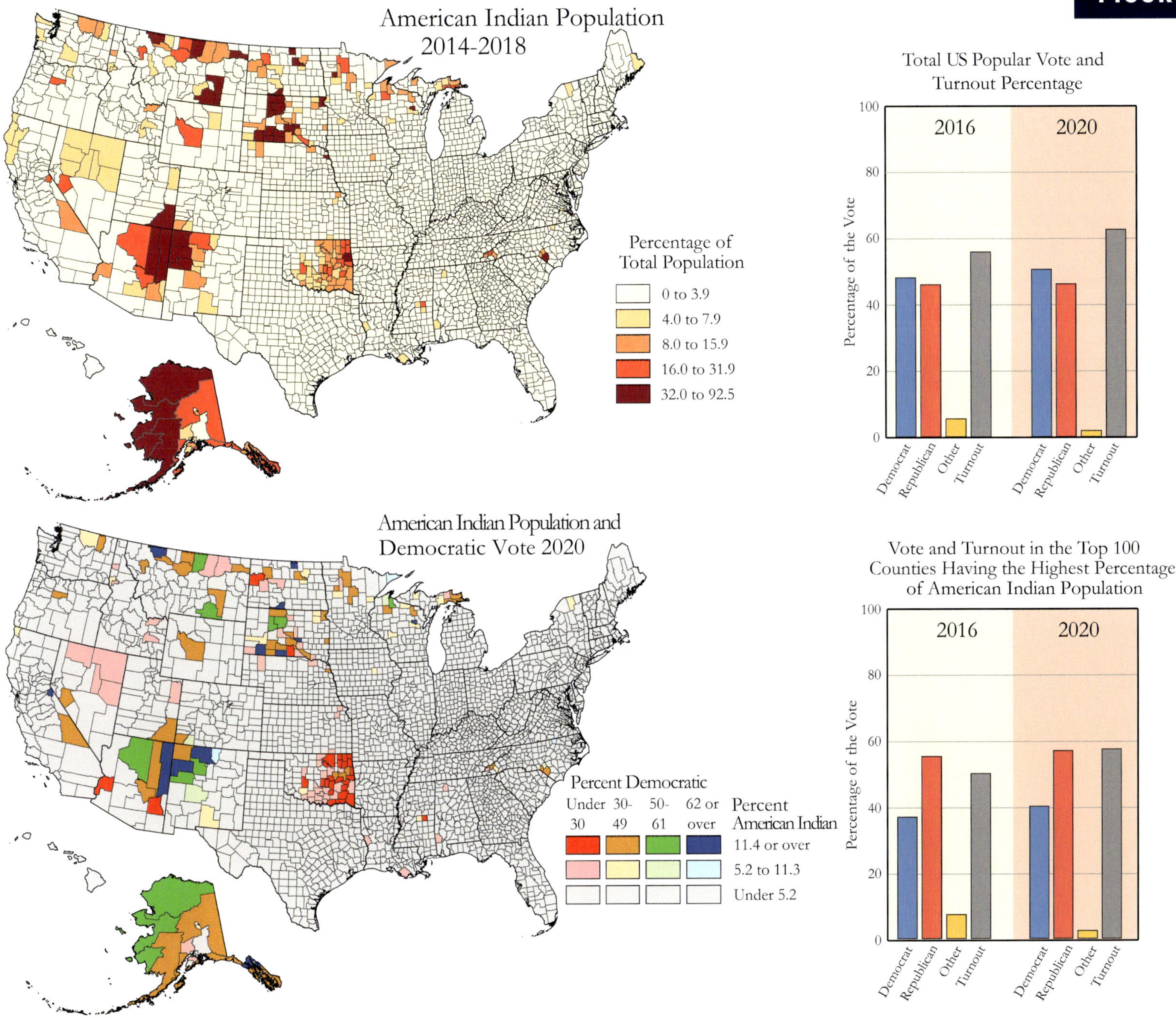
American Indian Population 2014-2018
Percentage of Total Population
0 to 3.9
4.0 to 7.9
8.0 to 15.9
16.0 to 31.9
32.0 to 92.5
Total US Popular Vote and Turnout Percentage
2016
2020
Percentage of the Vote
0
20
40
60
80
100
Democrat
Republican
Other
Turnout
American Indian Population and Democratic Vote 2020
Percent Democratic
Under 30
30-49
50-61
62 or over
Percent American Indian
11.4 or over
5.2 to 11.3
Under 5.2
Vote and Turnout in the Top 100 Counties Having the Highest Percentage of American Indian Population
2016
2020
Percentage of the Vote
0
20
40
60
80
100
Democrat
Republican
Other
Turnout

FIGURE 6.12

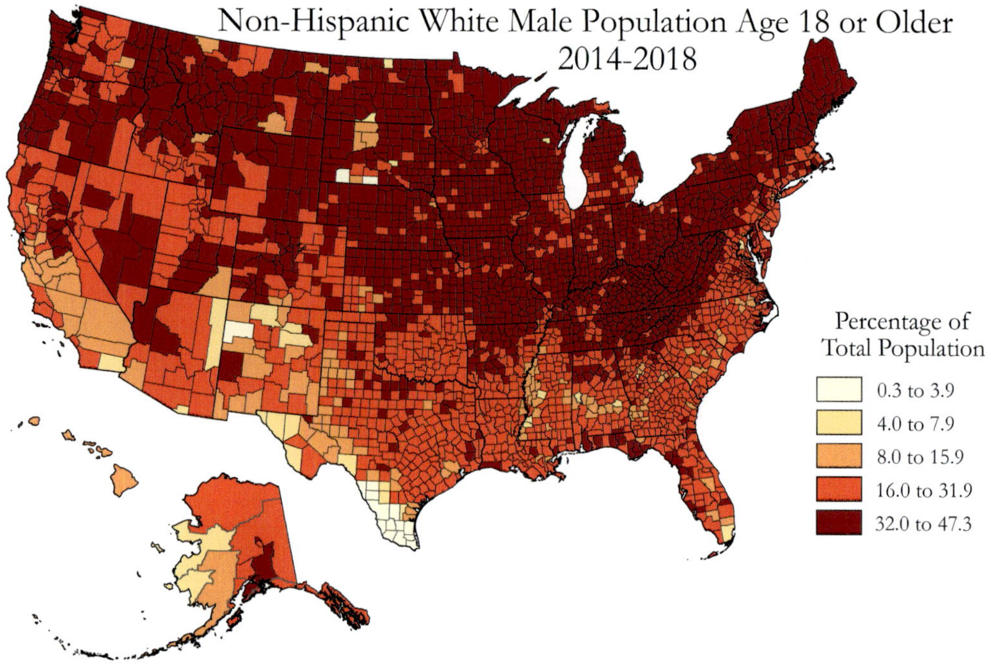
Non-Hispanic White Male Population Age 18 or Older
2014-2018
Percentage of
Total Population
0.3 to 3.9
4.0 to 7.9
8.0 to 15.9
16.0 to 31.9
32.0 to 47.3

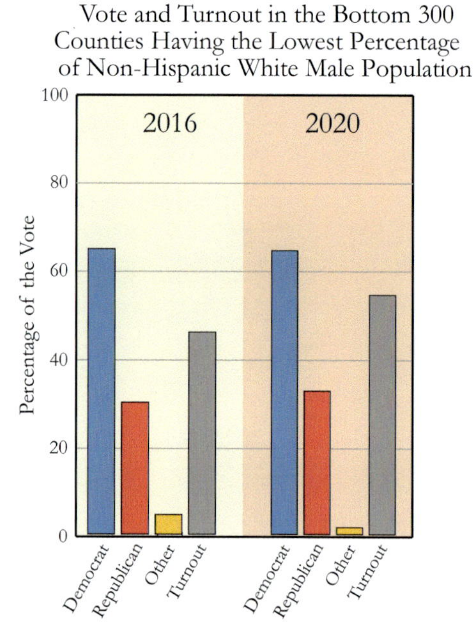
Vote and Turnout in the Bottom 300
Counties Having the Lowest Percentage
of Non-Hispanic White Male Population
2016
2020
Percentage of the Vote
0
20
40
60
80
100
Democrat
Republican
Other
Turnout
Democrat
Republican
Other
Turnout

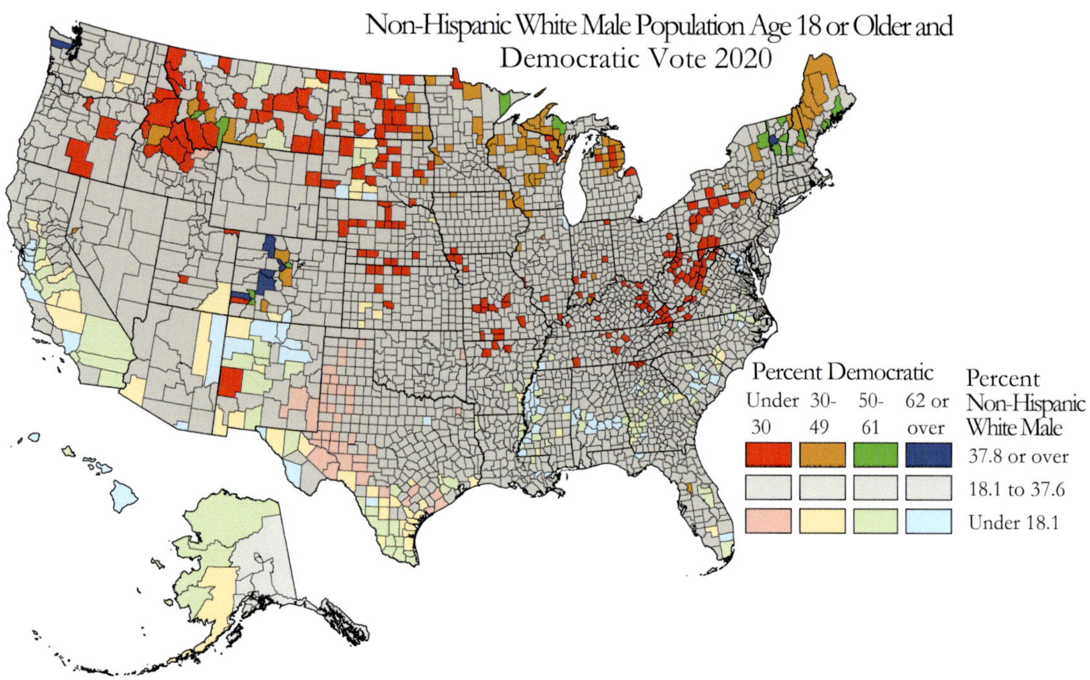
Non-Hispanic White Male Population Age 18 or Older and
Democratic Vote 2020
Percent Democratic
Under 30
30-49
50-61
62 or over
Percent Non-Hispanic White Male
37.8 or over
18.1 to 37.6
Under 18.1

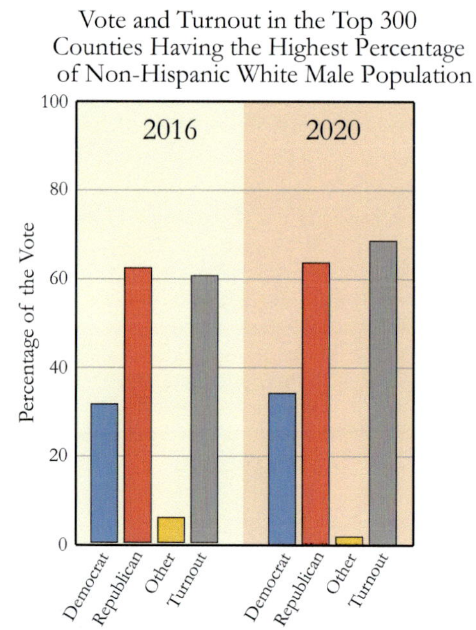
Vote and Turnout in the Top 300
Counties Having the Highest Percentage
of Non-Hispanic White Male Population
2016
2020
Percentage of the Vote
0
20
40
60
80
100
Democrat
Republican
Other
Turnout
Democrat
Republican
Other
Turnout

them by at least 70/30 percent. Conversely, counties in the West, along the US-Mexico border, and in parts of the Black Belt, with low percentages of white men, trended Democratic. The difference can be attributed at least in part to the higher tendency of white women voting Republican in the South and Texas, which increased the Republican percentage of the vote even in places with low percentage of white men.

EDUCATION

The relationship between education level and partisan preference is captured in the national exit poll data where college graduates supported Biden regardless of race/ethnicity or gender and a majority of those who did not graduate from college supported Donald Trump. However, there are significant regional variations in these patterns of support, which are captured in figures 6.13 and 6.14. Overall, these groups demonstrate contrasting patterns. For those counties with high percentages of the population with a BA and above, the preference is for the Democratic candidate by 24 percentage points, whereas for the three hundred counties with the highest levels of high school graduates, there is a very strong preference for Republican candidates. In 2016 a startling 70 percent of those counties voted for Donald Trump; in 2020 that gap widened slightly to 72 percent.

High School Graduates

Figure 6.13 shows the regional distribution of adults who stopped formal education after high school. The counties with the highest rates of adults with high school diplomas form a recognizable regional band from northern Pennsylvania west through the Midwest to northern Missouri and south through the Appalachians and lower Mississippi Valley to central Georgia and northern Louisiana. There are outliers in the northern Great Plains and Rockies as well as the entire west coast of Alaska. In terms of voting, these counties with high levels of high school graduates who did not go on to college form a distinct group of deep red (70 percent or more Republican) counties that overlap with counties that have high percentages of non-Hispanic white men (figure 6.12), manufacturing (figure 6.16) and mining (figure 6.15) employment, and low levels of college attainment. In contrast, the few counties in which a large percentage of the population has only a high school diploma but voted Democratic are either Native American–dominated counties (Alaska, Montana) or counties in the southern Black Belt from the Mississippi Delta to central South Carolina.

BA or Higher

In the top three hundred counties by percentages of BA graduates, Biden received 61.48 percent of the vote, against 36.57 percent support for Donald Trump. This represents a 4 percent increase in the Democratic vote since 2016, while the Republican vote increased by 0.04 percent. The population with a BA or higher, which tends to vote Democratic, is concentrated along the West Coast and in the Northeast, as well as in the central Rockies (figure 6.14). There are a large number of isolated counties in the highest class (40 percent or more adults with a BA or above) that represent either urban areas or college towns. In contrast to the high school map (figure 6.13), this map shows a strong relationship between populations with more than 24 percent BA graduates voting 70/30 percent for Biden. Places where high rates of college graduation correspond to high levels of Republican voting are concentrated in rural and suburban counties in red states, especially in areas with significant resources extraction industries (ranching and mining).

ECONOMY

The last group of variables assessed in this section are economic indicators including occupation, income, and employment status. Historically, agriculture and extractive industries have been associated with Republican voters, while manufacturing employment was associated with Democratic voting in the past (the Midwest "blue wall" of Pennsylvania, Ohio, Michigan, Illinois, and Wisconsin) through associations with union membership. More recently, however, the rejection of the Democratic party by the predominantly white and often male manufacturing workforce has led to a shift to Republican candidates. Service employment, associated with both very high and very low incomes, predominantly favors Democratic candidates, while median income, poverty, and unemployment demonstrate county-level relationships with both Republican and Democratic voters.

Agriculture, Forestry, Fishing, Hunting, and Mining

The counties that are dominated by primary industries show an unsurprising preference for Republican candidates, with the top three hundred counties voting Republican by 61 percent to 37 percent, although this represents a decrease of 2.21

FIGURE 6.13

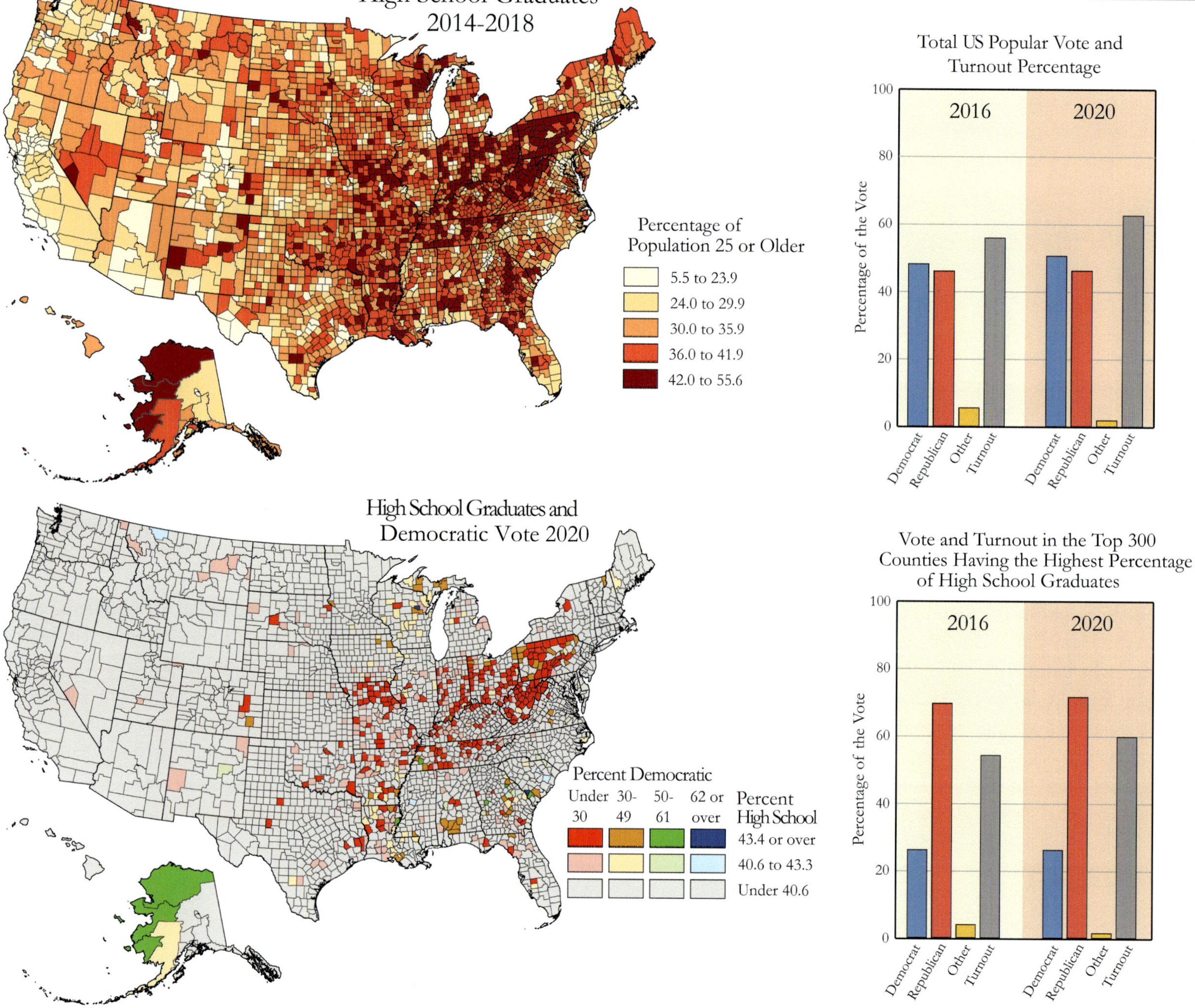
High School Graduates 2014-2018
Percentage of Population 25 or Older
5.5 to 23.9
24.0 to 29.9
30.0 to 35.9
36.0 to 41.9
42.0 to 55.6
Total US Popular Vote and Turnout Percentage
2016
2020
Percentage of the Vote
0
20
40
60
80
100
Democrat
Republican
Other
Turnout
High School Graduates and Democratic Vote 2020
Percent Democratic
Under 30
30-49
50-61
62 or over
Percent High School
43.4 or over
40.6 to 43.3
Under 40.6
Vote and Turnout in the Top 300 Counties Having the Highest Percentage of High School Graduates

FIGURE 6.14

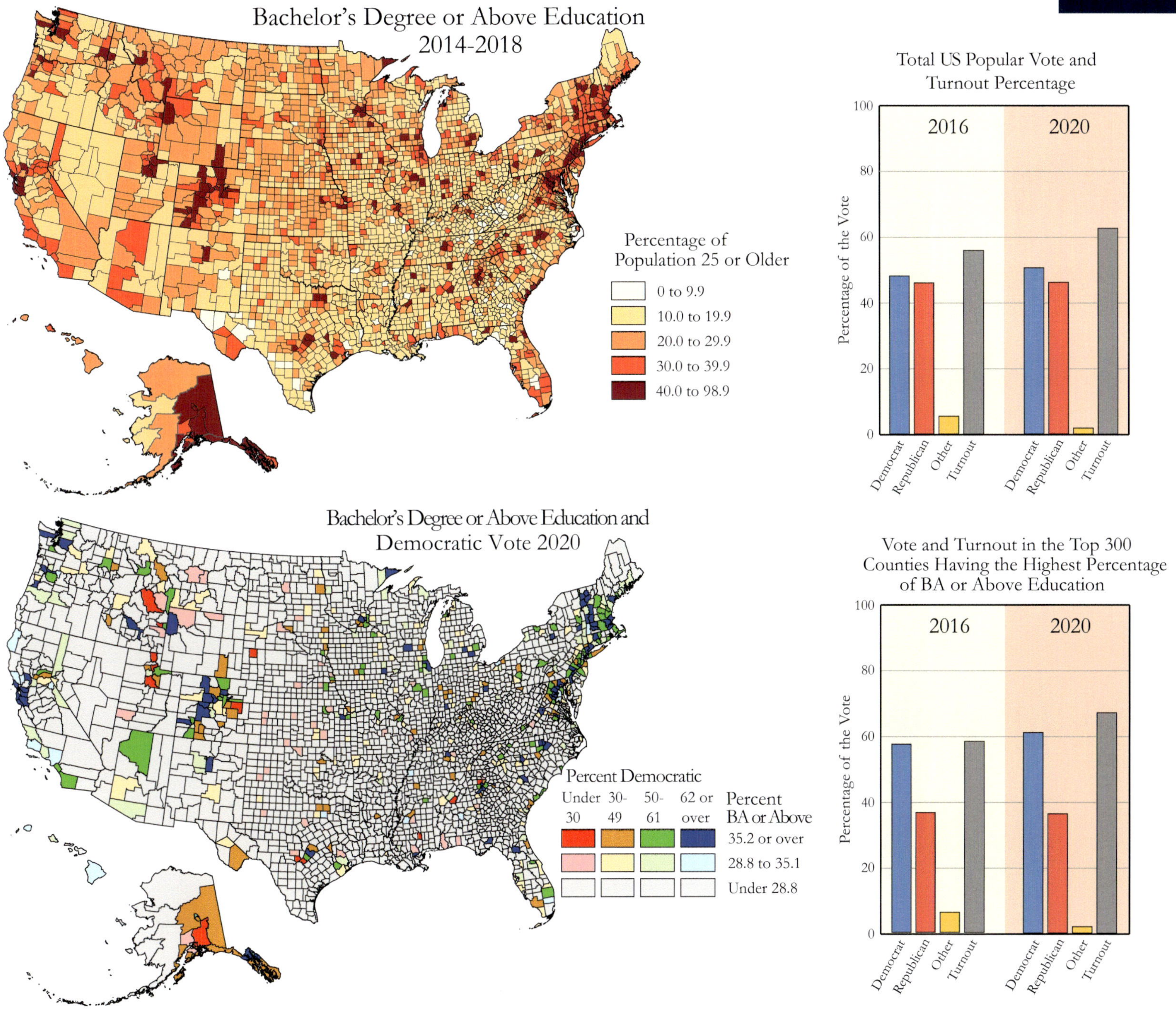

Bachelor's Degree or Above Education
2014-2018
Percentage of
Population 25 or Older
0 to 9.9
10.0 to 19.9
20.0 to 29.9
30.0 to 39.9
40.0 to 98.9
Total US Popular Vote and
Turnout Percentage
2016
2020
Percentage of the Vote
0
20
40
60
80
100
Democrat
Republican
Other
Turnout
Democrat
Republican
Other
Turnout
Bachelor's Degree or Above Education and
Democratic Vote 2020
Percent Democratic
Under 30
30-49
50-61
62 or over
Percent
BA or Above
35.2 or over
28.8 to 35.1
Under 28.8
Vote and Turnout in the Top 300
Counties Having the Highest Percentage
of BA or Above Education
2016
2020
Percentage of the Vote
0
20
40
60
80
100
Democrat
Republican
Other
Turnout
Democrat
Republican
Other
Turnout

FIGURE 6.15

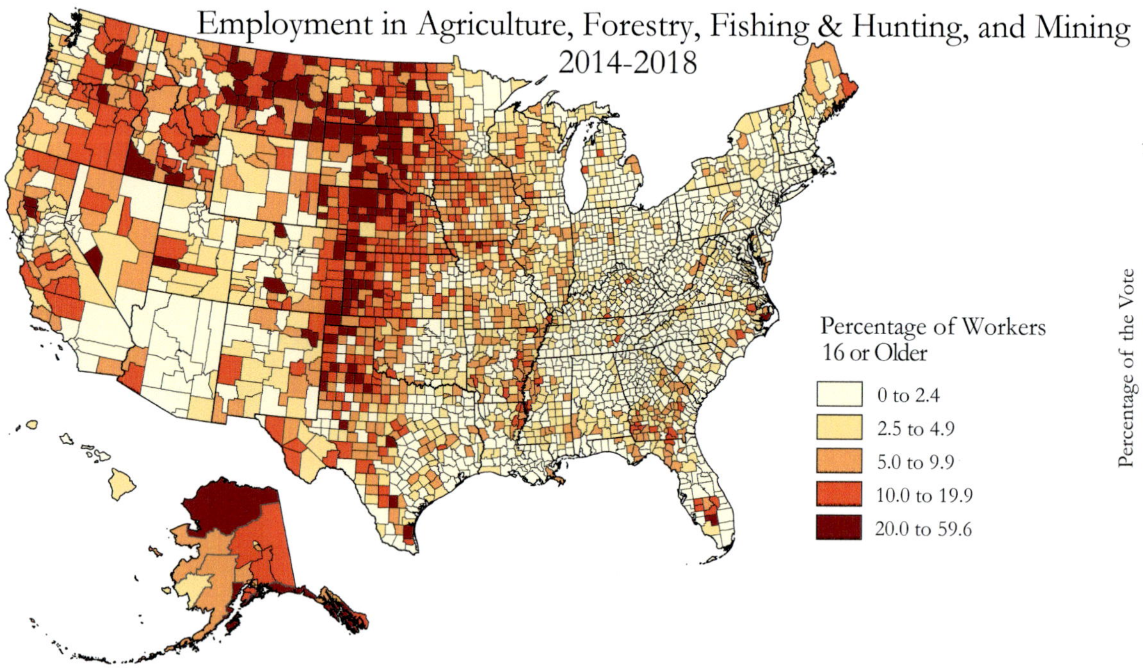

Employment in Agriculture, Forestry, Fishing & Hunting, and Mining
2014-2018
Percentage of Workers
16 or Older
0 to 2.4
2.5 to 4.9
5.0 to 9.9
10.0 to 19.9
20.0 to 59.6

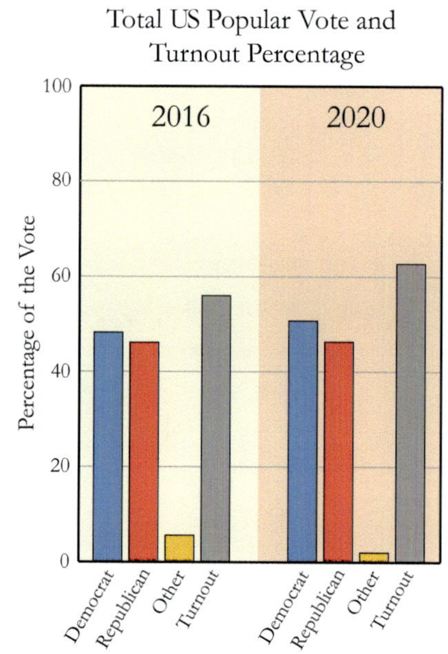

Total US Popular Vote and
Turnout Percentage
2016
2020
Percentage of the Vote
100
80
60
40
20
0
Democrat
Republican
Other
Turnout
Democrat
Republican
Other
Turnout

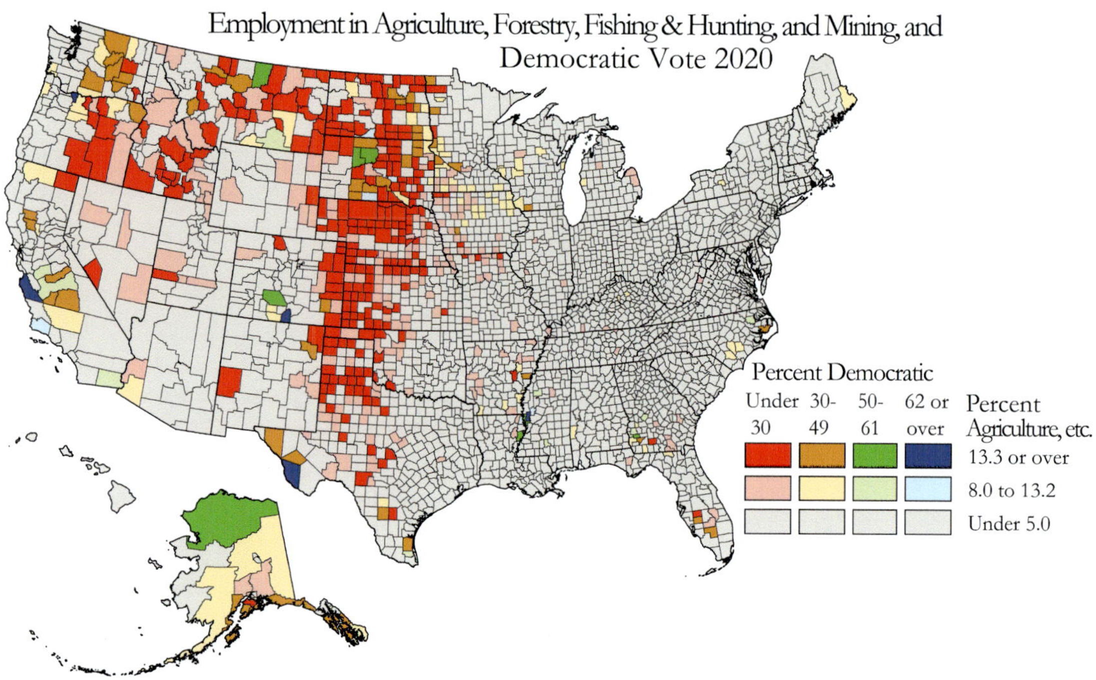

Employment in Agriculture, Forestry, Fishing & Hunting, and Mining, and
Democratic Vote 2020
Percent Democratic
Under 30
30-49
50-61
62 or over
Percent Agriculture, etc.
13.3 or over
8.0 to 13.2
Under 5.0

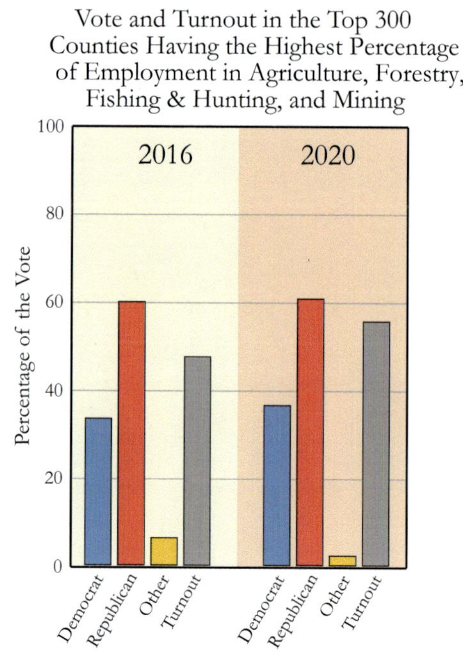

Vote and Turnout in the Top 300
Counties Having the Highest Percentage
of Employment in Agriculture, Forestry,
Fishing & Hunting, and Mining
2016
2020
Percentage of the Vote
100
80
60
40
20
0
Democrat
Republican
Other
Turnout
Democrat
Republican
Other
Turnout

FIGURE 6.16

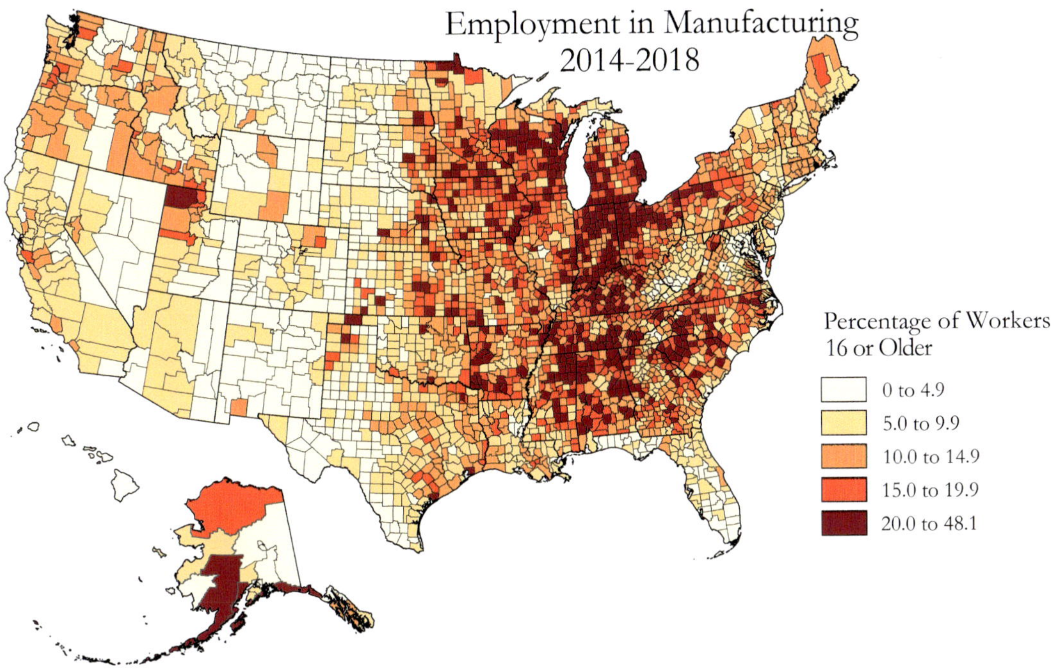
Employment in Manufacturing
2014-2018
Percentage of Workers
16 or Older
0 to 4.9
5.0 to 9.9
10.0 to 14.9
15.0 to 19.9
20.0 to 48.1

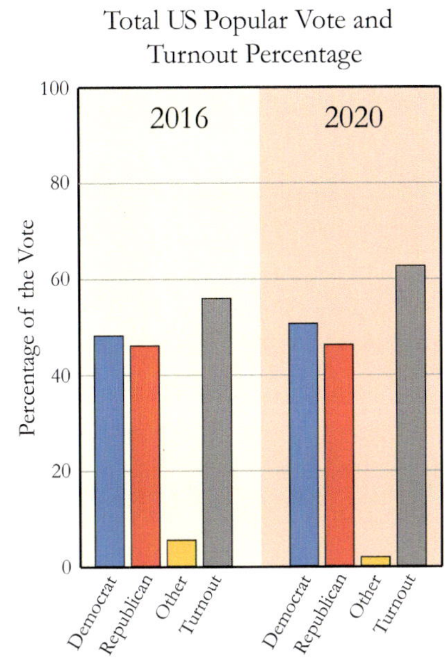
Total US Popular Vote and
Turnout Percentage
2016
2020
Percentage of the Vote
100
80
60
40
20
0
Democrat
Republican
Other
Turnout
Democrat
Republican
Other
Turnout

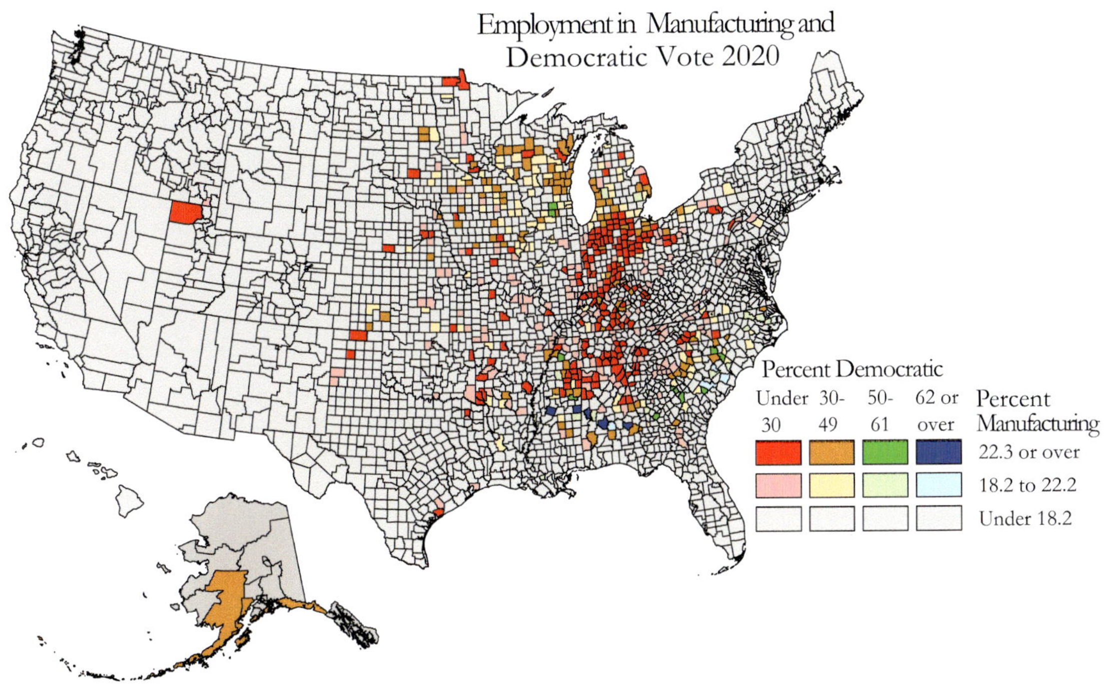
Employment in Manufacturing and
Democratic Vote 2020
Percent Democratic
Under 30
30-49
50-61
62 or over
Percent Manufacturing
22.3 or over
18.2 to 22.2
Under 18.2

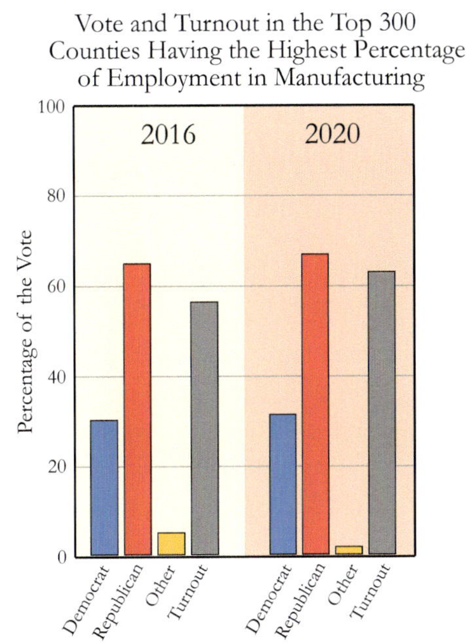
Vote and Turnout in the Top 300
Counties Having the Highest Percentage
of Employment in Manufacturing
2016
2020
Percentage of the Vote
100
80
60
40
20
0
Democrat
Republican
Other
Turnout
Democrat
Republican
Other
Turnout

percent in the 2016 Republican margin of victory. These counties are also largely rural counties, concentrated in the Great Plains, northern Rockies, and West Coast (figure 6.15). Of the more than 1,200 countries that have at least 4 percent of their employment in these employment categories, only fifty-four (4.3 percent) supported Biden, and almost all of those counties were either in California or had higher-than-average populations of color; the Black population of the Mississippi Delta and central Georgia; the Native American population of Montana, Alaska, New Mexico, and the Dakotas, and a scatter of US-Mexico border counties. Conversely, the great majority of these counties vote Republican, forming a distinct band from West Texas to North Dakota and west across the northern Rockies to eastern Oregon. This is also the region where county populations are less dense, older, and, with the exception of American Indian reservations, are predominantly non-Hispanic white.

Manufacturing

Manufacturing employment is concentrated in counties east of the Missouri River, particularly in the Midwestern corridor from Michigan/Wisconsin south to Alabama with only a handful of manufacturing counties in the Great Plains/South Central United States and, west of the Rockies, only Box Elder County, Utah (home of Thiokol/ATK) employs more than 20 percent of the population in manufacturing industries (figure 6.16). In terms of political affiliation, high employment in the manufacturing industry is strongly tied to Republican voting. Of the 755 counties with 17 percent or more of their employment in manufacturing only 39 (5.2 percent) voted for Biden and only 30 of the 891 (3.4 percent) counties with between 11 and 17 percent. In elections since Clinton's 1992 victory, manufacturing counties have trended Democratic; creating the Democrats' "blue wall" from New York to Minnesota (with the exception of Indiana, and occasionally Ohio). The 2016 election broke this pattern with Michigan, Pennsylvania, and Wisconsin all flipping Republican (albeit by very narrow margins), and it was critical to Biden's chances of winning more than 270 Electoral College votes that these states flip back to the Democrats in 2020. It is clear from these maps that, while these states did vote for Biden in 2020, that flip happened somewhere other than the manufacturing-based counties, where only eight of the almost 200 counties in that group (manufacturing employment of at least 17 percent) voted for Biden.

Management, Business, Science, and Arts

Unlike the other two employment classes, counties with large management, business, science, and arts (MBSA) populations tend to vote Democratic. In the 2020 election, Joe Biden won the top three hundred of these counties by 61.47 percent to 36.57 percent, an increase of almost 4 percent in the Democratic vote over the 2016 election. Counties with high levels of MBSA employment (more than 40 percent) are concentrated on the coasts, particularly in the Northeast and in urban areas of the West Coast and Rockies (figure 6.17). Metropolitan counties and counties with college towns also display high levels of MBSA employment. In terms of voting, there is a relatively even division; once again it is apparent that the more rural counties of the Great Plains and Rockies trend Republican, with some obvious exceptions in the Native American reservations and environmental amenity counties, while the urban and suburban counties of the Northeast and West Coast show a marked preference for Biden.

Unemployment

High levels of unemployment (6 percent or above) were a feature of a year in which the COVID-19 pandemic shut down service industries all across the country. However, by November many states, particularly in the South and Midwest, had removed restrictions on economic activities, and high levels of pandemic-related unemployment were primarily to be found in California and the Northeast. The national pattern of unemployment reflects both those pandemic-related activities and traditionally high-unemployment areas such as the Mississippi Delta and Appalachia, as well as Native American reservations and the Texas/Mexico border region (figure 6.18). The pattern of voting in high-unemployment regions reflects many of the same variables as other economic indicators. Nineteen percent (158 of 819) of these counties voted for Biden, primarily urban counties on the coasts and counties with large percentages of Native American, Black, and Hispanic populations. The remaining 81 percent of high-unemployment counties, largely rural and non-Hispanic white, voted for Donald Trump. These counties are clustered in Appalachia and rural Texas and Oklahoma, where the Democratic candidate received less than 30 percent of the vote. Overall, however, the counties with the highest levels of unemployment preferred Joe Biden by 12 percent, although this margin was 3 percent smaller than in 2016;

FIGURE 6.17

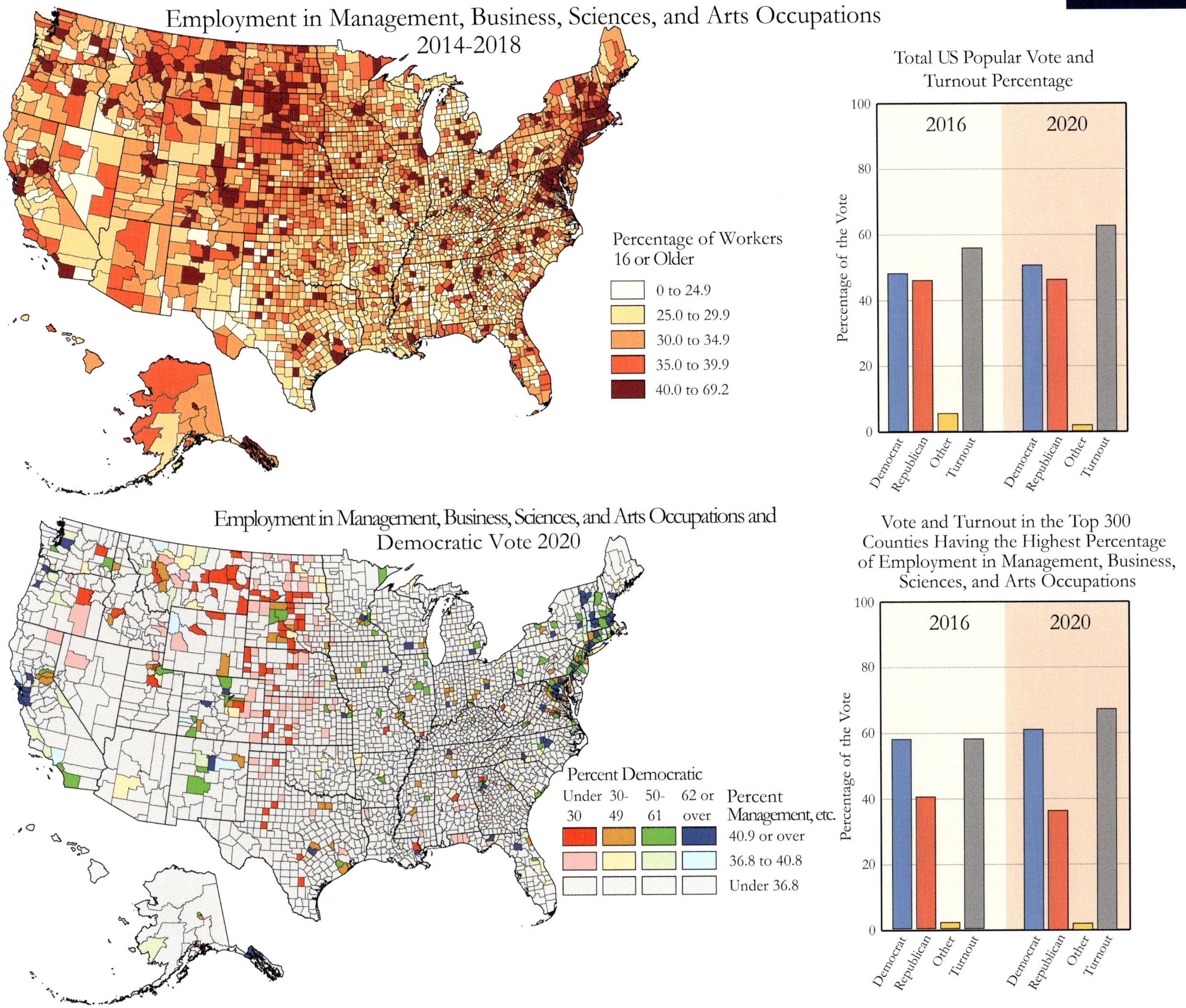

Employment in Management, Business, Sciences, and Arts Occupations 2014-2018
Percentage of Workers 16 or Older
0 to 24.9
25.0 to 29.9
30.0 to 34.9
35.0 to 39.9
40.0 to 69.2
Total US Popular Vote and Turnout Percentage
2016
2020
Percentage of the Vote
0
20
40
60
80
100
Democrat
Republican
Other
Turnout
Employment in Management, Business, Sciences, and Arts Occupations and Democratic Vote 2020
Percent Democratic
Under 30
30-49
50-61
62 or over
Percent Management, etc.
40.9 or over
36.8 to 40.8
Under 36.8
Vote and Turnout in the Top 300 Counties Having the Highest Percentage of Employment in Management, Business, Sciences, and Arts Occupations

FIGURE 6.18

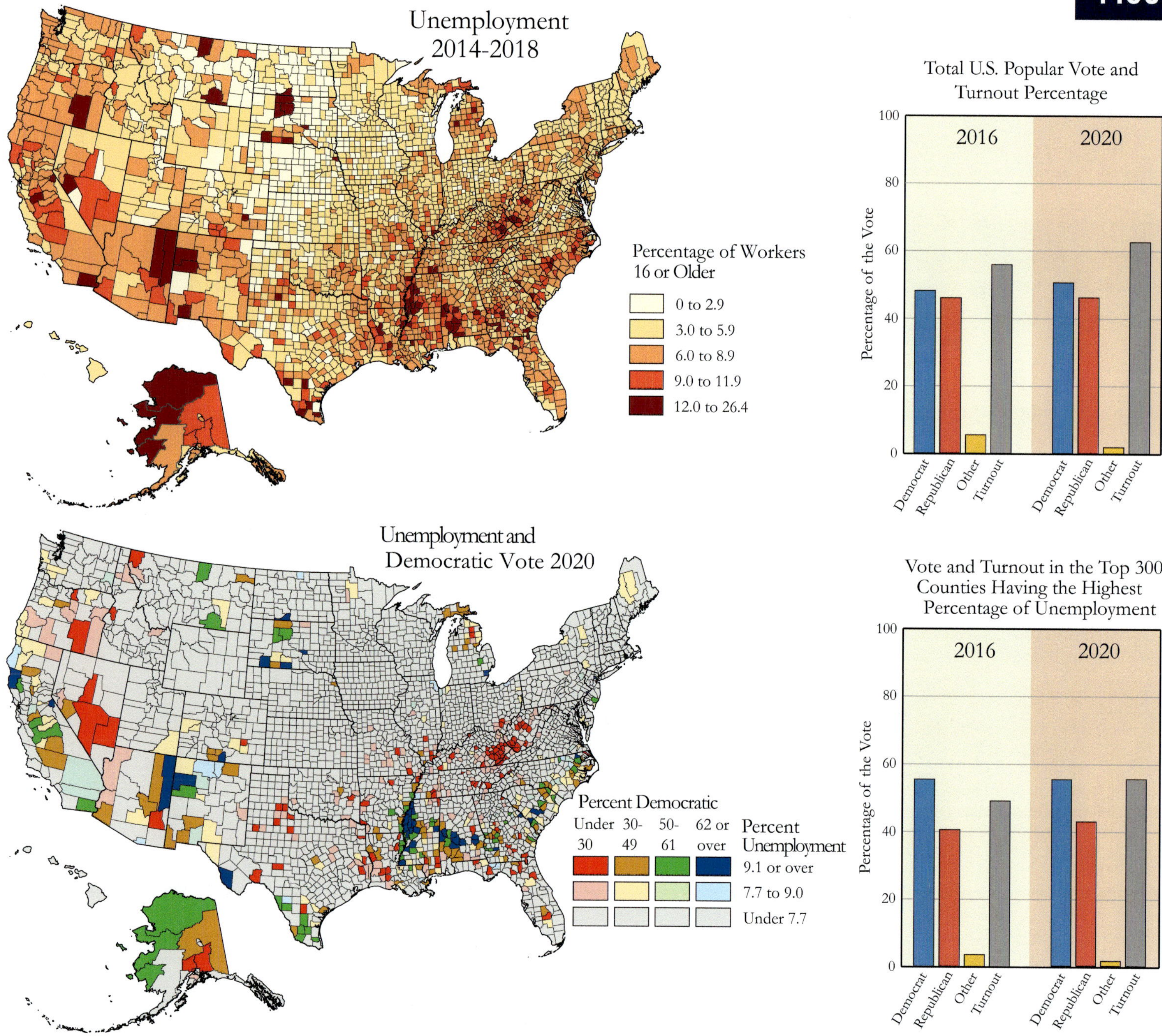
Unemployment 2014-2018
Percentage of Workers 16 or Older
0 to 2.9
3.0 to 5.9
6.0 to 8.9
9.0 to 11.9
12.0 to 26.4
Total U.S. Popular Vote and Turnout Percentage
2016
2020
Percentage of the Vote
100
80
60
40
20
0
Democrat
Republican
Other
Turnout
Democrat
Republican
Other
Turnout
Unemployment and Democratic Vote 2020
Percent Democratic
Under 30
30-49
50-61
62 or over
Percent Unemployment
9.1 or over
7.7 to 9.0
Under 7.7
Vote and Turnout in the Top 300 Counties Having the Highest Percentage of Unemployment
2016
2020
Percentage of the Vote
100
80
60
40
20
0
Democrat
Republican
Other
Turnout
Democrat
Republican
Other
Turnout

FIGURE 6.19

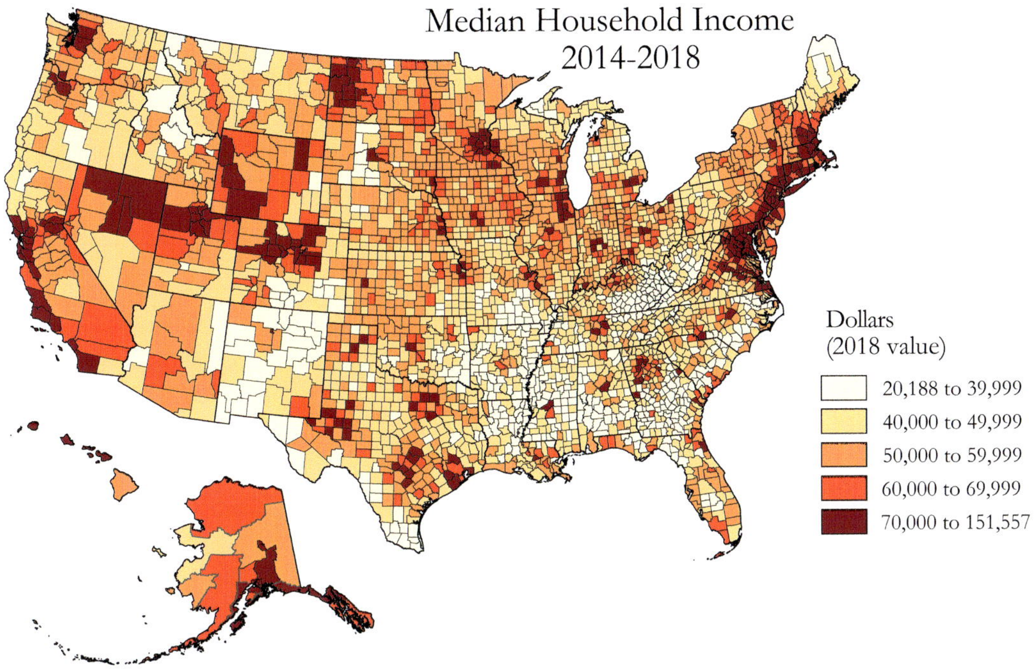
Median Household Income
2014-2018
Dollars
(2018 value)
20,188 to 39,999
40,000 to 49,999
50,000 to 59,999
60,000 to 69,999
70,000 to 151,557

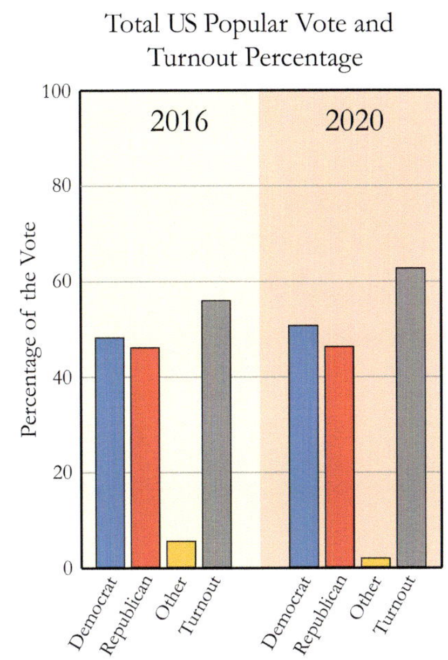
Total US Popular Vote and
Turnout Percentage
2016
2020
Percentage of the Vote
0
20
40
60
80
100
Democrat
Republican
Other
Turnout
Democrat
Republican
Other
Turnout

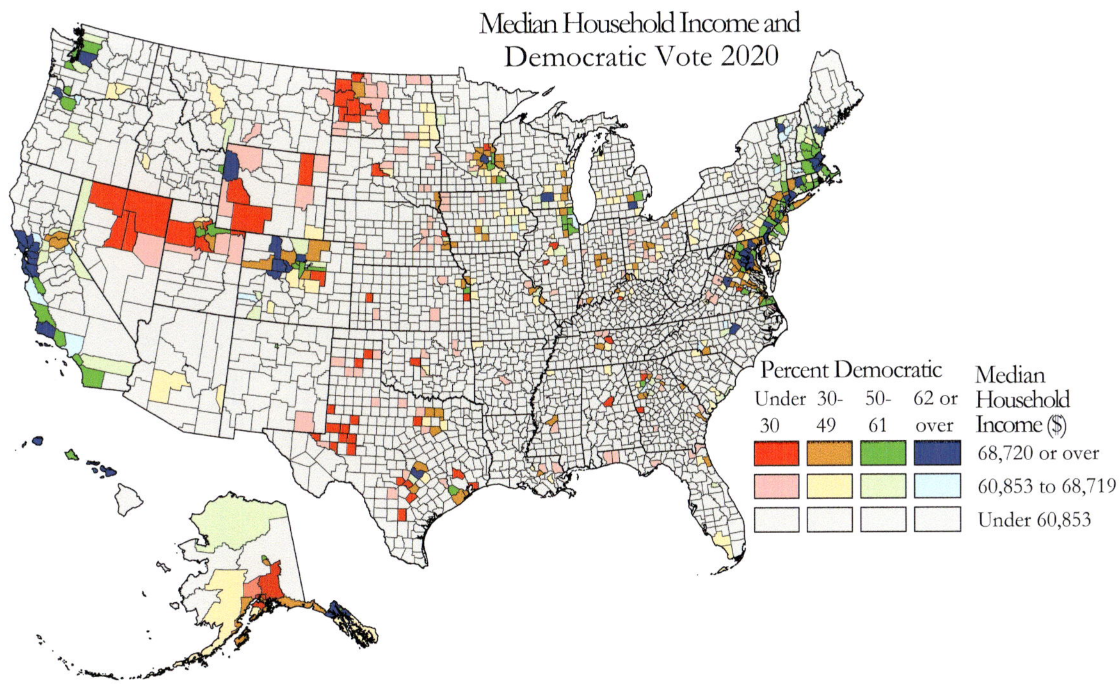
Median Household Income and
Democratic Vote 2020
Percent Democratic
Under 30
30-49
50-61
62 or over
Median Household Income ($)
68,720 or over
60,853 to 68,719
Under 60,853

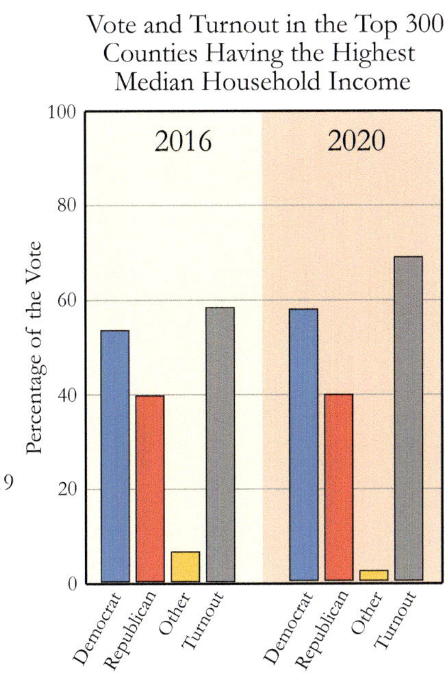
Vote and Turnout in the Top 300
Counties Having the Highest
Median Household Income
2016
2020
Percentage of the Vote
0
20
40
60
80
100
Democrat
Republican
Other
Turnout
Democrat
Republican
Other
Turnout

FIGURE 6.20

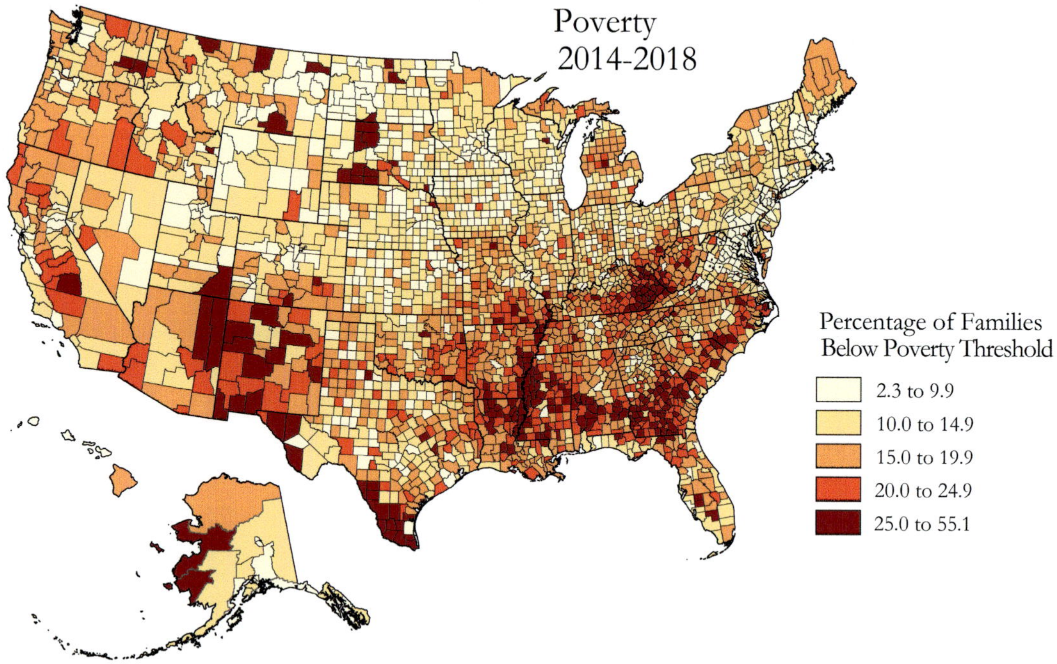
Poverty
2014-2018
Percentage of Families
Below Poverty Threshold
2.3 to 9.9
10.0 to 14.9
15.0 to 19.9
20.0 to 24.9
25.0 to 55.1

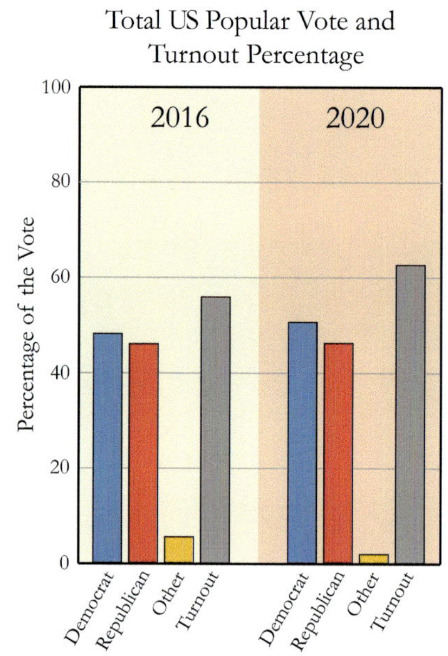
Total US Popular Vote and
Turnout Percentage
2016
2020
Percentage of the Vote
100
80
60
40
20
0
Democrat
Republican
Other
Turnout
Democrat
Republican
Other
Turnout

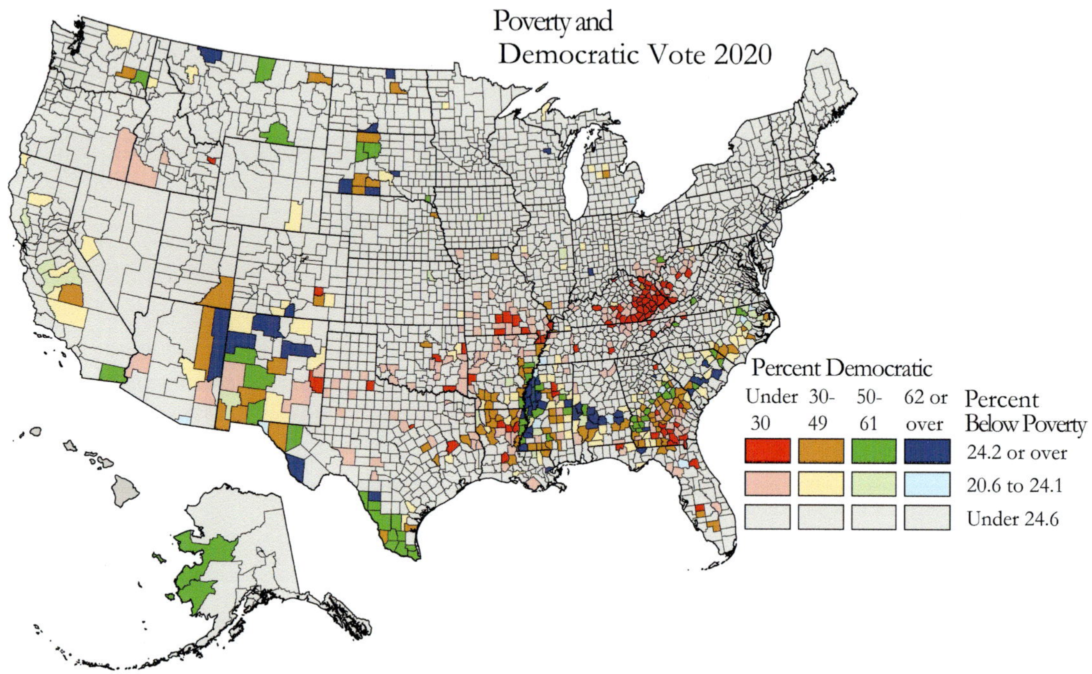
Poverty and
Democratic Vote 2020
Percent Democratic
Under 30
30-49
50-61
62 or over
Percent
Below Poverty
24.2 or over
20.6 to 24.1
Under 24.6

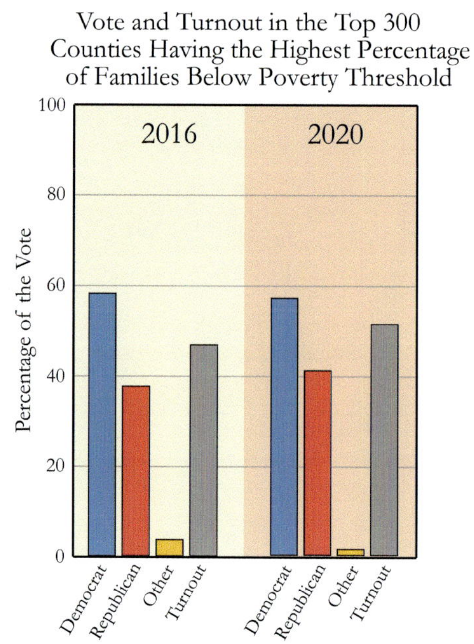
Vote and Turnout in the Top 300
Counties Having the Highest Percentage
of Families Below Poverty Threshold
2016
2020
Percentage of the Vote
100
80
60
40
20
0
Democrat
Republican
Other
Turnout
Democrat
Republican
Other
Turnout

the Republican vote in these counties increased from 40.75 percent to 43.02 percent.

Income and Poverty

High-median-income counties across the country voted for Joe Biden by a margin of almost 20 percent over the Republican incumbent (58 percent to 39 percent), an increase of 4 percent over the 2016 election. Geographically, these counties are concentrated in the metropolitan Northeast and along the West Coast, with an outlying area in the northern Rockies associated with extractive industries in North Dakota, Wyoming, and West Texas (figure 6.19). Unsurprisingly, high income in metropolitan areas is associated with support of the Democratic ticket, while the more central and rural high-income counties overwhelmingly supported the Republicans. The map of counties with significant levels of poverty (20 percent or more of the population living below the official poverty level) is almost a mirror image of the high-median-income map (figure 6.20). However, the highest-poverty counties demonstrate precisely the same partisan preference as the high-income counties. In 2020 Biden won the three hundred poorest counties in the United States by 57 percent to 41 percent, although this was a 4 percent decline in the Democratic vote from 2016. Poverty is widespread in primarily rural counties throughout Appalachia and the Southeast, in the counties with Native American reservations and along the southern border and in parts of the interior of the West Coast states. There are very few high-poverty counties in large metropolitan areas (Wayne County, Michigan) and only a few with over 16 percent of the population living in poverty (Los Angeles County, California; Harris County, Texas), suggesting that poverty is overwhelmingly a rural phenomenon in the contemporary United States. Within those rural areas the voting preferences break down largely along racial/ethnic lines. Counties that are poor and predominantly non-Hispanic white, in Appalachia, the Ozarks, the Ouachitas, and southern Georgia, as well as scattered counties in Texas and the Pacific Northwest trend Republican, while counties with higher levels of Black, Hispanic, and Native American populations (Mississippi Delta and Black Belt; southern border and Native American reservations) all voted at least 62 percent for Biden.

Across a wide spectrum of demographic and economic variables, across all 3,143 counties and county-equivalent units in the United States, it is possible to find a common, consistent pattern of voter preference. Republican counties can be defined by religion, race, and rurality. As the preceding text and the five following sections demonstrate, counties that are more rural, whiter, and more religious were much more likely to support Donald Trump over Joe Biden in the 2020 presidential election. Conversely, urban counties, minority-majority counties (or merely those with higher percentages of minority populations) and counties with lower levels of religious adherence all broke for Biden. There are other defining variables, of course; age is a factor in some counties, as is type of industry, but the most consistent explanatory pattern for voting at the county scale is rurality, race/ethnicity, and religion.

HISPANIC VOTING TRENDS AND THE 2020 PRESIDENTIAL ELECTION

RYAN WEICHELT

The terms Hispanic or Latino represent a historical amalgamation of Census policies aimed at quantifying population groups connected simply by language. In 1974, Census officials determined the use of speaking Spanish or having a "Spanish"-based last name was not unique enough to quantify groups as a distinct ethnicity. As a result, by the 1980 Census, Hispanics or Latinos could self-identify as part of this group. The resulting classification has been helpful at distinguishing this group across the United States, but it has also led to misunderstanding that Hispanic, like the term European, covers a vast array of cultures and distinct ethnicities. Therefore, Hispanic voters have long been seen as a homogenous voting bloc. Although many studies have identified spatial voting trends among Hispanic subgroups across the United States, the 2020 presidential election offered the most unique spatial representation of Hispanic voting patterns to date.

Hispanics comprise the largest minority population in the United States. Current US Census Bureau estimates show Hispanic populations comprise 18.5 percent of the total population, equating to an overall population of more than sixty-one million. Even though Hispanic voters have some of the lowest voter turnout rates, Pew Research estimates for the first time Hispanic voters will be the largest minority voting bloc, accounting for more than 13.3 percent of all votes, compared to 12.5 percent for African Americans (Cilluffo and Fry 2021). However, Hispanics are far from identical, spatially or ethnically. Figure 6.21 clearly identifies Hispanic populations are the largest along the southern US border. California (7.9 million), Texas (5.6 million), and Florida (3.1 million) have the largest total number of Hispanic eligible voters, though as a percentage of the total population of eligible voters, New Mexico tops the list with 43 percent followed by California (30 percent), Texas (30 percent), Arizona (24 percent), and Florida (20 percent; Noe-Bustamante, Lopez, and Krogstad 2020).

According to the 2019 US Census American Community Survey (ACS) data, Hispanic populations in the United States, either those that immigrated or were born in the United States, come from three main places: Mexico (62.3 percent), Puerto Rico (9.6 percent), and Cuba (3.9 percent). Additionally, 24.1 percent claim "Other Hispanic." This group includes many other Central and South American countries and groups such as "Tejano" or "Hispano" that claim no connections to a foreign country. Mexican ancestry populations are largest in the states along Mexican border but are also the most common throughout the Great Plains and Midwest as well as many large US urban areas. Puerto Rican populations are largest in New York City and in central Florida, while Cuban populations are la0rgest in Miami in southern Florida. "Tejano" populations are largest in southern Texas, with "Hispano" populations greatest in northern New Mexico and southern Colorado.

According to Pew Trust election data, the 2016 presidential election saw Hillary Clinton garner 66 percent of the Hispanic vote. This number was lower than 2012 and 2008, when Obama gained 71 percent and 67 percent of the Hispanic vote. Donald Trump managed to draw 28 percent of the Hispanic vote, 1 percent more than Romney in 2012, but lower than John McCain's 31 percent in 2008 (Krogstad and Lopez 2016). Although it is often believed immigration is the key issue among Hispanic voters, its importance among these voters continues to decline. The economy, health care, education, and other social issues all tend to poll higher for all Hispanic voters than immigration. Democrats hoped in 2016 Trump's racist statements toward Hispanics and his tag line of "Build the Wall" would see higher support for Clinton; this was not the case. Trump made a concerted effort to target conservative Christian Hispanic groups, which accounted for his higher-than-expected support among Hispanic voters (Khalid 2016).

As already stated, 2020 saw the largest Hispanic turnout on record. An estimated 16.6 million Hispanics cast votes across the country. Although exit polling of the 2020 presidential election was more difficult given the pandemic, several national polls yielded somewhat similar results. A national exit poll of more than fifteen thousand Hispanic voters conducted by Latino Decisions (2020) found 70 percent of voters supported Biden with 27 percent supporting Trump, while a UCLA (2021) study found 67 percent supported Biden and 32 percent supported Trump. No matter the polling, at the

FIGURE 6.21

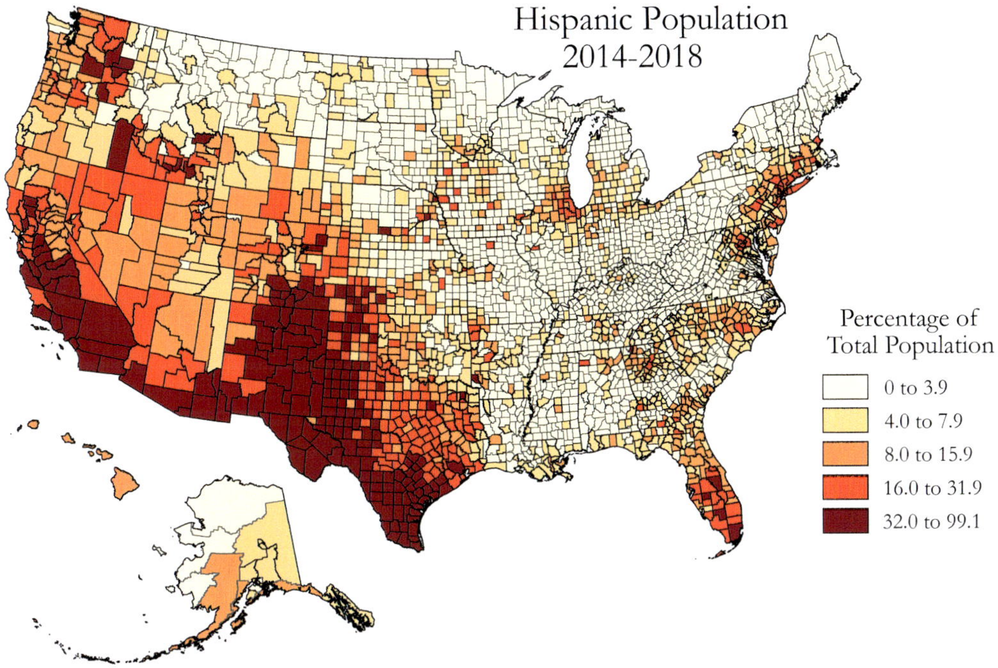

Hispanic Population
2014-2018
Percentage of
Total Population
0 to 3.9
4.0 to 7.9
8.0 to 15.9
16.0 to 31.9
32.0 to 99.1

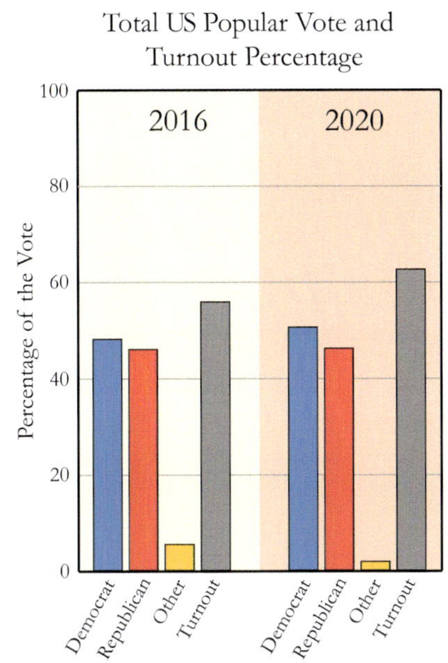

Total US Popular Vote and
Turnout Percentage
2016
2020
Percentage of the Vote
0
20
40
60
80
100
Democrat
Republican
Other
Turnout

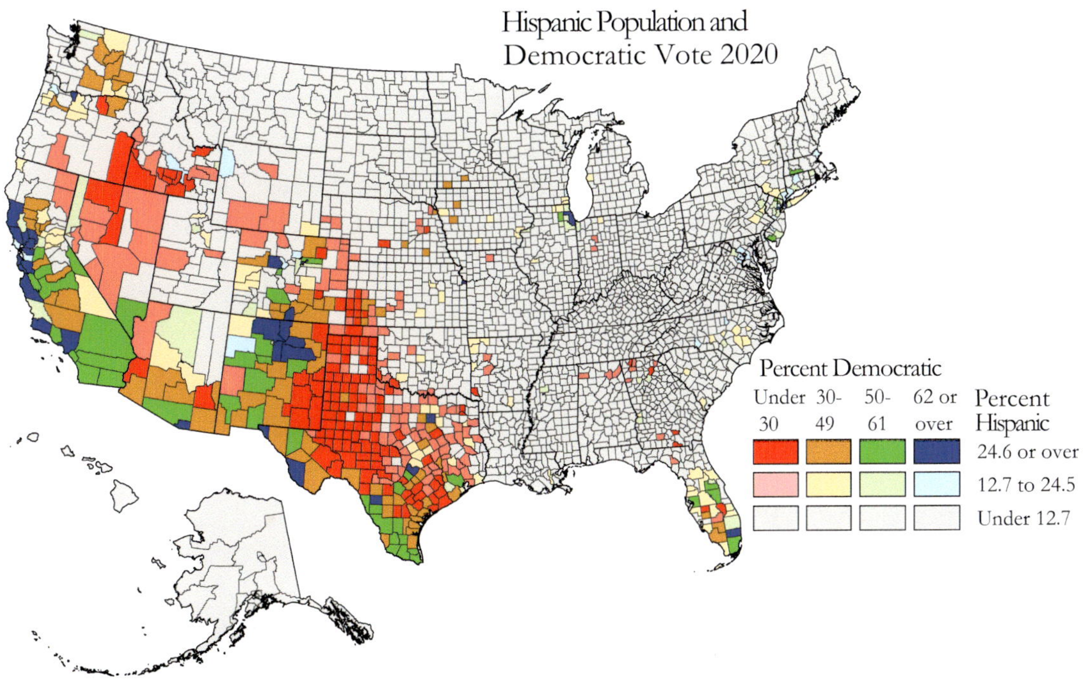

Hispanic Population and
Democratic Vote 2020
Percent Democratic
Under 30
30-49
50-61
62 or over
Percent Hispanic
24.6 or over
12.7 to 24.5
Under 12.7

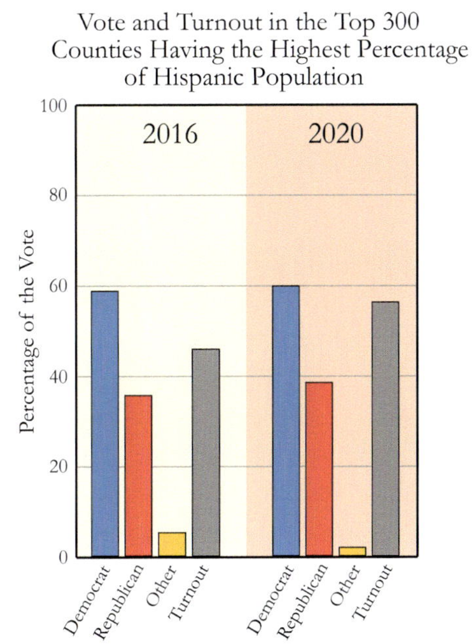

Vote and Turnout in the Top 300
Counties Having the Highest Percentage
of Hispanic Population
2016
2020
Percentage of the Vote
0
20
40
60
80
100
Democrat
Republican
Other
Turnout

state level, it was clear Hispanics played a varying and vital role in determining the outcome.

IMPORTANT ISSUES

For all Americans, COVID-19 was the main issue of the 2020 election. Nationally, Native American, African American, and Hispanic populations experienced proportionally higher rates of COVID-19 infections, hospitalizations, and deaths compared to white populations (Marshall 2020). An NBC *Wall Street Journal* poll found nearly 57 percent of Hispanics disapproved of Trump's handling of the COVID-19 pandemic and that the economic impacts of the pandemic were devastating for many Hispanic families and workers. The Latino Decisions poll (2020) found 25 percent of Hispanics lost their jobs and more than 42 percent had work hours or pay cut. These rates were much higher among younger Hispanic groups who more often work in service industries. As a result, 95 percent of respondents stated it was "important" (69 percent saying "very important") for a second round of stimulus to be passed.

Beyond COVID-19 and the economy, the nation grappled with the issues of racism and social justice. The deaths of Breonna Taylor and George Floyd at the hands of police influenced thousands of rallies and protests during the summer of 2020. Even before 2020, Trump became a lightning rod regarding issues of race. In the lead-up to his 2016 run, Trump would disparage all groups, but for Hispanics Trump's language describing immigrants as rapists and gangsters rung loudest. Based on the Latino Decisions poll (2020), more than 62 percent of Hispanics believe racism has increased from 2016 to 2020 under Donald Trump. The poll also indicated Spanish-speaking respondents were more likely to experience racism than non-Spanish-speaking respondents during the same period. This issue ranked as the fourth most important issue, above education and immigration reform and below the top three issues of the Coronavirus pandemic, jobs and economy, and health-care costs.

Even though Trump's support among Hispanics was low and the topic of immigration also polled low among Hispanic voters, immigration continued to be a favorite talking point for the president in the years leading up to 2020. Although Trump rarely missed a moment to reference the wall, his administration was facing a growing problem at the border itself. Since 2016, unaccompanied minors and family units arrested at the border were becoming increasingly more common. By 2019, family units comprised more than half (55.6 percent) of the 852,000 total border apprehensions and unaccompanied minors topped 76,000 apprehensions (Gramlich 2020). To deter the growing numbers, Trump ordered US Immigration and Customs Enforcement (ICE) to separate families with children held in detention centers along the US-Mexico border. By July 2019, NBC News shared an Associated Press report that nearly 5,400 children had been separated from their parents at the border (2019). Pictures of children locked in cages would highlight the immigration debate throughout the 2020 election cycle.

GEOGRAPHY OF ELECTION RESULTS

Ethnically, Biden's support among all groups of Hispanic voters was high except for Cuban voters in Florida. The Latino Decisions poll (2020) found Mexican and Puerto Rican voters held the highest level of support for Joe Biden at 74 percent and 70 percent, followed by Dominican voters at 61 percent. Biden support increased to 68 percent for Spanish-only speakers of all groups. More than 45 percent of Cuban respondents indicated support for Biden compared to 52 percent reporting support for Donald Trump. Dominican voters were the least likely to support Trump at only 21 percent.

Due to the close nature of the 2020 election across numerous states, Hispanic voters provided key votes for each candidate. Although the smallest in total population, Hispanic voters in Michigan and Wisconsin exhibited the highest percentage totals. According to the Latino Decisions poll (2020), Wisconsin, a state Biden won by only twenty thousand votes, saw the highest support of Hispanic voters at 77 percent to Trump's 22 percent. Although only comprising less than 5 percent of the total population, Hispanic voters in Wisconsin provided Biden with key votes, locking up his victory. Michigan polls found Hispanic voters supported Biden 76 percent to only 22 percent for Trump. Similarly high levels of support for Biden among Hispanics in Nevada, Arizona, and Georgia provided precious votes for Biden's razor-thin victories, solidifying Biden as the forty-sixth president of United States. The states with the largest Hispanic populations, California, New Mexico, and Colorado, all delivered a higher percentage of votes for Biden.

Florida and Texas provided Trump with his greatest support among Hispanic voters. While Democrats believed they could win Florida, assistance among Cuban American voters in Miami all but assured Trump would win his adopted home state of Florida. Support among Cubans was bolstered by the economic success felt before the pandemic and Trump's targeted message of Biden and Harris as socialists (Torres 2020). Outside of Miami, Trump's backing among Hispanics mirrored other areas of the country, but the large voting bloc of Cuban American voters in Miami-Dade County was too much for Biden to overcome. The effect was also felt down ticket, as Democratic incumbent Donna Shalala lost her congressional seat to a Republican and Cuban American candidate Maria Elvira Salazar.

Perhaps the greatest disappointment for Biden and Democrats came out of the state of Texas. Some late polling suggested Texas was potentially in play for Biden, but he would need high voter turnout in the largest urban areas and along the traditionally Democratic and largely Hispanic border counties of southern Texas. Biden accomplished some of these goals in the largest urban areas, but traditionally dependent Hispanic voters along the border did not turn out as needed and new voters broke for Trump. While no one reason explains the phenomena, some officials citied Democrats' tactic to not knock on doors due to the pandemic, economic successes in the areas before the pandemic, and stimulus monies provided some reasons for Trump's success among Hispanic voters in South Texas (Villa 2020). Besides Florida, Biden's support among Hispanic voters in Texas was the lowest in the nation. In the end, Democrats' dreams of flipping Texas was lost.

Nationally, Hispanic candidates continue to make slow gains. A 2017 National Association of Latino Elected and Appointed Officials (NALEO) study found at the national, state, and local levels nearly 6,600 Hispanics hold some sort of office. Although only 1,523 have a party affiliation, 85 percent (1,305) are Democrats. Furthermore, 63 percent (4,199) of all elected officials are male. Additionally in 2017, four members of the US Senate and thirty-four members of the House of Representatives identify as Hispanic. By 2021 the total number of Hispanics elected to both the Senate and House of Representatives increased to six and forty-one. Of the total forty-seven elected officials, twelve are Republicans with the remaining thirty-five Democrats. Even though there is slight progress among Hispanic elected officials, the total numbers elected are proportionately lower than the total population.

Hispanics now comprise the largest minority voting bloc in the country. Yet, like so many minority groups, Hispanics disproportionately felt the impacts of the COVID-19 pandemic. Nearly one-fourth of Hispanics eighteen and older stated they lost their jobs, while those eighteen to thirty-four saw that number increase to nearly 40 percent. Socially, Hispanic populations across the United States also experienced increased levels of racism, more so among those in Spanish-speaking homes. For these reasons alone, Democrats felt increasingly confident Hispanics would turn out in record numbers to support Joe Biden. While a record number of total Hispanic voters did turn out, the losses in Florida and declines in Texas clouded an election where Hispanics should be celebrated as key swing voters for Democrats. Polls indicated that in states with very close election results like Wisconsin, Nevada, Arizona, and Georgia, the large margin of support for Biden among Hispanic voters was vital to his victory. Even in Florida, where Trump gained ground with Cuban voters, Biden won other Hispanic groups at or above a two-to-one margin. Although Texas was the one blemish for Biden, the effect was only felt for the presidency. Democrats garnered normal vote margins down ballot in the heaviest-concentration Hispanic areas in the Lone Star State. Even though the number of elected officials that identify as Hispanic is disproportionate to the total population, Hispanic voters flexed their electoral muscles. The 2020 election demonstrated that Hispanic voters were key to Biden's historic victory.

REFERENCES

Associated Press. 2019. More than 5,400 Children Split at Border, According to New Count. *NBCNews.com*. https://www.nbcnews.com/news/us-news/more-5-400-children-split-border-according-new-count-n1071791 (last accessed June 4, 2021).

Cilluffo, A., and R. Fry. 2021. An Early Look at the 2020 Electorate. *Pew Research Center's Social & Demographic Trends Project*. https://www.pewresearch.org/social-trends/2019/01/30/an-early-look-at-the-2020-electorate-2/ (last accessed June 4, 2021).

Gramlich, J. 2020. How Border Apprehensions, ICE Arrests and Deportations Have Changed under Trump. *Pew Research Center*. https://www.pewresearch.org/fact-tank/2020/03/02/how-border-apprehensions-ice-arrests-and-deportations-have-changed-under-trump/ (last accessed June 4, 2021).

Khalid, A. 2016. Latinos Will Never Vote for a Republican, and Other Myths About Hispanics from 2016. *NPR*. https://www.npr.org/2016/12/22/506347254/latinos-will-never-vote-for-a-republican-and-other-myths-about-hispanics-from-20 (last accessed June 4, 2021).

Krogstad, J. M., and M. H. Lopez. 2016. Hillary Clinton Won Latino Vote but Fell Below 2012 Support for Obama. *Pew Research Center*. https://www.pewresearch.org/fact-tank/2016/11/29/hillary-clinton-wins-latino-vote-but-falls-below-2012-support-for-obama/ (last accessed June 4, 2021).

Latino Decisions. 2020. American Election Eve Poll 2020. *Latino Decisions.* https://latinodecisions.com/polls-and-research/american-election-eve-poll-2020/ (last accessed June 4, 2021).

Marshall, W. F. 2020. Why Are People of Color More at Risk of Coronavirus Complications? *Mayo Clinic.* https://www.mayoclinic.org/diseases-conditions/coronavirus/expert-answers/coronavirus-infection-by-race/faq-20488802 (last accessed June 4, 2021).

National Association of Latino Elected and Appointed Officials. 2021. At-A-Glance. *NALEO Educational Fund.* https://naleo.org/at-a-glance/ (last accessed May 23, 2021).

Noe-Bustamante, L., M. H. Lopez, and J. M. Krogstad. 2020. U.S. Hispanic Population Surpassed 60 Million in 2019, but Growth Has Slowed. *Pew Research Center.* https://www.pewresearch.org/fact-tank/2020/07/07/u-s-hispanic-population-surpassed-60-million-in-2019-but-growth-has-slowed/ (last accessed June 4, 2021).

Torres, N. G. 2020. "Invisible Campaign" and the Specter of Socialism: Why Cuban Americans Fell Hard for Trump. https://www.miamiherald.com/news/politics-government/election/article247233684.html (last accessed May 28, 2021).

UCLA Latino Policy and Politics Initiative. 2021. Latino Voters Were Decisive in 2020 Presidential Election. *UCLA.* https://newsroom.ucla.edu/releases/latino-vote-analysis-2020-presidential-election (last accessed June 4, 2021).

Villa, L. 2021. Donald Trump's 2020 Gains in South Texas Vex Democrats. *Time.* https://time.com/5946848/donald-trump-hispanic-latino-voters-texas/ (last accessed June 4, 2021).

THE BLACK VOTE

FIONA M. DAVIDSON AND RASHAUNA MICKENS HINTZ

For the last sixty years, from the civil rights era into the twenty-first century, the most consistent voting pattern evident in the United States is Black support of the Democratic Party. In fact, at no time since race data has been gathered on voting and party affiliation has a Republican candidate garnered more than 40 percent of the Black vote. In the early and middle part of the twentieth century, the partisan gap ranged from 40 percent (1936) to a low of 20 percent in 1956; then, in the wake of the Civil Rights Act (1964) and the Voting Rights Act (1965), Black support for the Democrats reached a high of 94 percent in 1964. Since then, the Black Democratic vote has never dropped below 80 percent, varying from a low of 83 percent in 1992, when 7 percent of the Black vote went to third-party candidate Ross Perot, to a record 95 percent in the 2008 election. While the Black vote for the Democrats dropped fractionally in 2012, that election saw the first and only time that Black turnout exceeded white turnout (Krogstad and Lopez, 2017).

CIVIL RIGHTS AND BLACK VOTING

Within a year of that record turnout, the Supreme Court's decision to eviscerate the Voting Rights Act by removing pre-clearance for voting rules changes at the state level led to a series of state-level decisions that have restricted voting rights, with a disproportionate effect on voters of color, especially Black voters. The timing of the *Shelby County vs. Holder* Supreme Court decision (2013) is intimately tied to the Republican fear of an energized, overwhelmingly Democratic, Black voter base that emerged in 2012. Those voter suppression measures have included restricting early voting, reducing voting hours, removing voting locations, restricting mail-in voting, instituting voter ID laws, and purging voter rolls (Center for American Progress, 2018), all measures that are technically "race-blind" but impact poorer, more marginalized communities far more than they do other voters.

These measures are a continuation of decades of white, largely Republican, pushback against the Voting Rights Act and culminated in the 2018 midterms where Stacey Abrams's narrow loss to Brian Kemp in the Georgia gubernatorial election, after Kemp as secretary of state had purged more than three hundred thousand voters from the Georgia voting rolls, highlighted the dangers of voter suppression. The result of that loss, as well as the activism engendered by the deaths of George Floyd and Breonna Taylor, was a revitalization of efforts to register Black voters and facilitate access to the polls for voters of color. Through organizations like Stacey Abrams's Fair Fight, as well as veteran voter rights organizations such as HeadCount, WhenWeAllVote and VotoLatino, voter registration for traditionally marginalized groups across the country from California to Georgia reached record levels in the run-up to the 2020 presidential election. High rates of registration did not result in record turnout, but Black turnout in 2020 (63 percent) was second only to white voter turnout (71 percent) and exceeded Black turnout in every election cycle other than the 2008 and 2012 Obama elections.

BLACK VOTING BY COUNTY

Tied to the twin historical circumstances of slavery and twentieth-century rural-urban migration, the Black vote displays one of the most distinctive geographical patterns in the United States (figure 6.22). Counties with high percentages of Black populations are clustered in two distinct locations, found in a broad belt of rural counties in the South, from the Mississippi Delta, through the fertile black earth crescent of Mississippi, Alabama, Georgia, the Carolinas, and Virginia, and in a series of metropolitan counties throughout the Northeast and Midwest. Except for Wayne County, Michigan (urban Detroit) the counties where Black residents make up more than 32 percent of the population form an almost continuous band from northern Louisiana to southeast Pennsylvania, passing through most of the major urban centers of the South and East, including Atlanta, Memphis, Washington, DC, and Philadelphia. Outside of this southern and eastern concentration there are relatively few counties with more than 4 percent Black population, and most of them are metropolitan counties in the industrial Midwest from Illinois to Massachusetts.

Given the strong relationship between Black voters and the Democrats, the crescent of high Black population/high Democratic voting is unsurprising throughout the Deep South and into the Carolinas and Virginia. However, because the highest class of Black population is only 31 percent, and because white voters in these southern states, especially in rural areas, vote as faithfully Republican as their Black neighbors vote Democratic, the high Black population/high Democratic vote counties are surrounded by a band of high Black population/high Republican vote counties. Overall, in the 2020 election the three hundred counties with the highest concentrations of Black voters supported Joe Biden by over 18 points more than the national average, a slightly greater gap than in 2016, and as turnout was up, the overall Democratic vote for those counties was also up from 62 percent to 63 percent, a gap that accounts for most, if not all, the increased turnout.

BLACK WOMEN IN THE 2020 ELECTION

It is important to remember that, despite the consistently high levels of support for the Democrats, Black voters are no more monolithic than any other group. While there is relatively little geographic variation in the Black vote—all counties with large Black populations demonstrate support for the Democrats—within the Black voting population there is one particularly important cleavage, the difference between male and female voting. In the 2020 election, Black men were less likely to vote for Joe Biden than were Black women by 14 points (79 percent to 93 percent), although it is important to note that Black women were twice as likely to vote as Black men (making up 8 percent of the total vote, while Black men accounted for only 4 percent). Much has been made of the increased Black vote for Trump, increasing by 3 percent for men and 4 percent for women, but except for the record Black vote for the Democratic Party in the Obama elections (2008 and 2012) these current numbers are entirely in line with Black voting over the last forty years. Since the early 1970s, Black men have historically been much more likely to vote Republican than have Black women; in 1972, 23 percent of Black men supported Nixon (Wright Rigeur, 2020). Support for the Republicans among Black men declined during the Reagan/Bush presidencies but rebounded again under Clinton and the second George Bush. Black women, on the other hand, are much more reliably Democratic, with their support for Democratic candidates never falling below 90 percent in the past half century.

BLACK CANDIDATES

The 2020 election resulted in the election of a record fifty-nine Black candidates to the US House and Senate, bringing Black representation in the House almost on par with the general population (12 percent versus 13 percent), while the Black population remains woefully underrepresented in the Senate (3 percent). Of those elected, the vast majority (94 percent) were Democrats, a higher percentage than for minority candidates as a whole, only 83 percent of whom were Democrats. The importance of Black women is once again apparent; 46 percent of the Black representatives in Congress are female, a far higher percentage than for any other group and vastly more than the 19 percent of white representatives who are female. Geographically, Black members of the House and Senate overwhelmingly represent coastal and urban areas. California, New York, Texas, Illinois, Florida, and Georgia are the only states with three or more Black representatives. The 2020 election also marked record success for Black candidates in state legislative elections, with Black representation reaching almost 10 percent for the first time, although a minuscule thirteen of those state representatives are Republicans. Across the country, Black state representation more closely resembles the distribution of the Black population, with Black representation in Georgia, Louisiana, Mississippi, and South Carolina exceeding 25 percent, and Maryland at almost 30 percent.

The importance of the Black vote, especially the activism of Black women, in the 2020 election cannot be overstated. In cities across the United States, especially in Philadelphia, Detroit, Milwaukee, and Atlanta, the movement to register and mobilize the Black population was critical to the narrow margin of Joe Biden's wins in Pennsylvania, Michigan, Wisconsin, and Georgia. The mobilization of the Black population in Georgia's runoff elections created the narrow Senate victories for Democrats Warnock and Ossoff in January 2021. However, it is this activism and these narrow victories that have instigated another round of voter suppression measures in state legislatures around the country. Since January 2021, Republican state legislators have introduced more than four hundred different measures to restrict early days, limit mail-in voting, reduce ballot drops, and even criminalize the provision of food and water to people

FIGURE 6.22

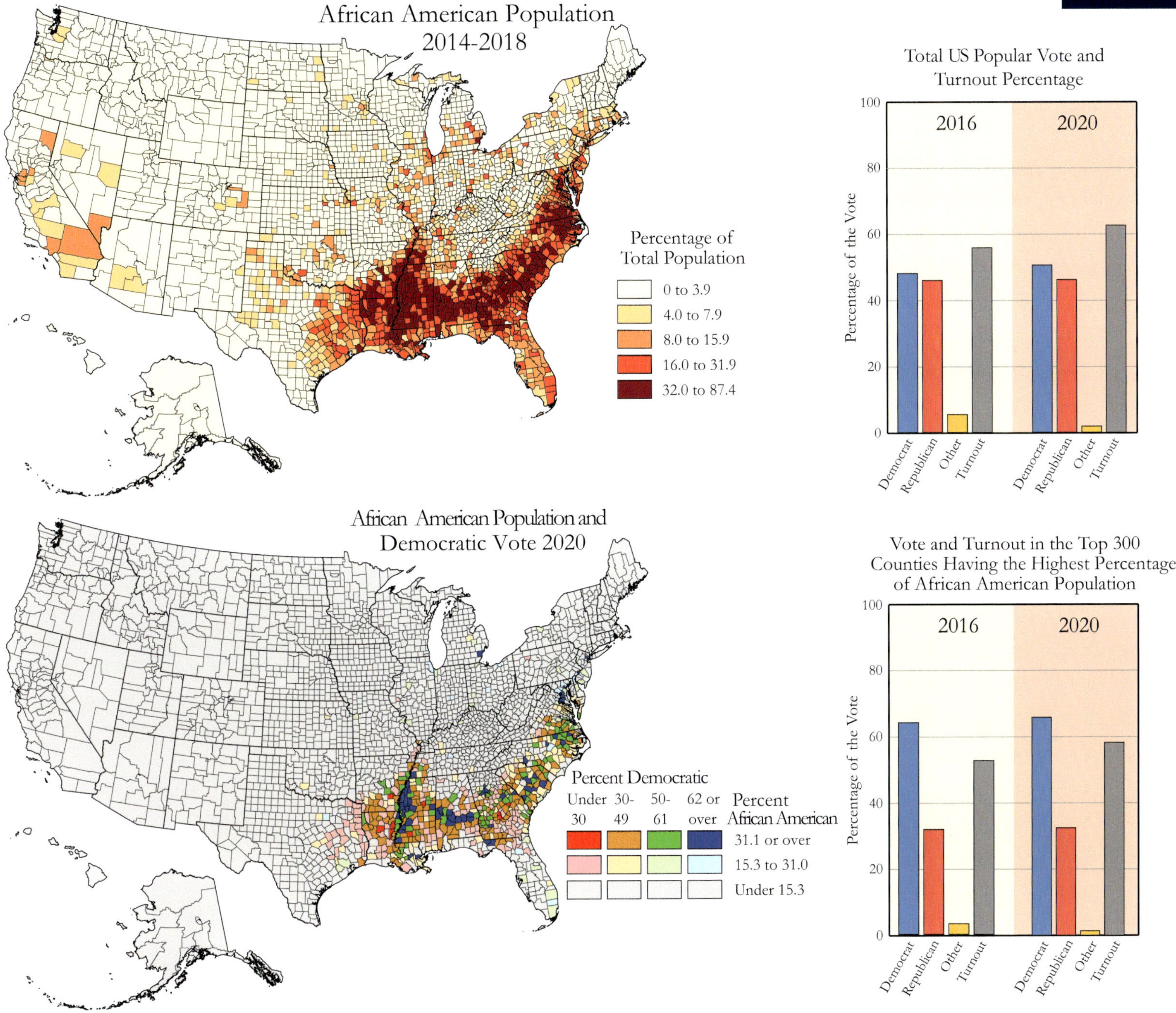
African American Population
2014-2018
Percentage of
Total Population
0 to 3.9
4.0 to 7.9
8.0 to 15.9
16.0 to 31.9
32.0 to 87.4
Total US Popular Vote and
Turnout Percentage
2016
2020
Percentage of the Vote
100
80
60
40
20
0
Democrat
Republican
Other
Turnout
African American Population and
Democratic Vote 2020
Percent Democratic
Under 30
30-49
50-61
62 or over
Percent
African American
31.1 or over
15.3 to 31.0
Under 15.3
Vote and Turnout in the Top 300
Counties Having the Highest Percentage
of African American Population

waiting in voter lines. While these measures are technically nonpartisan and race-blind, they will differentially impact populations who are urban, poor, marginalized, and minority, and above all, who vote Democratic. As Alice O'Lenick, the Republican chair of the Gwinnet County Board of Registrations and Elections said in January 2021 in reference to voting laws in the state of Georgia, "They don't have to change all of them, but they've got to change the major parts of them so that we at least have a shot at winning."

REFERENCES

Budiman, Abby. 2020. Key Facts about Black Eligible Voters in 2020 Battleground States. *Pew Research Center*. October 21, https://www.pewresearch.org/fact-tank/2020/10/21/key-facts-about-black-eligible-voters-in-2020-battleground-states/ last accessed June 18, 2021.

Center for American Progress. 2018. Voter Suppression During the 2018 Midterm Elections. https://www.americanprogress.org/issues/democracy/reports/2018/11/20/461296/voter-suppression-2018-midterm-elections/ last accessed June 15, 2021.

Center for American Women in Politics. 2020. Black Women in the 117th Congress. https://cawp.rutgers.edu/sites/default/files/resources/higher_heights_black_women_in_congress_fact_sheet_12.30.pdf last accessed June 10, 2021.

Jones, Martha S. 2020. Vanguard: How Black Women Broke Barriers, Won the Vote, and Insisted on Equality for All. New York: Basic Books.

Krogstad, Jens Manuel, and Mark Hugo Lopez. 2017. Black Voter Turnout Fell in 2016, Even as a Record Number of Americans Cast Ballots. *Pew Research Center*. May 12, https://www.pewresearch.org/fact-tank/2017/05/12/black-voter-turnout-fell-in-2016-even-as-a-record-number-of-americans-cast-ballots/ last accessed June 1, 2021.

Rigeur, Leah Wright. 2020. The Major Difference between Black Male and Female Voters. *The Atlantic*. November 21, https://www.theatlantic.com/culture/archive/2020/11/why-black-men-and-women-vote-so-differently/617134/.

ONE HUNDRED YEARS OF WOMEN'S SUFFRAGE AND THE 2020 ELECTION

FIONA M. DAVIDSON AND KIMBERLY JOHNSON MAIER

On August 18, 1920, Tennessee became the last state to ratify the Nineteenth Amendment to the US Constitution; when the amendment was certified on August 26, twenty-six million women were enfranchised at the stroke of a pen. However, for many millions of those women—African American, Native American, Hispanic, and Asian—that right was more theoretical than practical. One hundred years and sixty-nine days later, on November 3, 2020, Kamala Devi Harris, a child of Jamaican and Indian immigrant parents, became the forty-ninth vice president elect of the United States.

A BRIEF HISTORY OF THE US WOMEN'S SUFFRAGE MOVEMENT

In 1776, Abigail Adams wrote to John Adams, "In the new code of laws I desire you would remember the ladies and be more generous and favorable to them than your ancestors." Through their writing of the Constitution, the Founding Fathers laid the groundwork for determining who was and was not considered a citizen. These decisions ultimately determined the role women and women of color would have in the US political system. While Abigail Adams urged her husband, John Adams, to include women within the Constitution, they were not. Consequently, women and those of color were barred from political participation through informal and formal customs and laws. The goal of the women's suffrage movement was to provide a way for women to have equal political participation.

In 1848, a group of abolitionists gathered at Seneca Falls, New York, to discuss women's rights. The movement ultimately gathered steam into the 1850s with some divisions following the Civil War. For example, the Fourteenth Amendment (1868) extended constitutional protection to all men and the Fifteenth Amendment (1870) gave suffrage to Black men. In response, Susan B. Anthony and Elizabeth Cady Stanton created the National Woman Suffrage Association with the goal of advocating for a universal suffrage amendment in the US Constitution. The American Woman Suffrage Association, a pro-Fifteenth Amendment group, on the other hand, took a state-by-state approach. In 1890, the two groups united their efforts under the National American Woman Suffrage Association.

Over the course of the suffrage movement, women organized, paraded in the streets, picketed, and participated in hunger strikes. For example, on March 3, 1913, the day before President Wilson's inauguration, women organized and marched the inauguration parade route to show their exclusion from the democratic process. This same tactic has been utilized recently, most notably the Women's Marches in the days following the 2017 inauguration. States in the western part of the United States began extending the vote to women in the late 1800s, with Wyoming first in 1890. In 1920, women's suffrage was finally realized with the passage of the Nineteenth Amendment. Today, women play a significant role in the political process through their increased representation at all scales.

WOMEN IN THE 2020 ELECTION

This section of the atlas examines both the behavior of women as voters in the 2020 election and their fates as candidates at the national and state levels. National data on voting and turnout by gender was not available until Gallup began national exit polls in 1952 (Gallup Presidential Election Center) and at that time women tended to participate less than men and were more likely to vote Republican. By the 1960s, however, while women still tended to underperform in terms of turnout, the partisan gender gap had begun to emerge. In 1964 women preferred Johnson to Goldwater by 24 points and by 2 points more than men and in every election since then (with the exception of Carter/Ford 1976 and Clinton/Bush 1992, when third parties drew significant support from men) women have preferred Democratic candidates by anything from 1 percent (1972) to 12 percent (2016). In 2020, this preference continued, with the same 12 percent gap between male and female preference for the Democratic candidate; given the lack of third-party voting in this election, the Republican gap is also 12 percent. At the state level (figures 6.23 and 6.24) the most interesting change from 2016 is the consolidation of Republican states from Idaho to West Virginia—where in 2016 several of these states demonstrated a split in the vote (Republican win, but with Democratic majority from women voters) the 2020 vote shows Republican majorities from both men and women voters. Several other states, notably Colorado and Virginia, experienced a shift from a split vote (men Republican/women Democratic) to both voting Democratic, but most important are the states that shifted from Republican win/women Democratic to Democratic win/women Democratic. In all these states, the gender split did not change—a majority of men voted Republican, and a majority of women voted Democratic, but the relative balance between the two changed sufficiently that Wisconsin, Michigan, and Pennsylvania shifted from Trump in 2016 to Biden in 2020.

It is important to note that women are not a monolithic voting block and exit polls indicate that white women were not only more likely to vote for Trump than for Biden (55 percent to 44 percent), but that the Republican preference increased by 1 percent from the 2016 election. Conversely, Black and Latino women favored Biden by 90/9 and 69/30 respectively, and these differences have a startling regional pattern (figure 6.25). In states in the South in particular, that trend Republican but have large minority populations, the gap between white and (primarily) Black women is a striking 60 percent or more. The gap is almost as large in adjacent Republican states from Oklahoma to Ohio. Of all the states where there is at least a 40 percent gap between the voting preference of white and Black women, only Georgia and Michigan—by narrow margins—voted for Biden, and in both states the Democratic vote was largely driven by Black women in the metropolitan counties. The explanation of this regional gap in Democratic voting is complex, but relates to the ways in which women, particularly white women in the South, prioritize race versus sex in their self-identification (Maxwell and Shields 2019).

WOMEN VOTING BY COUNTY

Unsurprisingly, there is relatively little variation in gender distribution at the county level across the United States (figure 6.26), but a careful examination of county-level voting trends for women can be discerned in counties where women are over- or underrepresented, even by small margins. If we compare the top and bottom three hundred counties in terms of female representation (those with more than 52.6 percent women, or fewer than 47.4 percent women) there is a dramatic difference in the voting preference; the top counties preferred Biden over Trump by 26.22 percent (an increase of 1.35 percent over 2016) and the bottom counties voted for Trump by 65.26 percent to 32.92 percent (an increase of 0.47 percent over 2016). Counties in which women make up less than 47.3 percent of the population are unsurprisingly predominantly rural and concentrated in the West and Great Plains. These counties have high levels of employment in rural-based agriculture and extractive industries and, in keeping with national voting patterns for men and rural areas, the majority of these counties voted overwhelmingly Republican. There are relatively few counties where women are underrepresented that voted Democratic, but those that did are largely counties with large minority populations (South Texas, western Alaska) or environmental amenity counties like those of central Colorado, where the outdoor activities are leisure-based rather than extractive and attract a younger, less-conservative male population. Counties where women are overrepresented (52.6 percent or more) are rare in the West, and much more common in the South, particularly in the Lower Mississippi Valley and through the Black Belt to the Atlantic Coast. These counties are fairly evenly split between Democratic and Republican voting. Those that voted Democratic are counties with substantial minority populations, both rural (South Texas, the Black Belt, and Native American reservations) and urban (DeKalb County, Georgia, Wayne County, Michigan) reflecting the continuing trend for white women to favor Trump over Biden by 4 points. From southern Arkansas to South Carolina, rural and suburban counties with low levels of minority women demonstrate high, sometimes very high (70 percent+) levels of support for the Republicans.

Similar voting patterns are evident in counties where women are over- or underrepresented in the labor force (figure 6.27). High levels of female participation in the labor market are most prevalent in the Northeast, across the Midwest and upper Great Plains, and along the West Coast. In terms of voting, of the six hundred counties where more than 60 percent of women participated in the labor market, slightly more than half (59 percent) voted for Biden. Unsurprisingly, the counties with high female participation/Democratic vote are clustered in metropolitan cores throughout the country, as well as in exurban counties in the Northeast and environmental amenity counties in the Rockies, while counties with high female participation and Republican voting stretch from northern Nevada

FIGURE 6.23

Gender Gap 2016

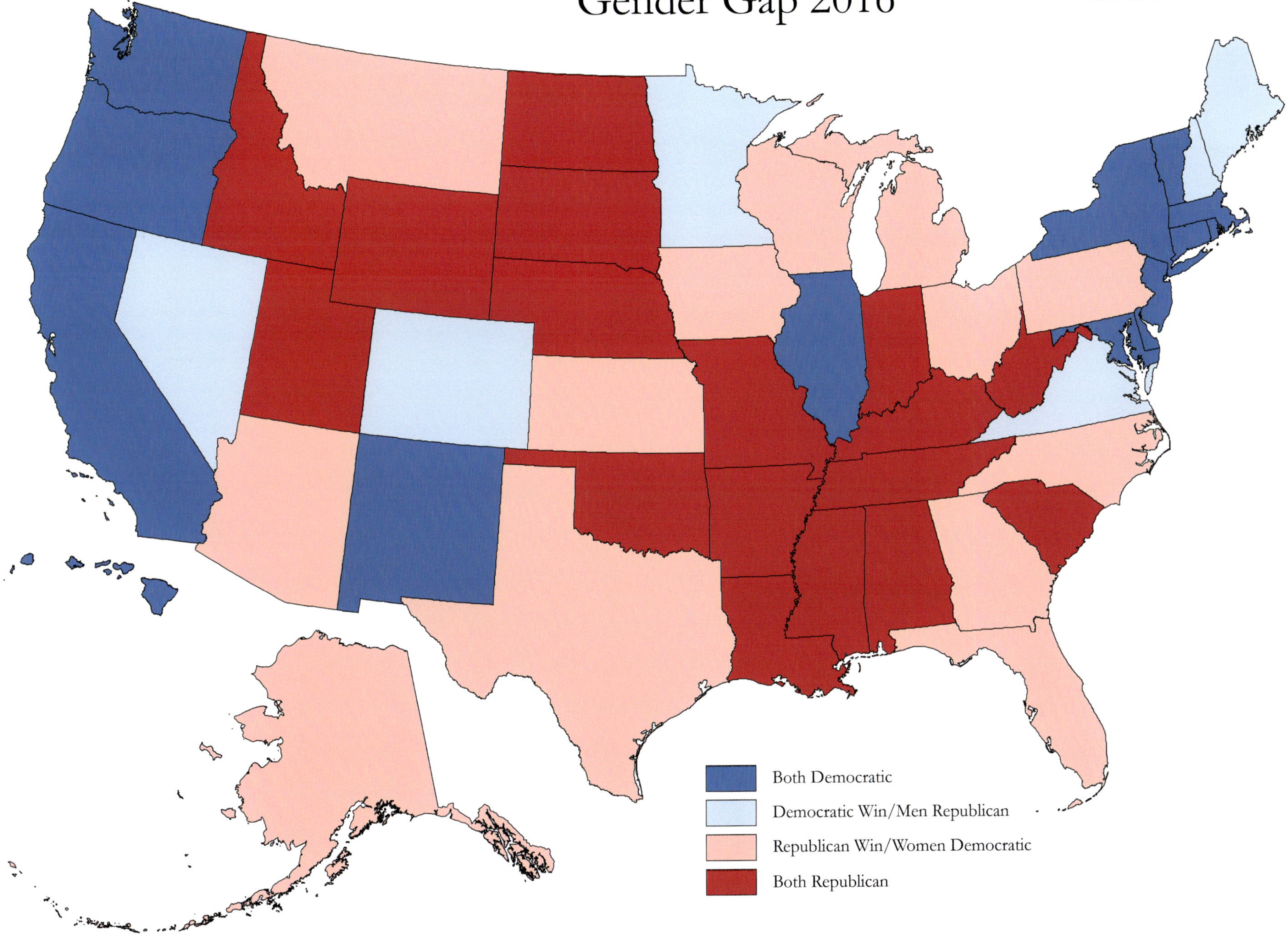

FIGURE 6.24

Gender Gap 2020

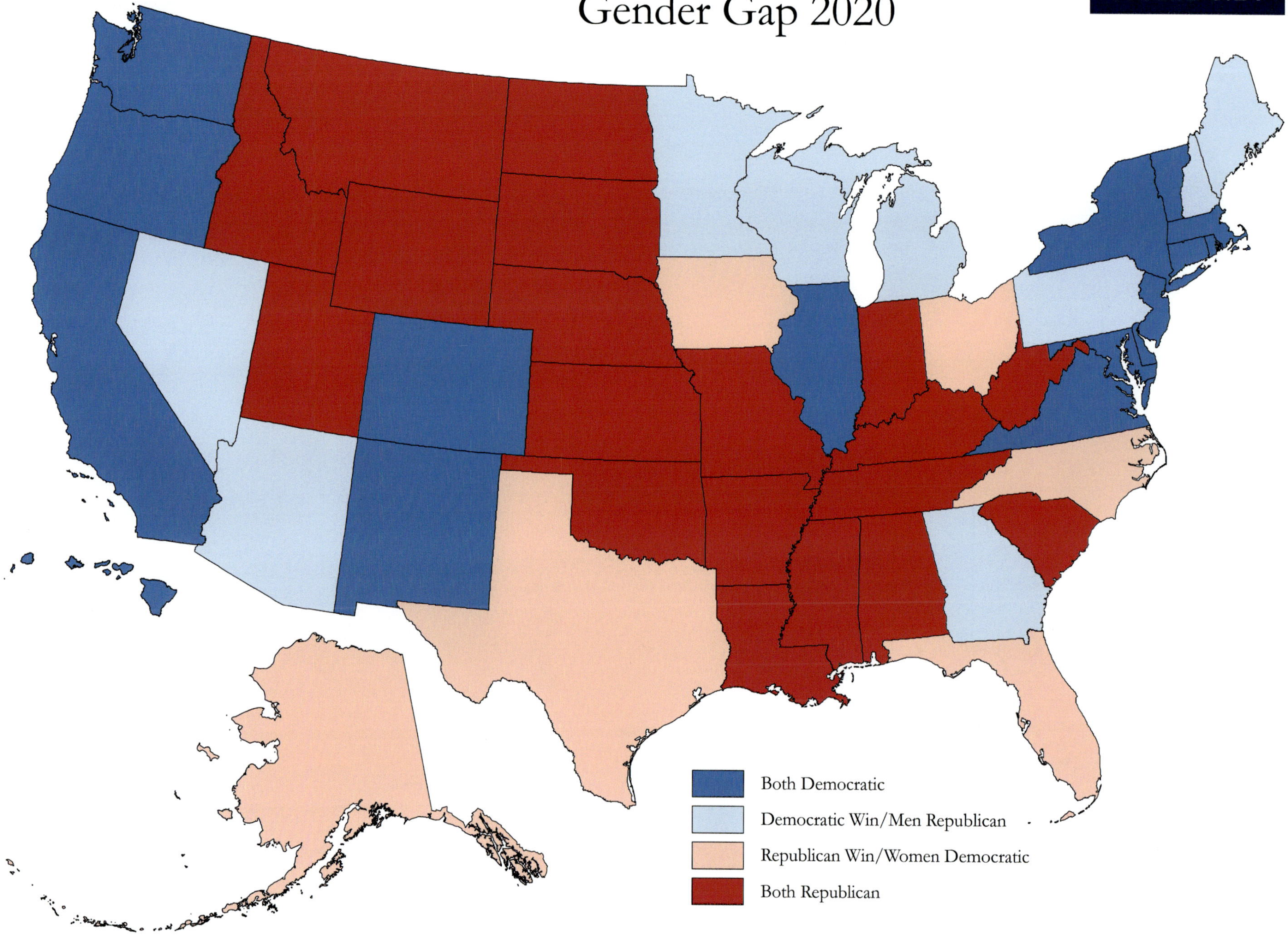

FIGURE 6.25

Race Gap for Democratic Women's Votes, 2020

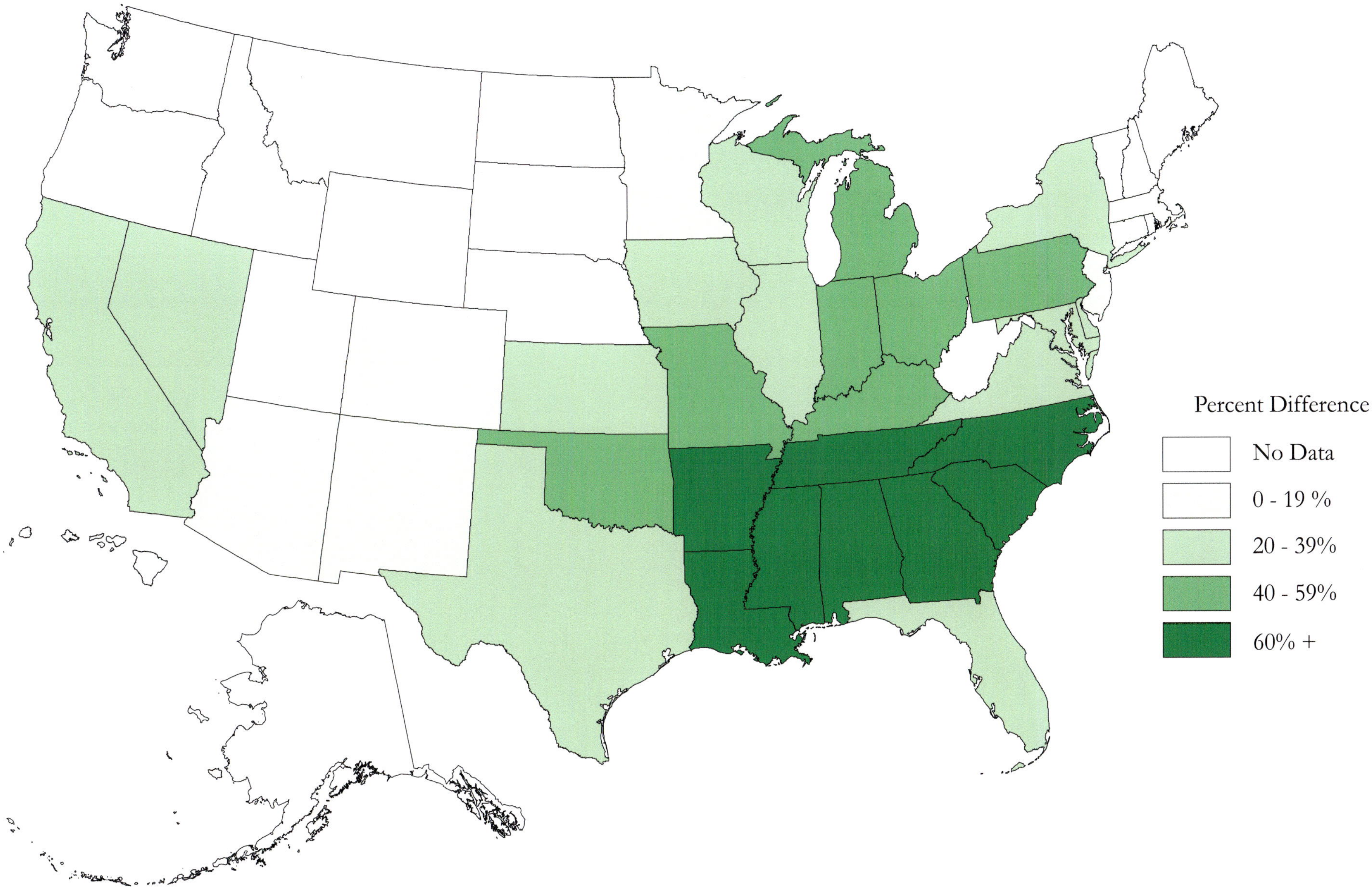

FIGURE 6.26

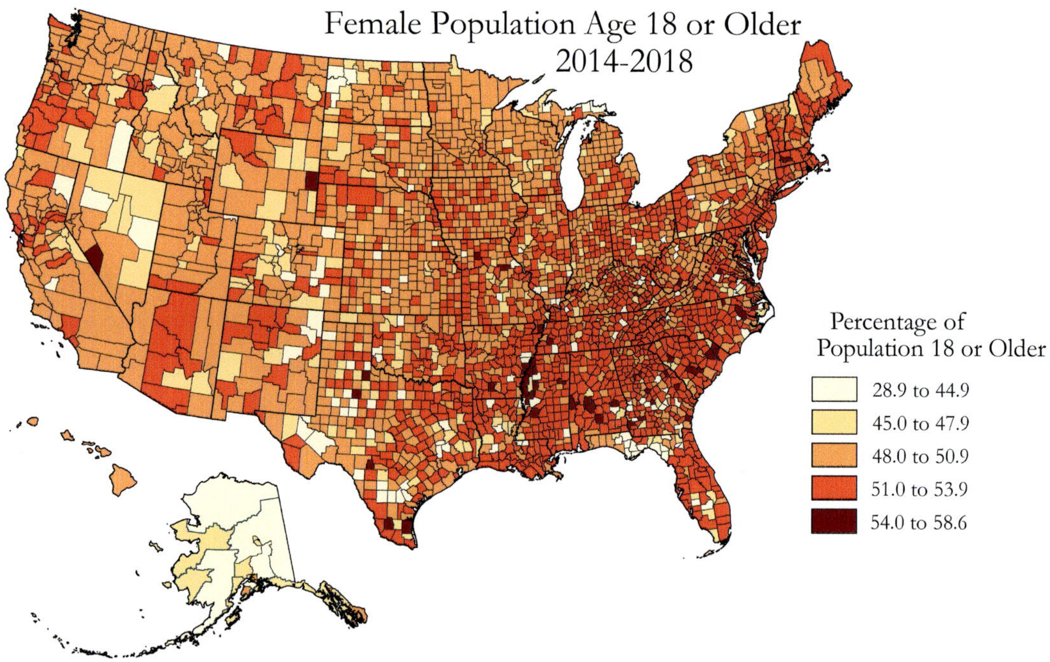

Female Population Age 18 or Older
2014-2018
Percentage of
Population 18 or Older
28.9 to 44.9
45.0 to 47.9
48.0 to 50.9
51.0 to 53.9
54.0 to 58.6

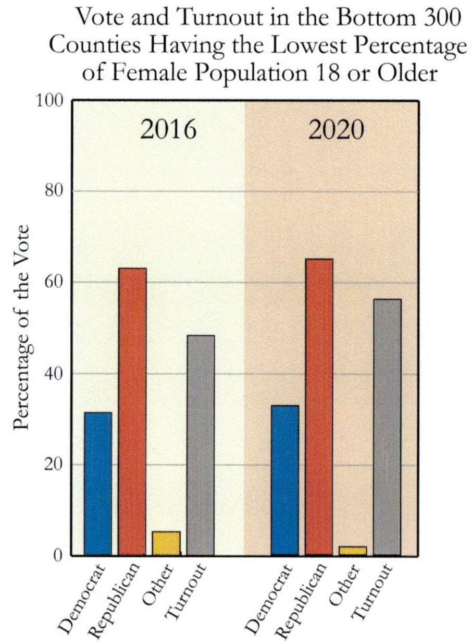

Vote and Turnout in the Bottom 300
Counties Having the Lowest Percentage
of Female Population 18 or Older
2016
2020
Percentage of the Vote
100
80
60
40
20
0
Democrat
Republican
Other
Turnout
Democrat
Republican
Other
Turnout

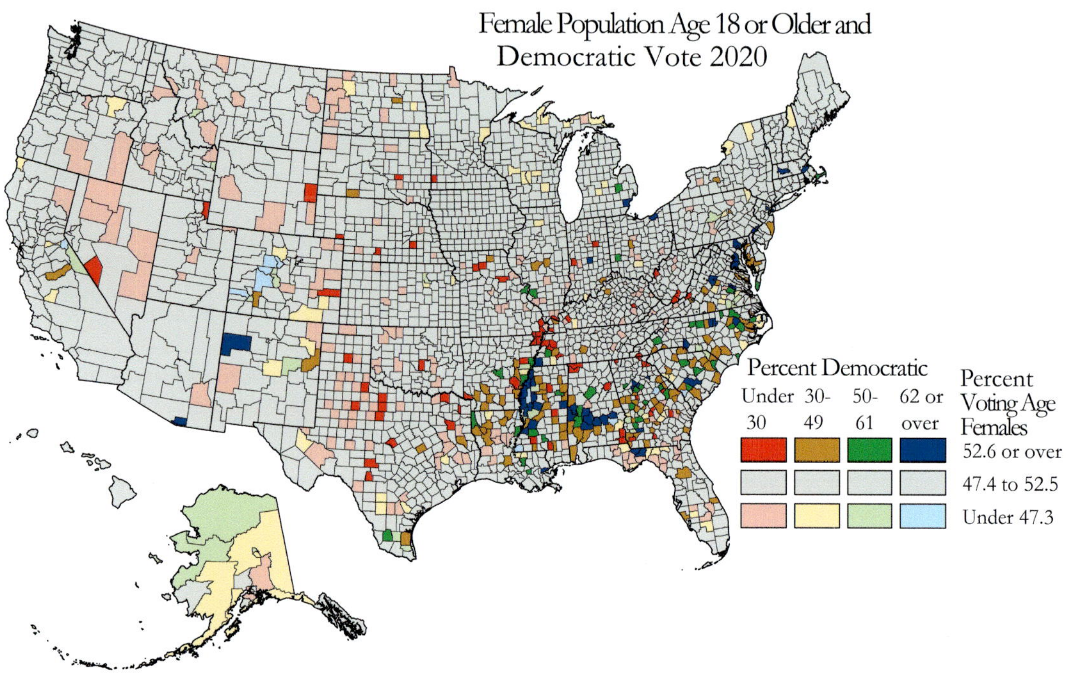

Female Population Age 18 or Older and
Democratic Vote 2020
Percent Democratic
Under 30
30-49
50-61
62 or over
Percent Voting Age Females
52.6 or over
47.4 to 52.5
Under 47.3

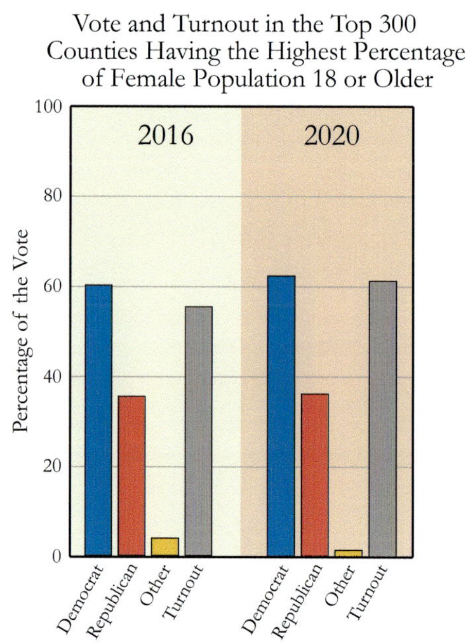

Vote and Turnout in the Top 300
Counties Having the Highest Percentage
of Female Population 18 or Older
2016
2020
Percentage of the Vote
100
80
60
40
20
0
Democrat
Republican
Other
Turnout
Democrat
Republican
Other
Turnout

FIGURE 6.27

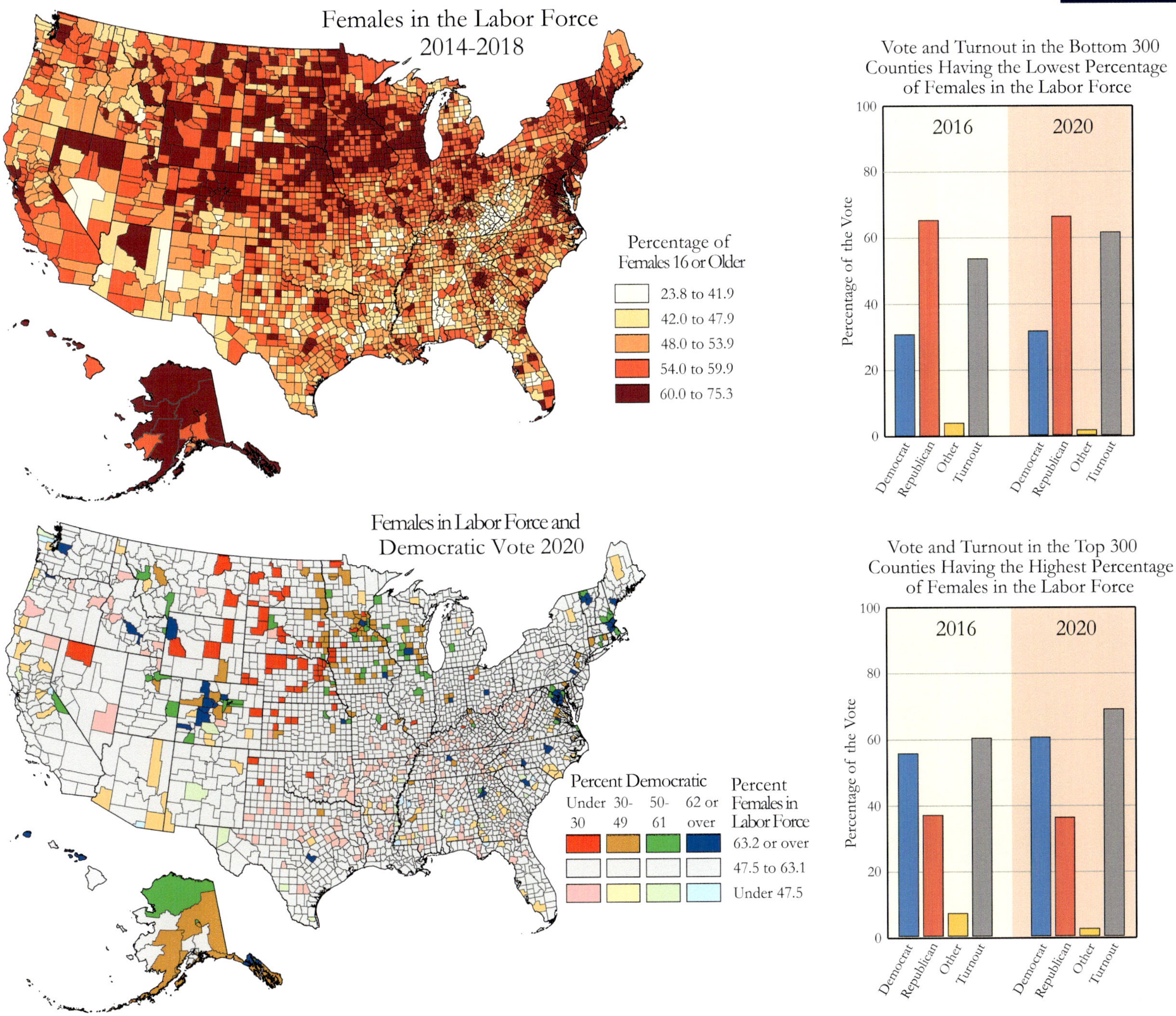
Females in the Labor Force
2014-2018
Percentage of
Females 16 or Older
23.8 to 41.9
42.0 to 47.9
48.0 to 53.9
54.0 to 59.9
60.0 to 75.3
Vote and Turnout in the Bottom 300
Counties Having the Lowest Percentage
of Females in the Labor Force
2016
2020
Percentage of the Vote
0
20
40
60
80
100
Democrat
Republican
Other
Turnout
Females in Labor Force and
Democratic Vote 2020
Percent Democratic
Under 30
30-49
50-61
62 or over
Percent
Females in
Labor Force
63.2 or over
47.5 to 63.1
Under 47.5
Vote and Turnout in the Top 300
Counties Having the Highest Percentage
of Females in the Labor Force
2016
2020
Percentage of the Vote
0
20
40
60
80
100
Democrat
Republican
Other
Turnout

FIGURE 6.28

Women in National Legislature, 2021

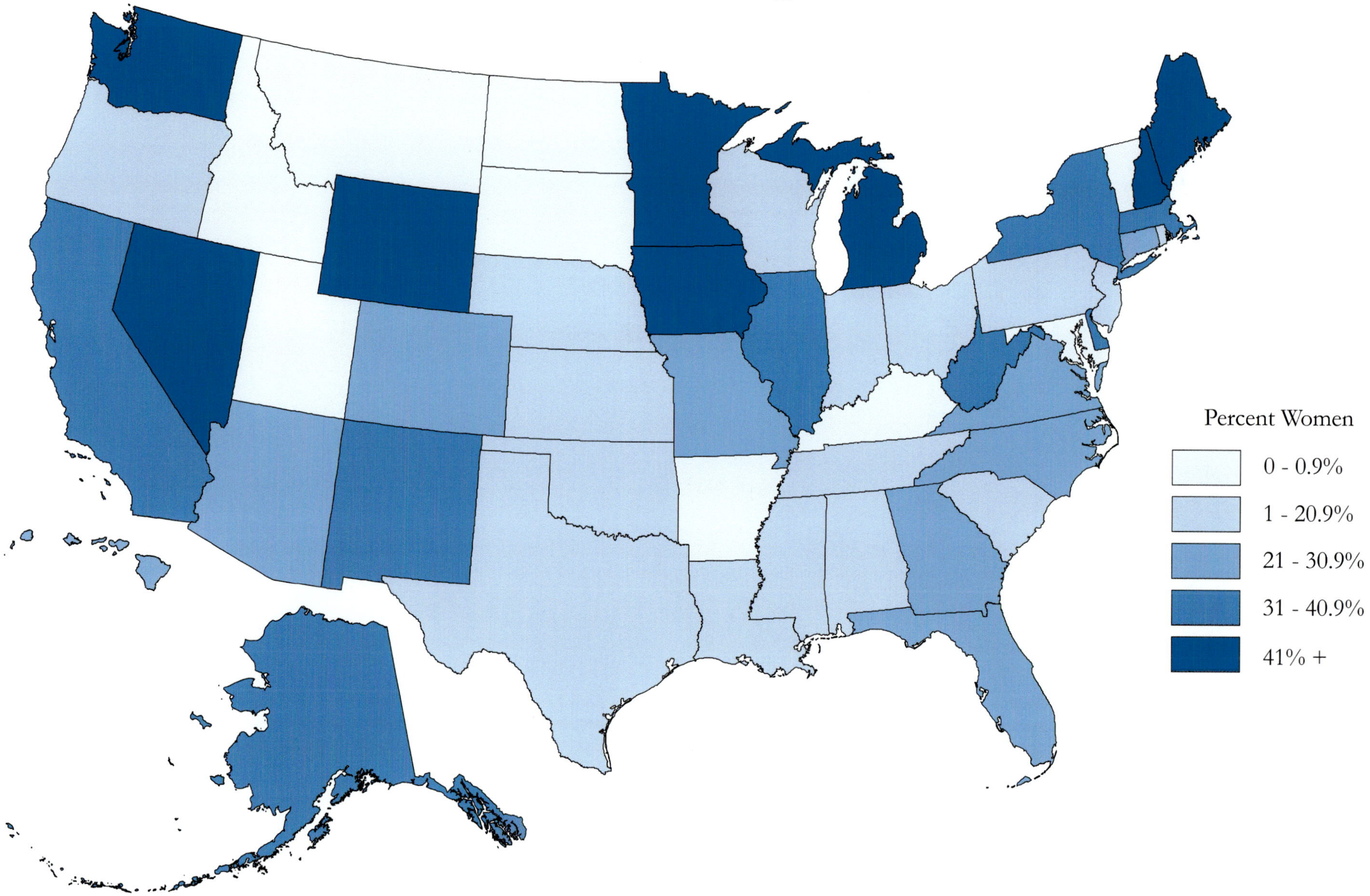

FIGURE 6.29

Women in State Legislatures, 2021

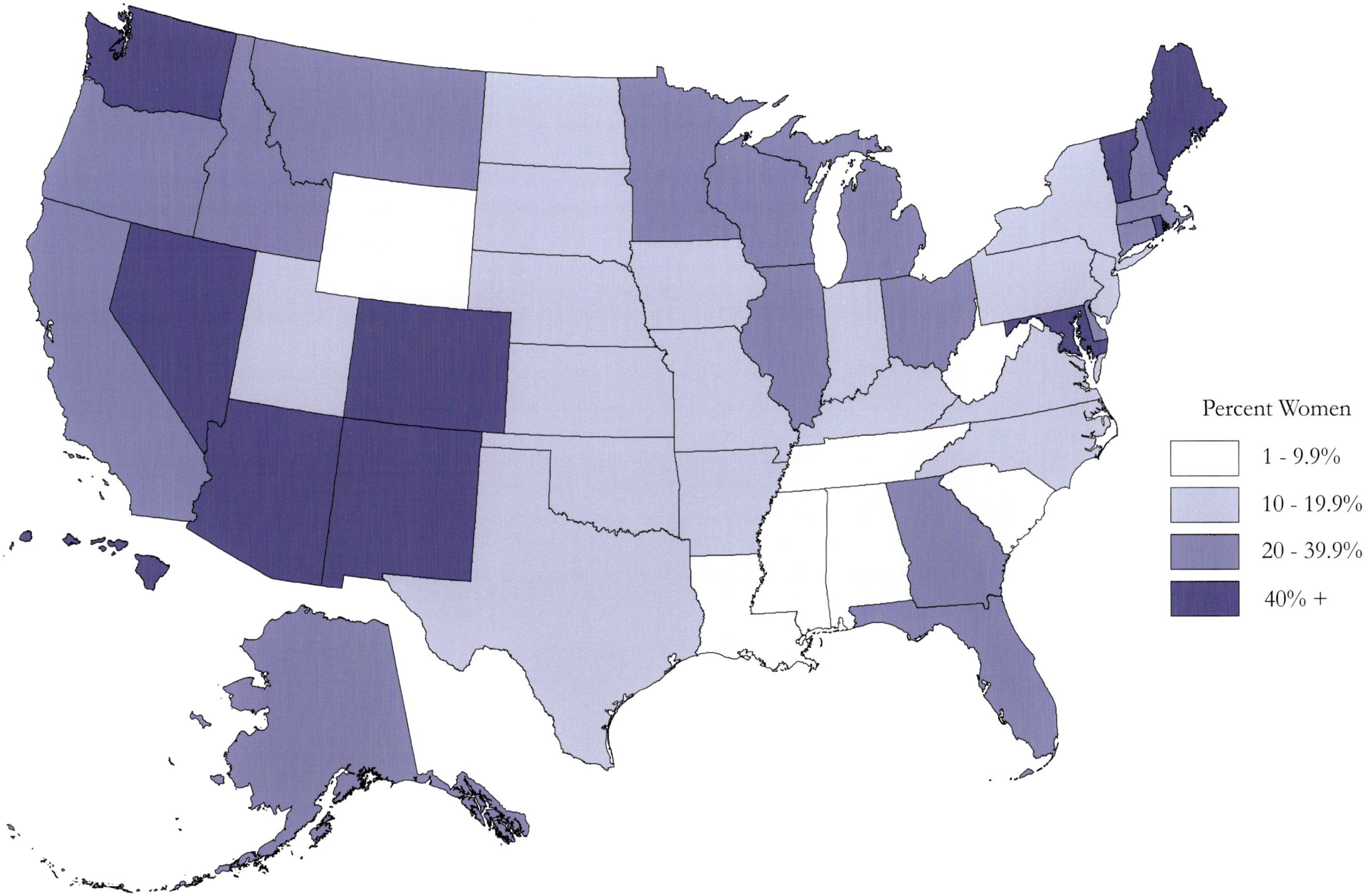

across the upper Great Plains to southern Wisconsin. These counties are overwhelmingly rural, white, and heavily dependent on agriculture, extractive, and (in Wisconsin) manufacturing industries. A lower rate of labor force participation for women is a feature of counties from West Virginia, through Appalachia, and into the South Central, with a cluster of low participation counties in the Interior West. The vast majority of these counties (97.2 percent) voted Republican, and the exceptions are almost all counties with large minority populations (South Dakota, New Mexico, Arizona, South Texas, and the Black Belt).

WOMEN AS CANDIDATES

While women have been active as voters since before the Nineteenth Amendment (many states permitted some, if not all, women to vote in local and state elections prior to 1921) widespread female participation as candidates is a much more recent phenomenon in the United States. At the national level, the first woman sworn into the House of Representatives was Jeannette Rankin (R) from Montana in 1917 and in the Senate, eighty-seven-year-old Rebecca Felton (D) of Georgia, who was appointed to fill a vacancy and served just twenty-four hours in 1922. The pattern of women being appointed to fill positions vacated by their husbands and fathers continued throughout the twentieth century and includes current Alaska senator Lisa Murkowski (R), who was appointed to fill her father's seat when he resigned in 2002, although she has since won three elections in her own right.

There was a great deal of press attention when the 2020 election represented a high point for women's candidacy for national office, with a record 583 women running for the House and 60 for the Senate (these records include both primary and general election numbers). However, 50 percent of these women lost their primaries, and more than 60 percent lost their general election runs. Overall, women are both less likely to run for national office (only 35 percent of the candidates for House and Senate elections in 2020 were women) and even less likely to win (27 percent of winning candidates were women) when they do run (Center for American Women in Politics 2020). Consequently, an examination of the current House and Senate indicates that, as of January 2021, women make up only 26 percent of the combined membership. At the state level, there are nine states with no female representation (one of which, Vermont, has never had a woman elected to national office) and only five states (Iowa, New Hampshire, Nevada, Washington, and Wyoming) with more women than men in their delegations (figure 6.28). There is no clear partisan pattern to this representation; while Democratic states like California and Washington have higher percentages of women in their delegations, so do Republican states with small delegations and popular, long-serving female candidates (Alaska, Iowa, Wyoming).

Women also made records at the state level, with 3,583 female candidates in the 2020 state legislative elections, representing 24 percent of the total. Of those women, 1,191 won their races, and currently women hold 28.92 percent of state legislative positions. Across the United States, northern and western states demonstrate the highest representation for women in state legislatures and southern states the lowest (figure 6.29). These states demonstrate a clearer partisan pattern than the national delegation. In seven states, six of which are in the South or Appalachia, women have less than 10 percent of their state legislative seats and they are dominated by Republican voters, whereas states with substantial female representation in their state legislatures (40 percent+) all voted Democratic in the 2020 presidential election.

Overall, the 2020 election saw a continuation of several long-standing voting patterns, with women preferring Democratic candidates and continuing to make inroads into both state and national representation. The election of Kamala Harris as vice president both reinforces this progress and also reminds us that women are not a monolithic voting bloc. As in 2016 and previous elections, there was significant divergence between the way white women and women of color voted; these variations are discussed in the chapter section on the Black vote.

REFERENCES

Center for American Women in Politics. https://cawp.rutgers.edu (last accessed June 9, 2021).

Gallup. US Presidential Election Center. https://news.gallup.com/poll/267905/gallup-election2020-center.aspx (last accessed June 14, 2021).

Maxwell, Angie, and Todd Shields. 2019. *The Long Southern Strategy*. Oxford: Oxford University Press.

Roesch Wagner, Sally, ed. 2016. *The Women's Suffrage Movement*. New York: Penguin Books.

Tetrault, Lisa. 2014. *The Myth of Seneca Falls: Memory and the Women's Suffrage Movement 1848–1898*. Chapel Hill: University of North Carolina Press.

RELIGION AND THE 2020 US PRESIDENTIAL ELECTION

FRED M. SHELLEY

For many years, religion has played an important role in predicting election outcomes. Outside of the South after the Civil War, Protestants were predominantly Republicans. Roman Catholics, many of whom were immigrants or descended from immigrants from primarily Catholic countries such as Ireland and Italy, were predominantly Democrats. Strong support from his fellow Catholics helped John F. Kennedy, the first Roman Catholic president, to win the 1960 election narrowly.

As the first evangelical Protestant president, Jimmy Carter got significant support among his fellow evangelicals in 1976. Since then, however, white evangelical Protestants have represented the core of Republican support in the South and have been an important source of Republican votes elsewhere. Meanwhile, many Catholics have drifted away from their traditional allegiance to the Democratic Party. This pattern held up in the 2020 election. According to the exit poll, 81 percent of white evangelical voters supported Donald Trump as compared to 18 percent for Joe Biden. Trump got 61 percent among all Protestants, including primarily Democratic Black Protestants and more moderate mainline Protestants. Catholics were split evenly, with 50 percent for Trump and 49 percent for Biden, who is a Catholic. On the other hand, people who claimed no religion voted 72 percent for Biden and 26 percent for Trump. These three groups represented 26 percent, 22 percent, and 21 percent of the electorate, respectively. Jewish and Muslim voters, although fewer in number, also supported Biden strongly.

What accounts for the strong support for Trump and previous Republican candidates, including candidates at the state level, among evangelicals? Issues driving this support are cultural, not economic. Prior to the early twentieth century, a key issue was prohibition, although this question became settled by the enactment of the Twenty-First Amendment, which repealed the earlier Eighteenth Amendment that banned the sale of alcoholic beverages, in 1933. Over the past several decades, two key issues that have driven evangelical support for the Republican Party have been LGBT rights, including same-sex marriage, and abortion. According to a 2019 survey by the Pew Research Institute, Democrats are much more likely to support the legality of same-sex marriage than are Republicans. However, there is evidence that this issue may disappear from the political scene. By 2019, nearly half of all Republicans and well over half of Republicans less than thirty years of age were supportive of same-sex marriage.

The question of abortion has remained significant in the nearly fifty years since the Supreme Court struck down state restrictions on abortion in *Roe v. Wade* in 1973. Both evangelicals and Roman Catholics were likely to oppose the court's decision, although today many Americans do not want the decision to be overturned. According to the 2019 survey, 61 percent of Americans believe that abortion should be legal in most or all cases. However, 62 percent of Republicans, including 77 percent of Republican conservatives, believe that abortion should be illegal in most or all cases.

In 2019, nine states enacted new laws restricting abortion access. Some were "heartbeat" laws in which an abortion would be banned after six weeks of pregnancy, at which time the heartbeat of the fetus could be detected. The sponsor of the Alabama law stated that an objective of proposing that law and ensuring its enactment was to force the US Supreme Court to address the abortion issue with an eye toward possibly overturning *Roe v. Wade*. Trump carried all of these states. On the other hand, six states enacted laws in 2019 that protected access to abortion. Biden carried all of these states (figure 6.30). Unfortunately, exit polls provide little insight into the geography of support for Biden and Trump based on religion at the state level. Exit polls are no longer conducted in every state, and in many states in which exit polls are conducted there are not enough adherents to particular religious traditions to allow for meaningful comparisons. For example, in Alabama 48 percent of respondents identified themselves as white evangelical Christians. Of these, 92 percent supported Trump, whereas 62 percent of Alabamians who did not identify themselves as white evangelical Christians supported Biden. As well, 91 percent of Alabama voters who believe that abortion should be illegal in most or all cases supported Trump. However, less than 10 percent of

Abortion Laws Enacted, 2019

FIGURE 6.30

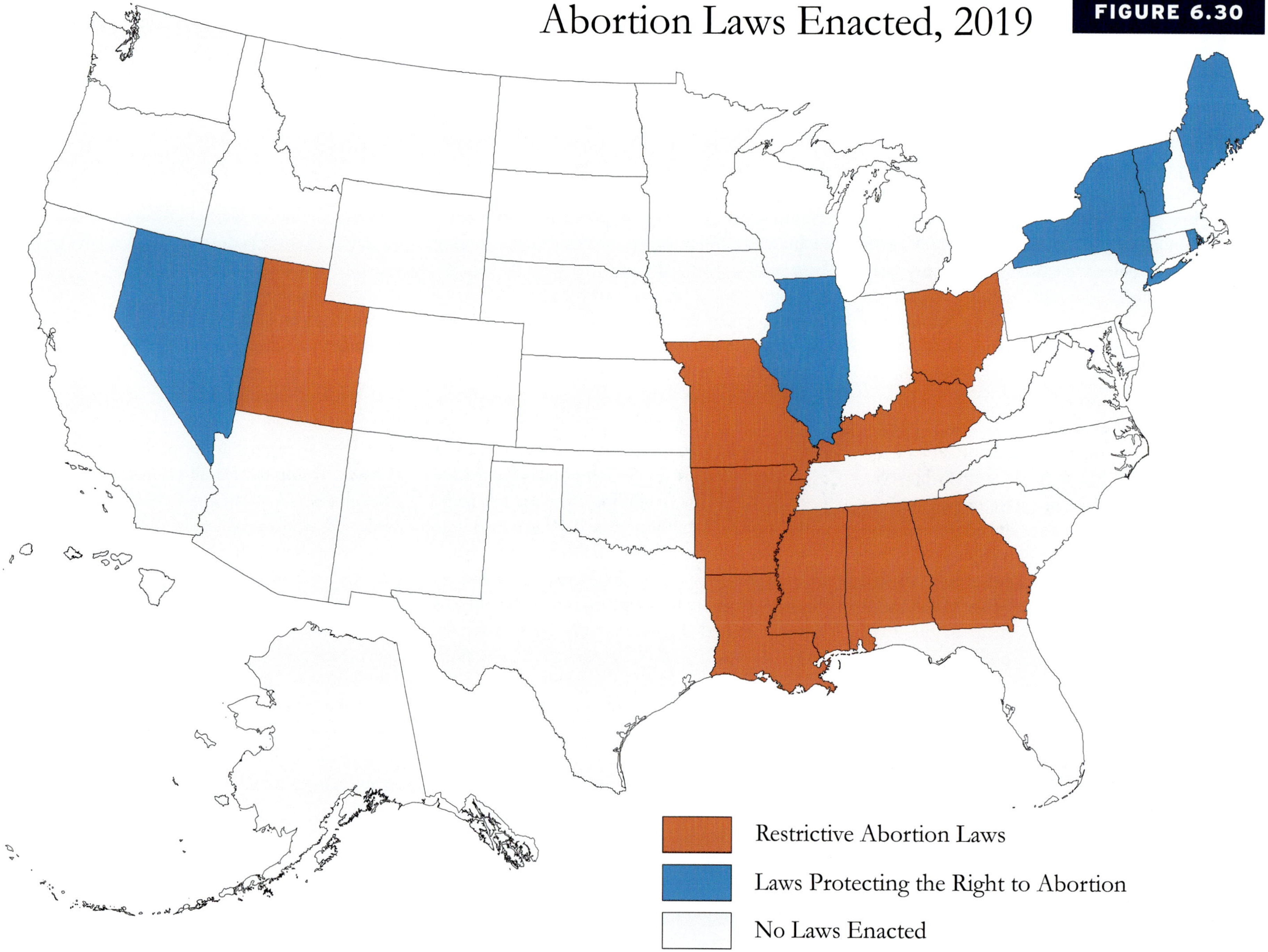

FIGURE 6.31

White Evangelicals, 2019

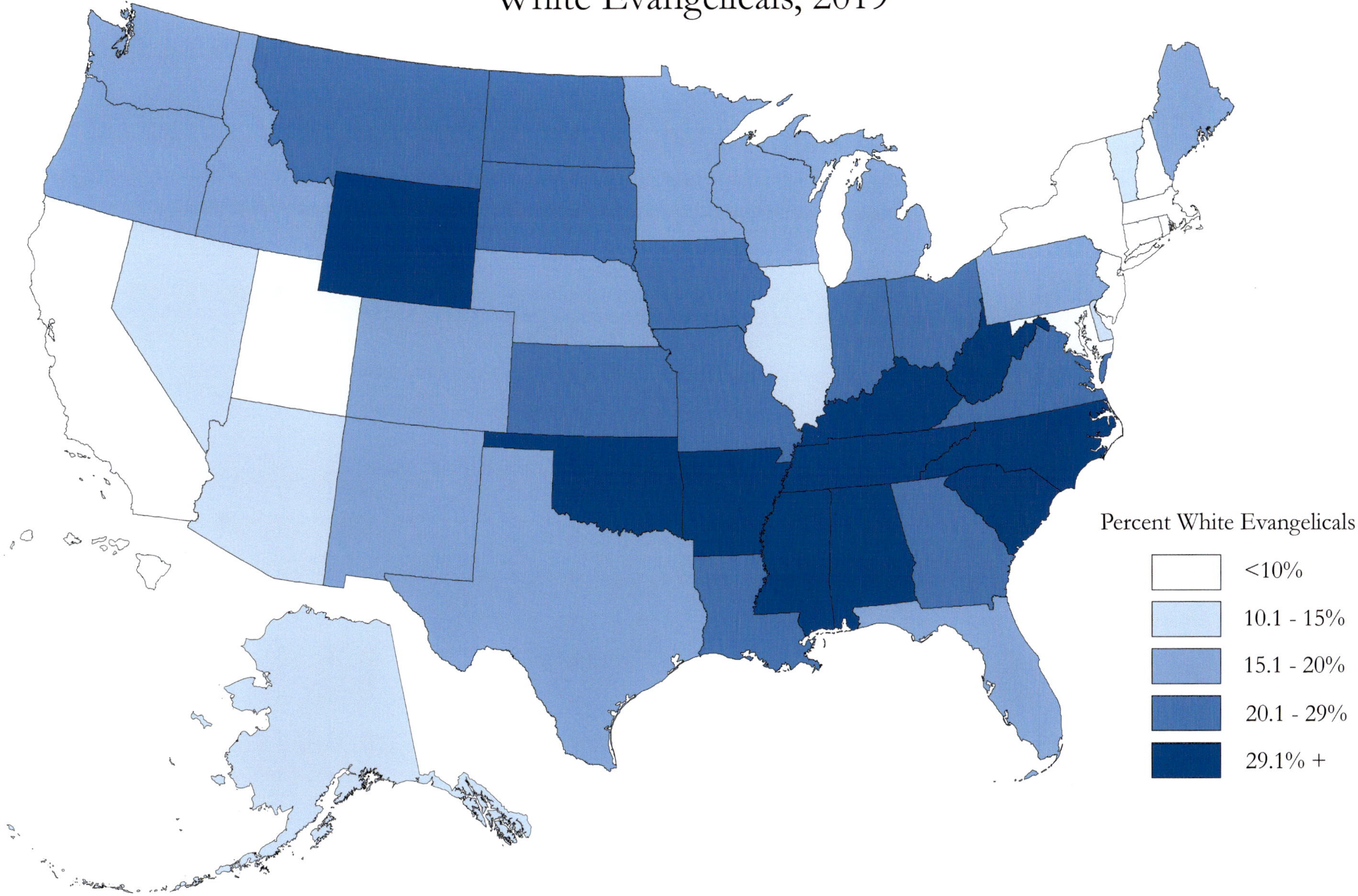

FIGURE 6.32

White Catholics, 2019

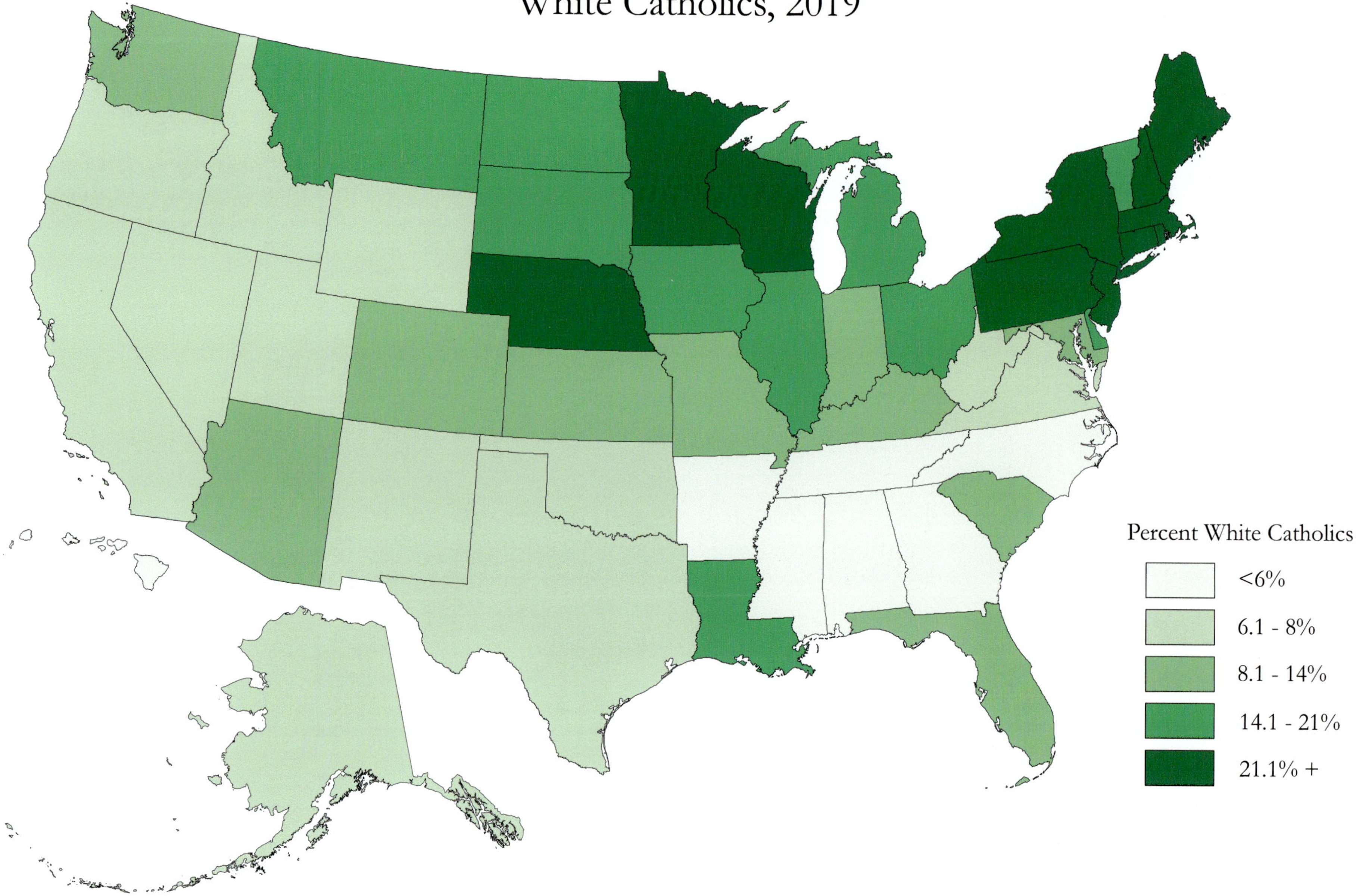

FIGURE 6.33

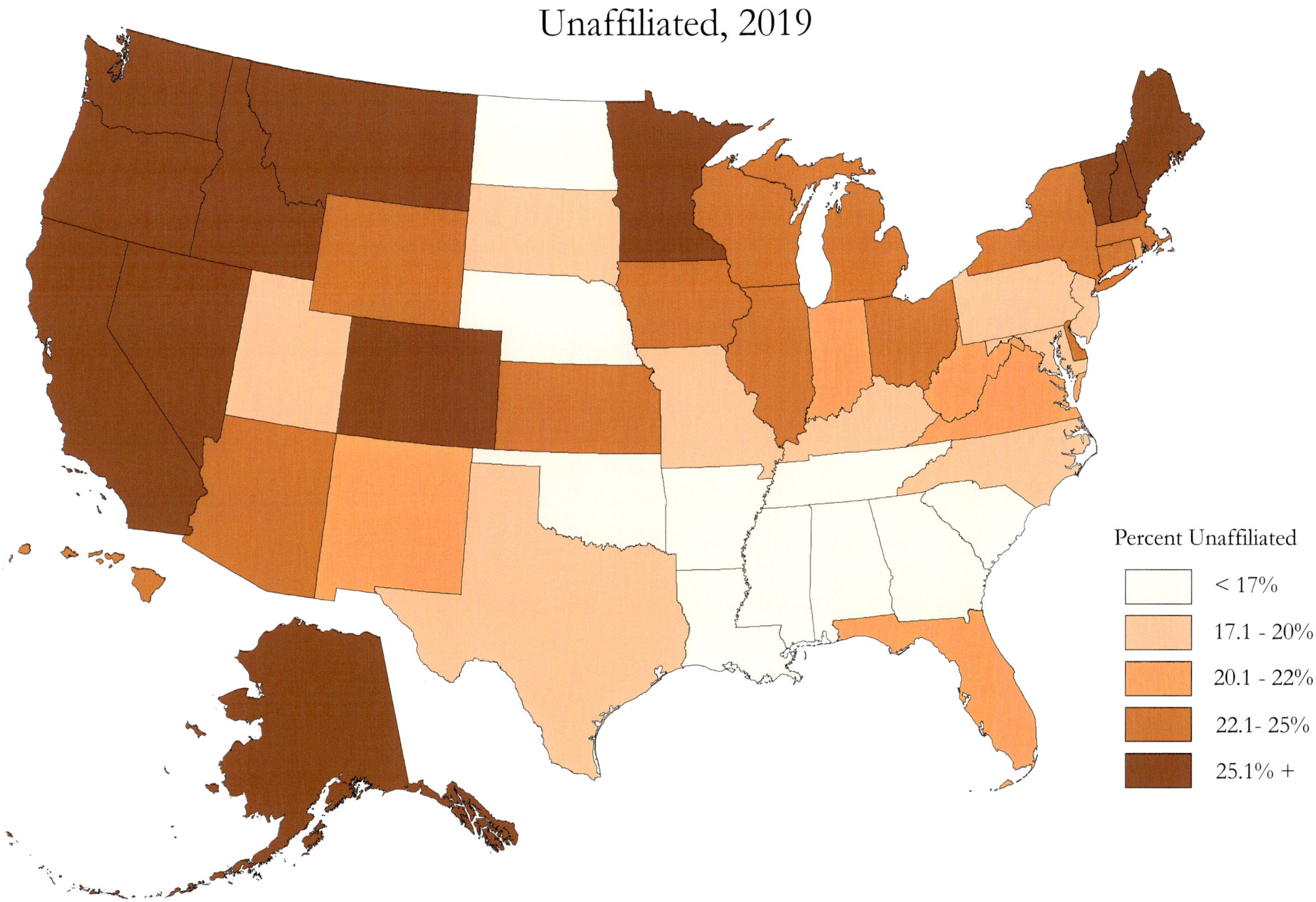
Unaffiliated, 2019
Percent Unaffiliated
< 17%
17.1 - 20%
20.1 - 22%
22.1- 25%
25.1% +

Alabama's residents are Catholics, so this comparison was not done.

However, general trends can be observed at the state level by comparing maps of religious affiliation with maps of election outcomes. According to results of a survey presented in the 2019 *American Values Atlas* published by the Public Religion Research Institute (PRRI), 15.2 percent of Americans described themselves as white evangelical Protestants. Across the United States, this percentage varied from 6 percent in Vermont and heavily Mormon Utah to a high of 35 percent in West Virginia (figure 6.31). The percentage of white evangelical Protestants exceeded 20 percent in eleven states, all of which were carried by Trump. Biden carried nine of the ten states with fewer than 10 percent white evangelicals with Utah, whose many Mormon residents are predominantly Republican, as the only exception.

According to the PRRI survey, 12 percent of voters identified themselves as white Catholics. These percentages ranged from 5 percent in Alaska to over 20 percent in Rhode Island, Massachusetts, and New Hampshire (figure 6.32). Biden carried eleven of the thirteen states with Catholic populations of more than 15 percent, losing only North Dakota and South Dakota among these states. However, Trump carried fourteen of the sixteen states with less than 10 percent Catholic populations, losing only Hawaii and Virginia among these states to Biden.

In a 2014 survey reported by the Pew Research Center, 23 percent of respondents reported having no religion, with the percentages varying from 12 percent in Alabama to 37 percent in Vermont (figure 6.33). Trump won eleven of the twelve states with less than 20 percent non-religious respondents, losing only New Jersey. On the other hand, Biden won eleven of the fourteen states with 27 percent or more non-religious respondents. Overall, religion continued to play a major role in the outcome of the 2020 election, although its importance may decrease given the tendency of younger Americans to drift away from formal, organized religion.

BIBLE BELT

DANIEL A. MCGOWIN AND GERALD R. WEBSTER

Although the term "Bible Belt" is ill-defined, especially as a geographic concept, it is often used in reference to the rural South with predominantly white evangelical populations. It is not uncommon during election cycles for reporters to refer to southern states as being part of the Bible Belt, such as Arkansas (Lynch 2017), Kentucky (Ambrose 2020), North Carolina (McCullough 2020), and South Carolina (Bohan 2007). That stated, such broad characterizations of southern states ignore regional diversity and changing demographics, as well as the existence of multiple "Bible Belts" (Grant 2016).

The origin of the term "Bible Belt" is credited to journalist H. L. Mencken's 1925 coverage of the Scopes "monkey trial" in Dayton, Tennessee (Webster et al. 2015, 1440). Although today the term is broadly applied to the South, we use an update by Brunn, Webster, and Archer (2011) to a previous spatial definition of the Bible Belt. The author of the previous definition, Charles Heatwole (1978), based the spatial location of the Bible Belt on the distribution of twenty-four Protestant denominations that believe in the literal interpretation of the Bible. Here we use 2010 data from the Glenmary Research Center to examine the top two hundred counties in terms of their proportion of adherents (with a minimum threshold of 43.3 percent) belonging to such denominations. In total there were twenty-nine million members of Bible Belt denominations in 2010, with nearly twenty million being members of the Southern Baptist Convention (Webster et al. 2015, 1442). The other large denominations include the Lutheran Church-Missouri Synod (2.2 million), Christian Churches and Churches of Christ (1.4 million), Seventh-Day Adventists (1.2 million), and Church of God in Cleveland, Tennessee (1.1 million; Webster et al. 2015, 1443).

This definition creates a discontinuous "belt" stretching from Virginia south to Alabama and west to Texas. There are several small concentrations within the belt, including in northern Alabama, western Kentucky, and northern Louisiana. However, the Bible Belt's "buckle" would today appear centered on the Oklahoma-Texas borderlands (Brunn, Webster, and Archer 2011). Although the Bible Belt is most associated with the South, the definition utilized here includes counties in some states outside of the region, namely Colorado, Illinois, Iowa, and South Dakota.

In 2020, the Republican Party (2016) opted to not adopt a new platform, choosing instead to continue using the 2016 platform. That platform, which reflects former President Donald Trump's vision, focuses on several issues that are of importance to many evangelicals, especially those in the South. Such issues include supporting "the public display of the Ten Commandments as a reflection of . . . our country's Judeo-Christian heritage" (12), school prayer, and being strongly antiabortion. These positions, along with Mr. Trump's appointment of three conservative justices to the US Supreme Court, resonate well with Bible Belt voters.

Former President Trump garnered 75.5 percent of the popular vote in Bible Belt counties compared to 24.5 percent for former Vice President Joe Biden (figure 6.34). For Mr. Trump, this number is an increase from the 73.4 percent that voted for him in 2016. The Democratic candidate also experienced a slight increase from Hillary Clinton's 23.4 in 2016. Despite the uptick in voting for Democrats, the overall trend reflects the further galvanizing of the Bible Belt vote for Republican presidential candidates, who received just over 66 percent of such votes in 2004.

In total, Mr. Trump won 197 of the 200 counties that comprise the Bible Belt. In ten of those counties, incumbent Trump received more than 90 percent of the vote, including more than 95 percent in Borden County, Texas. Most of these ten are in the "buckle" of the Bible Belt, with Winston County, Alabama (90.4 percent) and La Salle Parish, Louisiana (90.1 percent) falling outside of that core area. Additionally, Mr. Trump received more than 75 percent of the vote in 131 counties, and more than 60 percent in 184 counties. Mr. Biden's three wins in the Bible Belt is one more than Secretary Clinton won in 2016, although still two fewer than President Obama won in 2012. Biden won the city of Fredericksburg, Virginia (66.6 percent) and Marengo County, Alabama (50.4 percent), which were both won by the Democrat Clinton in 2016, as well as flipping Lynchburg city, Virginia (50.1 percent).

Bible Belt counties display several notable demographic tendencies. For example, Bible Belt

FIGURE 6.34

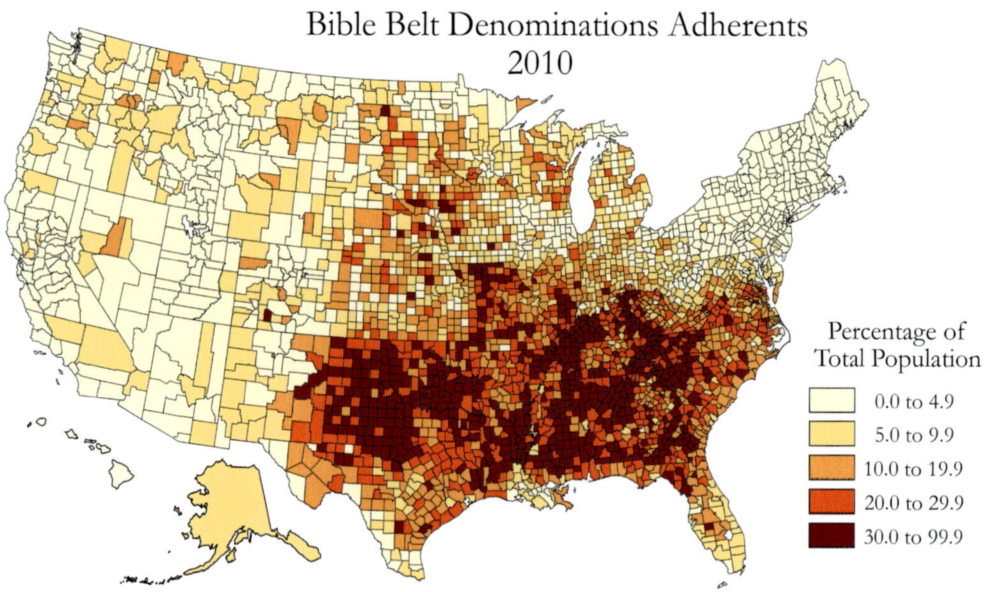
Bible Belt Denominations Adherents
2010
Percentage of Total Population
0.0 to 4.9
5.0 to 9.9
10.0 to 19.9
20.0 to 29.9
30.0 to 99.9

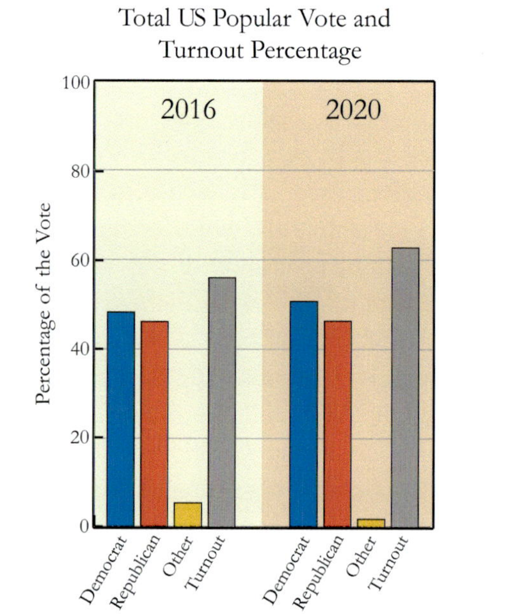
Total US Popular Vote and Turnout Percentage
2016
2020
Percentage of the Vote
0
20
40
60
80
100
Democrat
Republican
Other
Turnout

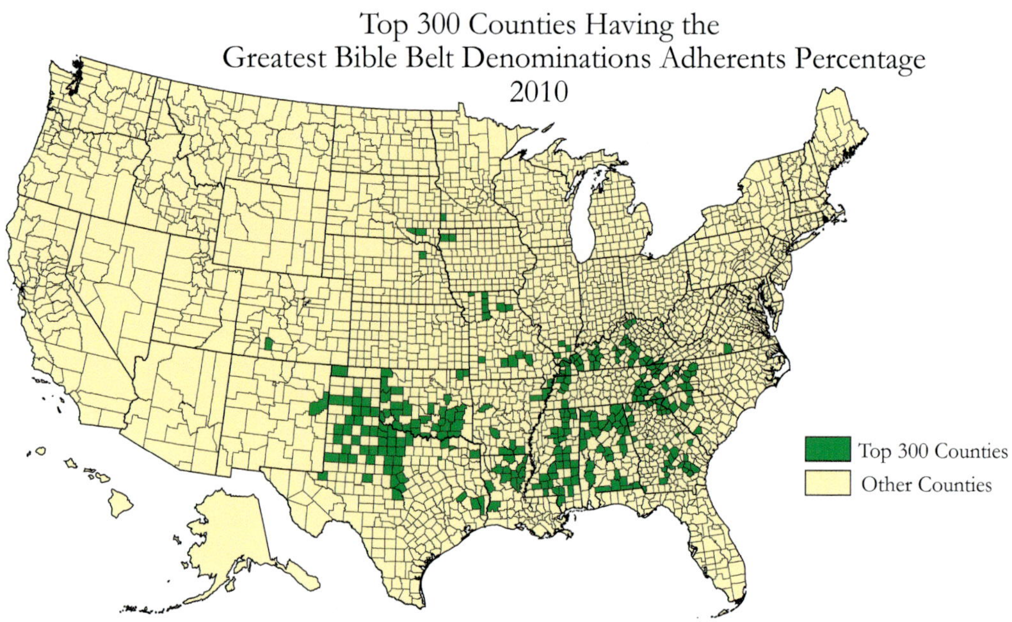
Top 300 Counties Having the Greatest Bible Belt Denominations Adherents Percentage
2010
Top 300 Counties
Other Counties

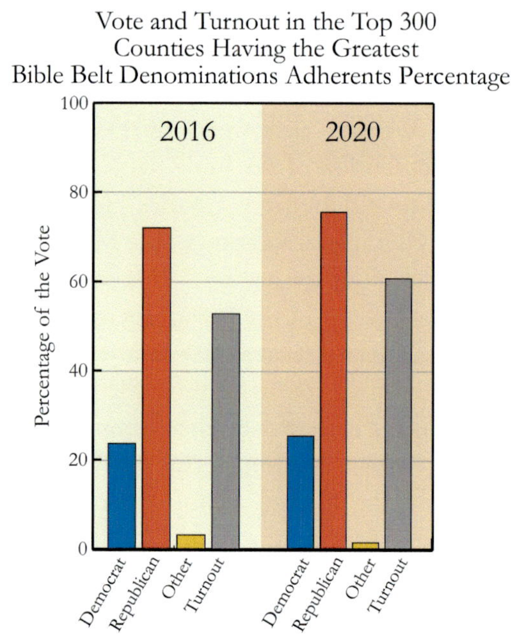
Vote and Turnout in the Top 300 Counties Having the Greatest Bible Belt Denominations Adherents Percentage
2016
2020
Percentage of the Vote
0
20
40
60
80
100
Democrat
Republican
Other
Turnout

counties have an average population of just over twenty-five thousand people compared to approximately one hundred thousand residents for the average county in the United States. Bible Belt counties also tend to be less densely populated than the rest of the United States—forty people per square mile versus the national average of eighty people per square mile. In terms of voting, counties with a higher percentage of votes for the Republican Party's nominee tend to have small populations. Indeed, seven of the ten Bible Belt counties with the highest percentage of votes for Mr. Trump have populations of less than ten thousand people, and six of those have a population of fewer than five thousand people. The populations of Bible Belt counties also tend to include fewer minority group members. Sixty-six percent of the US population is non-Hispanic white, but non-Hispanic whites constitute more than 80 percent of the population in the top two hundred Bible Belt counties. Finally, education levels lag in the top two hundred Bible Belt counties, with less than 15 percent of the adult population having a college degree, compared with more than 25 percent nationally.

Thus, these religiously conservative counties tend to be sparsely populated, largely composed of white residents with lesser levels of educational attainment, and they strongly support the Republican Party in presidential elections. As an example, of the 131 Bible Belt counties where Mr. Trump received more than 75 percent of the vote, eighty-nine have populations at least 80 percent white, and forty-five are more than 90 percent white. Conversely, in the sixteen counties where he received less than 60 percent of the vote, only two have white populations accounting for over 70 percent of the county—Clark County, Arkansas (70.4 percent) and Mineral County, Colorado (95.2 percent). It is also worth noting that the three counties won by Mr. Biden have populations that are relatively diverse, including Marengo County, which is 51.5 percent Black.

Given the above pattern, it is not surprising that Mr. Trump generally outperformed in Bible Belt counties compared to the percentage of votes he won statewide. For example, he won 77.8 percent of votes in Bible Belt counties in Alabama compared to 62.9 percent for the entire state. This is also true when comparing only the counties won by Mr. Trump. Removing counties lost by the incumbent in Alabama, Mr. Trump's victories in Bible Belt counties was greater than those in non–Bible Belt counties by a 78.6 percent to 67.3 percent margin. The trend holds true in fifteen of the seventeen states, with him underperforming in Bible Belt counties in Arkansas and Colorado. This pattern reflects the importance of Bible Belt counties in forming a solid base for any Republican nominee for president.

REFERENCES

Ambrose, G. 2020. In Elliott County, Ky., Democrats Went Big for Trump. *WFPL*. November 4, https://wfpl.org/in-elliott-county-ky-democrats-went-big-for-trump/ (last accessed March 24, 2021).

Bohan, C. 2007. Bible Belt State May Be Crucial in Election. *Reuters*. December 11, https://www.reuters.com/article/us-usa-politics-southcarolina/bible-belt-state-may-be-crucial-in-election-idUSN1044318320071211 (last accessed March 24, 2021).

Brunn, S., G. Webster, and J. C. Archer. 2011. The Bible Belt in a Changing South: Shrinking, Relocating, and Multiple Buckles. *Southeastern Geographer* 51(4): 513–49.

Grant, T. 2016. Trump Won Big in the Bible Belt (and Lots of Mini-Bible Belts outside the South). *Religious News Service*, April 21. https://religionnews.com/2016/04/21/evangelicals-trump-votes-counties-data (last accessed November 3, 2021).

Heatwole, C. 1978. The Bible Belt: A Problem in Regional Definition. *Journal of Geography* 77:50–55.

Lynch, S. 2017. Evangelicals in "Bible Belt" Solidly behind Donald Trump. *Irish Times*. August 14, https://www.irishtimes.com/news/world/us/evangelicals-in-bible-belt-solidly-behind-donald-trump-1.3185356 (last accessed March 24, 2021).

McCullough, M. 2020. In the Bible Belt It's Love Thy Neighbour, Not Thy Politics. *Sydney Morning Herald*, October 27, https://www.smh.com.au/world/north-america/in-the-bible-belt-it-s-love-thy-neighbour-not-thy-politics-20201025-p568cw.html (last accessed March 24, 2021).

Republican Party. 2016. 2016 Republican Party Platform. https://prod-cdn-static.gop.com/docs/Resolution_Platform_2020.pdf (last accessed March 24, 2021).

Webster, G., R. Watrel, J. C. Archer, and S. Brunn. 2015. Bible Belt Membership Patterns, Correlates and Landscapes. In S. Brunn, ed., *The Changing World Religion Map: Sacred Places, Identities, Practices and Politics*. New York: Springer.

7 CONGRESSIONAL ELECTIONS AND ROLL-CALL VOTES

A "BLUE WAVE": THE 2018 CONGRESSIONAL ELECTIONS

RYAN WEICHELT

For many decades, the party of newly elected presidents has struggled in the proceeding midterm elections. Bill Clinton saw Democrats lose control of the House in 1994 and Barack Obama lost the House to the Tea Party movement in 2010. The surprising election of Donald Trump in 2016 and the controversies that immediately followed the president provided Democrats a strong belief that 2018 could give them control of the House of Representatives for the first time since 2010. However, in the months leading up to the midterm elections, belief turned into reality and the question became not whether Democrats would take control of the House, but rather whether it would by a "Blue Wave" or "Blue Ripple."

After the 2010 midterm elections and the resulting redistricting period, some believed Democrats could never regain control of the House of Representatives. Republicans flipped sixty-four seats across the United States during the Tea Party wave, giving them a 242 to 193 advantage in the House. This would increase to a high of 247 to 188 after the 2014 midterm elections. With the conclusion of the 2016 elections, and even with Trump on the ballot, Republicans maintained a 241 to 194 advantage.

Donald Trump would lose the popular vote, but to the surprise of many was able to thread the needle and claim the presidency through the Electoral College. Democrats had to reflect on their failures. After the 2016 elections, Republicans controlled the House, Senate, and the White House for the first time since 2003. Many in the party feared that an all-Republican front would work to undo all that Democrats had accomplished under the leadership of President Barack Obama. Instead of taking advantage of their power, congressional Republicans stood idly by as leadership attempted to understand how Trump would impact the party. In late 2017, Trump himself would claim that he passed more legislation in his first year than any other president. The total number equated to ninety-six bills—the lowest number passed in the first year of one's presidency for all post–World War II presidents (Keith 2017). The only meaningful legislation passed between 2017 and 2018 was the Tax Cuts and Jobs Act of 2017.

Republican leadership quickly found that Trump would not easily be controlled. Moreover, media pressure, combined with an ongoing Mueller Investigation into Russian interference in the 2016 election, made it clear that Democrats and other anti-Trump voters would be energized heading into the 2018 midterms. Speaker of the House and former Republican vice presidential candidate Paul Ryan decided he would not seek reelection in 2018. It is not known whether his decision was the result of no longer wanting to engage with the Trump administration or sensing that his time as

FIGURE 7.1

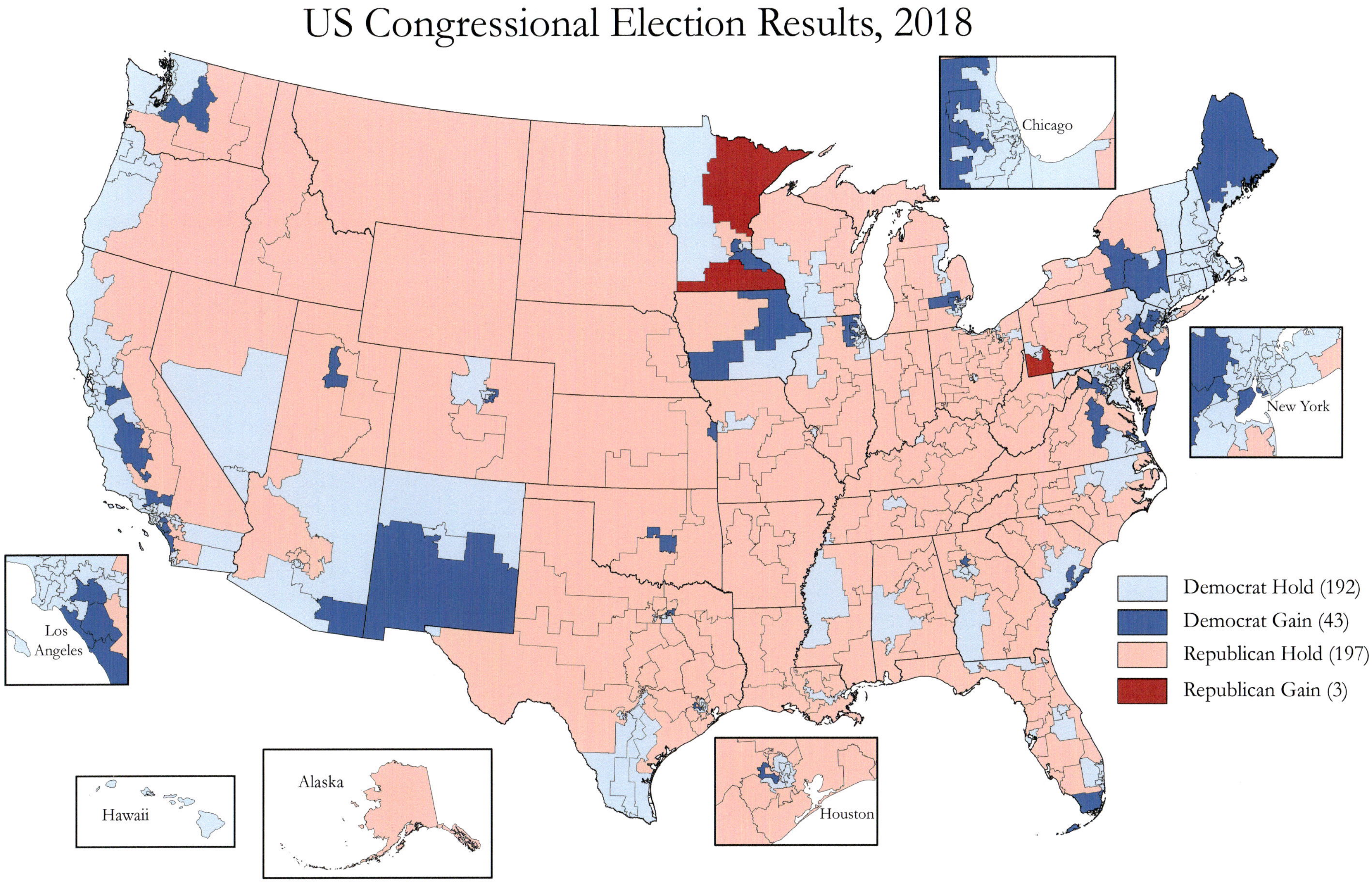
US Congressional Election Results, 2018
Chicago
New York
Los Angeles
Houston
Hawaii
Alaska
Democrat Hold (192)
Democrat Gain (43)
Republican Hold (197)
Republican Gain (3)

Speaker may be at an end. Political watchers predicted congressional Republicans would be punished at the polls for Donald Trump's behavior, but it was not clear to what extent.

Before the polls closed on November 6, political prognosticators estimated that thirty-five to forty-five seats could be won by Democrats (270toWin 2018). In the end, Democrats won forty-one seats, resulting in a 235- to 200-seat advantage. Figure 7.1 illustrates the results of the 2018 congressional elections. The largest gains were found on both coastal regions, especially among urban and suburban districts in and around New York City and Los Angeles. Surprising gains were found in Iowa, Oklahoma, and Utah. Republicans, for their part, picked up three seats previously held by Democrats: one in Pennsylvania and two in Minnesota.

District-level results reveal interesting partisan patterns. Among the most loyal Republican districts, approximately 70 percent saw an increase in Republican support compared to Trump's 2016 results. Most tended to be rural districts found throughout the United States, demonstrating Trump's increased support and newly found control of the Republican Party. The remaining districts, those won by both Democrats and Republicans, all saw increased Democratic support compared to Clinton's results in 2016. The greatest increases for Democrats were found in larger urban areas as well as in suburban areas across the United States (Watkins, Rebecca, and Yourish 2018).

The 2018 "Blue Wave" midterm elections introduced a variety of new faces to the American public representing all aspects of the political spectrum. For Republicans, 2018 helped elevate those congressmen who were Trump's most ardent supporters. Although Republicans like Jim Jordan from Ohio and Matt Gaetz from Florida had been in Congress previously, both the Mueller Investigation and continued attacks on Trump by the Left increased their profile among Trump supporters. Furthermore, newly elected Republicans found success by adopting a more pro-Trump stance compared to those running in 2016.

The "Blue Wave" for Democrats created a more diverse majority in the House than in any previous cycle. Some declared 2018 to be the "Year of the Woman"; women running for offices across the country reach an all-time high. In the House, 117 women were elected; Democrat women accounted for 100 of them. These victories also created several firsts for minority groups. In both Kansas and New Mexico, Sharice Davids and Deb Haaland became the first American Indian women elected to Congress. In Michigan, Rashida Tlaib became the first Palestinian American woman in Congress and in Minnesota, Ilhan Omar became the first Somali American elected to Congress. Both were also the first female Muslims elected to the House of Representatives. In New York, Alexandria Ocasio-Cortez became the youngest woman to be elected at twenty-nine and Ayanna Pressley became the first Black woman to be elected to Congress in Massachusetts (Cooney 2018). Moreover, Omar, Tlaib, Ocasio-Cortez, and Pressley were more progressive compared to many within the party. With control of Congress in hand, Democrats would select, for the second time, California Congresswoman Nancy Pelosi Speaker of the House, the only woman to ever serve in the role.

Although many suggested the "Blue Wave" could have been larger, no matter the size it became a defining election as a clear response to Donald Trump from both parties. Many Republicans struggled with their identity as Trump quickly worked to redefine the Republican Party. Some, like Speaker of the House Paul Ryan, left Congress, while others quickly took favor with the president's agenda. Results illustrated that in the most Republican of districts, voters had fully accepted Trumpism, and as a result those candidates who chose to endorse the president saw increased support even when compared to Trump's own margins in 2016. More moderate Republicans in suburban and urban areas clearly struggled. Those who were not defeated saw winning margins decrease to levels not seen in recent elections and those who lost were defeated by Democratic candidates.

Although diversity was a hallmark of the 2018 midterm elections, it also created tension within the Democratic Party. In the coming months freshman representatives like Alexandria Ocasio-Cortez would butt heads with moderate Democrats in her desire for a more progressive platform. Additionally, Ilhan Omar and Rashida Tlaib would also make controversial comments toward Israel, not only raising the ire of Democratic leadership, but also making them an easy target for Donald Trump to rally his growing base of Republican supporters. Although these headaches were minor, it did demonstrate the growing power of progressives within the Democratic electorate. The many nuances exposed across both parties by the 2018 midterms would help set the stage for the next two years in the run-up to the 2020 presidential election.

REFERENCES

Cooney, S. 2018. 2018 Midterm Elections Leads to Historic Wins for Women. *Time*. https://time.com/5323592/2018-elections-women-history-records/ (last accessed June 2, 2021).

Keith, T. 2017. Trump Signed 96 Laws in 2017. Here Is What They Do and How They Measure Up. *NPR*. https://www.npr.org/2017/12/27/573546883/trump-signed-96-laws-in-2017-here-is-what-they-do-and-how-they-measure-up (last accessed May 12, 2021).

Thomsen, J. 2017. Trump Vows to Repeal and Replace ObamaCare "Disaster." *The Hill*. https://thehill.com/homenews/administration/361693-trump-vows-to-repeal-and-replace-obamacare-disaster (last accessed May 18, 2021).

270toWin. 2018. House Election Forecast Maps. https://www.270towin.com/2018-house-election-predictions/ (last accessed May 2, 2021).

Watkins, D., K. K. Rebecca, L. Buchanan, and K. Yourish. 2018. Sizing Up the 2018 Blue Wave. *New York Times*. https://www.nytimes.com/interactive/2018/11/07/us/politics/how-democrats-took-the-house.html (last accessed June 3, 2021).

MEMBERSHIP, PARTY, AND CAUCUSES IN THE 116TH CONGRESS

ERIN H. FOUBERG

The 116th Congress (January 3, 2019 to January 3, 2021) marked a shift from Republican control of both the House and Senate in the 115th Congress (January 3, 2017 to January 3, 2019) to a split legislature. In the 2018 midterm elections, Democrats gained control of the House in a what has been coined a "Blue Wave." The 101 new members elected in 2018 not only shifted the House from a Republican to Democratic majority; they also brought more diversity to the House floor. Among the 101 new members, 42 were women; 24 were people of color; 2 were Muslims; 2 were Native American; more than 20 percent were millennials (born between 1981 and 1996); and nearly half were Gen X (born between 1965 and 1980).

The newly elected freshman class in the House helped propel the 116th Congress to the most diverse in US history. The racial composition of the House better reflected the racial diversity of the country. The 12 percent of House members who are Black and 1 percent who are Native American are on par with the proportions of Blacks and Native Americans in the country (Bialik 2019). The share of Hispanics (9 percent) and Asians (3 percent) in the House is half the share of both groups in the country—18 percent and 6 percent, respectively (Bialik 2019).

Figure 7.2 shows the party affiliation in the US House of Representatives in the 116th Congress: 233 Democrats; 195 Republicans, 2 Independents-Libertarians, and 5 vacant seats. Republicans held the majority in the Senate with 52 seats, compared to 46 Democrats and 2 Independents who caucused with Democrats. The youth of the large freshman class, with an average age of 47.9 years, helped pull the average age of members of the House in the 116th Congress down to 57.6 years, slightly lower than the 115th (57.8 years).

Elections for the 116th Congress reflected a growing dissatisfaction with President Donald J. Trump in some parts of the country. In the two years leading up to the 116th Congress, during the 115th Congress, average congressional approval ratings were 19.75 percent in 2017 and 19.83 percent in 2018. The large voter turnout in the midterm election helped shift congressional districts to Democrats. The *New York Times* reported that four of the twenty-five Democratic progressives in the freshman class won their seats in congressional districts that President Trump carried in the 2016 election. Another nineteen Democrats, not categorized by the *Times* as progressives, also won in congressional districts that President Trump carried.

After the midterm elections and the Democratic takeover of Congress, approval ratings rose to an average of 21.2 percent in the first year of the 116th Congress, averaging Gallup's monthly approval ratings over twelve months. The second year of the 116th Congress rated a bit higher with 22.3 percent congressional approval in 2020. However, in mid-December 2020, after the presidential election of Joseph Biden and before the January 6 insurrection at the Capitol and Biden's inauguration, congressional approval fell to only 15 percent.

Among the forty-two women elected new members of the House in 2018, thirty-eight were Democrats, and two of the newly Democratic women in the House, Alexandria Ocasio-Cortez (New York) and Ilhan Omar (Minnesota) were in the top ten most followed members of Congress on Twitter (Van Kessel et al. 2020). Ocasio-Cortez and Omar were not the most followed social media users in the House and Senate. In fact, they ranked third and tenth among all members of the House and Senate from both parties in the 116th Congress. According to Pew Research Center analysis, "voting members of the 116th Congress collectively produced more than 2.2 million tweets and Facebook posts in 2019 and 2020" (Smith and Shah 2021). The most-followed members of Congress were not the most prolific on social media, but their posts garnered the most engagements. The thirty members of the 116th Congress who had more than one million followers included five new members of Congress and accounted for "70 percent of followers, 71 percent of reactions and favorites and 65 percent of shares and retweets" (Smith and Shah 2021).

The 116th Congress set records for social media use, which was reflected in the kinds of legislation passed through the House before the 2020 presidential election. The Democratic-controlled House passed a total of 233 substantive bills and

FIGURE 7.2

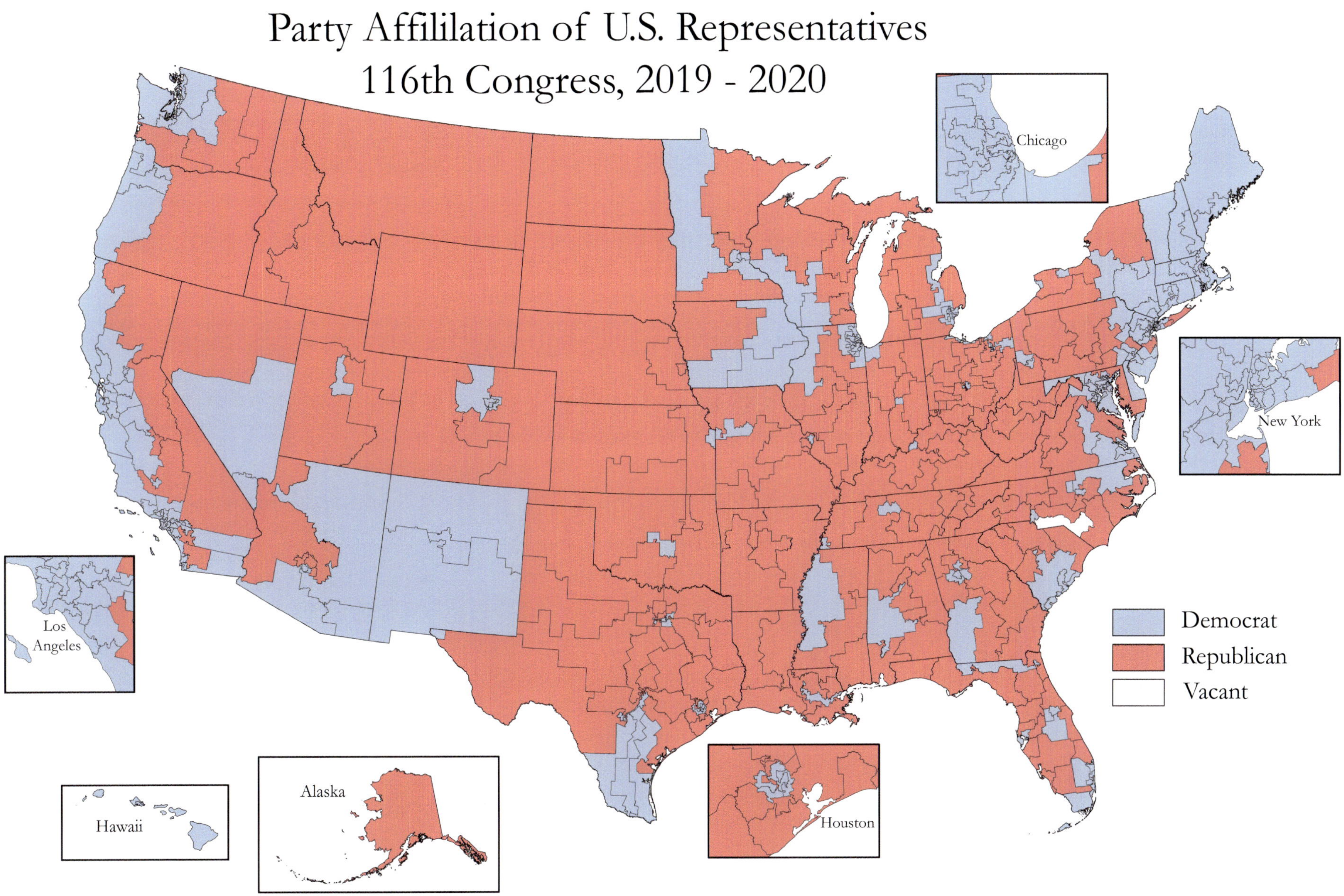
Party Affililation of U.S. Representatives
116th Congress, 2019 - 2020
Chicago
New York
Los Angeles
Houston
Alaska
Hawaii
Democrat
Republican
Vacant

111 ceremonial bills, according to Pew Research Center, making it "one of the least legislatively productive Congresses of the past five decades" (DeSilver 2021). For comparison, the 115th Congress passed 306 substantive and 136 ceremonial bills. The number of bills alone may not be the best gauge of productivity in Congress, though; the average length of bills is getting longer (DeSilver 2021). Regardless of whether the number of bills is a good measure of productivity, what is remarkable about the 116th Congress is how many bills were passed after the November 3, 2020, presidential election. Astoundingly, "More than four-in-ten bills that came out of the 116th Congress (151 of 344, or 44%) were passed in the final two months of its two-year term" (DeSilver 2021). Among the bills Congress passed in the two-month lame duck session were the second COVID-19 relief bill, which included a second stimulus check for many Americans and extended the eviction moratorium set in place earlier in the pandemic.

The relatively small number of bills and the topic of bills is likely a reflection of the Democrats taking control of the House in 2018 and the Democratic caucuses using the 116th Congress to set the tone for the 2020 presidential election. After the 2018 midterm election, the Congressional Progressive Caucus grew to ninety-six members in the House. Some saw the legislation coming out of the House as a "statement of Democrats' values ahead of the 2020 elections." (Brodey 2018). For example, the first piece of legislation in the 116th Congress was the "For the People Act," an expansion of the Voting Rights Act that quickly passed the House and was stalled in the Senate.

The New Democrat Coalition, a group of moderate Democrats, grew to 101 members during the 116th Congress and voted in 2019 to make it easier for the caucus to take positions on specific legislation (McPherson 2019). The Blue Dogs continued to function as a caucus in the 116th Congress, with twenty-four Democrats who identify as fiscally conservative. A bipartisan caucus established in 2017, the Problem Solvers Caucus is working to bring together representatives across the country to find "common ground on many of the key issues facing the nation." Their self-reported goal is to "create a durable bloc that champions ideas that appeal to a broad spectrum of the American people" (Problem Solvers Caucus).

Figure 7.3 shows Popular Vote for Winning Candidates for the 117th Congress. The darker red House districts had large majority votes for Republican candidates, and the darker blue House districts had large majority votes for Democrat candidates. Elections for the 117th Congress occurred with the 2020 presidential election. Despite progressive Bernie Sanders losing in the Democratic presidential primary to more moderate Joseph Biden, the progressive side of the Democratic party saw success in several congressional primaries and in the 2020 elections and eventually in the November elections. Many of the progressives elected to the 117th Congress represent cities and will likely push progressive causes on local issues, including affordable housing and transportation (Sisson 2020). With a smaller majority, the 220 Democrats in the 117th Congress will need to work together to move substantive legislation through Congress. The number of members of the Progressive Caucus declined in 2020, down to ninety in the House. With repeated statements that he would work across the aisle, President Biden may tap the Problem Solvers Caucus to move legislation through the 117th Congress.

REFERENCES

Bialik, K. 2019. For the Fifth Time in a Row, the New Congress Is the Most Racially and Ethnically Diverse Ever. *Pew Research Center*. https://www.pewresearch.org/fact-tank/2019/02/08/for-the-fifth-time-in-a-row-the-new-congress-is-the-most-racially-and-ethnically-diverse-ever/ (last accessed June 24, 2021).

Brodey, S. 2018. Will the Congressional Progressive Caucus Become the Freedom Caucus of the Left? *MinnPost*. https://www.minnpost.com/national/2018/12/will-the-congressional-progressive-caucus-become-the-freedom-caucus-of-the-left/ (last accessed June 22, 2021).

Congressional Research Service. 2020. Membership of the 116th Congress: A Profile. https://fas.org/sgp/crs/misc/R45583.pdf (last accessed June 24, 2021).

DeSilver, D. 2021. Nothing Lame about This Lame Duck: 116th Congress Had Busiest Post-Election Session in Recent History. *Pew Research Center*. https://www.pewresearch.org/fact-tank/2021/01/21/nothing-lame-about-this-lame-duck-116th-congress-had-busiest-post-election-session-in-recent-history/ (last accessed June 27, 2021).

Gallup. 2021. Congress and the Public. *Gallup.com*. https://news.gallup.com/poll/1600/congress-public.aspx (last accessed June 28, 2021).

McPherson, L. 2019. What's in a Position? This Is How Caucuses Show Their Strength. *Roll Call*. https://www.rollcall.com/2019/01/29/whats-in-a-position-this-is-how-caucuses-show-their-strength/ (last accessed June 22, 2021).

Problem Solvers Caucus. 2021. https://problemsolverscaucus.house.gov/about (last accessed June 28, 2021).

Sisson, P. 2020. At the City Level, Progressives Flex New Power. *Bloomberg CityLab*. https://www.bloomberg.com/news/articles/2020-11-05/progressives-score

FIGURE 7.3

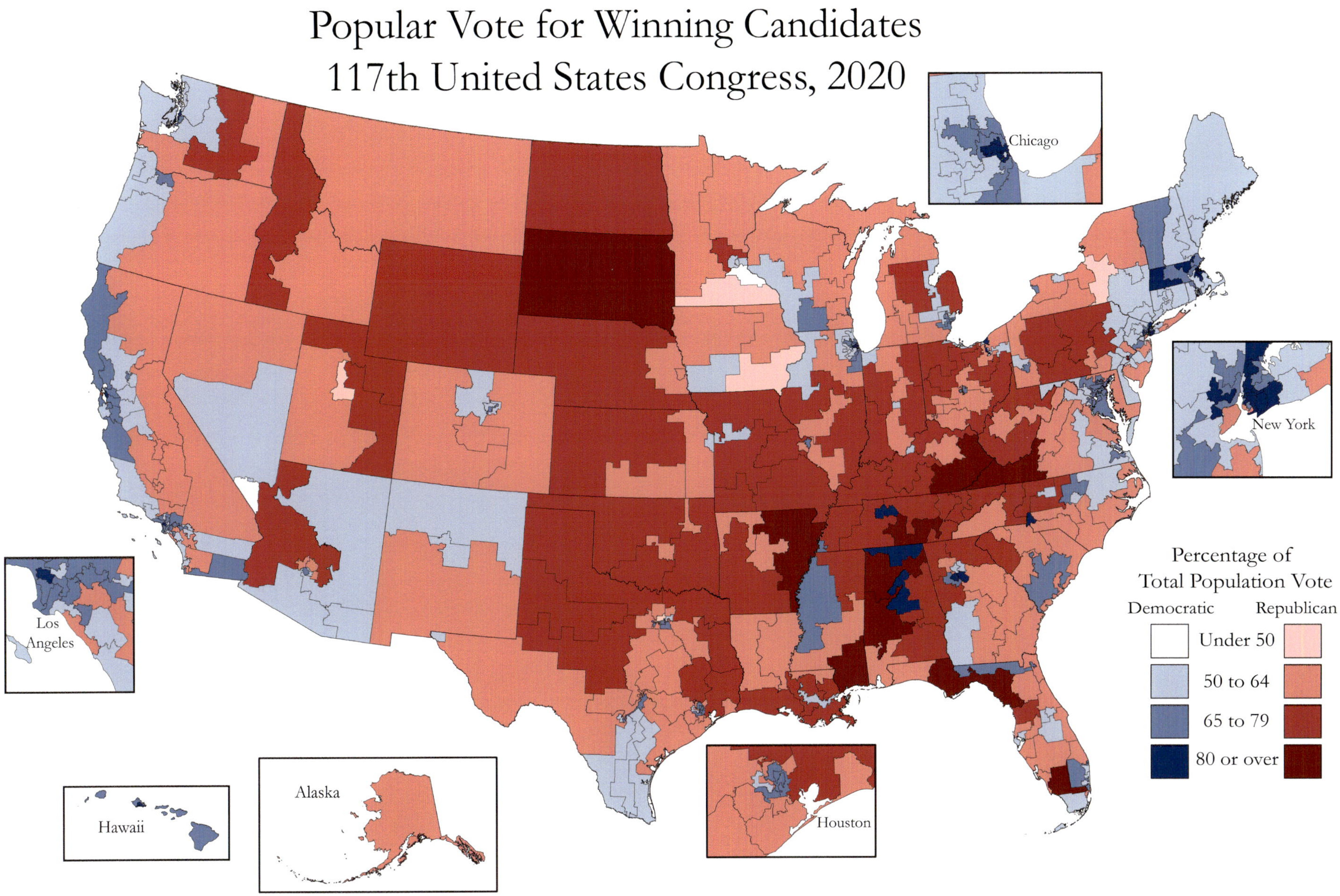
Popular Vote for Winning Candidates
117th United States Congress, 2020
Chicago
New York
Los Angeles
Hawaii
Alaska
Houston
Percentage of Total Population Vote
Democratic
Republican
Under 50
50 to 64
65 to 79
80 or over

-election-wins-in-u-s-cities (last accessed June 24, 2021).

Smith, A., and S. Shah. 2021. Though Not Especially Productive in Passing Bills, the 116th Congress Set New Marks for Social Media Use. *Pew Research Center.* https://www.pewresearch.org/fact-tank/2021/01/25/though-not-especially-productive-in-passing-bills-the-116th-congress-set-new-marks-for-social-media-use/ (last accessed June 21, 2021).

Van Kessel, P, R. Widjaya, S. Shah, A. Smith, and A. Hughes. 2020. Appendix A: Most-Followed Members of the 116th Congress. *Pew Research Center.* https://www.pewresearch.org/internet/2020/07/16/appendix-a-most-followed-members-of-116th-congress/ (last accessed June 21, 2021).

HOUSE ROLL-CALL VOTES IN THE 116TH CONGRESS

ERIN H. FOUBERG

The 116th Congress was in session during the last two years of President Donald Trump's term. Following the 2018 midterm elections, Democrats took the majority in the House and Republicans held the majority in the Senate. The 116th Congress was marked with fewer pieces of legislation passed and more social media posts than the 115th Congress, and much discussion in the House was signaling the voters and legislators to rally around the Democratic candidate for president in the 2020 election.

Using political scientists Keith Poole's and Howard Rosenthal's measure of political partisanship, the DW-NOMINATE metric, the 116th Congress can be compared the 114th, which was discussed in the *Atlas of the 2016 Elections*. The DW-NOMINATE metric compares roll-call votes on a two-dimension grid: one points to ideology from liberal to conservative and the other looks at regional differences. Poole and Rosenthal score every member of Congress who has served since 1789 on a score from –1 (liberal) to +1 (conservative), with 0 being moderate. Looking at Congress overall, the Republican party moved further right—to .51 in the 116th Congress, from the more moderate scores of .49 in the 115th Congress and .48 in the 114th Congress (Voteview 2021). The Democratic party moved closer to the center over the same time frame, landing at –.37 in the 116th Congress, from –.38 in the 115th and –.39 in the 114th.

At the end of the 116th Congress (2019–2021), on November 3, 2020, the House of Representatives had 197 Republicans, 232 Democrats, 1 Independent/Libertarian, and 5 vacancies. The Senate had 53 Republicans, 45 Democrats, and 2 Independent/Libertarians.

While the individual members of Congress are further apart politically, the US Senate and House remain closely divided. During the 1960s, Democratic majorities in the Senate and the House were sizable. However, "since then, the size of the majorities in both chambers has generally trended downward" (Schaeffer 2020). Reflecting the narrow division in Congress, Vice President Mike Pence cast the tie-breaking vote on thirteen pieces of legislation in the 115th Congress, when Republicans held the majority in both the House and Senate. Across all Congresses since the 88th (1963–1965), the most votes broken by a vice president before the 115th Congress was three (Schaeffer 2020). Roll-call votes in the 116th Congress reflect both the partisanship and the narrow margins in Congress. Many roll-call votes, whether passing or not, were basically split along party lines and others received generally universal support.

RUSSIAN SANCTIONS

The roots of the Russian sanctions resolution in June 2019 date back to H.R.3364, a bill from the 115th Congress supported overwhelmingly in the House (419–3) and Senate (98–2) in 2017. The original bill placed sanctions on Russia, Iran, and North Korea, specifically targeting Russia's undermining cybersecurity, government corruption, human rights abuses, aggression in Ukraine, and arms transfers to Syria. The bill gave Congress thirty days to reimpose sanctions against individuals if the president, through the US Treasury, proposed removing sanctions. President Trump reportedly considered vetoing the bill, but with huge, veto-proof majorities in both the House and Senate, Trump signed the bill (Greve 2019).

In January 2019, two months before special counsel Robert Mueller released "The Report on the Investigation into Russian Interference in the 2016 Presidential Election" (Mueller Report) to Attorney General William Barr and three months before the redacted version was released to the public, Congress considered whether to reimpose sanctions on a Russian oligarch through H.J. Resolution 30. The Trump administration moved to lift sanctions against Russian oligarch Oleg V. Deripaska and three companies through which he did business (Vogel 2019). This move invoked the 2017 law and gave Congress thirty days to counteract the Trump administration and reimpose the sanctions. The Democratically controlled House voted in favor of keeping the sanctions in place, in a largely symbolic vote just days after the Senate failed to get the sixty votes needed to keep the sanctions in place (Vogel 2019). The House and Senate members who voted to allow the sanctions to lift on the oligarch and his companies reportedly

FIGURE 7.4

H.J. Resolution 30: Disapproving the President's Proposal to Take an Action Relating to the Application of Certain Sanctions with Respect to the Russian Federation

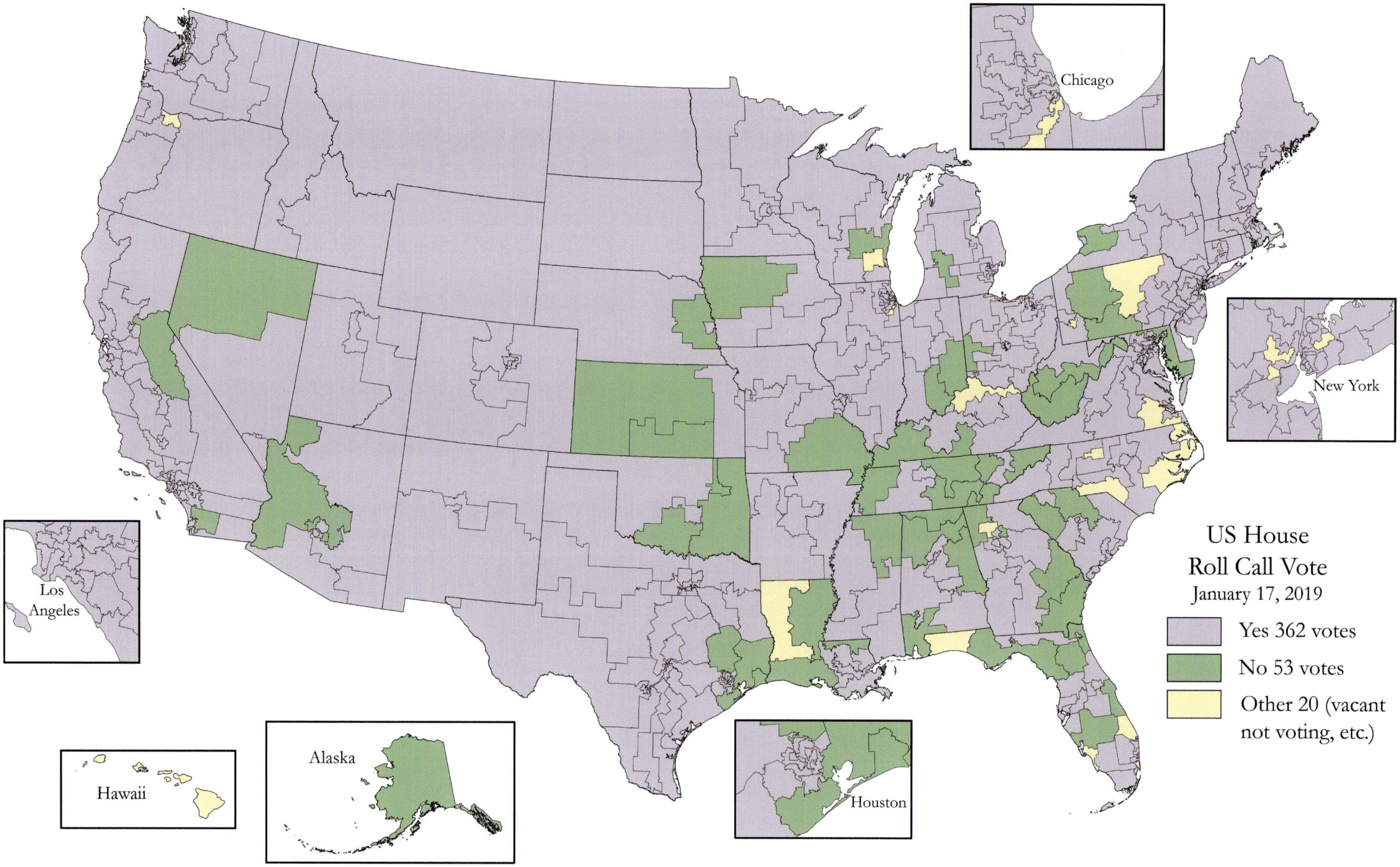

FIGURE 7.5

H.R. 3401: Emergency Supplemental Appropriations for Humanitarian Assistance and Security at the Southern Border Act, 2019

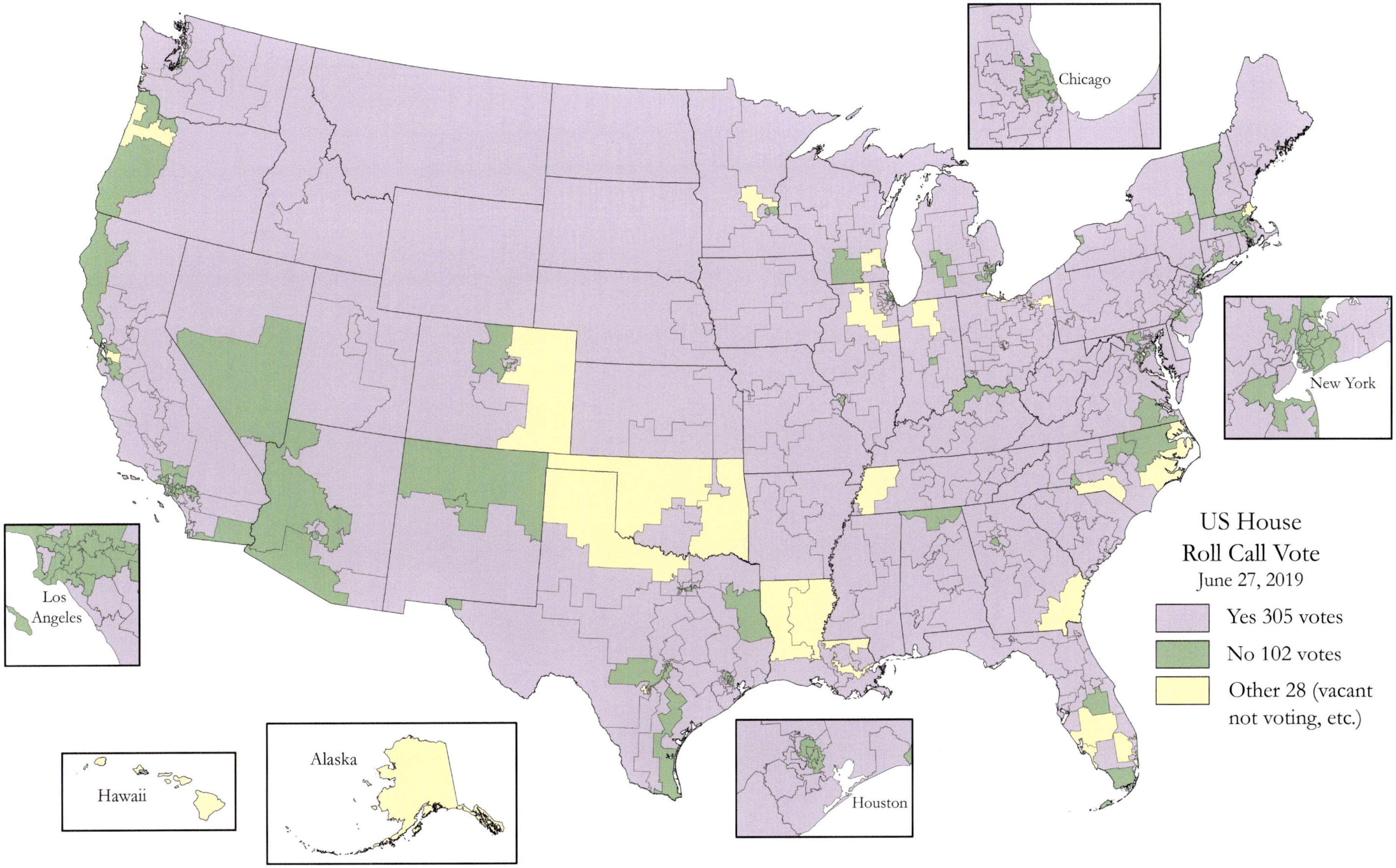

FIGURE 7.6

H.R. 5430: United States-Mexico-Canada Agreement Implementation Act

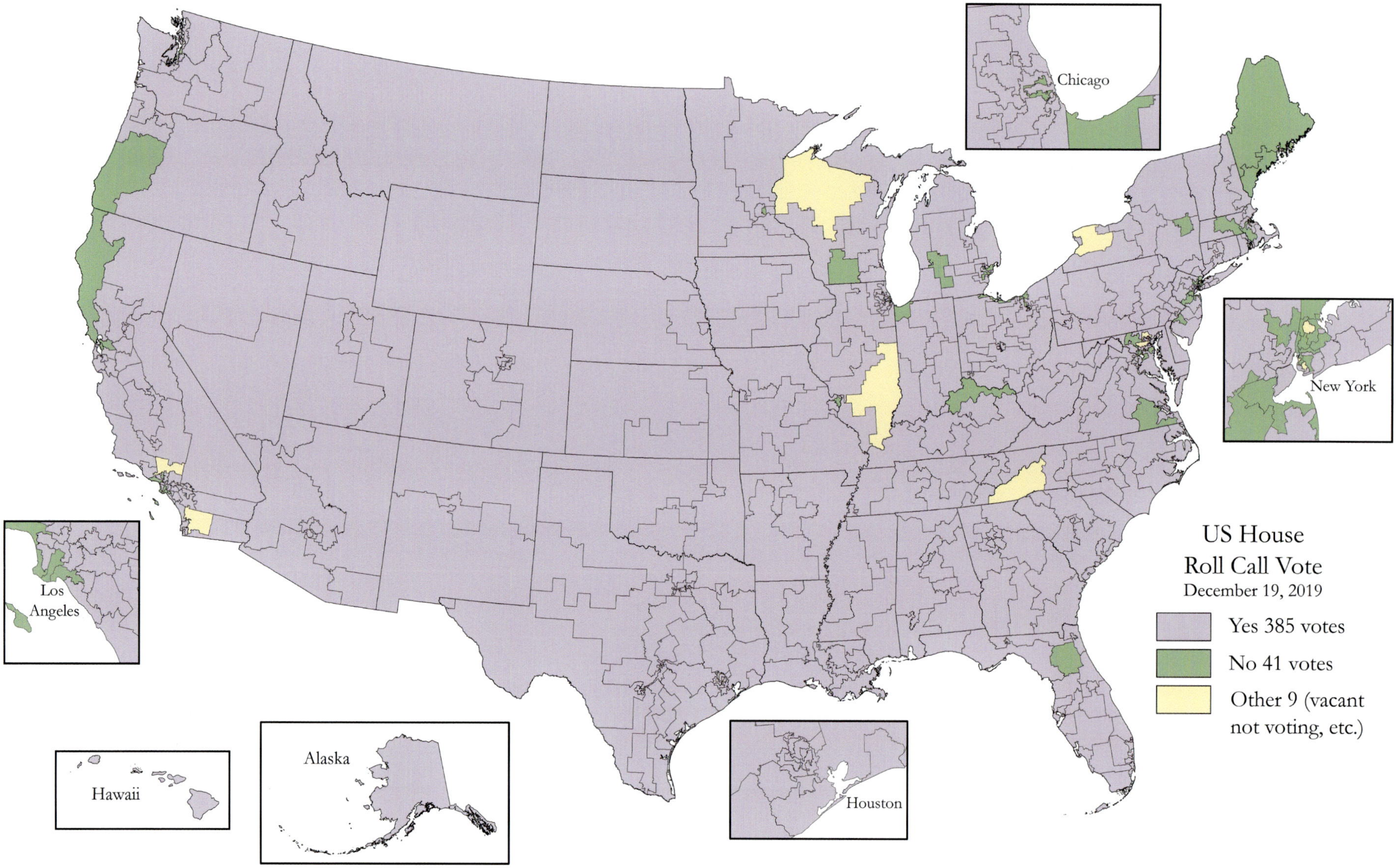

did so out of concern for the impact on the global aluminum industry (Zengerle 2019).

The distribution of House vote on January 17, 2019, shown in figure 7.4, had widespread support, with 362 votes in favor (226 Democrats and 136 Republicans), 53 voting against, and 18 not voting. All 53 voting against House Resolution 30 were Republicans. The pattern of No votes opposed party lines with 247 voting in favor (including two Democrats) and 186 voting against (all Democrats). In the Senate, eleven Republicans joined Democrats voting to reimpose the sanctions and not allow them to lift on Deripaska, a Putin ally. One argument was that no sanctions should be lifted against Russians at least until after the release of the Mueller Report.

EMERGENCY SUPPLEMENTAL APPROPRIATIONS

On June 27, 2019, the House voted in favor of H.R. 3401, to provide emergency supplemental appropriations for humanitarian assistance and security at the southern border. The House roll-call vote had a sizable majority, as shown in figure 7.5. Nearly all Republicans (176 to 7) and a majority of Democrats (129 to 95) voted in favor of the emergency appropriations. Prior to COVID-19, the border crisis continued to ramp up, with particular focus on the Trump administration's family separation policy and the news stories of children separated from their parents and living in cages under the control of the US Border Patrol. After six weeks of the zero-tolerance policy and backlash, the administration ended the controversial policy (Miroff 2020). Smugglers seized on the reversal and "began telling would-be clients that children were a passport into the United States" (Miroff 2020). A new wave of migration built up and "by May 2019 more than 144,000 migrants were taken into US custody amid a record surge of Central American families and children" (Miroff 2020).

By June 2019, Congress was under pressure to respond to the growing crisis at the border. The House passed a $4.5 billion bill on June 25, 2019, on a 230–195 vote after "last-minute changes were made to the legislation to sway progressive and Latino lawmakers who previously weren't on board" (Brufke 2019). House Resolution 3401, passed two days later, agreeing to a Senate amendment and the final form of the bill. That vote, mapped in figure 7.5, received even greater support, with 305 in favor and 102 opposed, as it gave more funding to the US Customs and Border Protection and other federal agencies operating on the border.

UNITED STATES-MEXICO-CANADA AGREEMENT

On December 19, 2019, a large majority in the House and Senate voted in favor of President Trump's signature legislation to replace the North American Free Trade Agreement (NAFTA). The United States-Mexico-Canada (USMCA) trade agreement received bipartisan support. In the Senate, 89 voted in favor of the new trade agreement, 10 voted against, and 1 did not vote. In the House vote on H.R. 5430, 385 voted in favor, 41 against, and 5 did not vote (see figure 7.6). In this vote, unlike the Russian sanction vote (H.J. Resolution 30), the opposition was from Democrats. Thirty-eight of the forty-one Nays were Democrats.

The trade agreement included concessions on "labor, environment, enforcement, and pharmaceutical provisions" that were requested by Democratic leaders of the House. "Those tweaks secured the backing of US labor, offering Democrats a pathway to supporting the agreement" (Kirby 2019). While work on negotiating USMCA began two years before the vote, the timing of the USMCA vote was tied to the timing of the Trump impeachment. The Mueller Report, released earlier in the spring of 2019, led to calls for impeachment of the president. Democrats in key swing states wanted and needed a political win before the recess. The December 2019 vote on USMCA provided that cover, as the Democrats announced their deal with President Trump and US Trade Representative Robert Lighthizer within an hour of announcing the articles on impeachment (Kirby 2019). The House voted on two articles of impeachment on December 18, 2019, and then voted overwhelmingly in favor of USMCA the next day.

REFERENCES

Brufke, J. 2019. House Passes $4.5 Billion Border Funding Bill. *The Hill.* https://thehill.com/homenews/house/450215-house-passes-45b-border-funding-bill (last accessed June 18, 2021).

Greve, J. 2019. The Daily 2020: The 11 GOP Senators Who Broke with Trump over Relaxing Sanctions against a Putin-Linked Oligarch. *Washington Post.* https://www.washingtonpost.com/news/powerpost/paloma/daily-202/2019/01/17/daily-202-the-11-gop-senators-who-broke-with-trump-over-relaxing-sanctions-against-a-putin-linked-oligarch/5c4002e61b326b3b88fef0be/ (last accessed June 16, 2021).

Kirby, J. 2019. The House Voted in Favor of the New Version of NAFTA after Democrats Secured

Changes to Labor and Pharmaceutical Provisions. *Vox*. https://www.vox.com/policy-and-politics/2019/12/19/21013178/usmca-trade-deal-passes-house-vote-approve-bipartisan-nafta-trump (last accessed June 15, 2021).

Miroff, N. 2020. "Kids in Cages": It's True that Obama Built Cages at the Border. But Trump's "Zero Tolerance" Immigration Policy Had No Precedent. *Washington Post*. https://www.washingtonpost.com/immigration/kids-in-cages-debate-trump-obama/2020/10/23/8ff96f3c-1532-11eb-82af-864652063d61_story.html (last accessed June 17, 2021).

Schaeffer, K. 2020. Slim Majorities Have Become More Common in the U.S. Senate and House. *Pew Research Center*. https://www.pewresearch.org/fact-tank/2020/12/01/slim-majorities-have-become-more-common-in-the-u-s-senate-and-house/ (last accessed June 17, 2021).

Vogel, K. 2019. 136 House Republicans Join Democrats in Vote Against Russia Sanctions Relief. *New York Times*. https://www.nytimes.com/2019/01/17/us/politics/trump-house-russia-sanctions.html (last accessed June 16, 2021).

Voteview. 2021. Parties: Congress at a Glance: Major Party Ideology. https://voteview.com/parties/all (last accessed June 17, 2021).

Zengerle, P. 2019. U.S. House Backs Sanctions on Russia's Rusal in Symbolic Vote. *Reuters*. https://www.reuters.com/article/us-usa-russia-sanctions/u-s-house-backs-sanctions-on-russias-rusal-in-symbolic-vote-idUSKCN1PB2D5 (last accessed June 17, 2021).

OVERVIEW OF THE 2020 SENATE ELECTIONS

RYAN WEICHELT

On January 5, 2021, the eyes of political watchers were focused on Georgia's two senate elections as control of the US Senate hung in the balance. After the November 3 election, Republicans were able to secure fifty Senate seats. Although Democrats were able to gain two seats, their count increased to forty-eight. If Democrats were able to secure both seats, it would bring their seat total to fifty, giving them control of the Senate since Democratic vice president Kamala Harris would serve as the tiebreaker. Although Joe Biden was able to defeat Donald Trump in November, Trump and his closest allies continued to sow discontent in the election process from November all through the final Senate election in January. As the 2020 election cycle ended in Georgia, it was clear Donald Trump and his own election successes and failures became a focus of the Georgia senate race.

The Democratic party's success in 2020 can be traced to the 2018 midterm election. For the first time since 2010, Democrats took control of the House of Representatives. While Republicans lost seats to the Democrats in the Senate, they were able to maintain a narrow lead. It was a strong midterm showing and their performance, combined with a growing discontent with the actions of Donald Trump before and during the pandemic, led many to believe that Democrats could not only win the White House but also gain back the US Senate.

Prior to the passage of the Seventeenth Amendment to the US Constitution in 1913, US Senators were selected by state legislatures. Upon the eventual ratification of the Seventeenth Amendment, the first direct election of senators occurred on November 5, 1918. Senators are elected for six-year terms and every two years, about one-third of the Senate seats are up for election. After the 2018 midterm elections, Senate Republicans held a narrow fifty-one- to forty-nine-seat advantage. Knowing which Senate seats would be up for grabs in 2020, Democrats were cautiously optimistic they could take back control.

In the lead-up to the 2020 elections, both parties were focused on seven seats in Arizona, Colorado, Alabama, two in Georgia, North Carolina, and Maine. Elections in both Arizona and Georgia were special elections owing to the vacancies filled in earlier years after the death of Arizona senator John McCain and the resignation of Johnny Isakson in Georgia in 2019. Colorado was viewed as an easy win for Democrats, while Republicans were confident former Auburn football coach Tommy Tuberville would unseat surprise 2018 special election winner, Democrat Doug Jones. However, there was not a clear front-runner in Arizona, North Carolina, Maine, or either Georgia senate races.

Independent of the candidate, it was clear nearly all elections in 2020 were both directly and indirectly about Donald Trump. The pressure to support or defy Donald Trump intensified during each of his years in office. Republicans struggled to find their footing in the turbulent political landscape Trump created. The 2018 "Blue Wave" illustrated that Republican support intensified in districts that ardently supported Trump in 2016, but it also highlighted declined GOP support in suburban and urban areas. This meant senators in swing states like North Carolina, Arizona, and Georgia were highly vulnerable. Moderate senators, like Maine's Susan Collins, had to walk a thin tightrope regarding Donald Trump by carefully diverging from the president while avoiding the president's ire.

Democrats' first order of business after taking the House was to investigate and impeach Donald Trump for his role in Russian interference of the 2016 Election. Although the Mueller Investigation would not formally indict the sitting president, Democratic leaders in the House and Senate believed there was enough to impeach and possibly remove Donald Trump from office. By the end of 2019, Donald Trump became only the third president to be impeached by the US House of Representatives, but not enough Republicans would break ranks to reach a two-thirds majority to remove him from office in the Senate.

It was clear heading into the February impeachment vote that Trump would not be removed, but it was not clear how moderate Republican senators, like Mitt Romney of Utah, Lisa Murkowski of Alaska, and Susan Collins would vote. Collins was up for reelection in 2020 and was the most vulnerable. Ultimately, only one Republican, Mitt Romney, would vote to remove Trump. The president was quick to chastise Romney in his "victory" speech, stating, "But you know, it's a failed presidential candidate, so things can happen when you fail so badly running for president" (Booker 2020). Susan Collins, however, received pushback from Democrats and moderates alike, who sharply criticized her support for Donald Trump. Already at odds with voters because of her vote to approve Supreme Court Justice Brett Kavanaugh, Collins's support among Maine voters declined as many made clear they believed she only cared about the party and not the law (Derysh 2020). These issues of loyalty would nag candidates across the spectrum throughout the 2020 election cycle.

Just a little more than a month after the impeachment trial, the COVID-19 pandemic would effectively shut the country down as political leaders in the United States and throughout the world struggled to mitigate the virus. Trump did little to ease tensions and consistently downplayed the danger the disease posed and lacked a coherent federal plan to combat the virus. In a rare moment of bipartisanship, representatives from both parties passed an economic stimulus package to ease the suffering of the initial wave of the pandemic. However, as spring gave way to summer, the partisan divide was quickly reestablished. Trump's most ardent supporters complained further stimulus was too expensive, creating headaches for Senate Republicans running for reelection, particularly in southern states that saw higher cases of COVID-19 compared to other areas of the United States. Furthermore, issues of race would also cloud 2020. The deaths of Breonna Taylor and George Floyd at the hands of police officers would bring race relations center stage. Although Trump and his most passionate legislative allies mocked Black Lives Matter protests, some believed Senate Republicans in places like Georgia and North Carolina could face strict backlash from the large Black populations found in their states.

Additionally, on September 18, 2020, Justice Ruth Bader Ginsberg lost her long battle with cancer. Even though Republicans had recently vowed never to select a replacement in an election year in the summer of 2016, they quickly moved forward to fill the vacancy. Their wish was to replace a staunch liberal with an equally staunch conservative, believing such a move would bolster support among their base in the upcoming election. Although it was hoped that moderates like Romney or Collins would break ranks, in the end, along strictly partisan lines, Republicans would approve Amy Coney Barrett as the next justice of the US Supreme Court only days before the November 3 election.

After polls closed on November 3, 2020, Democrats inched closer to controlling the Senate. Although it took nearly a week of counting votes, it was clear Joe Biden would defeat Donald Trump and Democrats would maintain control of the House, but the Senate was left undetermined. Figure 7.7 shows Democrats flipped Arizona and Colorado, but lost, as expected, in Alabama. Results in both North Carolina and Maine shocked some, but specific circumstances in each race explained how Republicans maintained each of the seats. Control of the Senate would come down to both Georgia runoff elections. Due to Georgia's election laws, if no one candidate gains more than 50 percent of the total vote in the initial election, a runoff election would occur between the top two vote-earning candidates. In both Senate elections on November 3, none of the candidates would reach the 50 percent threshold. In the end, both Democrats would win, giving each party fifty seats and with Vice President Harris, control of the Senate.

ARIZONA

After the passing of longtime Republican Senator John McCain on August 25, 2018, Martha McSally was appointed as his temporary successor until the next statewide federal election, November 3, 2020. McSally, who ran for the Senate in 2018 but lost to moderate Kristen Sinema, allied herself with Donald Trump in 2018 and again in 2020 (Sanchez 2018). Because this was John McCain's seat, however, the McCain family did not approve of McSally's initial appointment and were even more outspoken against her during the 2020 campaign as a result of her support for President Trump. Running against McSally was Democrat Mark Kelly, who was best known as the twin brother of fellow astronaut Scott Kelly, as well as the husband of former Arizona Congresswoman Gabby Giffords. As COVID-19 ravaged Arizona in August of 2020, Kelly campaigned on the need for more action to combat the virus and highlighted his stance as a strong moderate, while McSally struggled to find her footing as Donald Trump dragged her popularity down through his own controversies. In the end Kelly won the election by a larger margin than Joe Biden. His support was strongest in the Phoenix and Tucson areas, as well as in areas with larger Hispanic populations and Native American reservations found throughout the state.

FIGURE 7.7

2020 Senate Elections

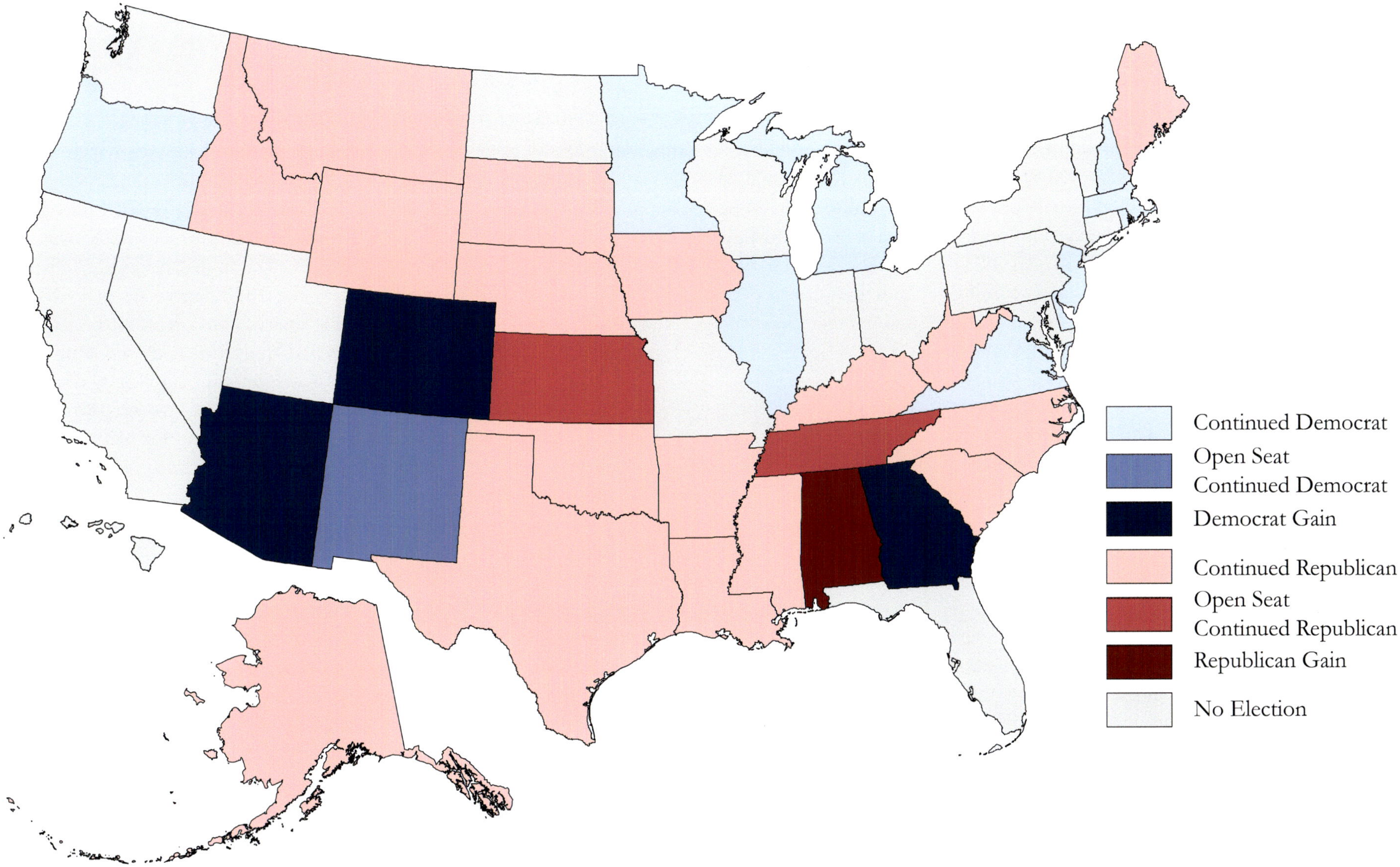

MAINE

Although Maine often elects Democrats for president, Republican Susan Collins has represented Maine since 1996. Her general moderate stance has been praised by many, but her recent and tepid support of Donald Trump left her vulnerable in 2020. The Democratic National Committee saw a chance to pick off the Republican seat and extended additional funds to Democrat and Speaker of the Maine House of Representatives, Sara Gideon. In 2019 Maine approved the use of a ranked choice voting system, allowing voters to rank candidates based on preference. If no candidate wins with a clear majority of 50 percent or greater, any first-place votes for candidates coming in third or lower are then transferred to the voter's second choice. It was believed that if third-party candidates could gain enough support, Gideon would be the most likely second choice of their ballots, providing her the necessary votes to win. In the end, Collins would exceed the 50 percent threshold by defeating Gideon 51 percent to 42 percent and the ranked choice system was not initiated.

NORTH CAROLINA

Leading up to the November 3 elections, polls indicated Democratic challenger Cal Cunningham had a chance to defeat the incumbent senator, Republican Thom Tillis. As was the case in nearly all elections in 2020, Donald Trump loomed large. Tillis was an ardent supporter of Donald Trump and gained his endorsement early in the election cycle. Due to the large Black population of North Carolina, race and defunding police became an early issue in the campaign. Tillis accused Cunningham of being silent on these issues, but Cunningham had made his opposition to defunding police clear in favor of reform. As September gave way to October, this campaign would spin into chaos. On October 3, Tillis indicated that he had tested positive for COVID-19 after attending what would be deemed a "super-spreader event," the celebration of Supreme Court nominee Amy Coney Barrett. On that same day, it was reported Cunningham had an extramarital affair. As polls closed, Tillis would keep his seat. It was clear the affair allegations impacted support for Cunningham throughout suburban areas of the state. He lost the election 48.7 percent to 46.9 percent, underperforming even Biden throughout the state and popular Democrat governor, Roy Cooper.

GEORGIA

Due to runoff laws in Georgia, winners of both Georgia senate seats would be determined on January 5, 2021. Nearly $1 billion was spent on campaign advertisements in this closely watched race. Trump and President-Elect Joe Biden would also make campaign stops for their respective parties' hopeful candidates in the weeks leading up to January 5. Both Republican incumbent senators, David Purdue and Kelly Loeffler, were ardent supporters of Donald Trump before and after the November 3 election. On the Democrat side, Jon Ossoff was widely known for his House election loss in 2017 and his ties to the late John Lewis. Reverend Raphael Warnock was a pastor at the Ebenezer Baptist Church in Atlanta and was handpicked by former gubernatorial candidate Stacey Abrams to run for office. Warnock's ability to speak and connect with Black voters made him a formidable candidate. In the end, both Democratic candidates won their elections by driving up voter turnout in traditional Black areas, as well as in urban, suburban, and college communities throughout the state. Republicans were less enthusiastic. Trump's attacks on Georgia's election processes and his absence on the January 5 ticket meant Republican turnout was lower compared to November 3, 2020. Although support was highest in the rural areas, decreased turnout in suburban areas proved to be too difficult for either Republican candidate to overcome. These elections provided Democrats the needed seats to take control of the Senate.

As with the 2018 midterm elections, Trump was central to the 2020 Senate election cycle. It was clear, that apart from a handful of Republicans, Trump had transformed the GOP into the Party of Trump, leading to complicated results. In swing states like Arizona and Georgia, the growing discontent with the president was just enough for senators to win their respective elections. Yet, in both Maine and North Carolina, candidates were able to survive even with their support for the president made public. It remains to be seen what kind of impact Trump will have on the 2022 midterm Senate races; however, given that he remains one of the most influential Republicans, Trump and Trumpism will continue to play a role.

REFERENCES

Booker, B. 2020. Trump Blasts Romney Over Impeachment Vote. *NPR*. https://www.npr.org/2020/02/06/803445433/trump-blasts-romney-over-impeachment-vote (last accessed June 22, 2021).

Derysh, I. 2020. Susan Collins Doesn't "Understand" Why Voters Turned on Her. Did She Forget Her Impeachment Vote? *Salon*. https://www.salon.com/2020/03/06/susan-collins-doesnt-understand-why-voters-turned-on-her-did-she-forget-her-impeachment-vote/ (last accessed June 22, 2021).

Sanchez, Y. W. 2018. In Arizona's Senate Race, Martha McSally Cozies Up to Donald Trump. *Arizona Republic*. https://www.azcentral.com/story/news/politics/arizona/2018/07/24/martha-mcsally-cozies-up-donald-trump-arizonas-senate-race/798813002/ (last accessed June 22, 2021).

ARIZONA SENATE ELECTION

WYATT PAJTASH

On August 25, 2018, the fate of the Senate majority for the 117th US Congress became much more competitive. It was on this day that the six-term senator, Republican John McCain, passed away from brain cancer. The death of McCain began a roller-coaster ride for Arizona's Class 3 Senate seat, which had been held consistently since 1969 by the Republican political giants of McCain and his predecessor, conservative icon Barry Goldwater.

Per Arizona law (Arizona Legislature 2021), vacancies of an Arizona Senate seat are to be temporarily filled by the appointment by the governor of an individual who belongs to the same political party as the outgoing senator. The appointee shall than serve until the next statewide election is held. Initially following McCain's death, Republican Governor Doug Ducey appointed McCain's former junior senator and minority whip of the Senate, Jon Kyl, who returned to the Senate after a lucrative tenure as a lobbyist. However, Senator Kyl stated he would resign from his post at the end of 2018, setting the stage for another appointment by Governor Ducey that would be influenced by the results of Arizona's other Senate election (Hansen and Sanchez 2018).

That successive appointee became the former air force pilot, Congresswoman Martha McSally. McSally had just come off a defeat for Arizona's other Senate seat, which was being vacated by Kyl's first successor to the Senate, Republican Jeff Flake. Senator Flake decided not to run for a second term, fearing a tough primary with his popularity declining in the Republican Party after being a vocal critic of President Trump and even ended up endorsing Joe Biden for president in August of 2020. Flake was succeeded by Democratic Congresswoman Kyrsten Sinema, who beat McSally by a 55,900-vote margin (Taylor 2018).

McSally ran her 2018 campaign for Senate with a similar rhetoric to that of President Trump, even after choosing not to endorse him in his 2016 election. This choice was likely the result of seeing Senator Flake's criticisms of the president ruin his chances of a second term and prompting tough primary challenges from candidates who had similar ideologies as Trump. This strategy proved fatal for McSally; Sinema became the first Democrat to win a US Senate election in Arizona since 1988. Sinema's campaign for Senate was perceived as moderate and created a campaign template for future Democratic candidates running in Arizona.

The loss of the Senate seat for Republicans in 2018 was coupled with defeats for the statewide offices for the superintendent of the public institutions and the secretary of state. This signaled a continued shift in the state's electorate to the Left after Trump only won the state by 3 percent in 2016 following Mitt Romney's nine-point win in 2012. Despite the loss and the changing electorate, Ducey saw McSally's name recognition as a key to keeping the Senate seat for Republicans.

Making matters worse for McSally's 2020 Senate campaign was that the Democrats recruited the popular Mark Kelly, who was a former astronaut and husband to former Arizona Congresswoman Gabrielle Giffords. Although both Kelly and Giffords were already in the public eye, they had been thrust into the national spotlight in 2011 after Giffords was shot and almost killed in an assassination attempt at one of her public events (Halloran 2011). After surviving a bullet to the head and a very tough recovery, Giffords and her husband became activists for gun safety, which become a central issue for Kelly's campaign. An important secondary issue for the Kelly campaign was heath care. Kelly was quick to point to the quality care his wife received in her long recovery as well as the care Kelly himself received as he battled skin cancer. Kelly would also lean on his role in NASA and his time in the navy, advocating for advancements in scientific research and infrastructure innovations. Like Sinema before him, Kelly ran as a centrist, attacking both McSally and Democrats for creating partisan gridlock (Sanchez 2018).

Outside of the presidential election, the special Senate race was not complicated by many outside figures; there were no third-party candidates running and neither McSally nor Kelly faced serious primary challenges. While polls showed a tight race in the early months of 2020, throughout the summer numbers started to lean toward Kelly. By Election Day, his polling averages were hovering around 50 percent with a 5 percent lead on the sitting Republican senator. McSally also fell behind Kelly in fundraising, with Kelly raising almost

FIGURE 7.8

Arizona US Senate Election Popular Vote Leader, 2020

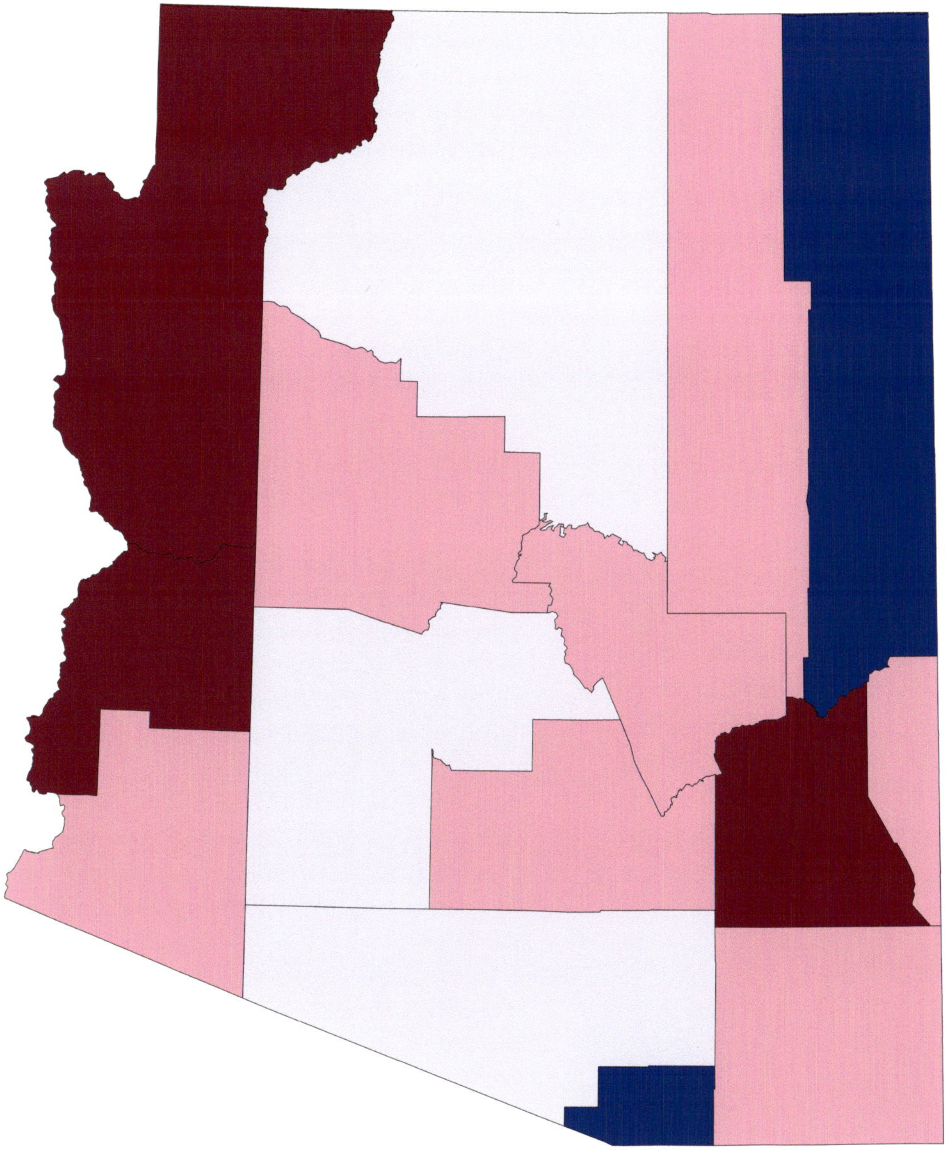

Percentage of Vote

Democrat

Kelly Majority
(50 to 64%)

Kelly Super Majority
(65% or more)

Republican

McSally Majority
(50 to 64%)

McSally Super Majority
(65% or more)

$100 million to McSally's $71 million. These realities had many political handicappers move Arizona's senate seat from "Toss-Up" to "leaning Democratic." *Sabato's Crystal Ball* was the first, labeling the seat as leaning Democratic as early as April 2020 and continuing through Election Day.

Senator McSally's 2020 campaign was almost a replica of her 2018 campaign; she clung to President Trump and his reelection campaign. Representing a border state, McSally promoted a secure border and was often heard discussing Trump's border wall. Her campaign also emphasized fighting back against China, especially claiming the United States was too dependent on China for prescription drugs and manufacturing while also blaming China for the coronavirus. Outside of her reliance on Trump, McSally's campaign tried to label Mark Kelly as too liberal for Arizona and ran several attack ads claiming he would "Do anything for a buck and say anything for a vote." As McSally lagged further behind Kelly in the polls and in fundraising, in an odd twist McSally ask her supporters to skip meals and donate the money they saved to her campaign (Montini 2020). This strange strategy had little effect.

As results came in on election night, Mark Kelly maintained a strong lead over McSally and as the votes slowly came in, the deficit was too large of McSally. On November 4, 2020, Mark Kelly was declared the winner. Despite the math clearly showing that McSally could not overtake Kelly, it took her nearly a week to concede to her challenger. Although true to form, McSally could not shake away from Trump. Her stalling to concede was likely influenced by protests to the Arizona elections by Trump himself and Republican supporters who stood outside the Maricopa County election center screaming "count the votes" as absentee ballots were slowly being counted (Knowles and Shepherd 2020). Overall, the final numbers appeared to be slightly closer than polls had suggested, with Kelly winning 51.2 percent of the vote to McSally's 48.8 percent equating to only a difference of 78,806 votes. Mark Kelly outperformed Joe Biden in every county and did an entire percentage point better than Biden statewide (see figure 7.8). The key factor in winning the state proved to be Kelly's ability to win Maricopa County, the home to the capital, Phoenix, and the most populous county in the state. Trump had won Maricopa County in his 2016 election victory. Kelly also excelled in the Tucson area and throughout the Native American reservations in the northern and eastern areas of the state.

As this was a special election, Mark Kelly was sworn in as a senator on December 2, 2020, which narrowed the Republican Senate majority to fifty-two to forty-eight for the remaining lame duck session of the 116th Congress and helped Democrats prevent possible nominations from President Trump, like that of Judy Shelton to the Federal Reserve Board (Dennis 2020). Senator Kelly's term runs until the end of the 117th Congress on January 3, 2023, meaning he will have a short turnover before having to run for a full six-year term in the 2022 midterms.

REFERENCES

Arizona Legislature. 2021. 16-222—Vacancy in the Office of United States Senator or Representative. https://www.azleg.gov/ars/16/00222.htm#:~:text=The%20special%20primary%20election%20shall,after%20the%20special%20primary%20election (last accessed April 24, 2021).

Dennis, S. T. 2020. Mark Kelly Sworn In to Senate, Shrinking GOP Majority. *Bloomberg.com*. https://www.bloomberg.com/news/articles/2020-12-02/mark-kelly-sworn-in-to-senate-shrinking-gop-majority (last accessed May 2, 2021).

Halloran, L. 2011. "Vitriol" Cited as Possible Factor In Arizona Tragedy. *NPR*. https://www.npr.org/2011/01/10/132764367/congresswoman-shot-in-arizona (last accessed April 25, 2021).

Hansen, R. J., and Y. Wingett Sanchez. 2018. Martha McSally Appointed to John McCain's Senate Seat. *Arizona Republic*. https://www.azcentral.com/story/news/politics/arizona/2018/12/18/martha-mcsally-named-doug-ducey-kyl-mccain-arizona-senate-seat-lost-sinema/2277884002/ (last accessed April 15, 2021).

Knowles, H., and K. Shepherd. 2020. Driven by Unfounded "Sharpiegate" Rumor, Pro-Trump Protesters Mass outside Arizona Vote-Counting Center. *Washington Post*. https://www.washingtonpost.com/nation/2020/11/05/arizona-election-protest-votes/ (last accessed April 22, 2021).

Montini, E. J. 2020. Sen. Martha McSally: "Fast a Meal" and Donate the Cash to Me. *Arizona Republic*. https://www.azcentral.com/story/opinion/op-ed/ej-montini/2020/08/21/sen-martha-mcsally-fast-meal-and-donate-cash-me/3417979001/ (last accessed May 10, 2021).

Taylor, J. 2018. Democrat Kyrsten Sinema Wins Arizona Senate Race, Flipping Second GOP Seat. *NPR*. https://www.npr.org/2018/11/12/667214148/democrat-kyrsten-sinema-wins-arizona-senate-seat (last accessed May 3, 2021).

AS MAINE GOES, NO GOES THE NATION: THE 2020 SENATE ELECTION

BENJAMIN FOREST

Once considered a political bellwether, Maine was the only state in 2020 that split its vote between the presidential and senatorial contests. While President Joe Biden won two of the state's four electoral votes with a comfortable 53 percent to 44 percent margin statewide, incumbent Republican Senator Susan Collins prevailed 51 percent to 42 percent over Democratic challenger Sara Gideon. (Presidents Biden and Trump split the two electoral votes awarded by Maine's congressional districts.) Independent candidates—progressive Lisa Savage and conservative Max Linn—captured 5 percent and 1.7 percent of the vote, respectively.

Collins's reelection was also surprising because she was considered vulnerable throughout the campaign and faced considerable opposition from Democratic groups after her votes to confirm Supreme Court Justice Brett Kavanaugh and to acquit President Donald Trump at his first impeachment trial (Rothenberg 2019; Sopko 2020). Famously, she declared that having "learned from this case," the president would be "much more cautious in the future" (Mazzei and Edmondson 2020). Collins's declaration that although she considered President Trump's conduct wrong but short of an impeachable offense is emblematic of her effort to position herself as a moderate Republican in a dramatically polarized national political environment. Her subsequent reelection—and vote to convict at the second impeachment trial in February 2021—reflects Maine's transition from a reliable national political barometer to an outlier. Nonetheless, local observers describe the state as "increasingly divided" with 2020 characterized by "fractured local elections" (Shepherd and Piper 2020). If such opinions are correct, Maine may soon return to its status as a national mirror.

Senator Collins entered the 2020 campaign as a four-term senator, first elected in 1996, after having been active in state-level politics since the mid-1970s, including running (unsuccessfully) as the state's first woman gubernatorial candidate in 1994 (Ballotpedia 2021b). She won reelection in 2002, 2008, and 2014 by increasingly large margins, prevailing in every county. Collins holds several positions at odds with the usual Republican Party ones, notably her support for abortion rights and the Affordable Care Act, and her voting record ranks her as one of the most liberal Republican Senators and an ideological centrist (GovTrack 2021). Her ideological ranking is similar to Maine's other senator, Independent Angus King (who caucuses with the Democratic party.)

Sara Gideon entered the 2020 campaign having served for four years as the Speaker of the Maine House of Representatives. She was first elected in 2012 after working in the private sector and becoming involved in local politics several years earlier. She won the Democratic Senate primary handily over two other women candidates, with more than 70 percent of the vote (Ballotpedia 2021a). Speaker Gideon prioritized liberal legislative positions, including Medicaid expansion and climate change, and received high positive rankings from labor and environmental groups (Maine AFL-CIO 2020; Maine Conservation Voters 2020). Nonetheless, her campaign emphasized her efforts to work with both parties, particularly her decision to intersperse the seating of State House representatives rather than having the two parties sit on opposite sides of the chamber.

Speaker Gideon's campaign attracted significantly more total contributions compared to Senator Collins (about $75 million compared to about $30 million), had a significantly higher proportion from small individual contributions (<$200), and from nearly every category of contributor other than business interests (Center for Responsive Politics 2020). The vast majority (over 93 percent) of contributions to both candidates came from out-of-state sources, indicating the intense national interest in the race, and independent spending from outside groups favored Gideon by about $10 million.

Despite her disadvantages in contributions and preelection polls, Senator Collins won by more than 70,000 votes, nearly a 9 percent margin (Piper and Shepherd 2020). Nonetheless, this result can be interpreted as a mark of either strength or weakness. The difference between support for presidential and senatorial candidates from different parties was among the highest of any state, showing that Senator Collins attracted an unusually high number of Biden supporters. Analysis immediately after the election speculated that polling missed both the prevalence of such ticket-splitting and the strong break of undecided voters to Senator Collins.

On the other hand, the size of her victory was the lowest since Senator Collins's initial election in 1996, reversed her trend of increasing winning margins, and—other than the 1996 election—was the first time since 1978 that the Republican Senate candidate failed to win a majority in every county. To a degree, the 2020 geographic pattern of voting (see figure 7.9)—with Speaker Gideon winning decisively in Portland's Cumberland County (54 percent to 38 percent)—resembles the 1996 election, where the Democratic candidate won the relatively urbanized Cumberland, Androscoggin, and Kennebec counties. In 2020, Speaker Gideon failed to win these latter two, but showed greater strength in coastal regions, winning Knox County by 1.9 percent and holding Senator Collins to her lowest margins in York (4.8 percent), Sagadahoc (3.2 percent), and Hancock (5.6 percent) counties.

Speaker Gideon's strength in Portland and South Portland, where she won by 46 percent and 30 percent respectively, may reflect an increasing urban-rural divide in the state. Other than a handful of small jurisdictions, she enjoyed her highest victory margin in Portland. Yet this may be more about a split between Portland and the rest of the state rather than urban-rural differences in general. Portland's voters also passed several progressive ballot measures that mirror Speaker Gideon's legislative priorities, including mandating new environmental standards and a $15/hour minimum wage (Shepherd and Piper 2020). In Maine's other large municipalities—Lewiston, Bangor (Senator Collins's home), and Scarborough, for example—the vote was split nearly evenly.

Beyond the geographical pattern of voting, surveys and exit polling reveal important divisions among Maine voters (Andre et al. 2020a, Andre et al. 2020b). As expected, support for the Senate candidates largely mirrored support for their party's presidential candidate, although exits polls indicated that favorable opinions of Collins were lower among Trump voters (70 percent) compared to favorable opinions of Gideon (90 percent) among Biden backers. Indeed, 95 percent of Trump supporters felt that the senator had not been sufficiently supportive of the president. Consequently, the key to the senator's victory may have been her relatively high rating among Biden supporters. Nearly 20 percent held a favorable view of Senator Collins, while only 4 percent of Trump partisans liked Gideon. This contrasts dramatically with partisan polarization at the national level, where respondents held 95 percent/5 percent favorable/unfavorable views of both presidential candidates. Thus, Senator Collins benefited from a relatively large partisan crossover vote.

As shown in figure 7.9, there is not a clear geographical pattern to the crossover vote. The two regions with the largest difference between Biden and Gideon were northern, lightly populated Aroostook and heavily urban Cumberland counties. The smallest differences were counties on opposite sides and the middle of the state, representing a mix of small and moderate populations (Oxford, Androscoggin, Piscataquis, and Washington).

The size of the senator's crossover vote is ironic given Republican opposition to Maine's new ranked choice voting system. In 2019, the state adopted a system that permitted voters to rank candidates in order of preference (Andrews, Shepherd, and Piper 2020). If no one won a majority initially, votes for the candidate with the lowest support would be dropped, and their second-choice candidate would be counted toward the total. Republicans opposed this method because they feared Left-leaning voters would list a progressive candidate like Lisa Savage first but mark a Democratic candidate as their second choice. Under the typical "first-past-the-post" system in the United States, votes for third parties are considered "wasted" because they (typically) reduce the total of the second-place candidate. Third-party candidates were widely blamed for Republican or Democratic losses in the 1992 (Ross Perot), 2000 (Ralph Nader), and 2016 (Jill Stein) presidential elections, for example. Under the ranked choice system, such voters can both express a preference for a third party and (potentially) support a winning candidate. Ultimately, however, the ranked choice system had no effect in 2020 because Senator Collins won an outright majority statewide.

REFERENCES

Andre, M., A. Aufrichtig, G. Beltran, M. Bloch, L. Buchanan, A. Chavez, N. Cohn, M. Conlen, A. Daniel, A. Elkeurti, A. Fischer, J. Holder, W. Houp, J. Huang, J. Katz, A. Krolik, J. C. Lee, R. Lieberman, I. Marcus, J. Patel, C. Smart, B. Smithgall, U. Syam, R. Taylor, M. Watkins and I. White. 2020a. Maine Exit Polls: How Different Groups Voted. *New York Times*. November 3.

———. 2020b. Maine Voter Surveys: How Different Groups Voted. *New York Times*. November 3.

Andrews, C., M. Shepherd, and J. Piper. 2020. Maine's Experiment with Ranked-Choice Voting Was Muted in 2020. *Bangor Daily News*. November 5.

FIGURE 7.9

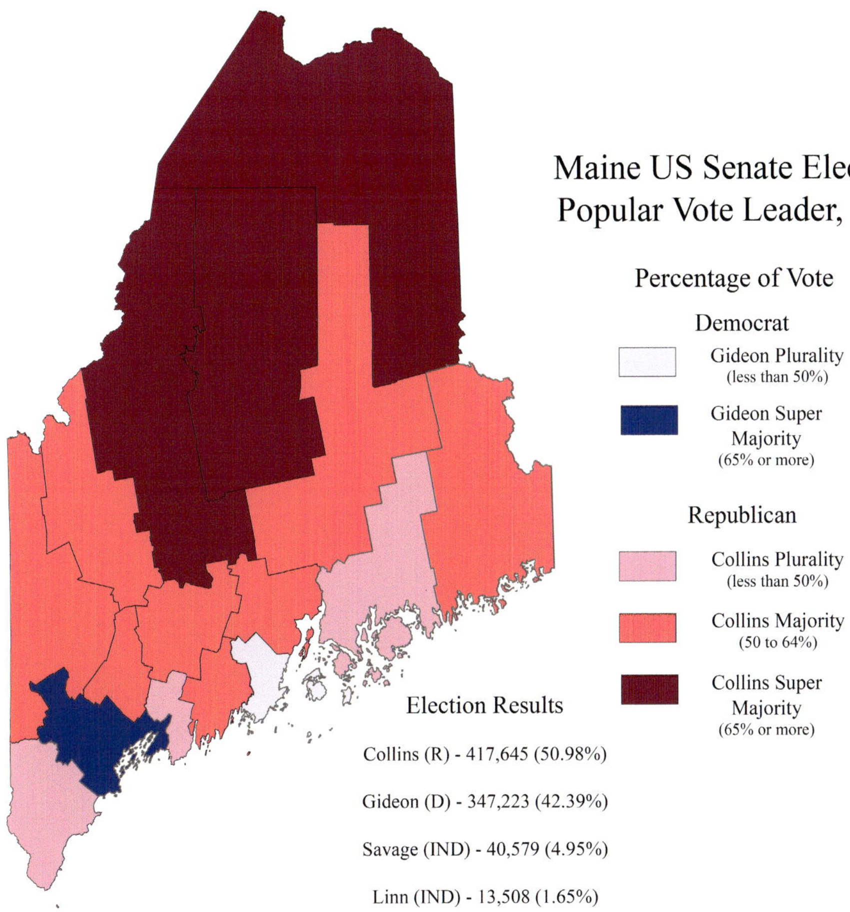
Maine US Senate Election
Popular Vote Leader, 2020
Percentage of Vote
Democrat
Gideon Plurality
(less than 50%)
Gideon Super
Majority
(65% or more)
Republican
Collins Plurality
(less than 50%)
Collins Majority
(50 to 64%)
Collins Super
Majority
(65% or more)
Election Results
Collins (R) - 417,645 (50.98%)
Gideon (D) - 347,223 (42.39%)
Savage (IND) - 40,579 (4.95%)
Linn (IND) - 13,508 (1.65%)

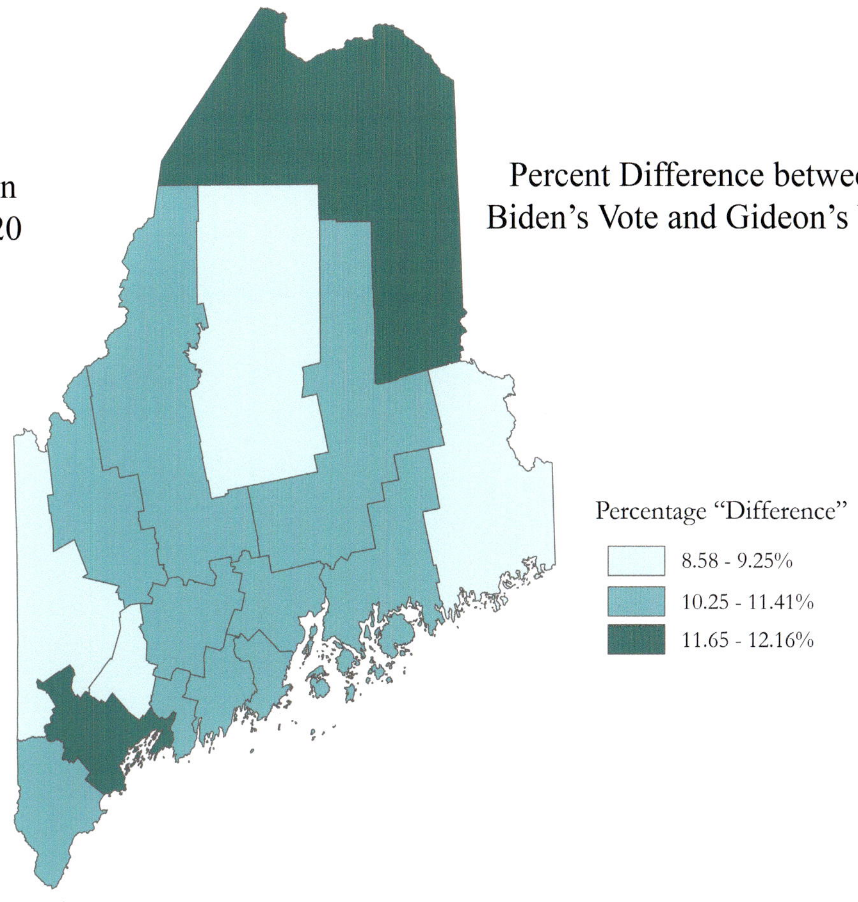
Percent Difference between
Biden's Vote and Gideon's Vote
Percentage "Difference"
8.58 - 9.25%
10.25 - 11.41%
11.65 - 12.16%

Susan Collins (Maine). 2021b. *Ballotpedia.* https://ballotpedia.org/Susan_Collins_(Maine) (last accessed May 1, 2021).

Sen. Susan Collins. 2021. *GovTrack.* https://www.govtrack.us/congress/members/susan_collins/300025 (last accessed May 3, 2021).

Sara Gideon. 2021a. *Ballotpedia.* https://ballotpedia.org/Sara_Gideon (last accessed May 7, 2021).

Legislative Scorecard. *Maine AFL-CIO.* http://maineaflcio.openscorecard.org/all-scores (last accessed May 6, 2021).

Maine Senate 2020 Race. 2020. *Center for Responsive Politics.* https://www.opensecrets.org/races/summary?cycle=2020&id=MES2 (last accessed May 7, 2021).

Maine Conservation Voters. 2020. 2020 Environmental Scorecard. https://www.maineconservation.org/assets/scorecards/2020-mcv-scorecard-web.pdf (last accessed May 6, 2021).

Mazzei, P., and C. Edmondson. 2020. Susan Collins Announces She Will Vote to Acquit Trump, Calling Conduct "Wrong." *New York Times.* February 4.

Piper, J., and M. Shepherd. 2020. Susan Collins Defied the Polls. Here's What They May Have Gotten Wrong. *Bangor Daily News.* November 5.

Rothenberg, S. 2019. Susan Collins Has a 2020 Problem. *Roll Call.* January 15.

Shepherd, M., and J. Piper. 2020. Maine's Most Interesting Local Races of 2020 Were all Over the Map. *Bangor Daily News.* November 4.

Sopko, G. 2020. Colby Poll Shows Statistical Dead Heat in Maine's U.S. Senate Race. *Colby News.* February 18.

THE 2020 MONTANA SENATORIAL ELECTION

KENNETH FRENCH

The 2020 Montana senatorial race between the incumbent Senator Steve Daines, a Republican from Bozeman, and Governor Steve Bullock, a moderate Democrat who grew up in Helena, was of national interest. Senator Daines is a fifth-generation Montanan who worked as vice president of RightNow Technologies in Bozeman and was elected in 2012 as Montana's lone congressperson (Daines 2021). His senatorial duties include work on the Finance committee, Energy and Natural Resources committee, Indian Affairs committee, and Banking, Housing, and Urban Affairs committee. Popular two-term governor Steve Bullock was the former attorney general of Montana, and he touted his success in bringing Republicans and Democrats together (Bullock 2020). Governor Bullock, due to term limits, could not run in the gubernatorial race in 2020. He focused his political efforts on the Senate seat after he was unsuccessful in gaining the Democratic nomination for president. The polling leading up to the November election indicated a tight race. However, the result of the 2020 Montana senatorial race was the comfortable reelection of Republican Senator Steve Daines (55 percent of the vote) over Governor Bullock (45 percent of the vote).

This Montana Senate seat was of national interest, especially for Democrats hoping for a historically Republican "red state" possibly turning "blue" and giving Democrats the majority in the Senate. Fundraising mirrored this national political interest. Democrats raised and spent $82 million for Governor Bullock, while Republicans spent $63 million for Senator Daines (Brown 2020)—by far the most money spent in any Montana election. According to the US Federal Election Commission, as of October 27, 2020, this race generated $96.6 million in third-party spending, with 82 percent of individual donations to Daines and 92 percent of individual donations to Bullock with out-of-state addresses (Dietrich 2020). Democratic fundraising increases were tied to anti-Trump sentiments and with the death of Justice Ruth Bader Ginsberg on September 18, 2020. Even though the Democratic senatorial candidate outspent the Republican candidate by nearly $20 million, the final result was still a victory for the Senate Republicans.

The main issues Montanans were concerned about were tied to public lands, health care, and political identity. Senator Daines cosponsored the Great American Outdoors Act that made federal royalties from oil and gas drilling go into conservation—an act popular in Montana and for environmentalists (Giroux 2020). Governor Bullock was seen as handling the COVID-19 shutdown well and was in favor of Obamacare. Senator Daines was able to walk a thin line on health care—being in lock step with Trump by voting to repeal Obamacare, while stating his support to protect preexisting conditions in the future. During the debates, Senator Daines tried to paint Governor Bullock as a puppet of the Democratic leadership of Senator Chuck Schumer and Congresswoman Nancy Pelosi. The strategy to cater to tribal politics was a success for Senator Daines in an independent-minded Montana.

The political geography of Montana, a state that consistently votes for Republican presidential candidates, does not seem complex. However, to "understand politics in Montana and the process of representation, one does not begin with people or politicians. One begins with *place* (italics in the original), because without place the rest does not—cannot—be made to make sense. How Montanans understand themselves, their representatives, their history, and their relationship to others—including the federal government—begins and ends with place." (Parker 2015, 13). There is an independent streak in Montana in which "split-ticket" voting is common and electing a pro-gun, moderate Democrat to the Senate would not be out of the ordinary. Nonetheless, the Republicans swept the presidential, senatorial, congressional, and gubernatorial races in 2020. Political observers of the Republican political dominance think the increasingly partisan national politics has taken over Montana's tradition of independence (Drew 2020).

As the map of election results illustrates (see figure 7.10), Senator Steve Daines won forty-seven of the fifty-six counties. Daines won Yellowstone County with 58.6 percent of the vote and Cascade County with 55.7 percent of the vote, home to the largest (Billings) and third largest cities (Great Falls), respectively. He also won the less-populated, rural counties by large percentages.

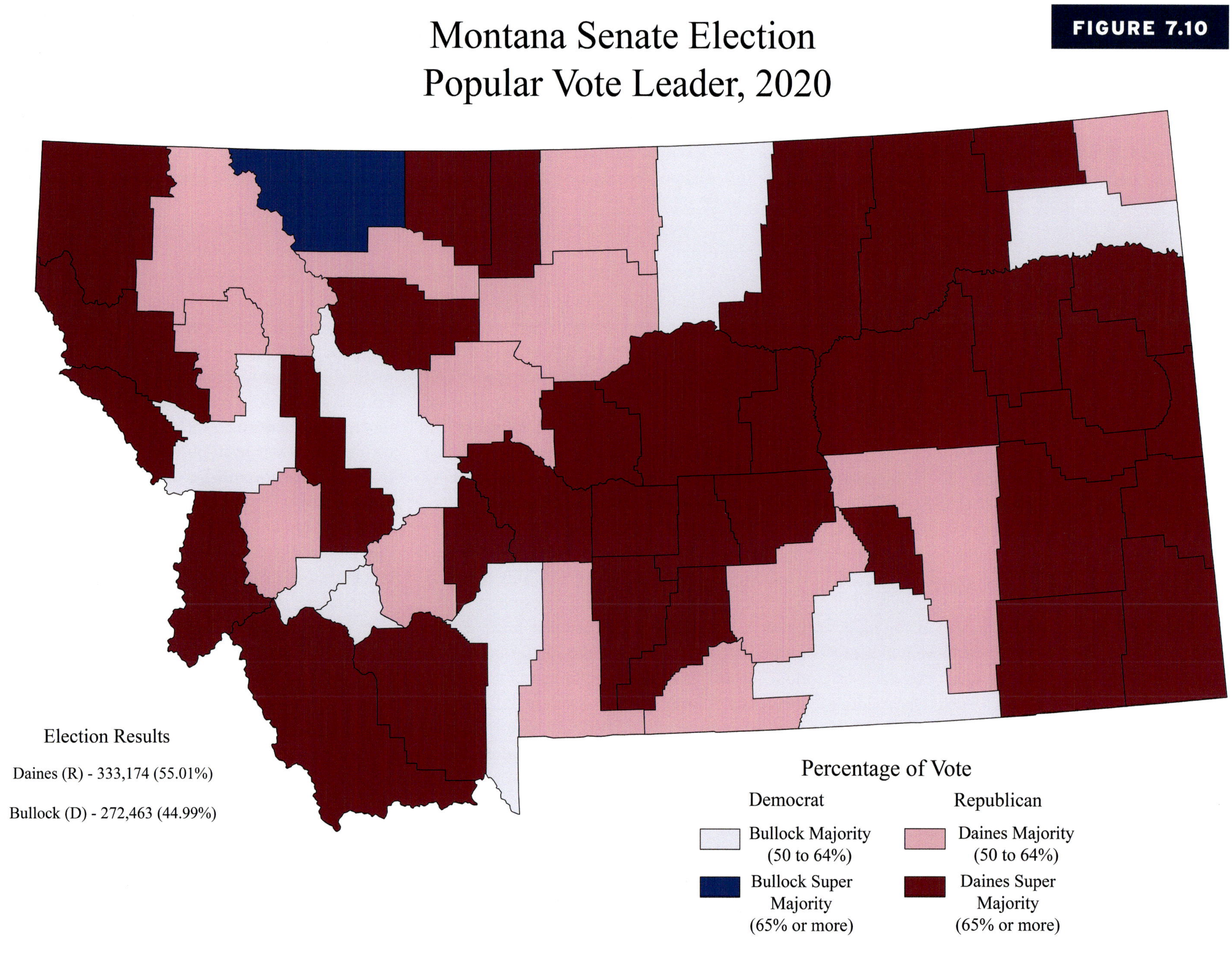
FIGURE 7.10
Montana Senate Election
Popular Vote Leader, 2020
Election Results
Daines (R) - 333,174 (55.01%)
Bullock (D) - 272,463 (44.99%)
Percentage of Vote
Democrat
Republican
Bullock Majority (50 to 64%)
Daines Majority (50 to 64%)
Bullock Super Majority (65% or more)
Daines Super Majority (65% or more)

For example, Senator Daines won 91 percent of the vote of large in area Garfield County, which accounted for a total of 736 votes. Governor Steve Bullock narrowly beat out Daines with 51.6 percent of the vote in Lewis and Clark County, the home to the capital city of Helena. Bullock did well in major college cities, with 64.4 percent of the vote in Missoula County and 55.4 percent of the vote in Gallatin County, locations of the University of Montana and Montana State University, respectively. Another Democratic center was in Butte, with 64.7 percent of the vote in Silver Bow County going to Bullock. The remaining Democratic strongholds were in counties with Native American reservations—Glacier County (Blackfeet Reservation) in the northwest, Roosevelt County (Fort Peck Reservation) in the northeast, and Big Horn County (Crow Reservation) in the south.

Overall, the 2020 Montana senatorial race between Mr. Daines and Mr. Bullock was not as close as the polling indicated. Democrats on the national stage were upset that they could not gain a "red" Senate seat by pushing a popular Democrat governor over the incumbent Republican senator. Political scholars will wonder if national interest in the 2020 Montana senate race, with the inclusion of tribal political identities and outside campaign funds, overcame the independent-minded voting populace of Montana. Is this a sign of future Republican dominance over all of Montana's political offices?

REFERENCES

Brown, M. 2020. Trump Noticeably Absent from Key US Senate Race in Montana. *ABC NEWS*. https://abcnews.go.com/Politics/wireStory/trump-noticeably-absent-key-us-senate-race-montana-73958019 (last accessed April 30, 2021).

Bullock, S. 2020. Meet Steve Bullock. *Bullock, U.S. Senate 2020*. https://stevebullock.com/meet-steve/ (last accessed April 30, 2021).

Daines, S. 2021. Meet Steve. *Steve Daines, United States Senator for Montana*. https://www.daines.senate.gov/meet-steve/biography (last accessed on April 30, 2021).

Dietrich, E. 2020. Where Is the Unprecedented Money Flooding Montana's Senate Race Coming From? Mostly out of State. *Billings Gazette*. https://billingsgazette.com/news/state-and-regional/where-is-the-unprecedented-money-flooding-montana-s-senate-race-coming-from-mostly-out-of/article_634e49a3-7dfa-5a5c-b6cf-ebcaddb61a4c.html (last accessed May 3, 2021).

Drew, M. 2020. Republicans Rout Caps Decade-Long Montana Political Shift. *Flathead Beacon*. https://flatheadbeacon.com/2020/11/04/republican-rout-caps-decade-long-montana-political-shift/ (last accessed May 3, 2021).

Giroux, G. 2020. Heated Montana Race Could Flip the Senate to Democrats. *Bloomberg Businessweek*. August 31, https://www.bloomberg.com/news/features/2020-08-31/could-montana-flip-the-senate-inside-the-race-between-daines-and-bullock (last accessed May 3, 2021).

Parker, D. C. 2015. *Battle for the Big Sky: Representation and the Politics of Place in the Race for the U.S. Senate*. Thousand Oaks, CA: Sage.

THE 2020 COLORADO THIRD CONGRESSIONAL DISTRICT ELECTION

REBECCA THEOBALD AND DWAYNE LILLER

While much of Colorado has turned toward the Democrats since 2008, the Western Slope and San Luis Valley remain more comfortable with conservative representatives; just over half of state representatives and senators in the Third District are Republicans. In June 2020, the incumbent five-term Republican, Scott Tipton (45.2 percent), was beaten in the primary by Lauren Boebert (54.6 percent), promoting her identity as a firearms owner running a small business. There was some hope among Democrats that Boebert's conspiratorial connections and the nation's increasing antipathy toward President Donald Trump would bring people to the polls to vote for Diane Mitsch Bush, a former state legislator and county commissioner who had run for the seat against Tipton in 2018. The race was framed on the extremes as being between "socialist" Mitsch Bush and "radical" Boebert, but the electors mirrored the voter registration patterns in the district as well as the continued division between rural and urban Colorado, reflected in changing demographics in mountain towns catering to tourists seeking recreation opportunities. Party registration numbers successfully predicted every 2020 congressional race in the state.

Despite significant financial advantages—the 2019–2020 contribution summary from the Federal Election Commission reported that Mitsch Bush had $5,055,863 in receipts compared to Boebert's $2,989,510—Mitsch Bush was never significantly ahead of Boebert in fall campaign season polls that had the Democrat winning. As with many other seemingly close races across the country, Boebert ended Election Day by winning 51.4 percent (220,634) over Mitsch Bush (45.2 percent—194,122), with another 14,563 votes to the Libertarian and Unity candidates. The polls were off, but Boebert's percent win when examining just the two main parties ranged from 23.9 percent to 84.3 percent among the twenty-nine counties, indicating significant division within the electorate.

Colorado's Third District has a varied geographic and political history, with its basic configuration dating from 1983 (98th Congress) and the current boundaries set during the most recent redistricting cycle in 2013. Over 106 years, the district has sent eight Democrats and seven[1] Republicans to Congress, although a Republican has held the seat since 2011. The district includes the Western Slope of Colorado bordering Utah from Wyoming to New Mexico at the corner of Arizona, stretching to the San Luis Valley to the east and including the Front Range population center of Pueblo as well as the cities of Grand Junction, Montrose, Durango, Steamboat Springs, and Glenwood Springs. By area, it is the ninth-largest congressional district comprising less than a whole state. According to American Community Survey 2019 estimates, the population was 756,566, with just under a quarter living in the urban areas of Pueblo or Grand Junction, another 20 percent living in cities with populations between five thousand and twenty thousand, and the remainder residing in small towns and rural areas. The Ute Mountain Ute and Southern Ute Indian Reservations lie within the Third Congressional District; just over two percent of the population is American Indian or Alaska Native, the largest percentage of any of Colorado's congressional districts. A quarter of the population identifies as Hispanic. The median age is 40.6 and the median household income is $59,973. The Cook Political Report Partisan Voting Index rated the District as R+6.

During the 2018 general election, Republicans had a strong advantage in most areas of the Third District. Democrats won 211 out of 485 precincts, with many of those nearer population centers, although pockets of support for each candidate were scattered throughout the region. Tipton had won his third term in 2014 with 68 percent of the vote, but in 2018 he beat Mitsch Bush with only 51.5 percent. An additional 83,730 people voted in 2020, an increase of 25 percent from 2018, consistent with turnout across Colorado. Examining registered voters from the two main parties, the split is Republican 54.98 percent to Democrat 45.02 percent—Boebert underperformed this split with 51.39 percent of the two-party vote while Mitsch Bush performed in line at 45.22 percent.

After the primary campaign, very little attention was paid to policy details (Ashby 2021). No debates took place between the two major candidates, in part due to coronavirus restrictions and in part due to disinterest in resolving scheduling conflicts. Boebert cast herself as a champion of President Trump and his policies, focusing on upholding gun rights, dismantling health-care funding, and decreasing business regulations. Mitsch Bush favored the Affordable Health Care plan, support for renewable energy, and "common sense gun control legislation" (VoteSmart 2020). Each of the candidates—Boebert at 95 percent and Mitsch Bush at 94 percent—relied primarily on individual contributors (FEC 2020). Colorado Public Radio reported "Democratic and Republican outside groups have each spent about $2.5 million attacking the other candidate" (Fish 2020). Anecdotal information indicates this race received a great deal of attention outside Colorado, but the consistency of the Republican ticket prevailed. With more Republicans (164,835) than Democrats (136,204) registered in the Third District, and more unaffiliated voters (203,195), the likelihood of a Democratic victory—even given the unpopularity of President Trump—was small.

The two previous election cycles resulted in Republican victors of 14.25 percent (2016) and 7.97 percent (2018) over Democratic competitors, with variable voter turnout in the last decade. The *Colorado Sun* reported that "Between Jan 1. and Nov. 1, the district added 13,746 registered Republicans compared to 9,771 Democrats and 24,446 unaffiliated voters" (Paul 2020). Contrary to patterns in more urban areas of Colorado, these unaffiliated voters do not necessarily tend to vote in more liberal directions. Unaffiliated voters, those not identifying with a political party (also referred to imprecisely as independent voters), now make up approximately 42 percent of the state's electorate, and analysts are still trying to figure out how to predict their voting patterns. Will they reflect the tendencies of the area where they live or does the unaffiliated badge provide license to break with the dominant party? Voters registering as unaffiliated may truly be independent or may be branding themselves as such while still toeing their preferred party's line.

The main component contributing to Boebert's victory was the existing voting pattern in the Third Congressional District. As figure 7.11 illustrates, Boebert was able to maintain the Republican strongholds of the Western Slope and San Luis Valley, while garnering enough votes in mountain areas and traditionally Democratic Pueblo to keep the seat in Republican hands. Economic instability also played a part. According to a map of federal unemployment data by the *Daily Yonder*, only five of the counties in the Third District lost fewer jobs on average in the past year than other parts of the country (Bishop 2021). Population and economic growth are relatively slow.

A confounding factor in this race was the presence of the controversial figure, President Trump, on the ballot. One way to see voter sentiments toward the incumbent presidential candidate is to compare Trump's vote percent in a two-party race to the various Republican congressional candidates. This "Trump effect" showed Trump garnering a smaller percentage of the two-party vote than the corresponding Republican congressional candidate in fifty-seven of sixty-four counties throughout Colorado. The measurement is Trump's percent win minus the corresponding Republican congressional candidate's percent win. Therefore, a strong showing by the congressional candidate will cause the Trump effect to be more pronounced (a larger negative number). This was the case in Districts Four and Five, the other two Colorado congressional districts won by Republicans, which displayed a higher Trump effect than District Three. Six of the seven counties where Trump outperformed the congressional candidate were in the Third District, suggesting either much higher support for Trump or less robust support for the Republican congressional candidate. Three of the 206 pivot counties (Ballotpedia 2016) in the United States are in the Third District, indicating some volatility in voter choice. A higher Republican congressional win percent than Trump implies that many voters split their tickets to vote for Joseph Biden while voting for Republicans down ballot. This was particularly evident in Districts Four and Five, which registered the largest differences in the state between Trump's percent of the two-party vote and the Republican congressional candidate's.

With strong campaign funding, a track record of public service, and previous experience in running for this office, one might have thought that Diane Mitsch Bush would be well poised to win in a competitive district hosting a significant number of unaffiliated voters. However, the labels of the voters do not necessarily reflect a significant change in their approach to policies and perceptions. Lauren Boebert's appeal to independence and rebellion was sufficient to win the confidence of people in larger cities, small towns, and rural areas throughout the district. She did not suffer from the Trump effect. Each county where Trump won Boebert also won.

In November 2021, the Colorado Supreme Court (2021) approved the map of eight congressional districts presented by the first Colorado Independent Redistricting Commission, generally agreeing the map was not perfect, nor did it please all stakeholders, but was a reasonable option. Whether the slightly revised map of the Third District protects incumbents or promotes competitive elections will be factors to examine in future analyses.

FIGURE 7.11

Colorado US House of Representatives District 3 Election Popular Vote Leader, 2020

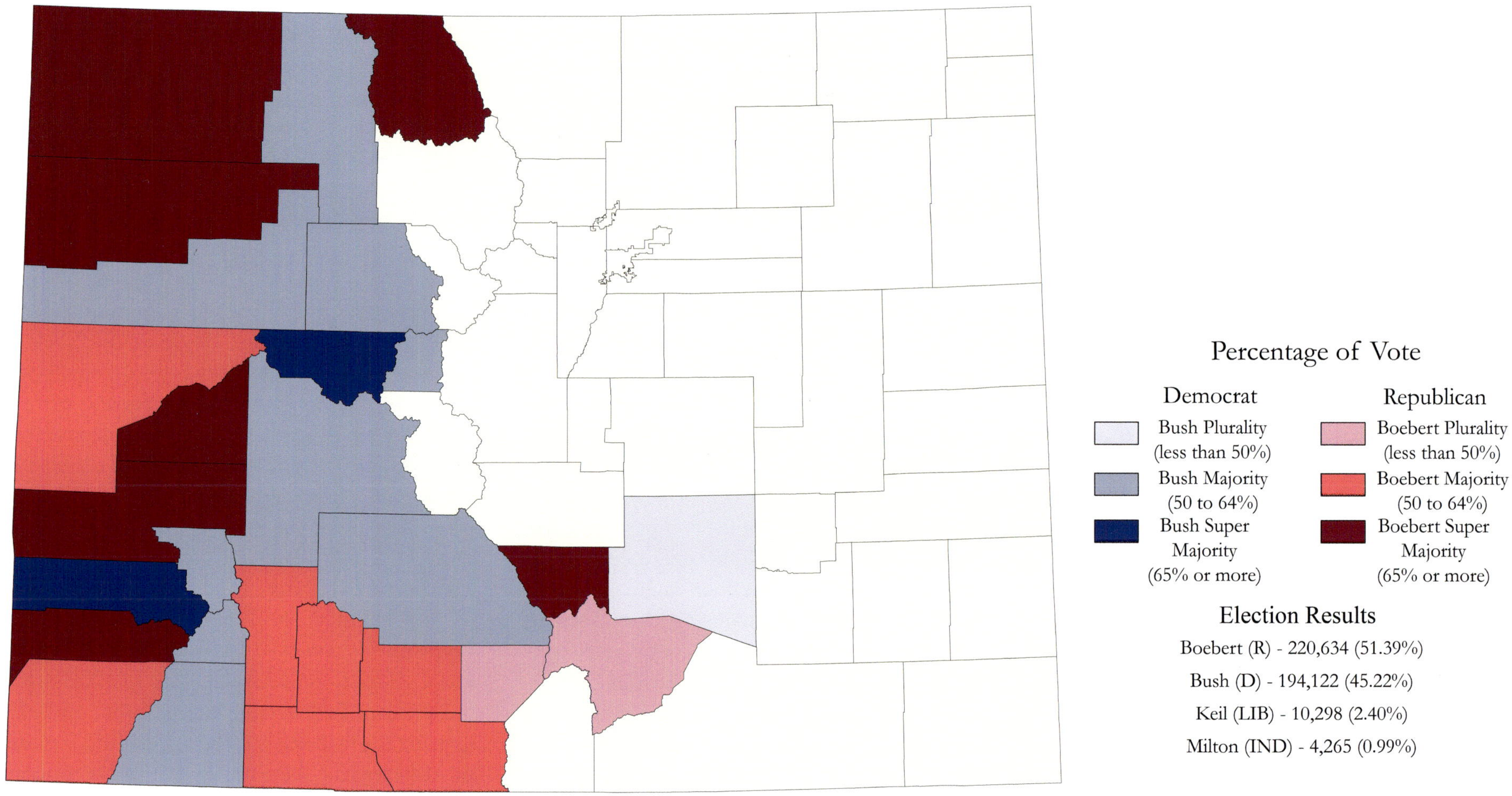

REFERENCES

Ashby, C. 2021. Race for 3rd CD is Tight, Poll Shows. *Daily Sentinel*. https://www.gjsentinel.com/news/western_colorado/race-for-3rd-cd-is-tight-poll-shows/article_7613553e-ddaf-11ea-8936-93fb370663a9.html (last accessed March 1, 2021).

Ballotpedia. 2016. Pivot Counties by State. https://ballotpedia.org/Pivot_Counties_by_state (last accessed March 1, 2021)

Bishop, B. 2021. Jobs Recovery Plateaus for Rural Counties in November. *Daily Yonder*. January 19, https://dailyyonder.com/jobs-recovery-plateaus-for-rural-counties-in-november/2021/01/19/ (last accessed March 1, 2021).

Colorado Supreme Court. 2021. 2021 CO 73–21SA208, In re Colo. Indep. Cong. Redistricting Comm'n. *Case Announcements*. Colorado Judicial Branch, Denver, Colorado. https://www.courts.state.co.us/Courts/Supreme_Court/Case_Announcements/Index.cfm.

FEC. Candidate Profiles. Campaign Finance Data. Federal Election Commission. https://www.fec.gov/data/candidate/H0CO03165/?cycle=2020 https://www.fec.gov/data/committee/C00707638/ (last accessed March 1, 2021).

Fish, S. 2020. Outside Spending Rushes into Colorado Race Between Lauren Boebert and Diane Mitsch Bush in Final Weeks. *Colorado Public Radio*. https://www.cpr.org/2020/11/02/outside-spending-rushes-into-colorado-race-between-boebert-and-mitsch-bush-in-final-weeks/ (last accessed March 1, 2021).

Paul, Jesse. 2020. 3 Numbers That Explain Republican Lauren Boebert's Victory in Colorado's 3rd Congressional District. *The Colorado Sun*. November 4.

VoteSmart. 2020. Issue Positions. https://justfacts.votesmart.org/candidate/political-courage-test/191290/lauren-boebert; https://justfacts.votesmart.org/candidate/political-courage-test/138740/diane-mitsch-bush (last accessed March 1, 2021).

NOTE

1. A Representative from Trinidad was elected after losing a seat in the previous election, so is counted twice over the 106 years of the district.

THE 2020 TEXAS TWENTY-THIRD CONGRESSIONAL DISTRICT ELECTION

CAMERON WINGREN

The 2020 Texas Twenty-Third Congressional District election was, like so many Texas Twenty-Third Congressional District elections prior, a hypercompetitive election fought right up until the very end. Unlike previous election cycles, however, three-term Republican representative Will Hurd had decided not to run for reelection. With no Republican incumbent, the Democratic Congressional Campaign Committee (DCCC) saw an opportunity to flip the district back under Democrat control. To do this, the DCCC tapped Gina Ortiz Jones, who had previously run against Will Hurd in 2018 and narrowly lost. At the same time as this, the National Republican Congressional Committee (NRCC) saw the importance of keeping the district under Republican control and invested heavily in the district as well. A field of nine Republican primary candidates led to a primary runoff election where Tony Gonzales was ultimately declared the winner to face Ortiz Jones in the general election. By Election Day, polling and race ratings had Ortiz Jones up and the district leaning Democratic. To the surprise of many in the DCCC, Texas Democratic Party, and Democrat leadership, Gonzales won an upset victory with 50.6 percent of the vote over Ortiz Jones's 46.6 percent and Libertarian Party candidate Beto Villela's 2.8 percent. Although Democrats were defeated in the district once again, this close result underscores the perennial competitiveness of Texas's Twenty-Third Congressional District.

Created after the passing of the Voting Rights Act of 1965, Texas's Twenty-Third Congressional District runs along much of Texas's border with Mexico. Nearly 58,000 square miles and encompassing parts of 29 counties, this district runs from eastern El Paso to western San Antonio. A predominantly rural district, save for a handful of small towns and county seats, it is difficult to campaign in. According to 2019 American Community Survey 1-Year Estimates, estimated total population of the district is 786,712, of which 69.6 percent are Hispanic. This district represents one of Texas's multiple Hispanic majority-minority congressional districts. Nearly 10 percent of potential voters in the district are naturalized citizens. Median household income in the district is $59,074, with 79 percent of individuals holding a high school diploma or equivalent and 23.9 percent of individuals holding a bachelor's degree or higher.

Texas's Twenty-Third Congressional District has a history of flipping between Democrat and Republican control, with control of the district swapping four times in the last two decades. Prior to the 2020 election, the district was represented by three-term Congressman Will Hurd, a moderate Republican known for his bipartisanship. In the 2016 election, Hurd was able to win reelection, but Hillary Clinton carried the district in the presidential election. This makes the Texas Twenty-Third Congressional District just one of five congressional districts represented by a Republican that Hillary Clinton won. In the 2018 election, Hurd once again won reelection, but by fewer than a thousand votes. With this electoral history in mind, it is no wonder the DCCC saw an opportunity to retake the district when Hurd announced he would not seek reelection in 2020.

Leading up to the 2020 election, Gina Ortiz Jones, who lost to Hurd in 2018 by fewer than a thousand votes, once again dominated the Democrat primary as she had in 2018. Ortiz Jones, an air force veteran who served as an intelligence officer in Iraq, previously worked at the Office of the US Trade Representative. Her key issues for her 2020 election campaign revolved around economic recovery from the COVID-19 pandemic, protecting/expanding health-care access, investing in education, and strengthening US relationships with allies to protect US interests, something for which she believes her prior experience left her well suited. For Republicans, nine primary candidates vied for the nomination, leading to a runoff election in which Tony Gonzales won by a mere forty-five votes (Office of the Texas Secretary of State 2020). Gonzales, while not a centrist in the swing district that is the Twenty-Third, still received Hurd's endorsement. A former US Navy cryptologist, Gonzales also worked as a Department of Defense legislative fellow and as an assistant professor at the University of Maryland system. His key issues for his 2020 election campaign were family values, strengthening security along the southern border, a strong national defense and foreign policy, and reducing government regulation of business. Finally, Libertarian Party candidate Beto Villela ran as the only third-party candidate on the ballot.

With hopes of flipping the district under Democrat control, the DCCC considered the Texas Twenty-Third as one of the top thirty-nine battleground districts in the 2020 election (DCCC 2020). With this, the DCCC also added Ortiz Jones to its Red to Blue program, which provides funding and guidance resources to candidates who are seeking election in Republican-controlled districts. Additionally, former president Barack Obama and presidential candidate Joe Biden endorsed Ortiz Jones. With this added attention and support, Ortiz Jones was able to raise $6,918,062 by Election Day (FEC 2020). At this same time, the NRCC added Gonzales to its Young Guns program, which supports candidates running for US House seats in battleground districts (NRCC 2020). Additionally, Gonzales received and welcomed the endorsement of President Donald Trump. By Election Day, Gonzales had raised $2,884,639 (FEC 2020). As the election grew near, polls and race ratings showed Ortiz Jones leading Gonzales. A Remington Research Group poll had Ortiz Jones up 45 percent to Gonzales's 43 percent, with 12 percent of voters undecided. On Election Day itself, two different race ratings had the district leaning Democratic while one race rating had the district tilting Democratic. As the polls closed on Election Day and results started to come in, Gonzales began leading Ortiz Jones by multiple percentage points. By 2:40 a.m. the next day, Gonzales was declared the winner and Texas's Twenty-Third Congressional District was retained firmly in the hands of Republicans.

Results from this election show the effect that the urban/rural divide as well as the Hispanic/non-Hispanic divide can have on deciding the outcome of an election. As expected, the region of El Paso County that falls into the Twenty-Third went Democrat and was won by Ortiz Jones with a supermajority (figure 7.12). Gonzales carried the area around San Antonio on the very eastern edge of the district, although only with a plurality (figure 7.12). This may seem surprising given the large, urban, Hispanic population in the area, but the inclusion of primarily non-Hispanic Republican strongholds to the north of San Antonio in the district negated the strength of the urban Hispanic. Rural counties within the Twenty-Third that lie between San Antonio and El Paso not directly on the US-Mexico border made up most counties won by Gonzales. While Ortiz Jones won some counties that lie along the US-Mexico border, counties with large Hispanic majorities, the interesting results were the multiple border counties won by Gonzales. These were the upset victories for Gonzales that Democrats did not expect; they had considered counties with a high Hispanic population to be easy wins for a Democrat candidate.

Various factors can be attributed to why this election was such an upset for the Democratic Party. One such factor is that Gonzales's message of traditional family values and strengthening the southern border simply played better with many in this rural border district. Ortiz Jones's message of expanding health care and education access among other things was well received, although still not enough to sway voters. Another factor, discussed in a Democratic Party postelection report, was that the Republican Party did a better job at outreach to gain new voters than the Democratic Party did in the lead-up to the election. Between the ruralness of the Twenty-Third District and the ongoing COVID-19 pandemic, reaching enough potential voters was a daunting task to be undertaken. Finally, a major factor as to why Ortiz Jones lost this race was that Texas's Hispanic population did not vote for Democratic candidates to the level that Democrats expected. While the Hispanic vote in this country remains heavily Democratic, something that helped secure victory for Biden in Arizona and Nevada, this was not seen to the same extent in Texas. This may be best summed up by a quote given to the *Texas Tribune* by Gonzales, in which he stated that the district is a "good example of how rural areas, particularly rural, heavily Hispanic areas, have not moved toward Democrats as quickly as the suburbs" (Svitek 2020).

With large financial support and many in the DCCC, Texas Democratic Party, and Democratic leadership expecting a victory for Ortiz Jones, all indications had Texas's Twenty-Third Congressional District election as a win for Democrats. This was not the case, however; Tony Gonzales was able to score an upset victory for Texas's Twenty-Third Congressional District and retain control of the district for the Republican Party. This election marks the fourth in a row that Republicans have won the district, although the closeness of this election once again highlights the competitiveness of the Twenty-Third. This competitiveness will most likely continue, making Texas's Twenty-Third Congressional District an election to watch in the upcoming 2022 midterms.

FIGURE 7.12

Texas US House of Representatives District 23 Election Popular Vote Leader, 2020

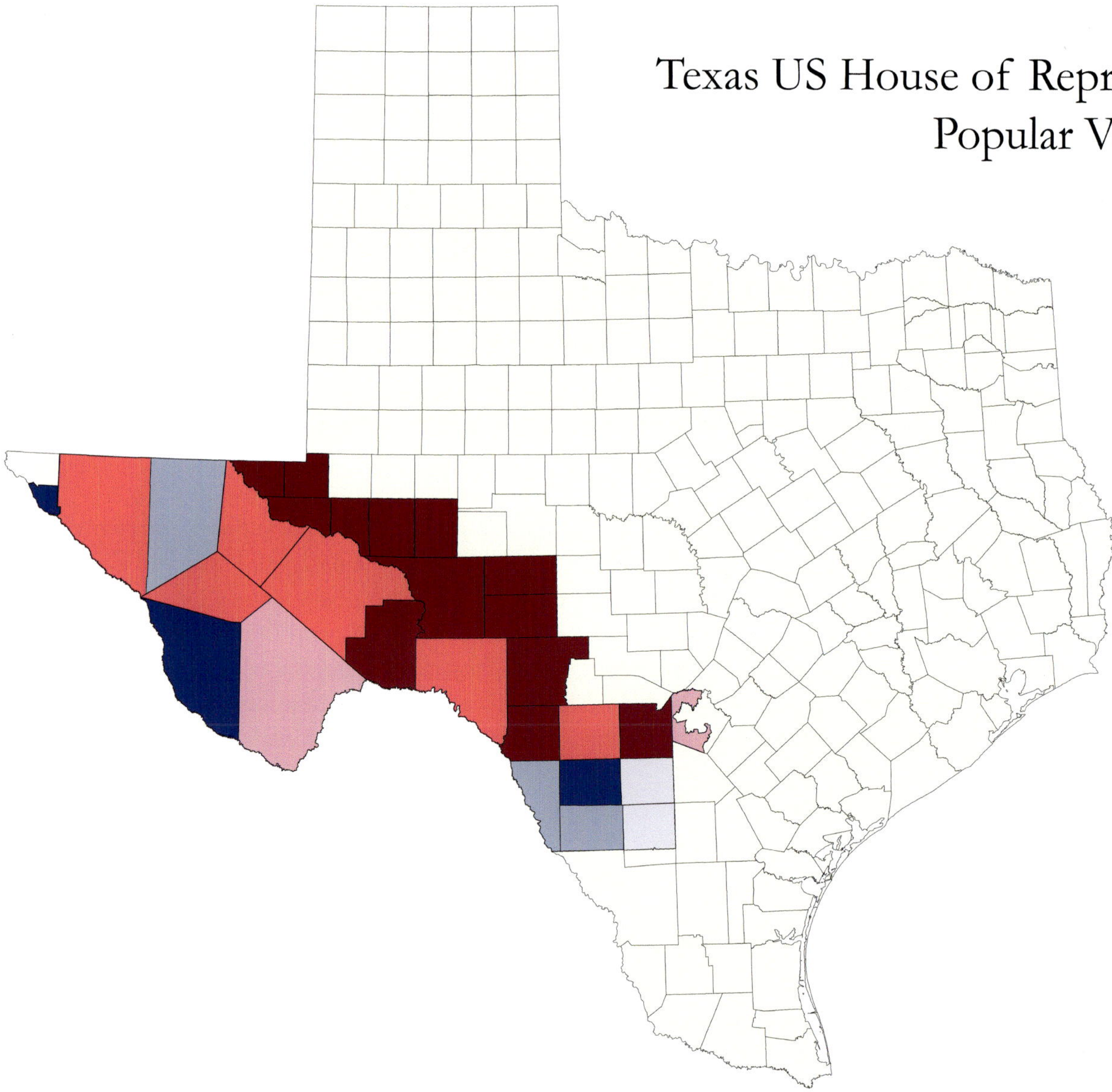

Election Results

Gonzales (R) - 149,395 (50.56%)

Jones (D) - 137,683 (46.60%)

Villela (LIB) - 8,369 (2.83%)

REFERENCES

DCCC. 2020. Expands Offensive Battlefield to 39 Districts. DCCC. https://dccc.org/memo-dccc-expands-offensive-battlefield-39-districts/ (last accessed March 17, 2021).

FEC. 2020. Election Data and Reporting Deadlines. Texas—House District 23. https://www.fec.gov/data/elections/house/TX/23/2020/ (last accessed March 24, 2021).

NRCC. 2020. Young Guns 2020. https://gopyoungguns2020.com/ (last accessed March 17, 2021).

Office of the Texas Secretary of State. 2020. Texas Election Results U.S. REPRESENTATIVE DISTRICT 23. https://results.texas-election.com/contestdetails?officeID=2002&officeName=U.%20S.%20REPRESENTATIVE%20DISTRICT%2023&officeType=FEDERAL%20OFFICES&from=race (last accessed April 17, 2021).

Svitek, P. 2020. Always a Battleground, Never a Landslide: Texas' 23rd Congressional District Stays Competitive Despite State's Political Shifts. *Texas Tribune*. https://www.texastribune.org/2020/10/29/texas-23-2020-gina-ortiz-jones-tony-gonzales/ (last accessed April 17, 2021).

US Census Bureau. 2020. 2019 American Community Survey 1-Year Estimates. https://www.census.gov/newsroom/press-kits/2020/acs-1year.html (last accessed March 12, 2021).

8 STATE ELECTIONS, LOCAL ELECTIONS, AND REFERENDA

PERSISTENCE AND CHANGE IN STATE AND LOCAL ELECTIONS, 2020

RICHARD L. MORRILL

Although media and voter attention may have been dominated by the race for president, the knowledgeable observer knows that the political power to govern depends as fundamentally on the entire set of offices, and especially on the votes for governor, senator, representative of the US House, and on the composition of state legislatures. It is the totality of these that defines the outcomes of liberal and conservative ideologies.

As in the 2016 election, the outcomes were basically divided, with small to marginal change in the composition of governors, the US Senate, the House of Representatives, and state legislatures, disappointing both parties. Overall, the message from voters is to reinforce a basic conservatism, an unwillingness to commit to consistent change.

GUBERNATORIAL CONTESTS

Only eleven states had gubernatorial races in 2020 (see figure 8.1). In the remaining thirty-nine states where no races were held, twenty are held by Democrats and nineteen are held by Republicans, the Democratic states tending to be the larger. Although the Democrats retained North Carolina and Washington, the loss of Montana was viewed as a failure to keep at least a balanced state. Republicans retained traditionally conservative states of Indiana, Missouri, North Dakota, Utah, and West Virginia; they also held on to New Hampshire and Vermont, disappointing liberal expectations.

Overall, this suggests a small rightward shift, despite the election of Biden. Because of the 2020 elections, Republicans now hold twenty-seven of the fifty possible gubernatorial seats, to the Democrats' twenty-three. Spatially, Democrats' gubernatorial strength is found in the far West Coast, the Northeast, and the Upper Midwest. Republicans' gubernatorial strength is found in the interior and southern states of the country.

SENATE CONTESTS

No Senate races were held in sixteen states, including California, New York, Ohio, Pennsylvania, and Florida. Republicans retained winning seats in nineteen states including moderate Maine, North Carolina, and Iowa. Democrats held on to vital seats in ten states that were not necessarily strongly Democrat, such as New Mexico and Virginia. Alabama moved to Republican sentiments. Overall, the Senate results were not supportive of a new Democrat presidency, but rather a pattern of "middle of the road" caution.

While Democrats certainly had successes, such as winning the runoff elections in Georgia (Ossoff

FIGURE 8.1

Persistence and Change for Gubernatorial and US Senatorial Elections

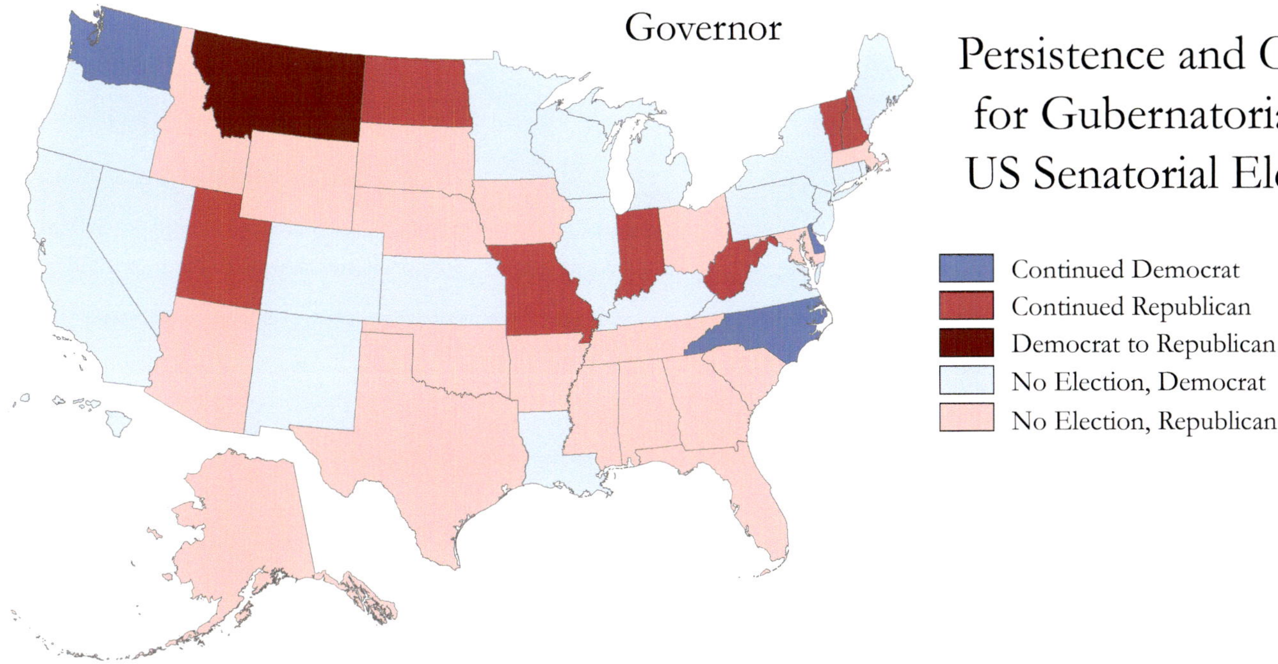

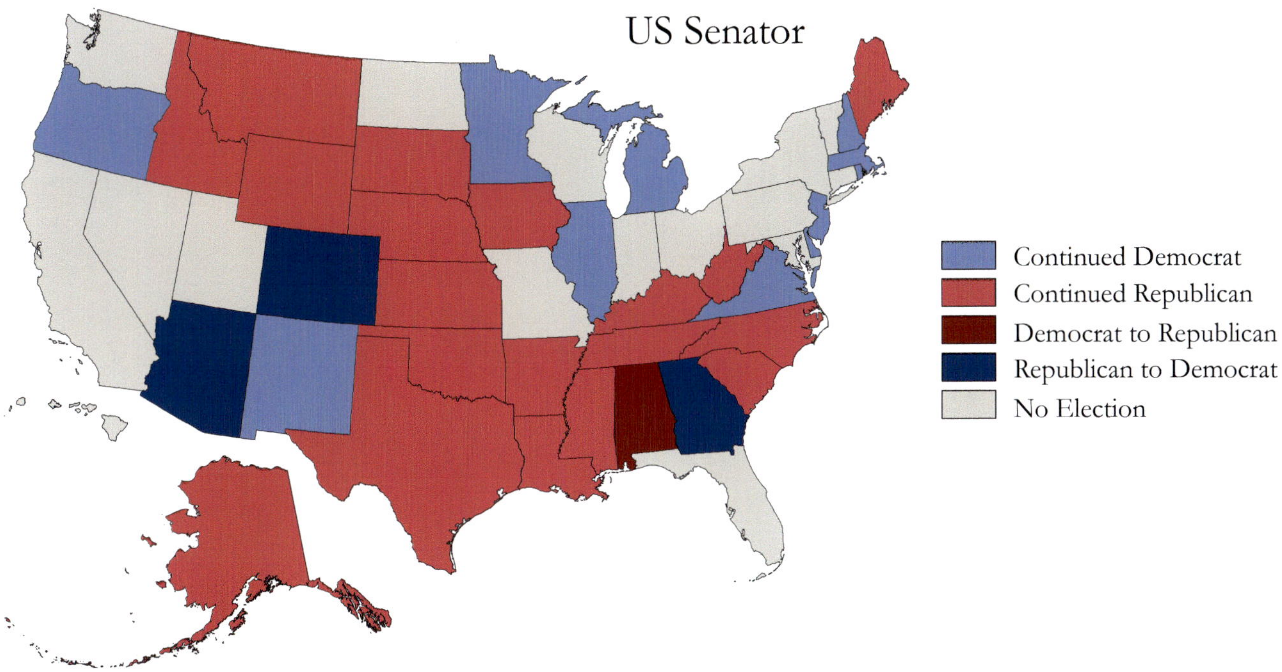

FIGURE 8.2

Persistence and Change for US House and State Legislatures

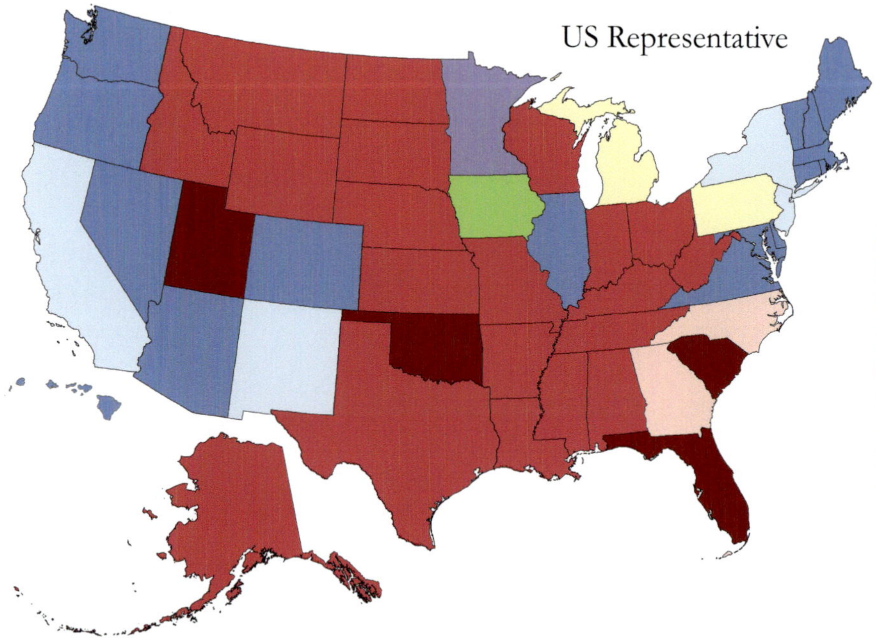

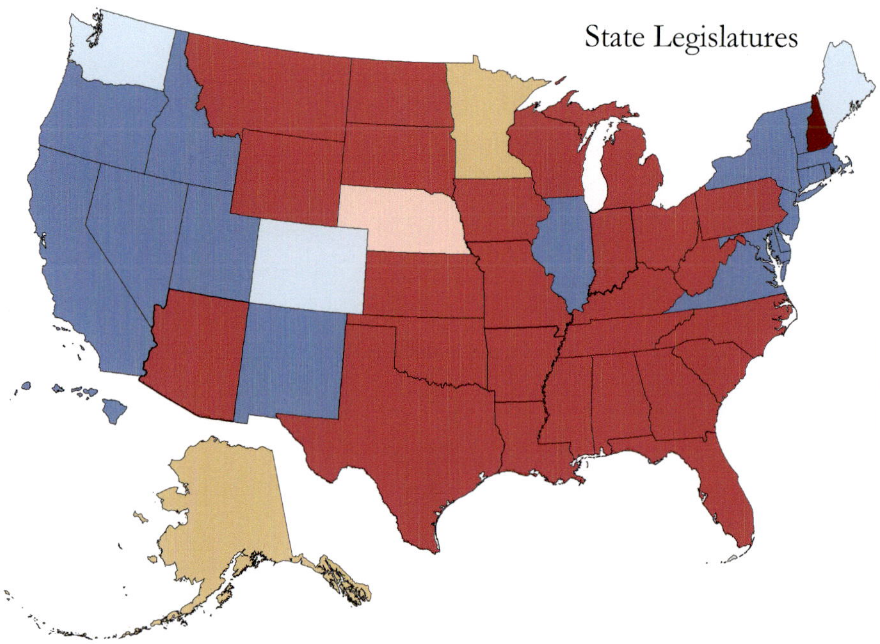

FIGURE 8.3

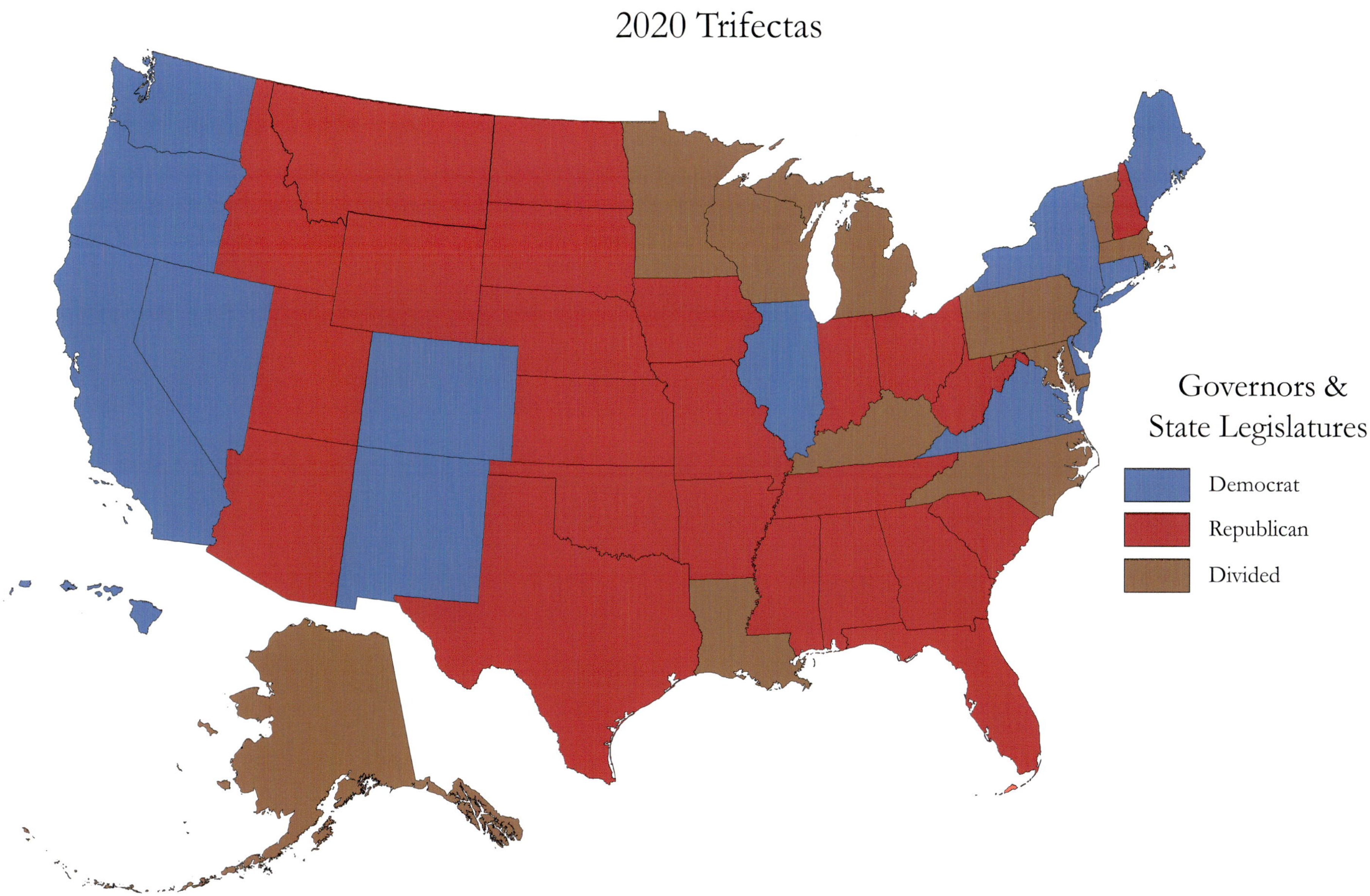
2020 Trifectas
Governors &
State Legislatures
Democrat
Republican
Divided

over Perdue and Warnock over Loeffler), Mark Kelly winning the Senate seat in Arizona against Martha McSally, and the Hickenlooper victory in Colorado, Republicans also had their own successes. For example, Collins winning reelection in Maine, Thom Tillis's victory in North Carolina over Cal Cunningham, and Joni Ernst maintaining her seat in Iowa. Like gubernatorial strength, Democratic strength is located on the West Coast, the Southwest, the Northeast, and the Great Lakes, while the Republicans remain strong in the South, much of the Great Plains, and the Interior West.

THE 116TH CONGRESS, HOUSE OF REPRESENTATIVES

The pattern of uncertainty, with small marginal changes, held true of the House races as well. Democrats held on to a majority of seats in sixteen states, including "moderate" Arizona, Colorado, Nevada, Maine, and Virginia, and gained seats in California, New Mexico, Minnesota, Michigan, New York, and New Hampshire. Republicans similarly gained in Utah, Oklahoma, Iowa, South Carolina, and Florida, as well as adding to their majority in Georgia and North Carolina, again indicating an overall pattern of indecision to commit strongly to one party.

While Democrats maintained control of the House of Representatives, Republicans managed to narrow the Democratic majority to a handful of seats. In tallying all House members, Democrats hold the majority of seats in twenty states, and Republicans hold the majority in twenty-seven. Three states, Michigan, Minnesota, and Pennsylvania, are tied. Figure 8.2 illustrates the balance of Democratic versus Republican control. As is the case with senators and governors, Democratic support is strongest in the Northeast and far West, with Republicans dominating the southern and interior states.

STATE LEGISLATURES

Thirty states have legislatures totally controlled by Republicans, while only twelve states are Democrat controlled (see figure 8.2). Only two states are divided, Minnesota and Alaska. Six states, including Washington, Colorado, Maine, Illinois, Maryland, and Delaware moved from divided to both Democratic controlled, a small Democratic shift overall.

Consequently, similar patterns as described earlier can be seen in the geography of party control of state legislatures. Of the twelve houses controlled by Democrats, all but Nevada, New Mexico, Idaho, Utah, Colorado, and Illinois are found on the east and west coasts. Of the thirty states where Republicans hold both houses, all are found within the South and interior states of the country, with exception of New Hampshire.

TRIFECTAS

Trifecta results if one party controls three entities: the governorship and both houses of the legislature. Trifectas are a key measure of political imbalance of power. Republicans have such control in twenty-three states, Democrats in thirteen, while power is divided in the remaining eleven states (see figure 8.3). As with the preceding measures, the pattern continues the conclusion of caution in conceding too much power to one party.

MISSISSIPPI CHANGES STATE FLAG FROM CONFEDERATE INSIGNIA TO MAGNOLIA

JORDAN BRASHER

According to the Southern Poverty Law Center, at least 168 Confederate symbols were removed from public and private spaces in 2020 alone (Southern Poverty Law Center 2021). Almost all of them were removed in the latter half of the year, in the aftermath of George Floyd's murder by police in Minneapolis, Minnesota on May 25, 2020. Key to the demands of the Black Lives Matter uprisings—which swept not only the nation but the world in response to the killing—has been the removal of Confederate symbols from public spaces. Monuments, memorials, and place names honoring Ku Klux Klansmen, enslavers, and Confederates have been taken down at a historically unprecedented rate. Of the 168 Confederate symbols removed in 2020, 94 were Confederate monuments, nearly twice the 58 removed during the preceding four years combined (Southern Poverty Law Center 2021). New demands to remove racist symbols—especially those associated with the Confederate States of America (1861–1865)—arrived on the heels of the 2015 Charleston Massacre, the 2017 Charlottesville tragedy, and the 2020 murder of George Floyd amid growing unrest around state-sanctioned and extralegal murders of unarmed Black women and men.

It is within this context that in the November 2020 election the state of Mississippi voted on whether to retain or change its state flag, which since 1894 had included the infamous Confederate battle emblem in the upper left corner. The flag had been subject to a referendum for change in a 2001 statewide vote when a (white) majority voted to keep the flag as is (Klinker 2002; Leib and Webster 2012; Orey 2004; Orey et. al 2011). Yet the 2020 election would change that. Among the pressures to change the flag in 2020 included the National Collegiate Athletic Association (NCAA)'s Southeastern Conference, which warned that it could prohibit conference championships from taking place in the state if the emblem was not removed (Suss 2020a). Indeed, collegiate athletes and coaches at Mississippi State University, the University of Mississippi, and elsewhere played an important role in exerting public pressure to change the flag. For example, Mississippi State University star football player Kylin Hill tweeted that he was no longer willing to represent any Mississippi state institutions on the field until the flag was changed (Horka 2020). In June 2020, not long after Floyd's murder, more than forty-six Mississippi athletic directors and college coaches gathered on the state capitol steps to give impassioned speeches on the need to change the flag (Suss 2020b). Given the outsized role that collegiate sports play in southern life, identity, and politics, these events likely contributed to the groundswell of public support for changing the state flag.

Figure 8.4 shows the vote breakdown on Ballot Measure 3, the referendum on the Mississippi state flag. The vote reveals much about the deeply racially divided nature of one of the poorest states of the former Confederacy. A comparison with figure 8.5 shows that counties voting in favor of the flag design change mostly correspond with the majority-Black counties in the upper Mississippi Delta region in the western part of the state, as well as its major majority-Black capital city of Jackson. Additionally, majority-Black areas in the east-central part of the state bordering Alabama in Lowndes and Noxubee counties, for example, also voted to adopt the new flag. A majority-Black county named—perhaps somewhat ironically—for the former president of the Confederacy, Jefferson Davis, was one of few counties in an otherwise mostly white and Republican south-central and southeastern Mississippi to both vote yes to adopt the new flag and for Joe Biden in the 2020 presidential election.

Yet figure 8.5 perhaps also reflects the slowly shifting attitudes of white southerners toward Confederate symbols. Some majority-white counties in north central Mississippi and southern and southeast Mississippi that voted for Trump in 2020 also voted with slim majorities to change the state flag. Some of those counties include Lafayette, Yalobusha, DeSoto, Lee, Chickasaw, and Montgomery counties in the north and Harrison, Forrest, Rankin, and Scott counties in the south. A Quinnipiac University national poll (2020) found in June 2020 that a majority of Americans support removing Confederate monuments from public spaces, a major break from research over the last thirty years. It remains to be seen whether the nearly 150 Confederate symbols that remain in the

FIGURE 8.4

Mississippi: Ballot Measure 3, State Flag Referendum
2020 Election

African American Populations
for Mississippi Counties

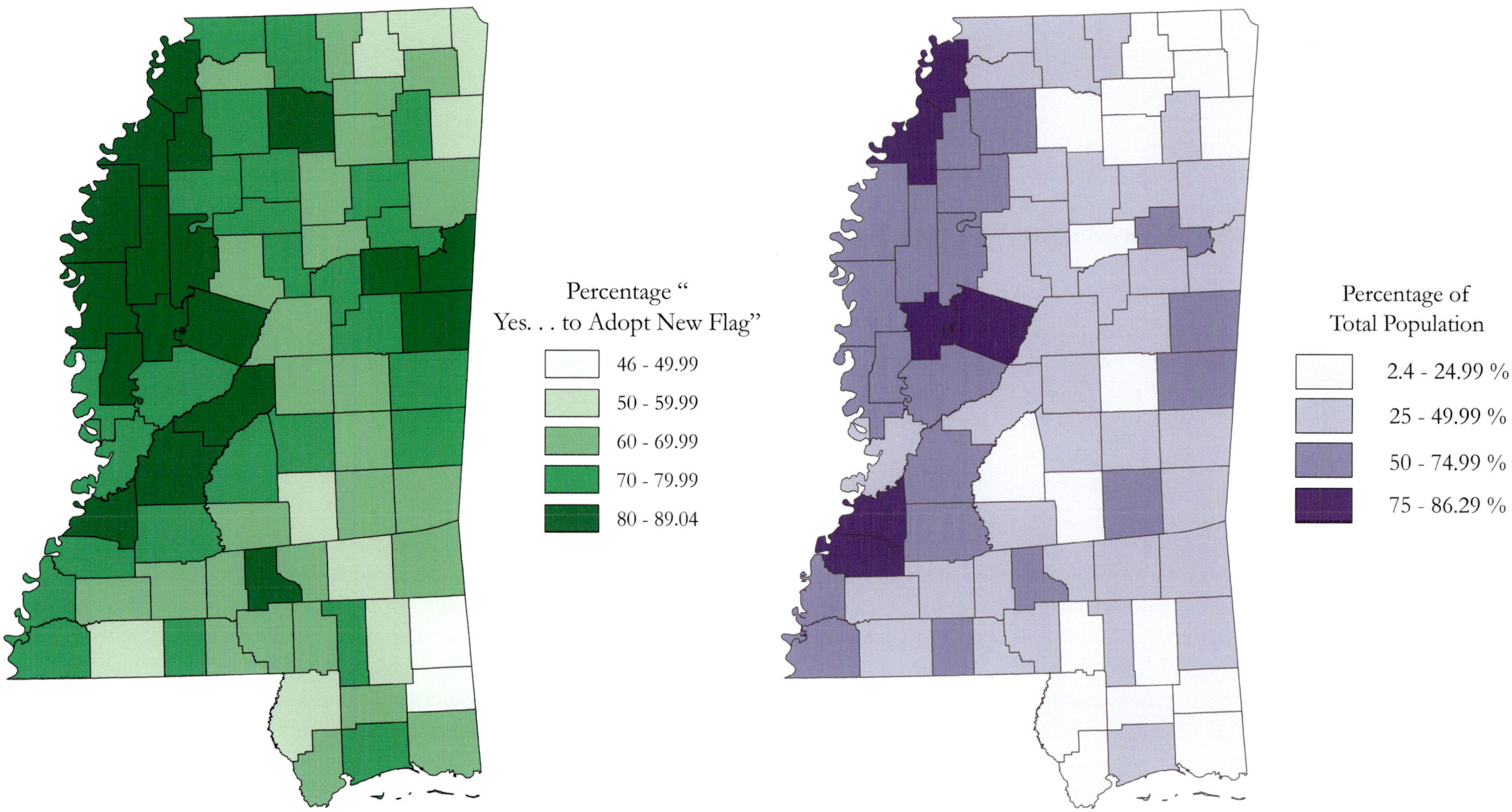

FIGURE 8.5

Mississippi: House Concurrent Resolution 79, State Senate and House Vote to Allow Flag Referendum

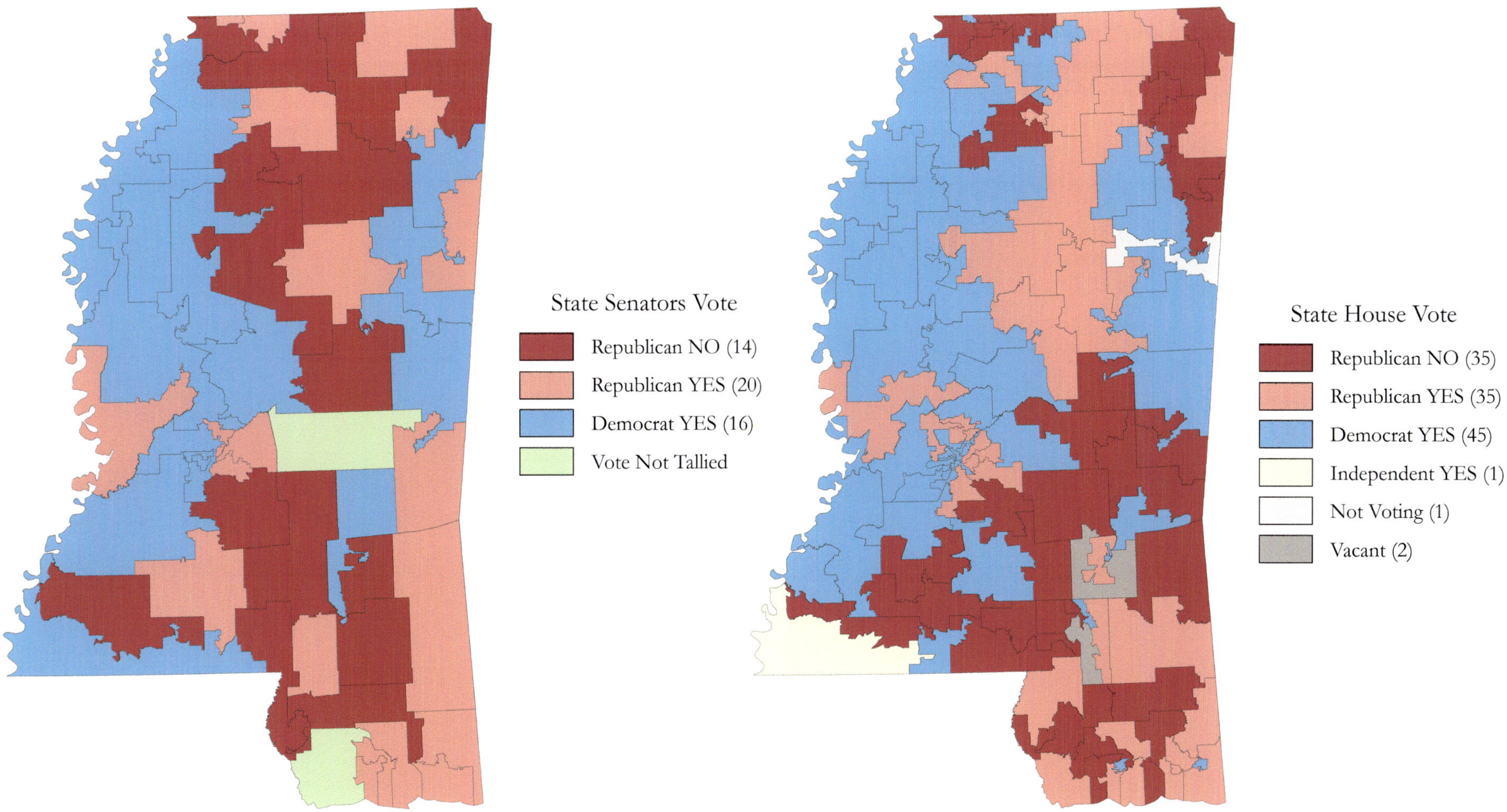

form of monuments, memorials, and place names across the state of Mississippi—and the more than 1,700 that remain across the nation (Southern Poverty Law Center 2021)—will be changed, too.

REFERENCES

Horka, T. 2020. Mississippi State RB Kylin Hill Says He Won't Represent "This State Anymore" If Flag Not Changed. *Mississippi Clarion Ledger*. June 22, https://www.clarionledger.com/story/sports/college/mississippi-state/2020/06/22/mississippi-state-running-back-kylin-hill-mississippi-flag-confederate-symbol/3237053001/.

Klinker, P. A. 2002. Waving the "White" Flag: Racial Voting Patterns in the 2001 Mississippi Flag Referendum. *Political Science & Politics* 34(3):647.

Leib, J. I., and G. R. Webster. 2012. Black, White or Green? The Confederate Battle Emblem and the 2001 Mississippi State Flag Referendum. *Southeastern Geographer* 52(3):299–326.

Orey, B. D. 2004. White Racial Attitudes and Support for the Mississippi State Flag. *American Politics Research* 32(1):102–16.

Orey, B. D., L. M. Overby, P. K. Hatemi, and B. Liu. 2011. White Support for Racial Referenda in the Deep South. *Politics & Policy* 39(4):539–58.

Quinnipiac University Poll. 2020. 68% Say Discrimination against Black Americans A "Serious Problem," Quinnipiac University National Poll Finds; Slight Majority Support Removing Confederate Statues. https://poll.qu.edu/national/release-detail?ReleaseID=3663.

Southern Poverty Law Center. 2021. SPLC reports over 160 Confederate symbols removed in 2020. February 23, 2021. URL: https://www.splcenter.org/presscenter/splc-reports-over-160-confederate-symbols-removed-2020.

Suss, N. 2020a. Commentary: Give Sports Credit; Athletes Played a Huge Role in Mississippi's Flag Change. *Mississippi Clarion-Ledger*. June 29, https://www.clarionledger.com/story/sports/2020/06/29/mississippi-flag-change-ole-miss-mississippi-state-athletes-commentary/3276230001/.

———. 2020b. Here Are the 46 Mississippi College Coaches Who Appeared at Legislature for Change to State Flag. *Mississippi Clarion Ledger*. June 25, https://www.clarionledger.com/story/sports/2020/06/25/mississippi-flag-sports-coaches-administrators-lobbying/3256703001/.

GEORGE FLOYD AND VOTER ATTITUDES AND TURNOUT IN MINNESOTA

JOHN HEPPEN AND MATTHEW MILLETT

The presidential election was approximately five months after the murder of George Floyd. Joseph Biden carried Minnesota by a healthy margin despite Trump doing better than expected in 2016, coming within a little over 44,000 votes of catching Hillary Clinton. Trump and Biden both made campaign stops in Minnesota. George Floyd died on May 25 in Minneapolis and in the days after, Minneapolis and neighboring St. Paul were rocked with protests, arsons, and property damage. Thousands of people protested during those days. The National Guard was called in and curfews were extended throughout the Minneapolis and St. Paul metropolitan area. Fired Minneapolis police officer Derek Chauvin was convicted of the murder of George Floyd on April 12, 2021 and sentenced on Friday, June 25.

The murder of George Floyd prompted calls for change from across the political spectrum. City Council members of Minneapolis even called for the abolition of the Minneapolis Police Department, as did Congresswoman Ilhan Omar, who represents Minneapolis. Even President Trump tweeted his condolences. Trump and Republicans throughout the campaign used calls to abolish the police against the Democrats and reaffirmed their commitment to law and order and support for law enforcement. The question is, did the George Floyd murder and the subsequent debate about racism and law enforcement impact the election and turnout in Minnesota? Figure 8.6 shows that voter turnout increased at the county level throughout urban and rural Minnesota. Hennepin County is home to Minneapolis and first-ring suburbs; neighboring Ramsey County is home to St. Paul and first-ring suburbs. All counties bordering those two core counties also constitute the Twin Cities Metropolitan area (figure 8.7)

While it is difficult to attribute the increased voter turnout throughout the metropolitan area and the state, there is evidence that the death of George Floyd, the Black Lives Matter movement, and the debate over race and policing played an important role in the 2020 election. In a poll of Minnesota voters conducted in late September 2020, eight hundred likely voters were asked about the presidential election and the issues important to them. The question, "Which one of the following public policy areas is most influential in deciding your vote for president?" was asked. For Biden voters, 28 percent said "racial justice and equality" was the most influential ahead of coronavirus at 20 percent (Minnesota Poll 2020). Statewide, among all likely voters "economy and jobs" led with 23 percent and "racial justice and equality" was second with 15 percent (Minnesota Poll 2020).

In another important question, likely voters were asked, "Do you think the death of George Floyd was an isolated incident, or a sign of broader problems in the treatment of Black people by Minneapolis police?" This question was broken down geographically. For residents of Hennepin and Ramsey County, 69 percent agreed that it was a broader problem; for the rest of the metropolitan area it was 47 percent for a plurality and statewide it was 52 percent, reflecting the population dominance of Hennepin and Ramsey counties (Minnesota Poll 2020).

Exit polling of Minnesota voters by the *New York Times* asked voters, "Which one of these five issues mattered most in deciding how you voted for president?" The results placed racial inequality at 14 percent, which was surpassed by the economy at 32 percent, the coronavirus pandemic at 19 percent, and health-care policy at 17 percent (*New York Times* 2020). Of the 14 percent who placed racial inequality as mattering most, 90 percent of them voted for Biden.

Another way to gauge the impact of the George Floyd murder on voting is to look at the voter turnout at a local level. George Floyd was murdered within State Senate District 62 (SD 62). In 2016 and 2020, there were state senate elections for that senate district. In 2016, 33,655 votes were cast for state senate and in 2020 37,200 votes were cast in SD 62, an increase of close to four thousand votes. At the more local level, George Floyd was killed at the intersection of 38th and Chicago in Minneapolis, which is near Ward 8 Precinct 8. In 2016 a total of 2,350 votes were cast for Clinton or Trump. In 2020, Ward 8 Precinct 8 saw 2,728 votes cast for Biden or Trump with Biden winning 90 percent of the vote, an increase of 378 votes. While it is difficult to obtain survey data at that level, there was clearly an increase in

FIGURE 8.6

Percent Difference in Voter Turnout of Voting Age Population 2020 to 2016, Minnesota Counties

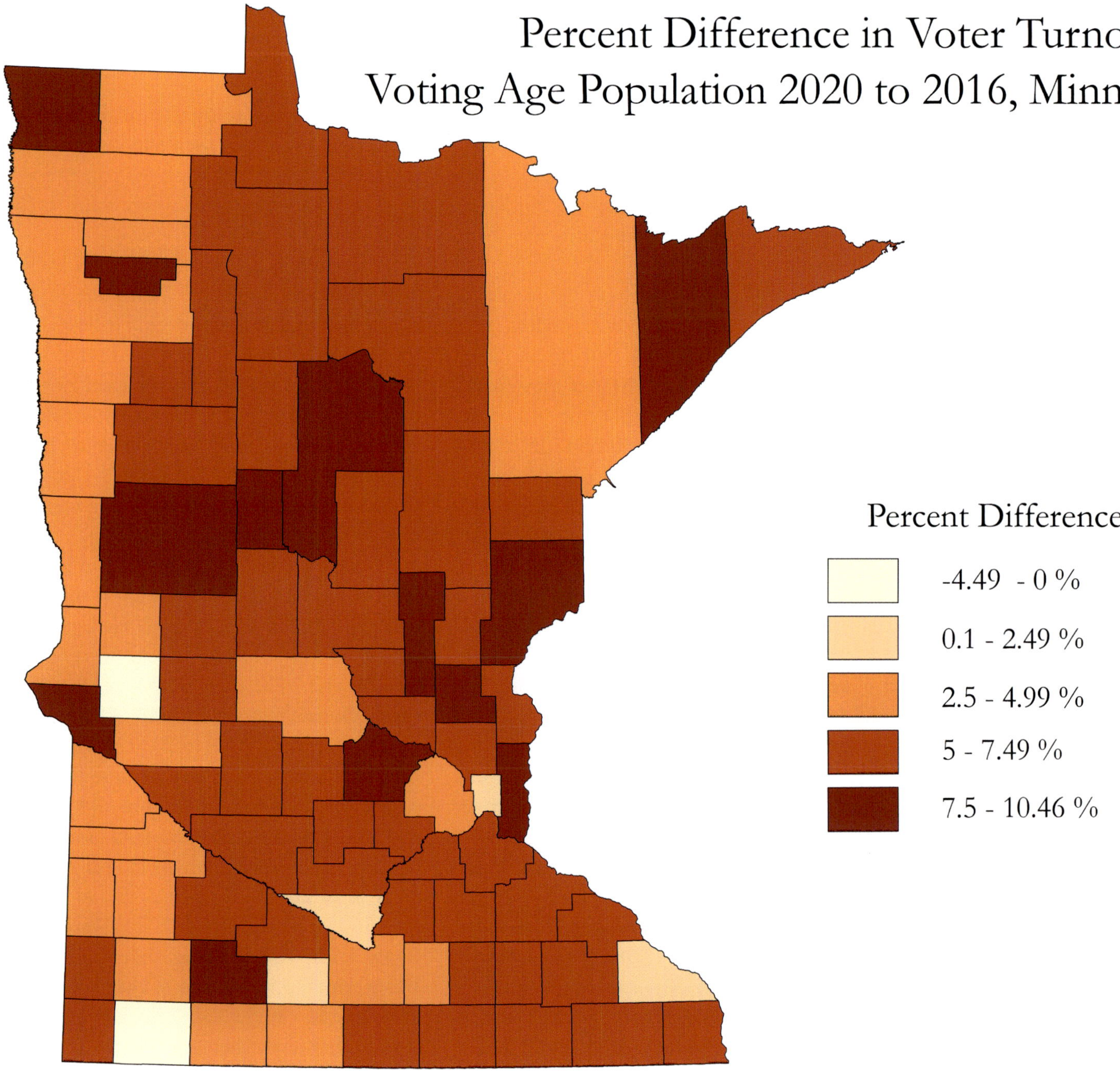

FIGURE 8.7

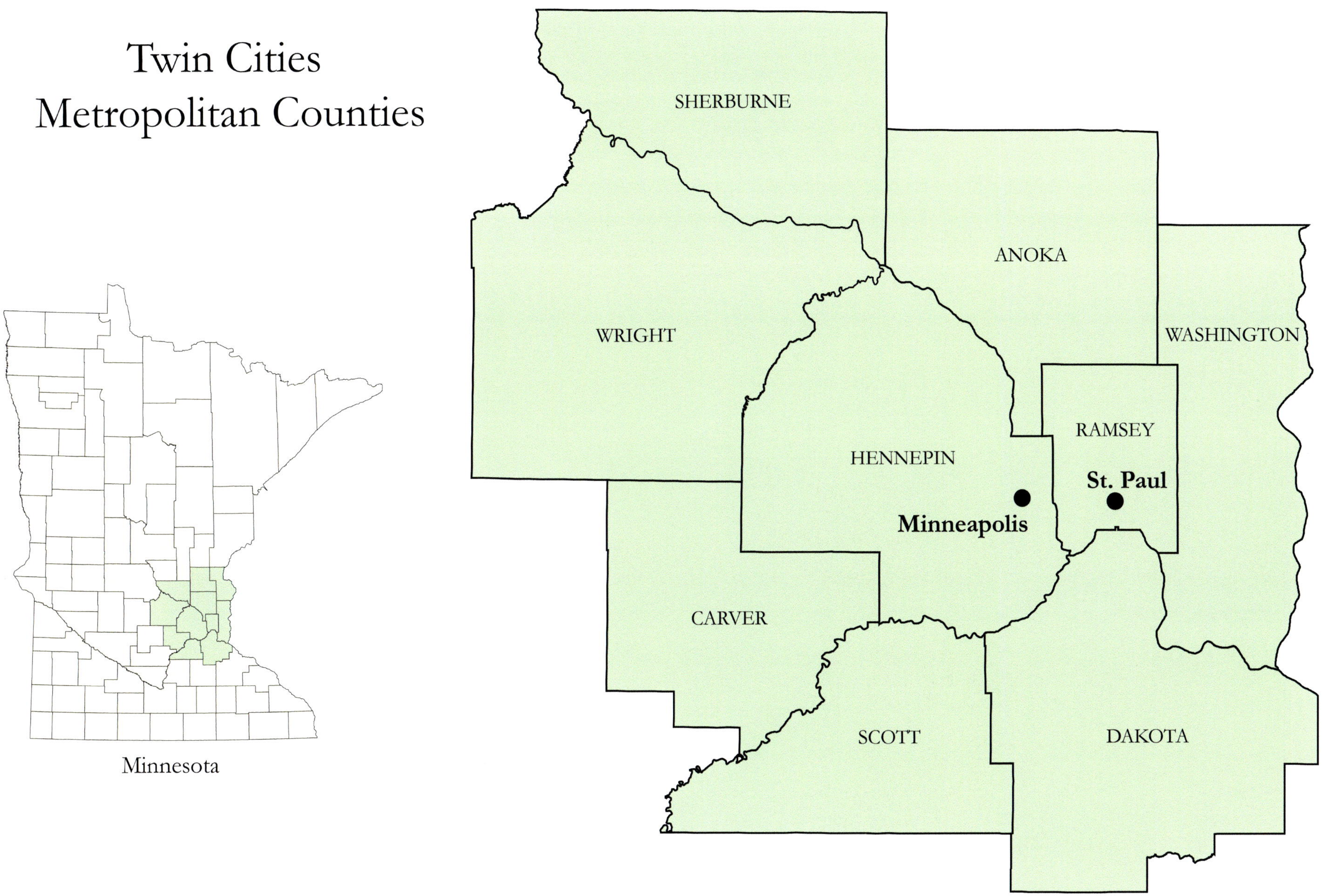
Twin Cities
Metropolitan Counties
Minnesota
SHERBURNE
ANOKA
WRIGHT
WASHINGTON
RAMSEY
HENNEPIN
St. Paul
Minneapolis
CARVER
SCOTT
DAKOTA

voter participation in the locality where the murder occurred. Coupling that with statewide and regional survey data of all voters and Biden voters, it is a reasonable hypothesis that the murder of George Floyd led to increased voter interest and turnout in Minneapolis.

REFERENCES

Minnesota Poll. 2020. September 30, https://www.startribune.com/the-minnesota-poll/468458743/.

New York Times. 2020. Minnesota Exit Polls: How Different Groups Voted. https://www.nytimes.com/interactive/2020/11/03/us/elections/exit-polls-minnesota.html

NOTE

1. Voting Data for Minneapolis is available from the Minnesota Secretary of State at https://electionresults.sos.state.mn.us/136 (2020) and https://electionresults.sos.state.mn.us/20161108 (2016) (last accessed June 24, 2021).

THE 2020 ELECTION IN MONTANA: A TRIUMPH FOR TRUMPISM–BUT FOR HOW LONG?

LARRY KNOPP

The 2020 election in Montana can only be viewed as a triumph of Trumpism. Not only did Donald Trump carry the state by more than 16 points; populist Republicans who embraced him won the US Senate and House races much more easily than expected, as well as every statewide executive office on the ballot. Even an extreme gun-rights measure that was widely expected to fail instead prevailed (albeit narrowly). Republicans gained the governorship for the first time in sixteen years and simultaneous control of the governorship and state legislature for the first time in eighteen, expanding their majorities in both houses of the state legislature in the process. The only arguably non-conservative (but still populist) outcome was the success of a pair of marijuana legalization measures, a result that was replicated in two other deep-red states (South Dakota and Mississippi—see "Marijuana Legalization on the Ballot in 2020" below).

An outcome of this magnitude was not anticipated, either by pundits or pollsters. Biden's 16-point loss to Donald Trump was by any measure huge, although it marks a slight improvement over Hillary Clinton's 20-point loss. However, the absence of a Green Party candidate on the ballot in 2020 appears to account for much of this difference, with the remainder appearing to come from a slight drop-off in support for the Libertarian candidate. Popular outgoing Democratic governor Steve Bullock, meanwhile, was expected to at least give incumbent Republican Senator Steve Daines a run for his money but ended up losing by 10 points. Democratic US House candidate Kathleen Williams, who had run a close but losing campaign for the state's open House seat in 2018, lost to Trump-backed State Auditor Matt Rosendale (who had lost a high-profile race for US Senator just two years earlier) by an even larger 12 points. And controversial and combative outgoing US Congressman Greg Gianforte, who only narrowly defeated Williams in 2018 for the House seat, won the governorship by 8 percent.

Paradoxically, just two years earlier Montana arguably repudiated Donald Trump (if not Trumpism) when it reelected Democratic US Senator Jon Tester, despite Trump making several trips to the state on behalf of Rosendale. Moreover, Tester for the first time in three elections garnered an absolute majority of the vote. More broadly, Montana has historically been one of the more "purple" of the Rocky Mountain states, due to a strong labor history and class-based politics associated with mining and timber industries in the west and along the Great Northern/BNSF "Hi-Line" railroad in the north. A strong streak of anti-corporate populism and support for public lands (and access to public lands for recreation) have added to its more moderate political pedigree. Even after 2020 it remains less Republican than its immediate neighbors (Idaho, Wyoming, and the Dakotas). But in recent decades the class-based politics of Montana have diminished in influence while culture war politics, especially around environmental issues, have become more influential. This shift has clearly benefited populist Republicans.

Geographically, the most prominent divide historically has been east-west, with ranchlands of the east leaning Republican and the mining and timber-oriented mountain west leaning Democratic (farmers in both regions have historically been a swing constituency). A racial divide between whites and Native Americans has been a secondary divide, with counties containing Indian Reservations leaning Democratic. Now, however, the partisan divide is more urban-rural and cultural than regional, while the white–Native American divide has widened (even though Native American voters have never been monolithic in support of Democrats). New voters are also a factor, given Montana's rapid rate of growth in recent years. While frequently stereotyped as disproportionately wealthy and liberal, in-migrants most likely constitute a partisan wash overall, with Democrats prevailing in some areas and Republicans in others. There is perhaps a slight advantage to Democrats overall, since the fastest growth by far has been in now strongly Democratic Gallatin County, home to Montana State University and a gateway to Yellowstone National Park. But in-migrants are both Democratic and Republican, conservative and liberal. Ravalli County, just south of Missoula County (home to the University of Montana and often the state's most Democratic), has become a center of populist ultra-conservatism and has a high in-migration rate. The same is true

FIGURE 8.8

Montana Presidential Election Percent Democrat, 2020 Election

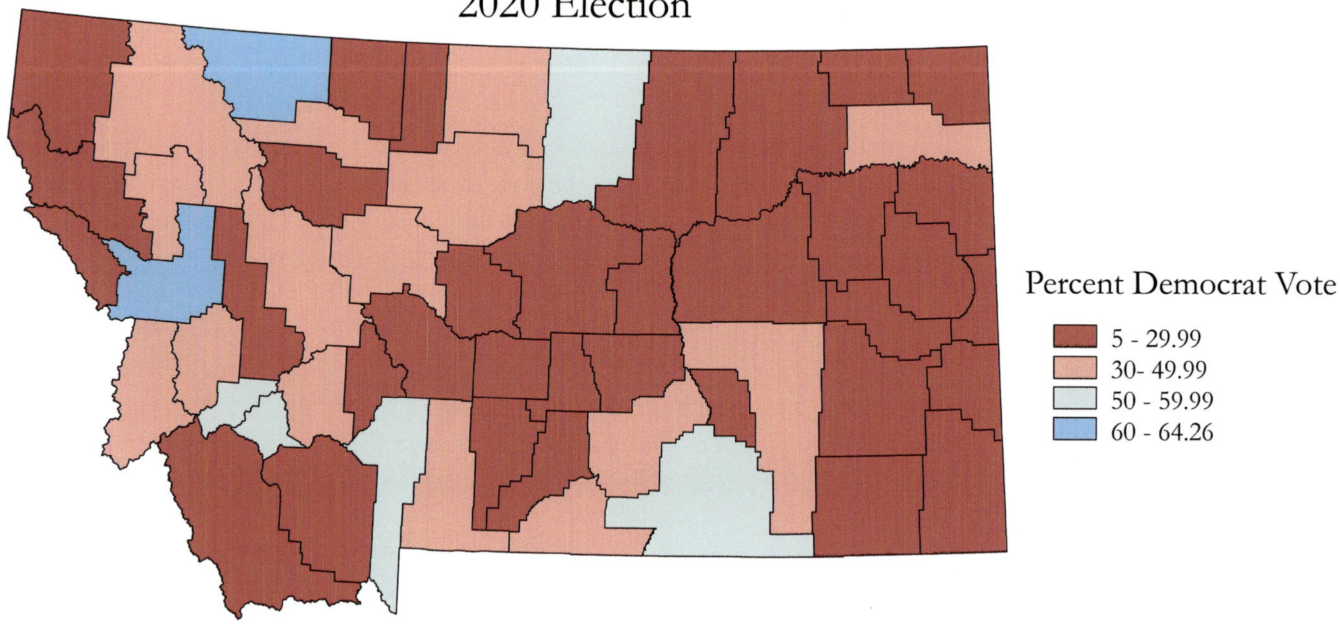

Montana Gubernatorial Election Percent Democrat, 2020 Election

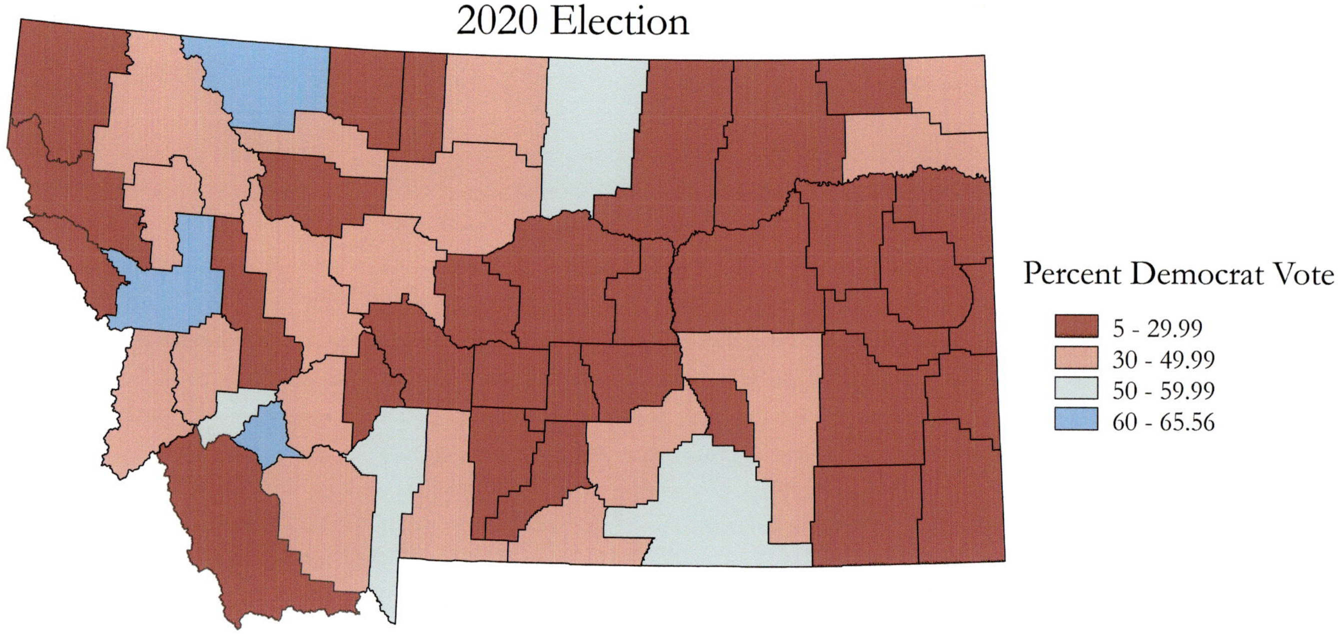

of Flathead County, which contains the small city of Kalispel and is both a gateway to Glacier National Park and popular for recreation Flathead Lake (although the strongly Democratic ski resort town of Whitefish is a Democratic outlier). Predominantly white rural counties are overwhelmingly Republican, regardless of region.

Indeed, there is now a remarkable consistency in partisan voting patterns geographically in Montana. Gallatin and Missoula counties are both anchored by university towns that have become heavily Democratic. Silver Bow and Deer Lodge counties, which contain the once-populous and influential but now declining mining towns of Butte and Anaconda, respectively, remain stubbornly Democratic due to their strong labor histories and traditions. Heavily Native American Glacier County, just east of Glacier National Park, is solidly Democratic—sometimes the most Democratic in the state—while Big Horn and Roosevelt counties (in the southeast and northeast, respectively) are roughly half Native American and lean Democratic. Differential turnout rates between whites and Native Americans, along with the contingencies of particular campaigns and elections, mean that neither county is a lock for Democrats, however. Roosevelt, for instance, voted narrowly Democratic in the 2020 Senate and House races but narrowly Republican for president and governor (figure 8.8).

Lewis & Clark County, home to the state capital of Helena in west-central Montana, matches Roosevelt County in this respect. Like many state capitals, Helena has trended Democratic in recent decades. But rural parts of the county tend to be solidly Republican. The result is that Lewis & Clark is a swing county that on balance leans Democratic, especially in state and local elections. Its voters opted for Trump in both 2016 and 2020 (by 7 and 4 points, respectively), for Democrats in all the other high-profile 2020 races, and for Republicans in low-profile down-ballot and uncompetitive races. Lake County in the northwest (containing most of Flathead Lake and much of the Flathead Reservation) and Park County (a second gateway to Yellowstone National Park just east of Bozeman) are rapidly growing and increasingly expensive environmental amenity counties that can also be competitive (although not in 2020). Their Republican lean in 2020 suggests that heavy turnout (at 81.3 percent statewide, the highest since 1972) may have benefited Republicans. Blaine County, along the Hi-Line in the north, is closely divided racially and very competitive. It is the only county to flip from Republican to Democratic in the 2020 presidential election, while none flipped the other way.

Overall, then, the 2020 election in Montana must be seen as at least a temporary triumph for Trumpism. But it is unclear how durable this triumph will prove to be. Republicans clearly benefited from Trumpism, but their office holders remain controversial and polarizing. Moreover, the trajectory and sustainability of the culture wars that have fueled Trumpism's popularity in Montana (as in the rest of the country) are uncertain, given that Trump's own future influence, along with the long-term impact of the state's changing economy and in-migration, are also far from clear.

THE 2020 NORTH CAROLINA PRESIDENTIAL AND GUBERNATORIAL ELECTION

KATIE WEICHELT

The 2020 North Carolina gubernatorial and presidential election results proved that voters still elect candidates not strictly based on party lines and ideology, but rather the complex expectations of each office. The election outcomes also revealed the advantages incumbents will have in any election cycle. The Democratic gubernatorial incumbent, Roy Cooper, with his practical and mild-mannered style juxtaposed against the Republican presidential incumbent, Donald Trump, with his more aggressive and confrontational nature, would lead many to conclude that both candidates could not be successful in the same election cycle. Yet North Carolina gave a majority of their votes to both candidates and not for the first time. This split result has become even more notable in an era of increasing political polarization.

According to polls in the lead-up to the 2020 election, Donald Trump held a small lead over his opponent, Joseph Biden—a lead that was so slight, some political watchers placed the race in the category of toss-up. For this reason, North Carolina was one state in which political watchers were inclined to give their attention. Indeed, Donald Trump's own behavior in the lead-up to the election seemed to indicate that the state was in play. According to Chris Cooper (2020), Trump tweeted about North Carolina more than 140 times between 2016 and the 2020 election. His diligence paid off. In the end, Trump again claimed the state's electoral votes, proving that Democrats have more work to do in securing the presidential vote.

A look at the geographic distribution of the presidential vote reveals a continuation of electoral trends seen in previous election cycles. From 2012 to 2016, both Democrat- and Republican-leaning regions moved further away from the center (figure 8.9). In 2020, Biden received majorities in counties making up the Research Triangle, the Charlotte region, and the Black Belt in the northeast. Trump received a majority of vote share primarily in counties in the western region of the state and the southeastern region. The 2020 election cycle also produced an increasingly more entrenched urban-rural divide. Counties in the Research Triangle, a longtime Democratic stronghold, gave an even larger portion of their votes to Biden compared to Clinton in 2016. In Wake County, for example, Clinton took just over 57 percent of the vote total while Biden captured more than 62 percent. Moving southwest to the Charlotte area, we also see increased Democratic support in this urban region that was once much more friendly to Republicans. Ever since Charlotte's Mecklenburg County flipped for Kerry in 2004, the Democratic presidential candidates have claimed a bigger share of the vote total in nearly every election cycle. In 2016, Clinton captured just over 62 percent of the vote while Biden took 66 percent (Leip 2016, 2020). The Black Belt region, as was previously mentioned, also tends to vote Democratic, although they differ from the Triangle and the Charlotte area in that it is more rural. The region is also home to more non-white residents than the state average (Weigel 2020). Unlike the Research Triangle and Charlotte, the region saw an erosion of Democratic support in the presidential race in 2016 compared to 2012, and these trends continued in 2020 with Biden getting even less of the vote share than Clinton in many counties, although he was still able to capture a majority of the vote share (Leip 2020). Trump increased his vote share in many Republican-leaning counties. One analysis found that of the twenty-three counties that gave Trump at least 70 percent of the vote share in 2016, twenty-one increased their support for him in 2020 (McCreary 2020).

Exit polls showed that Trump was significantly more popular with voters over sixty-five years of age. Biden was more successful with younger voters, in particular those eighteen to twenty-nine years old. Biden also attracted more voters who were college graduates; these two cohorts are predominant in counties voting for Biden. Voters without college degrees tended to break for Trump. Trump also proved to be more popular among white voters at a rate of two to one. However, among non-white voters Biden was successful at a rate of four to one (*New York Times* 2020).

Although the coronavirus pandemic dominated the headlines for much of 2020, this was not the most pressing issue for those in North Carolina. According to exit polling, only 14 percent of the state's voters viewed the coronavirus

FIGURE 8.9

North Carolina Presidential Election Popular Vote Leader, 2020

Election Result

Trump (R) - 2,758,775 (49.93%)

Biden (D) - 2,684,292 (48.59%)

Jorgensen (LIB) - 48,678 (0.88%)

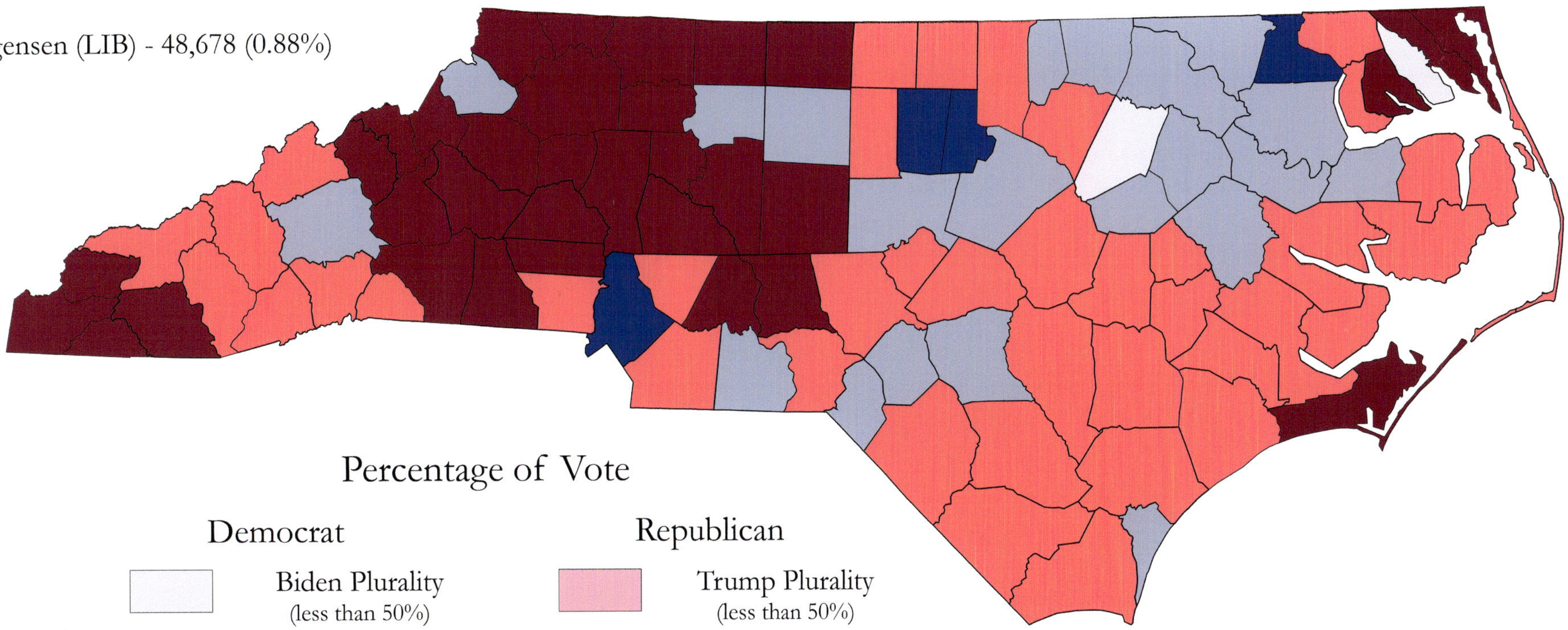

Percentage of Vote

Democrat	Republican
Biden Plurality (less than 50%)	Trump Plurality (less than 50%)
Biden Majority (50 to 64%)	Trump Majority (50 to 64%)
Biden Super Majority (65% or more)	Trump Super Majority (65% or more)

pandemic as the most important issue in deciding how they voted for president. The economy (35 percent) followed by racial inequality (21 percent) were the most important issues in deciding their presidential vote. More than 52 percent of voters believed that the US efforts to contain the coronavirus pandemic were going very well or somewhat well. The large majority of those who believed that efforts were going well ultimately cast their vote for Trump (*New York Times* 2020).

North Carolina voters would give a majority of their vote share to another candidate in the election cycle. Roy Cooper, the Democratic incumbent, secured a second term against Republican Dan Forest. While the winner of the presidential race would take several days to declare, Cooper was declared the winner the morning following Election Day due to the large lead over Forest and would ultimately gain a larger portion of the popular vote at 51.2 percent compared to Trump's 50.1 percent. Cooper's victory in 2016 was plagued by accusations from his opponent McCrory of voter fraud and requests for recounts; however, Cooper's majority on election night 2020 was too large for his opponent to challenge.

Roy Cooper is not the first Democrat to win the governor's seat while the Republican candidate claimed a larger share of the vote in the presidential election cycles. Indeed, North Carolina has a strong trend of doing just that. Since 1992, Democrats won the governor's seat in seven out of eight elections. During the same election cycles, the state has awarded their Electoral College votes to the Republican presidential candidates seven out of eight times. The exception was Barack Obama in 2012.

Like Biden, exit polls show that Cooper was more popular among younger voters, non-whites, and those who lived in more urbanized regions such as the Raleigh-Durham region and the Charlotte area (CNN 2020). Cooper, indeed, received a super majority (more than 65 percent of the popular vote) in Mecklenburg County, Wake County, Durham County, and Orange County (figure 8.10). Forest was more popular in rural counties and with those who, unsurprisingly, also gave a majority of their vote share to Donald Trump. However, Cooper was also able to claim an additional four counties where Trump also claimed the majority of the votes: Scotland, Martin, Granville, and Lenoir. However, a look back at the recent voting patterns of these counties shows a history of non-loyalty to any party. All four counties went for Obama in 2012 and all, with the exception of Scotland, flipped to Trump in 2016. Of these same four counties, only Scotland and Granville went for Cooper in 2016, but Martin and Lenoir saw a majority of their votes cast for Forest in the same cycle. Elsewhere in the state, Cooper actually lost Jackson County, in which he had previously received the majority of votes in 2016; however, he made up for it by picking up and gaining a bigger share of the vote in several counties in urban areas (Leip 2016, 2020).

Cooper himself, in an interview with *The Atlantic*, could not articulate what made him successful in two consecutive governor's races while Trump also claimed victory in the cycle (Dovere 2021). However, some have attributed the candidate's success to how he steered the state through the coronavirus pandemic without too much confrontation with Trump (Dovere 2021). A *New Republic* article published prior to the November election noted the stark difference in the two gubernatorial candidates' personalities. While Cooper is known for being mild and uncontroversial with a style that appealed in both rural and urban environments, Forest was known for his usage of more extreme rhetoric, including railing against the usage of masks, unwillingness to run as a more traditional Republican focused on fiscal responsibility and more moderate on social issues (Dovere 2021, Martin 2020). Overall, North Carolina has a history of electing governors focused more on education, infrastructure, and the economy than on national hot-button issues (Funk 2020). As Rob Christensen, a scholar of North Carolina politics, states in an article in the *Ashville Citizen Times*, "North Carolina voters seem to be looking for someone to make sure we have decent roads and schools when voting for governor, not ideology" (Boyle 2020). Cooper's style and focus on local issues may serve him well as the state moves on from the 2020 election. He will have his challenges. The legislature stands with a Republican majority, although not so much that they may override the governor's veto. Whether Cooper remains a popular governor throughout his second term remains to be seen. It also remains to be determined whether the state will continue with its history of refusing to vote strictly along party lines. What is certain is that the state will continue to be one of interest to political watchers.

REFERENCES

Boyle, J. 2020. Boyle Column: How Can North Carolina Go for Trump and Cooper in the Same Election? *Asheville Citizen Times*. https://www.citizen-times.com/story/news/local/2020/11/27/boyle-column-how-can-nc-go-trump-and-cooper-same-election/6421604002/ (last accessed June 14, 2021).

FIGURE 8.10

North Carolina Gubernatorial Election Popular Vote Leader, 2020

Election Result

Cooper (D) - 2,834,790 (51.52%)

Forest (R) - 2,586,605 (47.01%)

DiFiore (LIB) - 60,449 (1.10%)

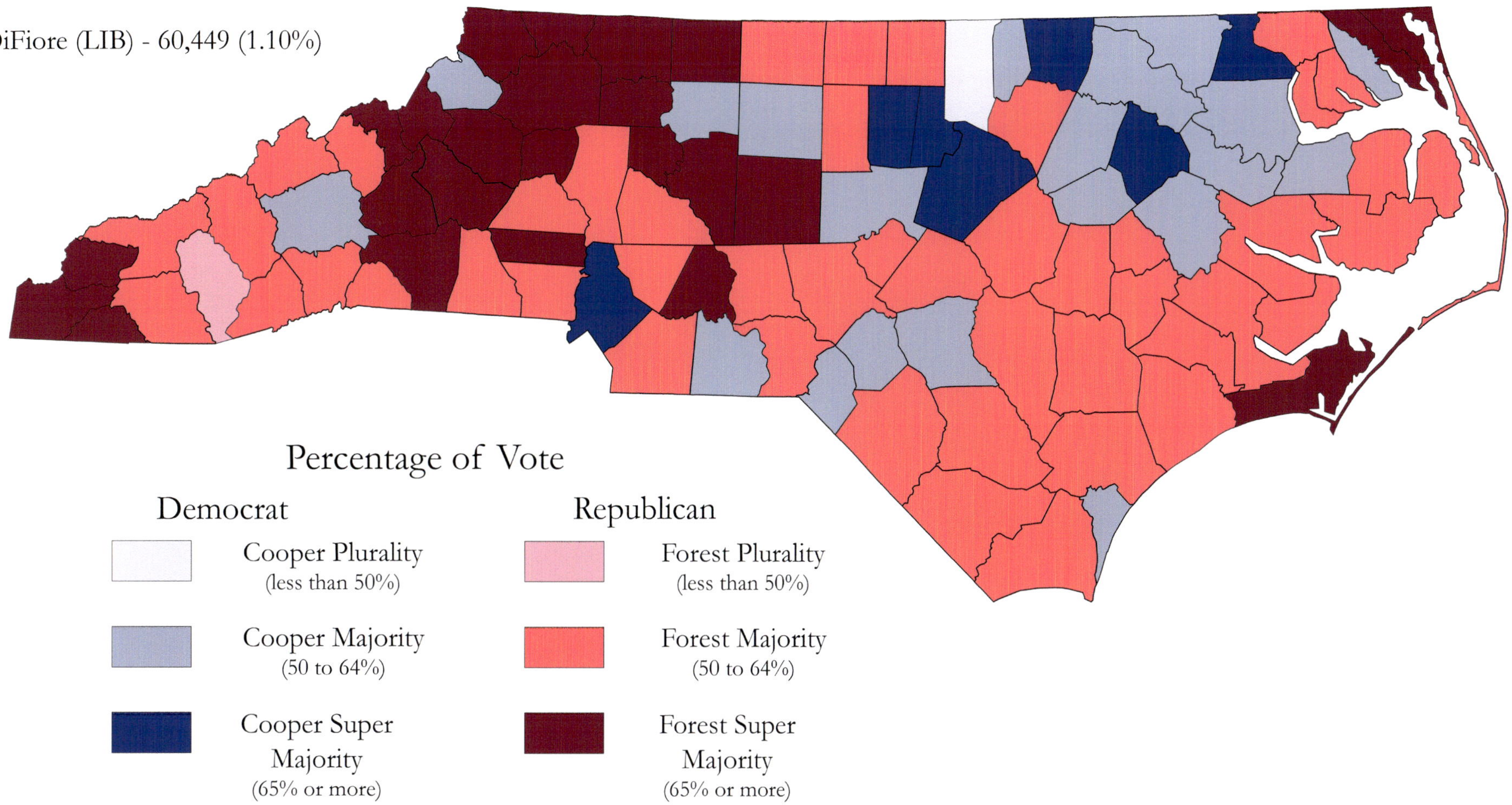

CNN. 2020. North Carolina 2020 Gubernatorial Exit Polls. *CNN.* https://www.cnn.com/election/2020/exit-polls/governor/north-carolina (last accessed June 14, 2021).

Cooper, C. 2020. Primary Primers: Why North Carolina May Be One of the Most Important States in the 2020 Election. *USAPP.* https://blogs.lse.ac.uk/usappblog/2020/08/25/primary-primers-why-north-carolina-may-be-one-of-the-most-important-states-in-the-2020-election/ (last accessed June 14, 2021).

Dovere, E.-I. 2021. What Does This Man Know That Other Democrats Don't? *The Atlantic.* https://www.theatlantic.com/politics/archive/2021/02/what-north-carolina-and-roy-cooper-can-teach-democrats/617874/ (last accessed June 14, 2021).

Funk, T. 2020. Trump and Cooper Both Win NC? How Does That Happen? *Charlotte Observer.* https://www.charlotteobserver.com/news/politics-government/election/article246964987.html (last accessed June 14, 2021).

Leip, D. 2016. 2016 Presidential General Election Results—North Carolina. *Dave Leip's Atlas of U.S. Presidential Elections.* https://uselectionatlas.org/RESULTS/index.html (last accessed June 14, 2021).

———. 2020. 2020 Presidential General Election Results. *Dave Leip's Atlas of U.S. Presidential Elections.* https://uselectionatlas.org/RESULTS/ (last accessed June 14, 2021).

Martin, N. 2020. The Legend of Roy Cooper. *New Republic.* https://newrepublic.com/article/159930/legend-roy-cooper (last accessed June 14, 2021).

McCreary, J. 2020. Reddest of NC's Counties Turned Even Redder in This Election, Data Show. *CBS17.com.* https://www.cbs17.com/news/your-local-election-hq/reddest-of-ncs-counties-turned-even-redder-in-this-election-data-show/ (last accessed June 14, 2021).

New York Times. 2020. North Carolina Exit Polls: How Different Groups Voted. 2020. *New York Times.* https://www.nytimes.com/interactive/2020/11/03/us/elections/exit-polls-north-carolina.html (last accessed June 14, 2021).

Weigel, D. 2020. The Six Political States of North Carolina. *Washington Post.* https://www.washingtonpost.com/graphics/2020/politics/north-carolina-political-geography/ (last accessed June 14, 2021).

THE GEOGRAPHY OF WISCONSIN ELECTION RESULTS AND THE 2020 PRESIDENTIAL ELECTION

RYAN WEICHELT

The 2016 election solidified Wisconsin as a true "purple" state. Having not voted for a Republican presidential candidate since 1984, Wisconsin surprised the nation by electing Donald Trump by approximately twenty thousand votes. Although much can be attributed to a poor campaign by Hillary Clinton and an obvious disdain for the former secretary of state, Donald Trump struck a chord with many voters in the Badger State. By 2020, Wisconsin would once again play the role as a key battleground state as the election results looked like 2016—except this time, Trump would lose.

To understand the 2020 result in Wisconsin it is necessary to revisit the 2018 gubernatorial election and the 2020 primary. By 2018, a virtual "Blue Wave" crashed over the United States as Democrats took back the House of Representatives. In Wisconsin, the "Wave" also hit, but due to obvious gerrymandering at all levels, only the sitting governor, Scott Walker, got wet. Walker had been a lightning rod for Wisconsin politics as he rose to power in the Tea Party era of 2010. Although Walker was an early darling of conservatives for the 2016 nomination, Trump attacked early and often during the initial days of the primary. Walker was the first to bow out, but in the coming months, Walker would echo much of Trump's rhetoric. In both an obvious backlash to Trump and the fact Trump was not on ballot himself, Walker was defeated by a quirky state superintendent of schools, Democrat Tony Evers. Evers was able to gain enough support in key areas across the state to win a by a slim margin. These same patterns would continue in the COVID-19-strained 2020 primary and state supreme court election in April. Liberal judge Jill Karfosky easily defeated her opponent, conservative and Scott Walker–appointed incumbent judge, Brian Kelly, by not only replicating Evers's performance but expanding support throughout much of the state.

Other events in Wisconsin during the months leading up to the November 3 election would bring Wisconsin into the national spotlight. The April 7 primary election would enhance the debate about the legitimacy of mail-in voting. The unnecessary shooting of Jacob Blake by Kenosha police officers garnered the national attention of both campaigns as Trump and Pence as well as Biden and Harris would campaign in Kenosha in the days following the shootings (Brewster 2020). Furthermore, the virus itself would see Wisconsin emerge as the national hotspot in the weeks leading up to November 3. Starting in the eastern half of the state, COVID-19 would spread westward, forcing both hospitals and morgues to capacity (Jannene 2020). Debates regarding mitigation of the virus would only heighten partisan divides across the Badger State.

The spatial patterns of voting provide rich data for understanding how Biden was able to win Wisconsin. In general, the state can be divided into four distinct electoral areas: northern Wisconsin, western Wisconsin, the I-94 urban corridor stretching from Madison to Milwaukee, and the Fox River Valley of eastern Wisconsin. These regions provide a spatial representation of all electoral landscapes: rural farming communities, small-town manufacturing, large urban areas, minority populations, and suburban communities. The 2020 election results for the Badger State are tied to these geographies, which were also replicated in many areas across the United States.

The northern Wisconsin region is a largely white and rural landscape dotted by dairy farms and small-town manufacturing. Although a former Democratic stronghold for decades, the 2010 elections signaled a drastic shift to the Right for this region. The western Wisconsin region stretches from the suburban areas of the Twin Cities along the Mississippi River to Prairie du Chien. The area is home to five university towns, but throughout the rural landscape numerous dairies and poultry farms can be found. Coupled with farming, this area also includes manufacturing centers all employing larger-than-expected Hispanic populations. Although Democrats continue to hold the Third Congressional District, election results have seesawed back and forth between the parties during the decade.

The remaining two regions represent the largest population areas in the state. The I-94 corridor, stretching from Madison to Milwaukee, represents ardent supporters for both parties. Dane County, the home to Madison, is ground zero for the Wisconsin Democratic Party. This county represents the fastest-growing population in the state, and high Democratic turnout in recent elections have washed away gains by Republicans in rural areas across the state. Moving eastward, the suburban counties around Milwaukee, known as the WOW counties (Waukesha, Ozaukee, and Washington) have long been conservative strongholds. Because it is comprised of educated and wealthy white populations, the GOP has long depended on its high voter turnout to win statewide elections. Finally, Wisconsin's largest city, Milwaukee, is also Wisconsin's most racially diverse. Democrats have long depended on support from the large African American populations found throughout the central and northern areas of the city.

The remaining area, the Fox River Valley, is a fast-growing and economically changing area. The Fox River Valley follows the Fox River from Green Bay, to Appleton, to Neenah, and along Lake Winnebago to Oshkosh. The entire area was once dominated by the paper industry and manufacturing continues to be the largest employer throughout the area. Yet as paper plants and other manufacturing facilities slowly closed, an influx of a younger and more educated service-based workforce has slowly begun to replace the population of the region. The result of this population change has triggered a slow but steady increase in Democratic support in each successive election.

The results of the 2020 election in Wisconsin were remarkably similar to the 2016 presidential election. Like Clinton in 2016, Biden held a comfortable lead in the polls heading into November 3, but in the end only about 20,500 votes separated the two candidates. The four regions provide overwhelming evidence as to how Biden was able to accomplish what Clinton could not. When Tony Evers defeated Scott Walker in 2018, rural voters throughout Wisconsin did not show the same enthusiasm as they had in 2016. In 2020 rural voters throughout the northern Wisconsin and western Wisconsin regions were once again enthusiastic for Donald Trump. Figures 8.11 and 8.12 show the 2020 results and comparisons between 2020 and 2016 the voting district level. It was clear Trump would be able to replicate and, in some areas, expand on his 2016 success in rural sections of the state. This was especially clear in the northern and western voting districts of Wisconsin.

Where Trump struggled, compared to 2016, was in the WOW counties as well as in the Fox River Valley. Decline in support for Trump in the suburban areas was largely based on decreased support from educated voters and women upset with Trump's stance on race. Perhaps one of the largest surprises was the decreased support for Trump in the state's most trusted GOP strongholds, Waukesha and Ozaukee counties. To the north, in the Fox River Valley, figure 8.12 shows similar declines in Republican support, especially in the larger urban areas. Although race relations certainly contributed to this decline, COVID-19 played a larger role. As October progressed, COVID-19 infections exploded across eastern Wisconsin and especially in the Fox River Valley area (Santhanam 2020). Trump's eventual diagnosis with the virus and his obvious disregard contributed to the loss of precious votes for the incumbent president. In the end, his decline in support in these two key areas were the major reasons for his loss.

For Democrats, results across the state represented a mixed bag. The continued decline in support in the northern region was not surprising, but Trump's success in the western Wisconsin region was surprising. Although longtime Democratic Congressman Ron Kind was able to keep his seat in the Third Congressional District, his victory was the closest in his career as he narrowly defeated a proud Trump supporter, Derrick Van Orden. As rural Republicans flexed their muscles for Trump, Dane County Democrats easily countered any of their gains. Dane County saw record turnout as well as Democratic support for Joe Biden, who won the state with a surprising 76 percent of the vote. Figure 8.11 clearly shows this, as does Biden's increased support throughout the WOW counties and into the Fox River Valley. Yet Biden was unable to inspire African American voters throughout Milwaukee, Racine, and Kenosha to cast their votes. Turnout in these areas was on par with if not below 2016 numbers.

In the end, Wisconsin's electoral regions provide strong spatial results, helping to explain the 2020 election. Compared to the Democrats' surprising wins in the state in 2018, it is clear rural voters are highly motivated when Trump is on the ballot. Polls once again were unable to capture Trump's strong support among rural white voters. Yet, Trump's weaknesses were exposed among educated voters. The decline in support throughout the largest urban and suburban areas highlighted voters' frustration with Trump's behavior, his stand on race relations, and his ignorance toward COVID-19. Biden's increased support in Dane County, coupled with higher turnout in the other areas, were able to negate Trump's gains in rural areas across the state to win back Wisconsin for the Democrats.

FIGURE 8.11

Percent Biden Vote for the 2020 Presidential Election for Wisconsin at the Voting District Level

Joe Biden (D) - 1,630,866 (49.45%)

Donald Trump (R) - 1,610,184 (48.82%)*

Third Party Votes - 56,991 (1.73%)

*Denotes Incumbent

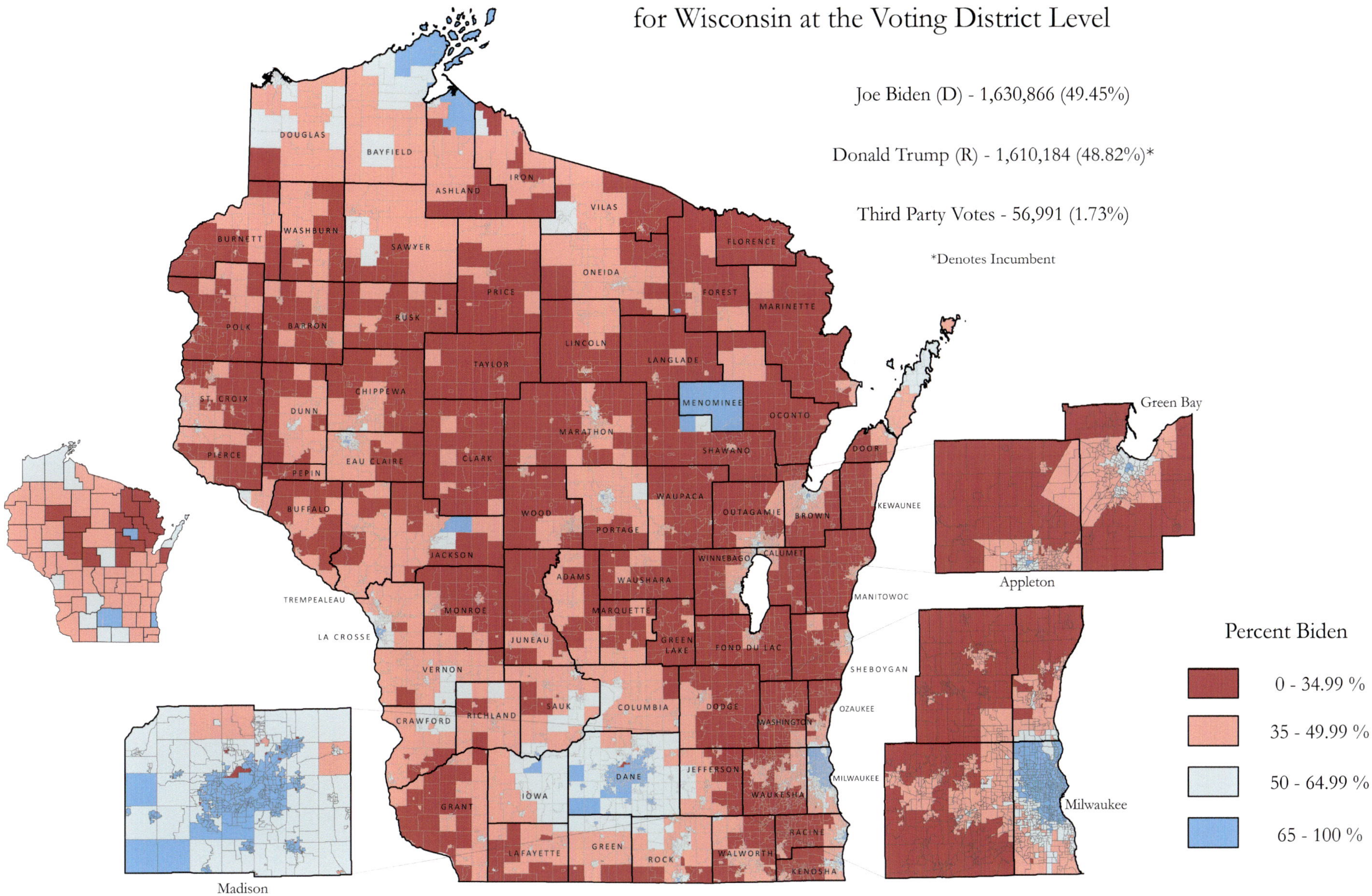

FIGURE 8.12

Percent Difference in Democratic Vote for the 2020 and 2016 Presidential Elections for Wisconsin at the Voting District Level

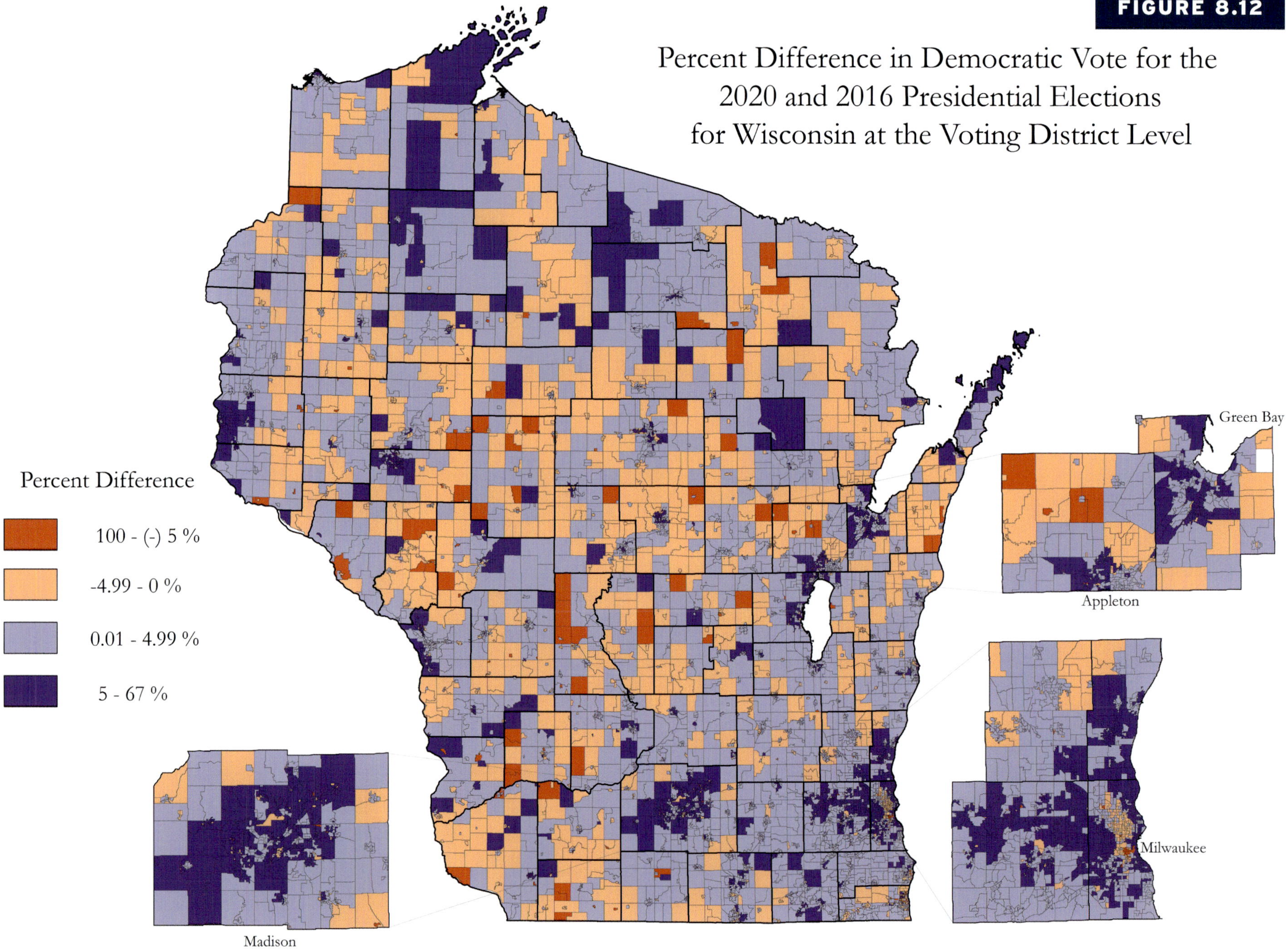

REFERENCES

Brewster, A. 2020. Wisconsin 2020 Election Results: Biden Projected to Win, but Trump Plans to Ask for Recount. *CBS News*. https://www.cbsnews.com/live-updates/wisconsin-election-results-2020/ (last accessed June 1, 2021).

Jannene, J. 2020. WI Daily: October Starts with Single-Day COVID-19 Case Record. *Urban Milwaukee*. https://urbanmilwaukee.com/2020/10/01/wi-daily-october-starts-with-single-day-covid-19-case-record/ (last accessed June 2, 2021).

Santhanam, L. 2020. Amid COVID-19 Pandemic, Wisconsin Voters Choose Biden over Trump. *PBS*. https://www.pbs.org/newshour/politics/amid-covid-19-pandemic-wisconsin-voters-choose-biden-over-trump (last accessed June 2, 2021).

MARIJUANA LEGALIZATION ON THE BALLOT IN 2020: THE JUGGERNAUT CONTINUES

RICHARD L. MORRILL AND LARRY KNOPP

Four states' voters considered some form of marijuana legalization in 2020: Arizona, Mississippi, Montana, and South Dakota (figures 8.13 to 8.16). Voters in Arizona, Montana, and South Dakota voted on citizens' initiatives to legalize recreational use while those in South Dakota also voted on a citizens' initiative legalizing medical use. Mississippi voters chose between two versions of medical use legalization only: a less restrictive citizens' initiative and a more restrictive alternative placed on the ballot by the state legislature (a quirk in state law required a vote on the question of medical use legalization first, then a choice between the two alternatives). Montana's vote also technically involved two measures: a constitutional amendment allowing the issue to be considered and a legal minimum age for use set, and a citizens' initiative that accomplished the actual legalization. In all cases legalization was approved. Legalization of recreational use easily passed in Arizona (60 percent), Montana (58 percent to allow consideration and 57 percent to legalize), and South Dakota (54 percent). Medical use was legalized by even larger margins in Mississippi and South Dakota (69 percent and 70 percent, respectively), with the less restrictive version easily defeating the more restrictive one in Mississippi (74 percent–26 percent).

These results in three deeply conservative and one "swing" states make clear that opposition to marijuana legalization is no longer a wedge issue working in either Republicans' or Democrats' favor. While legalization was generally more popular in Democratic than Republican counties, there is no evidence that these ballot measures had coattail effects that benefited Democrats. Nor is there evidence that Republican-led opposition—including on the parts of the popular governors of Arizona and South Dakota, the successful Republican candidate for governor of Montana, and many Republican state legislators—led to Republicans being punished. All that can be safely said about any correlation with partisan voting is that in most cases the highest votes against legalization tended to be in sparsely populated, heavily Republican rural counties.

Issues of race, ethnicity, and local contingency complicate the picture. In Mississippi, South Dakota, and Montana, there was a fairly consistent pattern of majority non-white counties voting more strongly for legalization than majority-white counties. This was particularly true in Mississippi, where the pro-legalization vote was strongest in counties with Black majorities. But these also tended to be the most heavily Democratic counties, and the "yes" vote was strong across the state. So conclusions about the relative roles of race/ethnicity and partisanship in that state should be made with caution.

In South Dakota and Montana, counties with Native American majorities (or near-majorities) also voted more heavily in favor of legalization than those with lower Native American populations, although this pattern was not universal. In Montana, all four majority Native American counties (Glacier, Big Horn, Roosevelt, and Blaine) voted very strongly for legalization. But so did major population centers with low Native American population (the main exception being Yellowstone County, which contains the state's largest city, Billings, and voted only narrowly for legalization). Opposition was highest in low-density, sparsely populated, rural counties. One of these, Pondera, has a substantial Native American minority (14.3 percent). In South Dakota, two Native American–majority counties (Jackson, which includes portions of the Pine Ridge Reservation and Mellette, which includes part of the Rosebud Reservation) actually voted narrowly against legalization. However, the results in these three seemingly anomalous counties most likely reflect a racial divide in voting on the matter; all three are very small, sparsely populated rural counties containing parts of reservations. The two South Dakota counties containing the majority of the Pine Ridge and Rosebud reservations (Oglala Lakota and Todd) voted heavily in favor. Counties containing the bulk of the Cheyenne River and Crow Creek reservations (Dewey, Ziebach, and Buffalo in South Dakota) did as well. However, Corson County, which contains South Dakota's portion of the Standing Rock Reservation, voted yes but at a slightly *lower* rate than the state as a whole. This is interesting in part because Standing Rock became a *cause célèbre* among progressives across North America in 2016–2017, when tribal

FIGURE 8.13

Arizona: Proposition 207 Marijuana Legalization Initiative
2020 Election

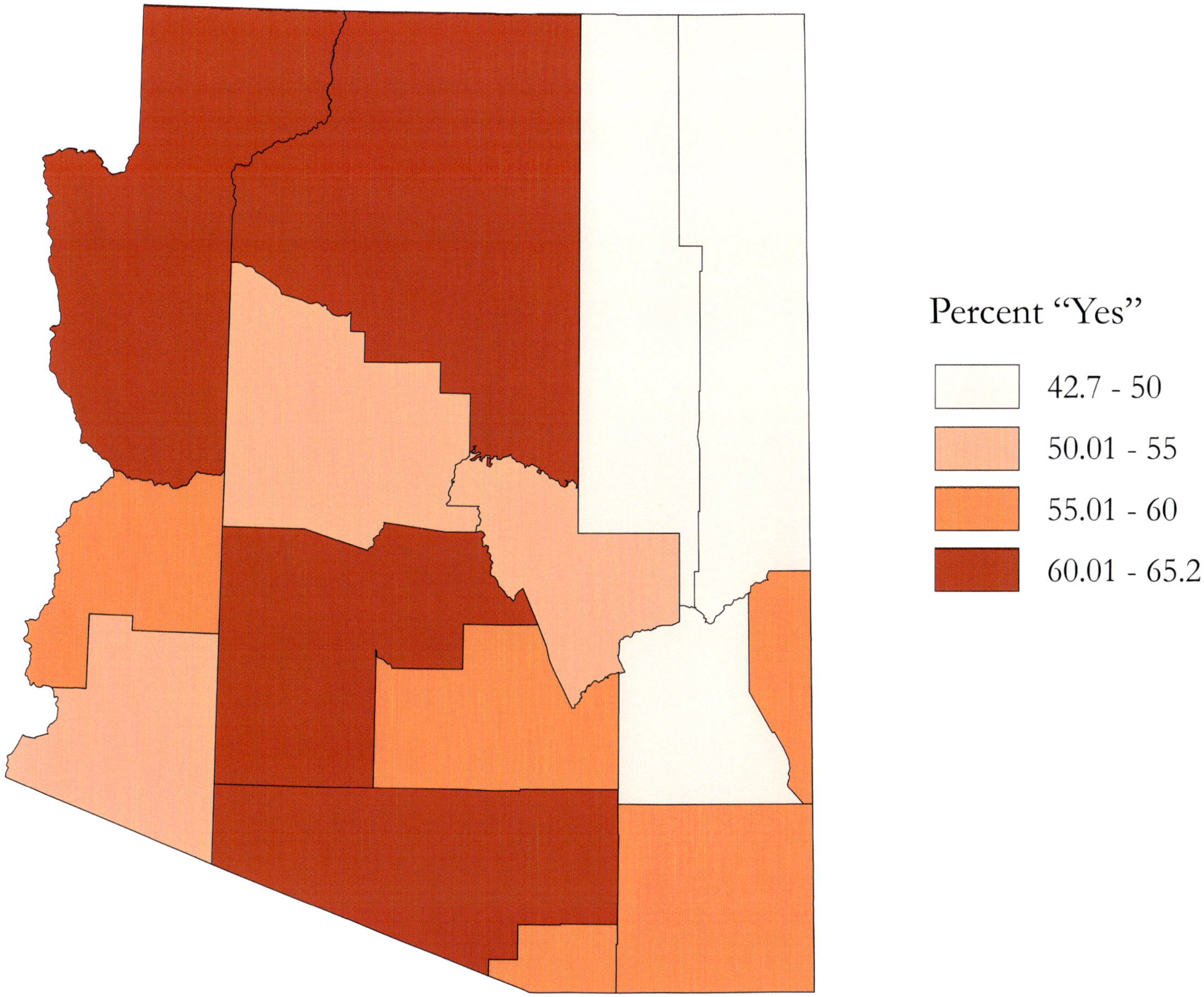

FIGURE 8.14

Mississippi: Marijuana Legalization Measure 2020 Election

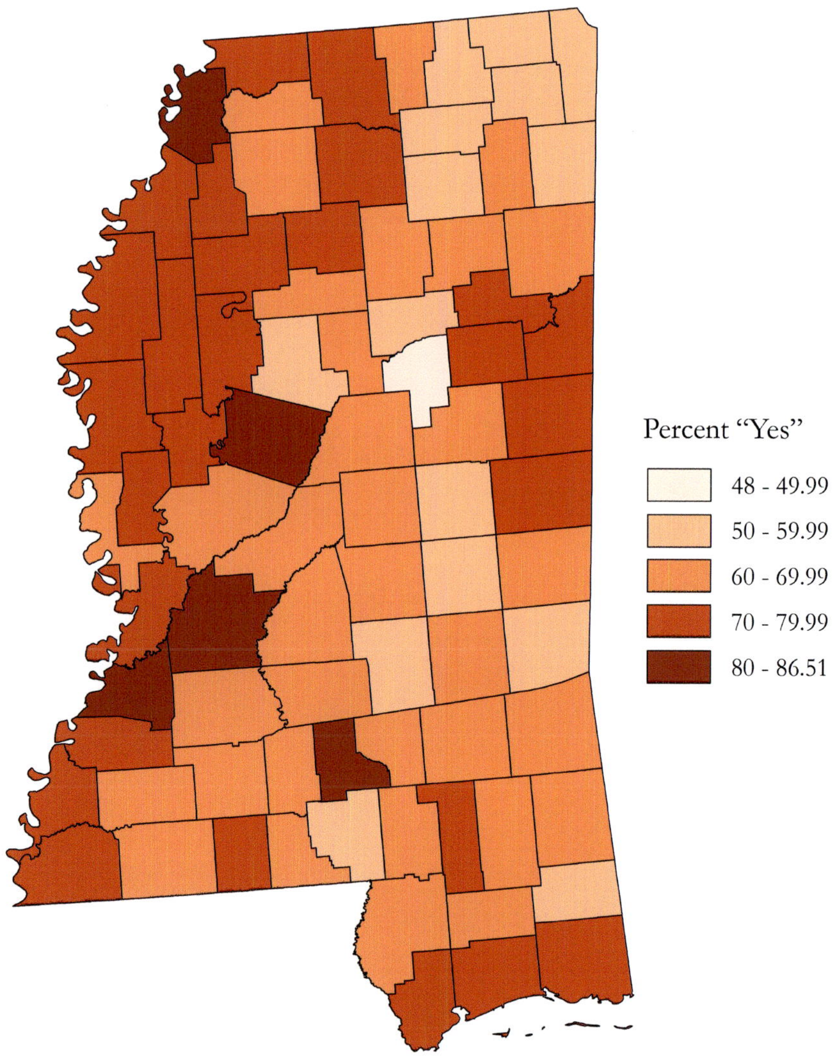

Mississippi: Marijuana Legalization "Initiative 65" 2020 Election

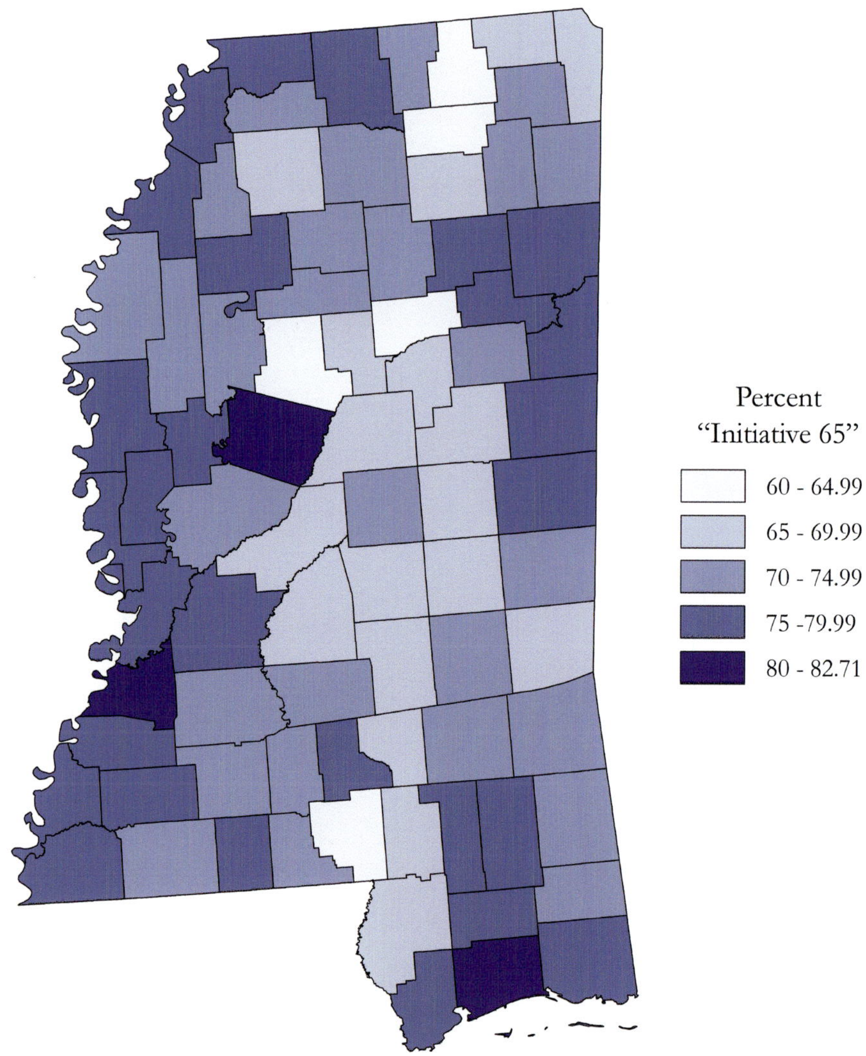

FIGURE 8.15

Montana: Marijuana Legalization "Initiative I-190"
2020 Election

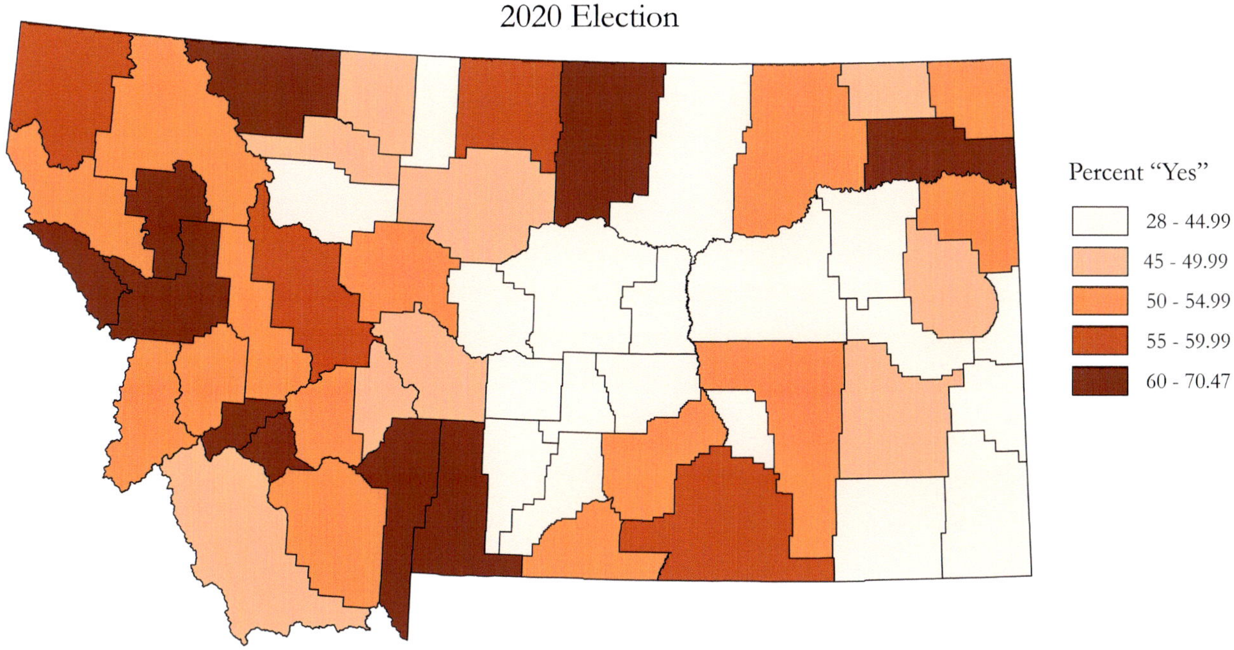

Montana: Allow Legal Age for Marijuana "Constitutional Amendment C-118"
2020 Election

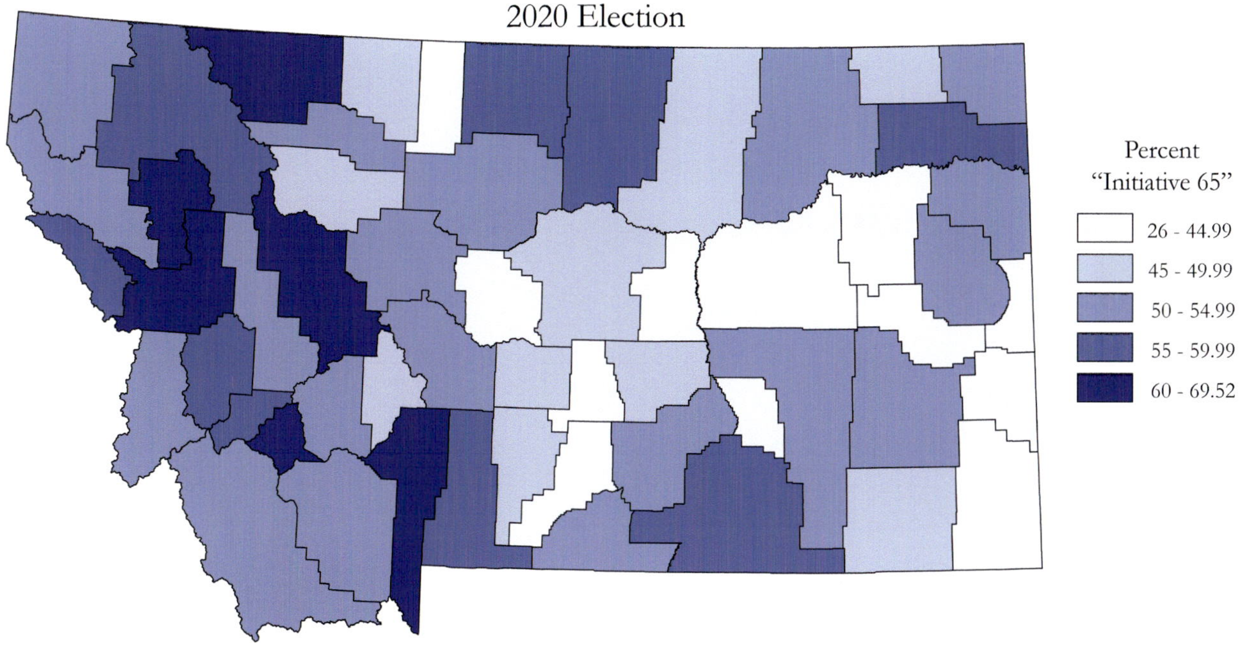

FIGURE 8.16

South Dakota: "Amendment A" to State Constitution, Legalizing Marijuana
2020 Election

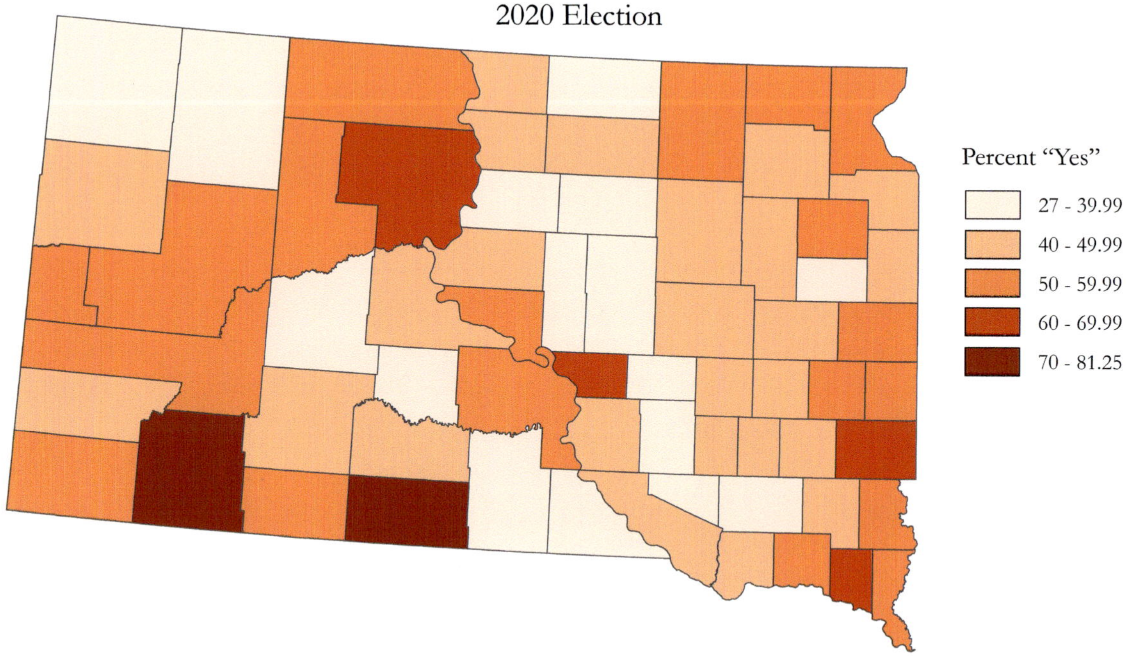

South Dakota: "Initiative 26" Legalizing Marijuana for Medical Use
2020 Election

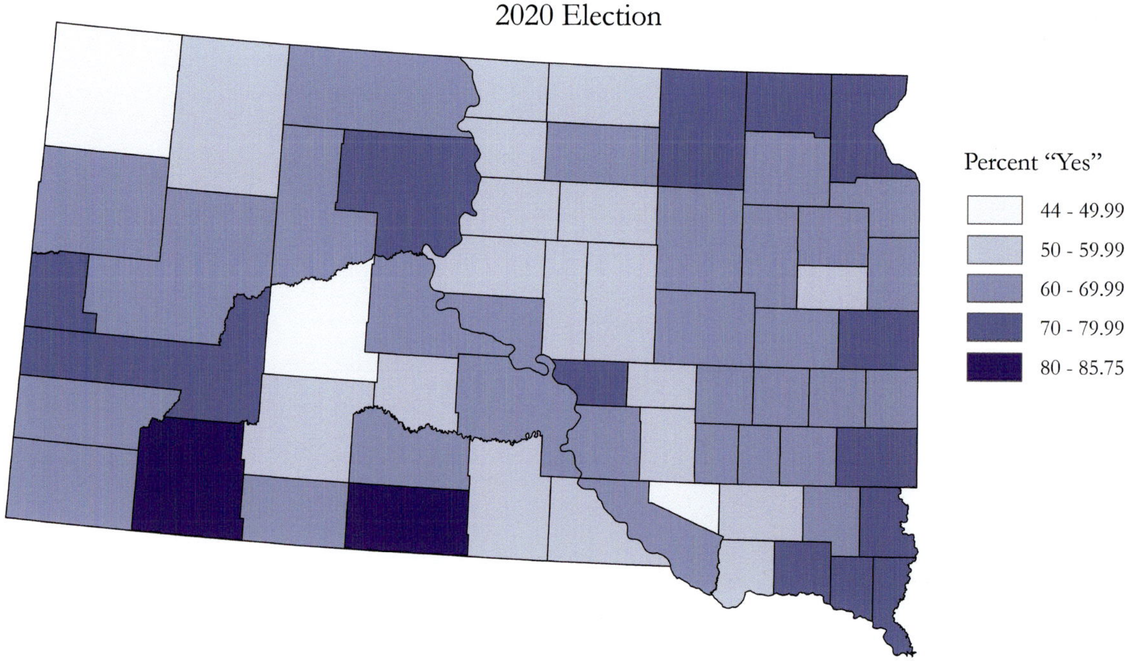

members and leaders, along with Native American and progressive activists from across North America, staged a months-long sit-in against construction of the Dakota Access oil pipeline at the North Dakota edge of the reservation. So in this particular case, at least, Native American political activism, and the support it drew from progressives, did not correlate disproportionately with support for marijuana legalization.

Contrary to Montana and South Dakota, Arizona counties with larger Native American populations tended to vote *against* legalization. Indeed, of the three counties that produced majorities against legalization, one (Apache) is dominated by the Navajo Reservation and a second (Navajo) has a near-majority Native American (and minority Anglo) population. One important factor in these two counties appears to have been active opposition to legalization on the part of the Navajo nation's leadership (which arguably made the difference in defeating a similar legalization measure in 2016).

Anglo versus Hispanic or Latino voting on the issue, meanwhile, is difficult to discern based on county-level data (and virtually impossible in Mississippi, Montana, and South Dakota, where Hispanic and Latino populations are quite small relative to the United States as a whole). But of the three Arizona counties with the highest percentages of "yes" votes (Coconino, Maricopa, and Pima) it is worth noting that one (Coconino) is substantially more Anglo than the state as a whole, one (Pima) substantially more Hispanic and Latino, and one (Maricopa, home to the Phoenix metro area) roughly mirrors the statewide racial and ethnic profile. The third county to vote against legalization, meanwhile (Graham, after majority Native American Apache and Navajo), also roughly mirrors the statewide profile. So it appears likely that in Arizona, at least, there is little difference between the way Anglos and Hispanics or Latinos voted on the issue.

What appears to characterize pro-legalization voting across all four states more than either partisanship or race/ethnicity is a combination of population density and economic/occupational/educational profile. Counties containing major population centers, strong economies, and colleges or universities all tended to support legalization, often strongly. Where one or more of these conditions did not exist, support tended to decline. There are, of course, exceptions (Montana's Yellowstone County, for example, the state's most populous, voted only narrowly for legalization). But overall the movement toward marijuana legalization, which began in the 1990s around medicinal use and the 2010s around recreational use, appears now to have become a mainstream, bipartisan, multiracial, multiethnic juggernaut. Opposition is weaker than ever and concentrated primarily in sparsely populated rural areas or areas (such as the Navajo Reservation in Arizona) where local contingencies play an outsized role.

FLORIDA'S AMENDMENT 2 AND THE 2020 PRESIDENTIAL ELECTION

THOMAS CRAIG

As has been the pattern over the past few decades, Florida played the role of swing state in the 2020 presidential election. While the state was eventually won by Trump, who captured more than half of the state's popular votes, statewide measures proved to be not as aligned with this result (figures 8.17 and 8.18). Namely, Florida's Amendment 2, which sought to increase the minimum wage from $8.56 in 2020 to $15.00 by 2026, passed by a significant margin, with 60.8 percent of the population voting yes.

This was the first time since 2004 that Florida voted on a minimum wage ballot measure, which also was approved by 71 percent of the vote. The 2020 amendment was specifically designed to increase the minimum wage in the state, beginning in September 2021, to $10.00, and then increasing an additional dollar every subsequent year, ending at $15.00 in September of 2026. While seven states had previously passed bills to increase minimum wage incrementally to $15.00, Florida become the first state to do so by ballot measure.

Trump was able to carry the state of Florida for a second time by expanding his percentages from 49 percent in 2016 to 51.2 percent in 2020. In most cases, Trump increased his winning percentage in every county he won in 2016, including important pivot counties like Jefferson, Monroe, and St. Lucie. Indeed, beyond the Democratic strongholds around Miami, Orlando, and Tallahassee, the only pivot county that Biden was able to flip was Pinellas County. Yet, despite only twelve counties having a majority voting in favor of Biden, thirty-two counties approved Amendment 2. Put another way, statewide nearly 6.4 million voted in favor of raising the minimum wage, while Trump received only about 5.7 million votes.

Unsurprisingly, all the counties that Biden won also voted to approve the progressive measure, and in most of cases, voted over +5 percentage points higher. That said, a couple of counties' approval percentages were slightly lower than Biden's winning percentage—Gadsden (-1.6 percent) and Leon (-3.4 percent)—although these counties still passed the measure with more than a 60 percent approval.

When it came to the four pivot counties (Jefferson, Monroe, St. Lucie, and Pinellas—the first three of which Trump won), three voted in favor of raising the minimum wage, and by wide margins. Monroe (58.1 percent), Pinellas (59.7 percent), and St. Lucie (58.4 percent) all voted yes by at least 16 points.

But it was not only pivot counties that swayed. Several counties where the vote went significantly toward Trump also voted in favor of Amendment 2. For instance, Sumter County, located in the central part of the state between Orlando and Tampa, voted for Trump with 67.9 percent. However, when it came to the measure, it was almost a 50-point swing, with 56.5 percent voting yes.

Similarly, in Collier, Marion, and Martin counties, all which elected Trump with about 62 percent of the respective votes, also voted yes for Amendment 2 by more than 55.5 percent. Even in counties that voted heavily for Trump, the swing in favor of yes was substantial, such as in Taylor (76.5 percent for Trump, 44.0 percent for Amendment 2, 41.8-point swing) and Walton (75.4 percent for Trump, 45.2 percent for Amendment 2, 42.2-point swing) counties. The largest shift in the state was found in rural Holmes County, located in the panhandle north of Panama City. Here, while Trump collected a staggering 89.1 percent of the vote, 36.4 percent voted to raise the minimum wage, nearly a 52-point swing.

Understandably, not all counties that Trump won followed suit with larger numbers approving the measure. Madison County, for instance, voted in favor of Trump at 59.4 percent but did not pass the amendment with 47.9 percent voting yes, about a 15-point swing. That said, in forty-nine of the fifty-five counties Trump won, the swing in favor of Amendment 2 was at least 25 points.

Moving forward, the swinging nature of national *and* state elections or measures seems to certainly be the hallmark for Florida. Understandably, we can expect the state to remain on the national and local radar for political pundits to argue about the swaying ways of Florida's votes.

REFERENCES

Ballottpedia. 2020. "Florida Amendment 2, $15 Minimum Wage Initiative (2020). https://ballotpedia.org/Florida_Amendment_2,_$15_Minimum_Wage_Initiative_(2020) (last accessed June 25, 2021).

FIGURE 8.17

Percent Democratic Vote
2020 Presidential Election

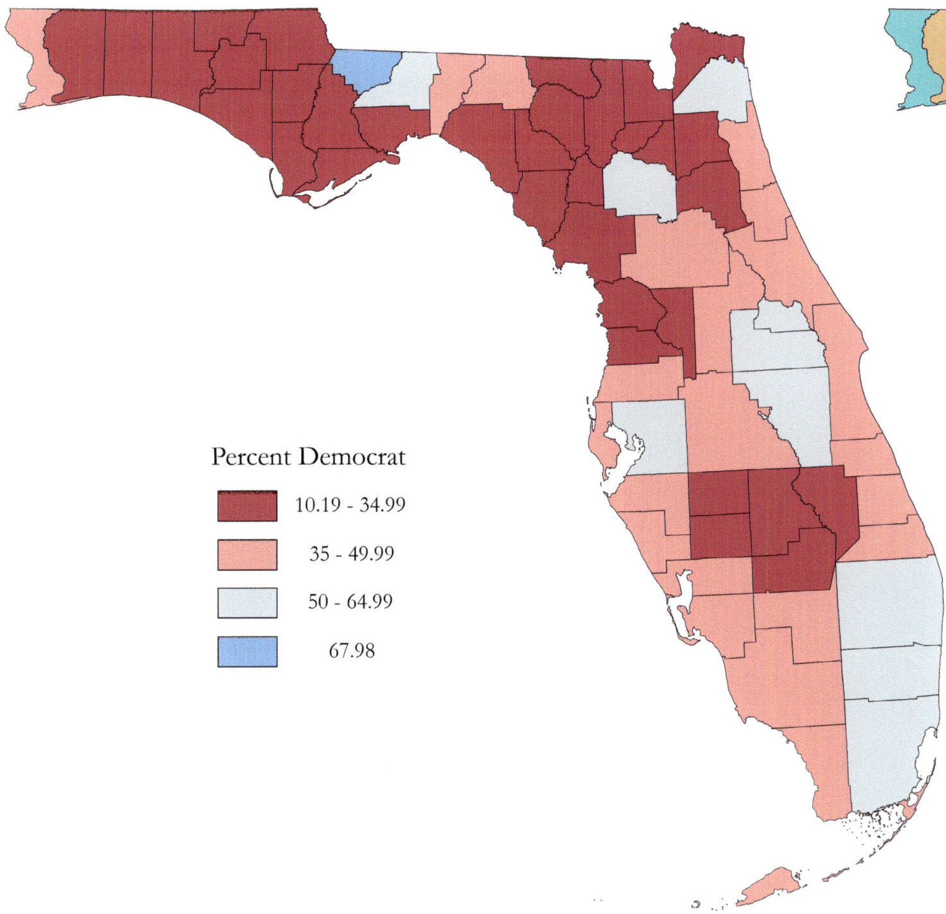

Amendment 2: Increase Minimum Wage to $15
Percent "Yes" Vote, Florida 2020 Election

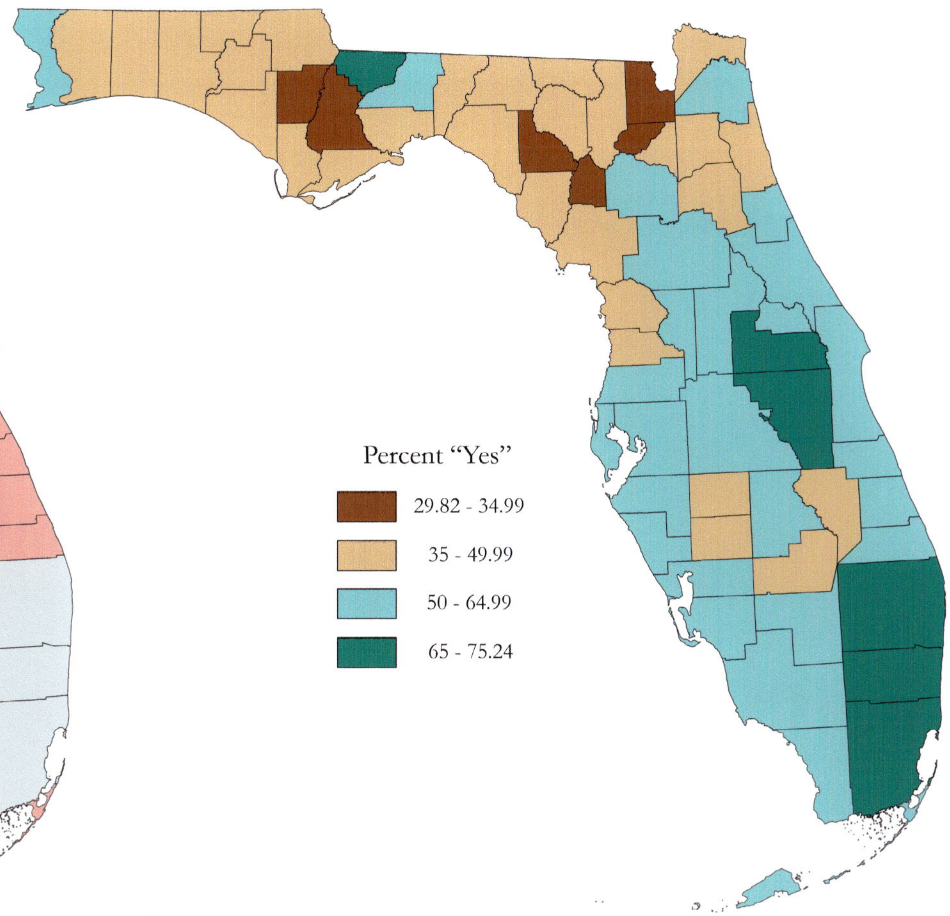

Percent Difference of Biden Vote and Amendment 2

FIGURE 8.18

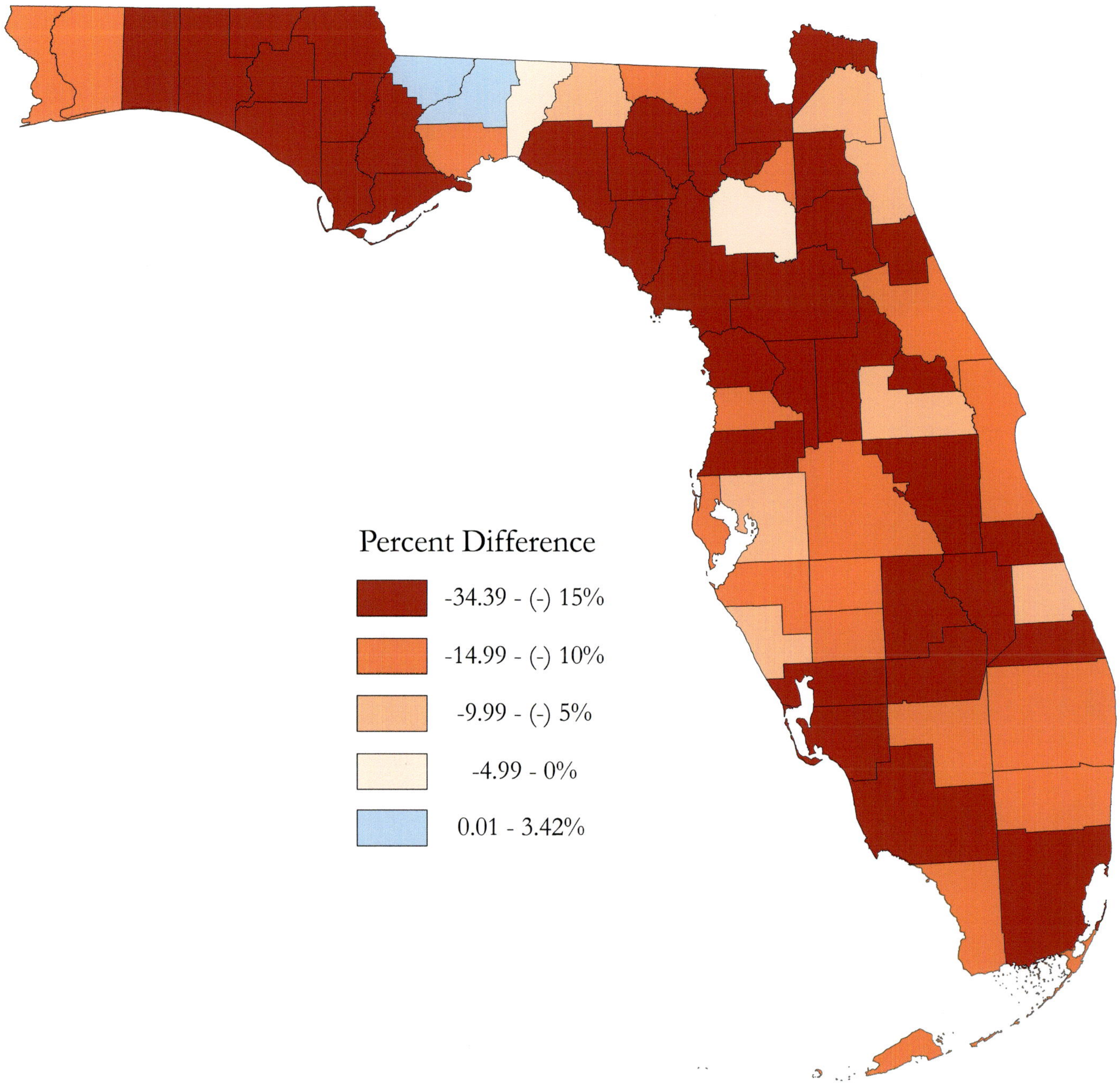

9 POSTELECTION EVENTS

GEORGIA RUNOFFS: DEMOCRATS TAKE THE SENATE MAJORITY WITH DOUBLE VICTORIES

J. MILES COLEMAN

Georgia was unique among the thirty-five states that hosted senatorial contests in 2020; it was the only one to see a "double-barreled" situation, where both of its seats were up. Georgia's contests were settled in January 2021 runoffs. For political observers, the runoffs punctuated an already volatile election cycle.

SETTING THE STAGE

Going into the 2020 cycle, Georgia was set to have at least one election; Senator David Perdue, a Republican who was initially elected in 2014, was scheduled to face voters. Still, this "regular" election would ultimately come down to a unique state legal quirk: aside from presidential contests, Georgia law requires that statewide officeholders be elected with a majority of the vote (Justia 2019). In other words, if Perdue or the eventual Democratic nominee placed first in November, but only with a plurality, a runoff would ensue.

The genesis of the state's second contest was in August 2019. That month, then-Senator Johnny Isakson, a Republican, announced that he would step down at the end of the calendar year. As governor, fellow Republican Brian Kemp was given the task of appointing Isakson's replacement; his designee would assume office in January 2020, a special election would be held that November, and the winner would serve the balance of Isakson's term. Importantly, special elections in Georgia are held under Louisiana-style jungle primary rules: all candidates, regardless of partisan labels, run on the same ballot. If no one clears 50 percent of the vote, then a runoff is triggered.

When Isakson's seat opened, then-President Trump implored the governor to appoint then-Representative Doug Collins, a strident conservative from rural northeastern Georgia. Instead, Kemp appointed Kelly Loeffler, a wealthy businesswoman who could self-fund a potentially expensive campaign. Perhaps Kemp calculated that Collins was too far right to win statewide and that Loeffler's profile as successful businesswoman would sell better in the electorally critical suburbs.

THE CAMPAIGNS

For the regular election, Jon Ossoff won the June Democratic primary outright, with 53 percent. Although Ossoff had no elected experience and was only thirty-three years old, he was well known in both Democratic circles and in the Atlanta metro. In 2017 he was the Democratic nominee for a nationally watched special election in the state's Sixth Congressional District. As a candidate, Ossoff stressed his work as a staffer for Representative John Lewis, a civil rights icon (Lewis died in July 2020).

For the special election, the all-party jungle primary would be held concurrently with the November general election. Although Doug Collins was snubbed for the Senate appointment, the Trump ally entered the race anyway. Early on, Collins floated internal polls that suggested GOP voters favored him over Loeffler (Bowden 2020). Adding to Loeffler's woes were allegations of insider trading in early 2020, although she was eventually cleared. In an ironic twist, considering the reasoning behind her appointment Loeffler's solution was to outflank Collins on the right. Perhaps sensing that Democrats would eventually consolidate behind one candidate, her short-term goal became beating out Collins for the second runoff berth. In one of the cycle's most memorable ads, Loeffler claimed to be "more conservative than Attila the Hun."

Although he was not the immediate frontrunner, Raphael Warnock emerged as the top Democrat in the special election. As pastor of Martin Luther King's former church, Atlanta's Ebenezer Baptist Church, Warnock was known for his powerful sermons. Beyond his role as a pastor, Warnock was also aided by nationally known and former Democratic gubernatorial candidate Stacey Abrams, who recruited Warnock for the position.

NOVEMBER TO JANUARY

On Election Day 2020, then-former Vice President Joe Biden carried Georgia by 11,779 votes over Trump, making it the closest margin in the country. However, the picture down ballot was more mixed for Democrats.

Given the Senate special election's large field of almost two dozen candidates, there was no question that the Senate race would go to a runoff. Warnock finished first, taking 33 percent of the vote, while Loeffler placed next with 26 percent and Collins finished with 20 percent.

In the regular Senate election, Perdue fell just short of a majority, necessitating a runoff. He took 49.7 percent, while Ossoff drew 48 percent, and the Libertarian candidate, Shane Hazel, who was given airtime during the debates, claimed just over 2 percent. Still, it was a potentially ominous sign for Democrats that Ossoff received almost one hundred thousand fewer raw votes than Biden.

With both contests unsettled, political observers looked to January 5, 2021, a day that would feature dual Georgia runoffs. As if the stakes were not already high, control of the Senate would hinge on the outcome. As a result of the November senatorial elections, Democrats would hold at least forty-eight seats; if Democrats won both Georgia seats, the chamber would be tied fifty-fifty, but Vice President Kamala Harris's tie-breaking vote would provide them control.

Spending for the runoffs was immense. The Center for Responsive Politics found that about $830 million was spent between the two races (Evers-Hillstrom 2021). In addition, then-President-Elect Biden made multiple stops in Atlanta during the runoff campaign while Trump held rallies in rural Georgia.

THE RESULTS

In the end, Democrats won both seats, and by relatively comfortable margins. Ossoff defeated Perdue by 55,000 votes, while Warnock's margin over Loeffler was an even larger 93,000 votes. What went wrong for Republicans?

First, in terms of voter mobilization, Democrats simply outmaneuvered Republicans. Throughout the 2020 campaign, Trump encouraged his supporters to vote on Election Day, while Democrats tended to vote early. This dynamic held in the runoff. In a number higher than November's 19.5 percent, roughly 30 percent of runoff votes were cast on Election Day. Perdue won these voters by 27 percent, a margin larger than Trump's 23 percent. But Democrats built up a superior lead in the early vote. Early in-person voters seemed to be the key group: while Biden lost them 53 percent to

TABLE 9.1. Georgia: November vs. January Vote by Method

	November 2020				January 2021			
Vote Method	Biden	Trump	D – R	% Share	Ossoff	Perdue	D – R	% Share
By Mail	64.5%	34.3%	30.3%	26.3%	67.7%	32.3%	35.4%	24.2%
Early In-Person	46.4%	52.7%	-6.3%	53.9%	50.6%	49.4%	1.3%	46.2%
Election Day	37.6%	60.2%	-22.6%	19.5%	36.4%	63.6%	-27.2%	29.4%
Provisional	63.5%	34.4%	29.2%	20.0%	68.6%	31.4%	37.3%	20.0%
Total	49.5%	49.3%	0.3%		50.6%	49.4%	1.2%	

FIGURE 9.1

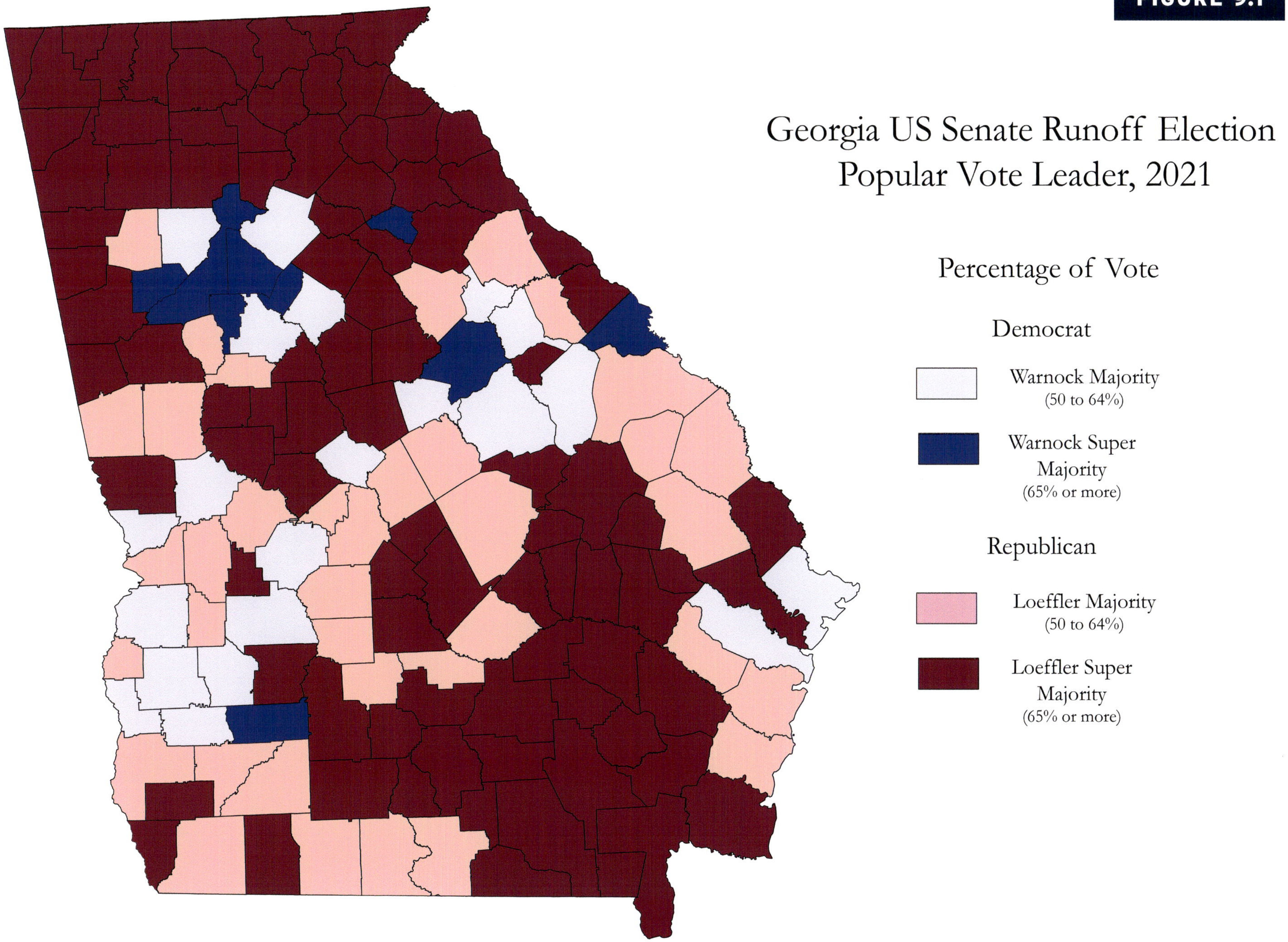
Georgia US Senate Runoff Election
Popular Vote Leader, 2021
Percentage of Vote
Democrat
Warnock Majority
(50 to 64%)
Warnock Super
Majority
(65% or more)
Republican
Loeffler Majority
(50 to 64%)
Loeffler Super
Majority
(65% or more)

FIGURE 9.2

Georgia: Percent Difference between Warnock and Biden 2020 Election and 2021 Special Election

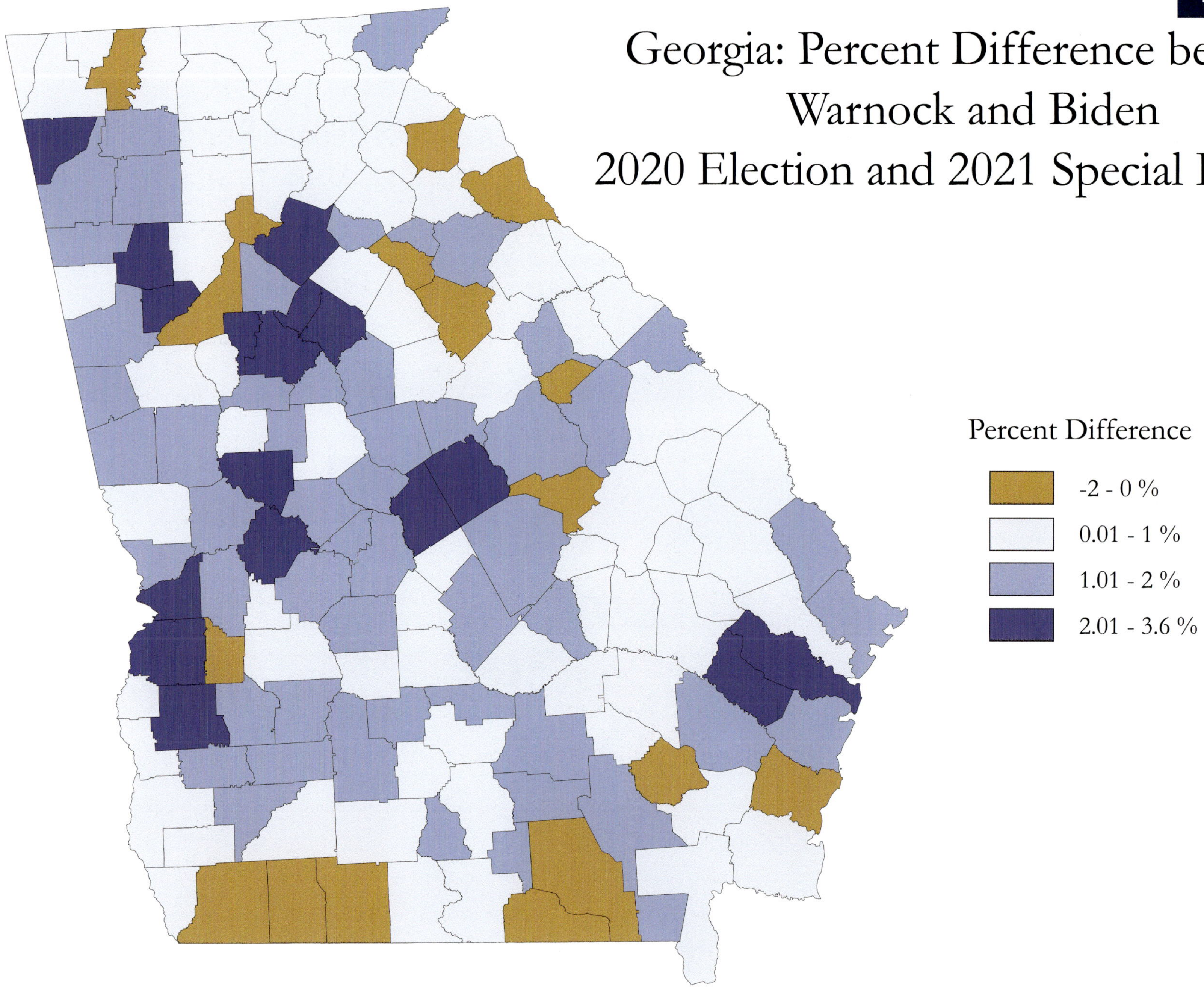

46 percent in November, Ossoff carried them by almost 51 percent to 49 percent. Table 9.1 (Georgia Secretary of State 2021) demonstrates that the contours of the special election were similar to the November general election.

Voter turnout, for a runoff, was astonishingly robust. Only about 10 percent fewer raw votes were cast in January than November. For context, in a senatorial runoff after the 2008 November elections, turnout fell by close to 50 percent. Still, the slight drop from the presidential result was not evenly distributed across the state. GA-14, a safely Republican seat in the rural northwest, saw the largest turnout dip, 15 percent, while most of the Democratic-leaning Atlanta metro districts saw only single-digit percentage drops.

Since his November loss, Trump repeatedly claimed that the results (in Georgia and other close states) were rife with fraud. This belief perhaps convinced some Republicans in rural Georgia to refuse to partake in a supposedly rigged system. After the election, some Republican operatives also blamed legislative infighting for dampening enthusiasm on their side when then-Senate Majority Leader Mitch McConnell was reluctant to go along with Trump's call for $2,000 stimulus checks.

In parts of rural Georgia, there was evidence that Republican messaging boosted Democratic turnout. The Black Belt is a largely rural geographic swath that cuts across most states in the South (its name refers to the soil, not the people). By composition, Georgia's portion of the region is just over 40 percent African American. Furthermore, adherence to local Black churches played a critical role in the area, boosting turnout for Democrats. In their attempts to frame Warnock as a "radical liberal," Republicans ran ads featuring fiery snippets of his sermons, which were sometimes taken out of context. As figures 9.1 and 9.2 show, some of Warnock's biggest gains over Biden came in Black Belt counties. Warnock also saw increases compared to Biden in many of the state's urban areas. Warnock improved most over Biden in Clayton County, which is majority Black and is home to the busy Hartsfield-Jackson Atlanta International Airport. Warnock's strength with Black voters also seemed to buoy Ossoff's performance.

In the Atlanta area, trends with suburban voters were illustrated in the Sixth Congressional District. This area is in the city's northern suburbs and is comprised of many college-educated white voters. Compared to 2016, this area flipped from Trump to Biden. In his runoff election, Ossoff carried the area by 5 points yet had lost the area by the same margin in 2017. Ironically, Warnock's margin was a few points better, although this may have had more to do with his opposition, where Loeffler's trek to the right seemed to help against Collins. Most likely her populist tone made her less appealing in the suburbs. Although neither Ossoff nor Warnock could match Biden's 10 percent spread in GA-6, it seems that Democrats can still count on a critical mass of college whites in elections.

LOOKING FORWARD

Because of their historic wins in Georgia, Democrats took the Senate, and enabled President Biden to come into office with control of Congress. Looking to 2022, Warnock's seat will again be up; key goals for his campaign will surely be inspiring turnout in the Black Belt while holding on to Trump-era suburban gains.

REFERENCES

Bowden, J. 2020. Internal Collins Poll Suggests He Holds Huge Lead Over Incumbent Sen. Loeffler in Georgia Special Election. *The Hill*. https://thehill.com/homenews/campaign/491459-poll-rep-collins-holds-huge-lead-over-incumbent-sen-loeffler-in-georgia (last accessed March 3, 2021).

Evers-Hillstrom, K. 2021. Georgia Senate Races Shatter Spending Records. *OpenSecrets News*. https://www.opensecrets.org/news/2021/01/georgia-senate-races-shatter-records/ (last accessed April 24, 2021).

Georgia Secretary of State. 2021. *Elections*. https://sos.ga.gov/index.php/elections (last accessed May 13, 2021).

Justia. 2019. 2010 Georgia Codec—Number of Votes Required for Election. *Justia Law*. https://law.justia.com/codes/georgia/2010/title-21/chapter-2/article-12/21-2-501 (last accessed May 14, 2021).

HOUSE OF REPRESENTATIVES VOTES ON CERTIFYING ARIZONA AND PENNSYLVANIA ELECTORAL COLLEGE RESULTS, JANUARY 6 AND 7, 2021

KENNETH C. MARTIS

Wednesday, January 6, 2021, is one of the most momentous days in American political history. On that day a riotous crowd pushed aside outer police barriers surrounding the US Capitol, stormed the Capitol, broke through doors and windows, overwhelmed law enforcement, and occupied the building. Among the places seized were the House of Representatives and Senate Chambers and offices of members of Congress, including the office of Speaker of the House Nancy Pelosi. Members of Congress and staff were ushered into safe rooms to be protected from the crowd. More than 140 law enforcement officers were injured in the ensuing violence and later one died (Schmidt and Broadwater 2021). One of the intents of the mob was to stop the constitutionally mandated congressional Electoral College certification of Joe Biden as winner of the 2020 presidential election.

The events leading up to the Capitol takeover began before, during, and after the November 3, 2020, presidential election. Accusations of voter fraud, ballot harvesting, non-citizen voting, computer manipulation, and deliberate miscounting were made by the Donald Trump presidential campaign, several national and state Republican leaders, a number of conservative commentators, and numerous far right conspiracy groups. The litigation that followed was substantial. In the end more than sixty lawsuits were filed. Highly targeted were the states of Arizona, Wisconsin, Michigan, Pennsylvania, Nevada, and Georgia. In all cases the suits were dismissed or dropped for lack of evidence, or deemed meritless, and in some cases determined frivolous. In several of the targeted states, the final vote certifications were made by Republican local and state election officials. The prime example is Georgia. Here Joe Biden won the state by a narrow 0.26 percent. The Republican secretary of state certified the final numbers as accurate and free of massive voter fraud, and the Republican governor confirmed (Brumback 2020). In addition, many of the dismissed lawsuits were adjudicated by Republican-appointed judges.

The Twelfth Amendment of the Constitution directs each state to transmit the Electoral College results: "sealed to the seat of the government of the United States, directed to the President of the Senate; the President of the Senate shall, in the presence of the Senate and House of Representatives, open all the certificates and the votes shall then be counted; the person having the greatest number of votes for President, shall be the President." The date of the certification of the 2020 presidential election was set for January 6, 2021, President of the Senate, Vice President Mike Pence, presiding.

In spite of the lack of evidence of widespread electoral fraud, a January 6, 2021, Washington, DC, protest rally was called for by President Trump after the December 13, 2020, Electoral College state votes confirmed a Biden win. The rally was bolstered online and through social media by numerous pro-Trump, conservative, and conspiracy groups. The "Save America March" rally took place south of the White House at the Ellipse with President Trump giving a one-hour speech. Vice President Mike Pence was admonished by the president for not doing his duty to overturn the election. The rally participants then marched to the Capitol and at approximately 1:00 p.m. a large number of them broke through the outer police barriers.

FIRST CERTIFICATION ROLL-CALL VOTE: ARIZONA

At about the time of the breaching of the outer Capitol barricades, Vice President Pence and US Senators entered the House Chamber to join with the Representatives to certify the Electoral College results (Congressional Record 2021). The roll call of the states began with Alabama and Alaska; their electoral votes were approved. A motion was made at 1:12 p.m. to challenge the Arizona results. At this point the vice president and Senate retired to their own chamber to debate the objection. Almost exactly one hour later the rioters breached the Capitol building itself. At this point the debate was suspended in both chambers and the vice president, members of Congress, and staff were evacuated into safe rooms.

What followed was an afternoon of a well-documented well-photographed takeover of the

FIGURE 9.3

House of Representatives Votes on Certifying Arizona and Pennsylvania Electoral College Results, January 6 and 7, 2021

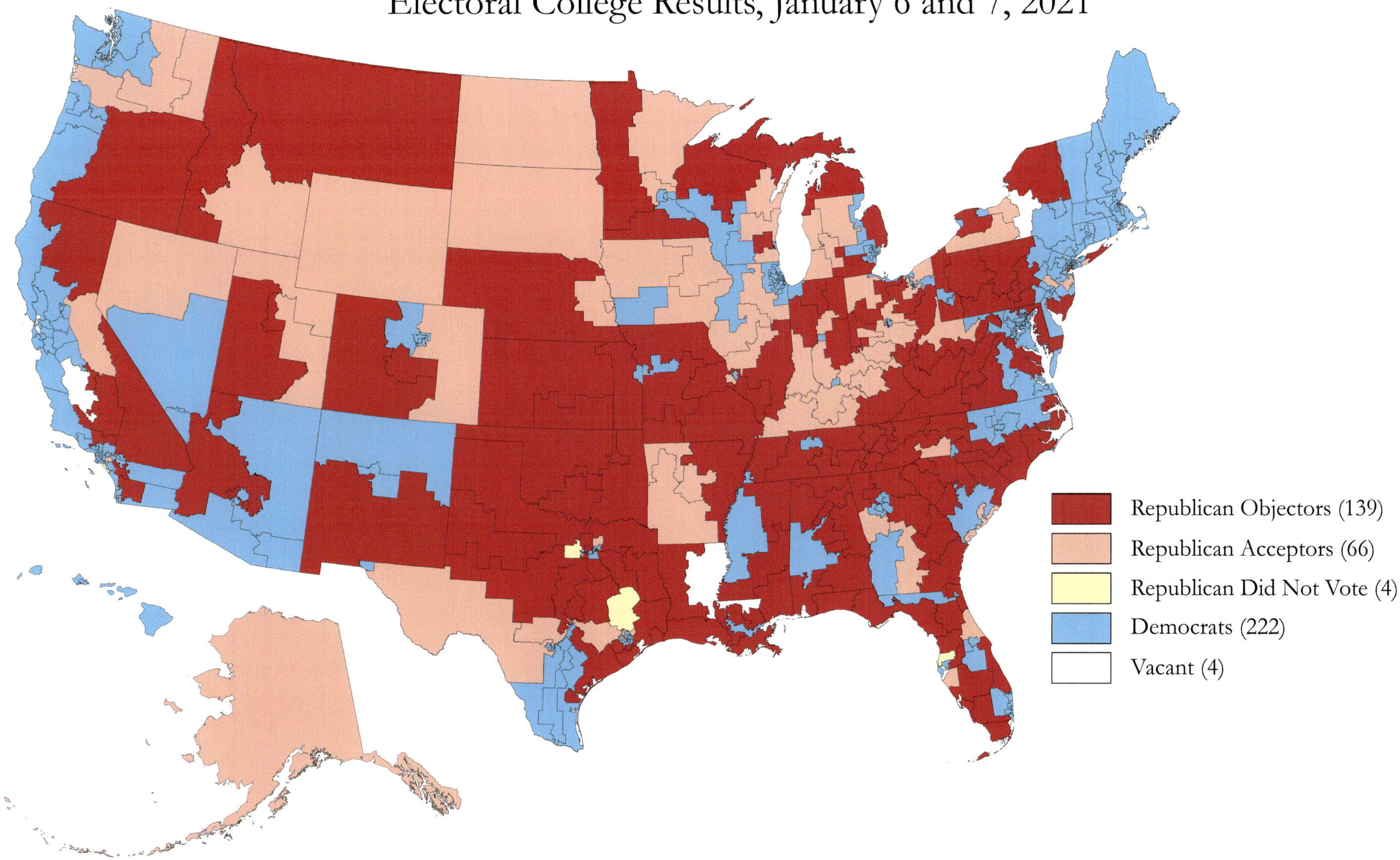

US Capitol Complex. The violent intent of many rioters was put forward in social media before and during the occupation. Reinforcements from the Capitol Police, DC Metropolitan Police, and DC National Guard arrived in late afternoon. A DC curfew was put into effect at 6:00 p.m. By 7:30 p.m. the Capitol was declared secure. Both Republican and Democrat leaders agreed the constitutional process should continue that day. The Senate reconvened and debated the challenge to the Arizona Electoral College results. By a count of 93–6 the Senate voted down the objection. The House of Representatives also reconvened and debated the Arizona results and by a vote of 303–121 dismissed the objection. All 121 votes to overturn the Arizona results came from Republican members; however, 83 Republicans voted to accept the results certified by the state. This vote came at 11:08 p.m. (Roll Call Vote 10 2021).

SECOND CERTIFICATION ROLL-CALL VOTE: PENNSYLVANIA

After the certification of Arizona, the roll call of states continued into the early hours of Thursday, January 7, 2021. The next and final objection by congressional Republicans concerned the results of the Pennsylvania election. The Senate turned down the Pennsylvania objection by a vote of 92–7. After considerable heated debate, inspired by the events of the day, the House also turned down the Pennsylvania objection by a vote of 282–138. In this case 64 Republicans joined with 218 Democrats to accept the results submitted by the state of Pennsylvania. The House vote came at 3:08 a.m. (Roll Call Vote 11 2021). Approximately fifteen minutes later Vice President Pence affirmed that Joe Biden won the 2020 presidential election by a 306–232 Electoral College vote.

ROLL-CALL CERTIFICATION MAP

After the November election the movement to challenge the results of the presidential election split the Republican Party, both nationally and in Congress. When it came time to finally certify the election on January 6, the divide among Republicans was most evident in the House of Representatives. As noted above, more Republicans objected to the Pennsylvania results than Arizona's. In addition, a number of Republicans were absent or did not vote on one or the other roll call. All Democrats present voted in both roll calls to accept the state-submitted results. However, like the Republicans, several Democrats were absent or did not vote on one or both of the roll calls (Kondik 2021). At the time of the certification there were four vacant seats in the House.

Figure 9.3 maps the combined Arizona and Pennsylvania roll-call votes on January 6 and 7, 2021. The 222 Democratic members of the House are colored blue. The Republicans are divided into three categories: 139 members voting to object to at least one of the state certifications (Objectors); 66 members voting to accept state certifications (Acceptors); and four members not voting. Almost one-third of House Republicans voted to accept the Electoral College results in Arizona and Pennsylvania. These Acceptors came largely from the Midwestern states: Ohio, Indiana, Illinois, Iowa, Minnesota, Wisconsin, and Michigan. Though few in number, the one other Acceptor cluster is in the Interior West and two adjacent Dakota representatives from the upper Great Plains. Most notable in this group is Liz Cheney of Wyoming, third-ranking House Republican. Republicans from Appalachia, the Southeast, and southern Great Plains were solidly in the Objector category. Pennsylvania is noteworthy as the only northern state where the entire Republican delegation was unified in objection. One general trend was that Republican Objectors tended to come from safe Republican districts, while most Republican Acceptors tended to come from more competitive districts (Kondik 2021). However, there were outliers in both directions, Representative Cheney being one example.

AFTERMATH

The events of January 6, 2021, came as a shock to the American body politic. The riot demonstrated the extreme polarization in US politics. The vast majority of political leaders from all regions, levels, and party condemned the mob violence. That being said, many conservatives and conspirators (including former President Trump) claim months after the last litigation that the election was stolen (Colvin 2021). These false claims play on the fears and beliefs of many Americans. Middle-class suburban and woman voters have tended toward the Democrats in recent elections. The actions and images of the rioters, and false claims of election fraud, may have intensified this trend. Data shows voters switching party registration in the weeks following the riot (Wilson 2021). The electoral consequences of January 6 in the 2022 congressional and 2024 presidential elections and beyond are yet to be determined.

REFERENCES

Brumback, Kate. 2020. Georgia Officials Certify Election Results Showing Biden Win. *APNews*. November 20, https://apnews.com/article/georgia-certify-election-joe-biden-ea8f867d740f3d7d42d0a55c1aef9e69.

Colvin, Jill. 2021. Trump Airs Old Election Grievances in Return to Rally Stage. *APNews*. June 26, https://www.usnews.com/news/politics/articles/2021-06-26/trump-targeting-gop-impeachment-voter-at-ohio-revenge-rally.

Congressional Record. 2021. Proceedings and Debates of the 117th Congress, First Session. Wednesday, January 6, 2021, Vol. 167, No. 4, H76.

Kondik, Kyle. (2021). How House Republicans Voted on Electoral College Disputes. University of Virginia Center for Politics. January 12, https://centerforpolitics.org/crystalball/articles/the-objectors-versus-the-rejecters.

Roll Call Vote 10. 2021. On Agreeing to the Objection, January 6, 2021, 11:08 pm—117th Congress, 1st Session, House of Representatives. https://clerk.house.gov/Votes/202110?Page=17.

Roll Call Vote 11. 2021. On Agreeing to the Objection, January 7, 2021, 3:08 am—117th Congress, 1st Session, House of Representatives. https://clerk.house.gov/Votes/202111?Page=17.

Schmidt, Michael S., and Luke Broadwater. 2021. Officers' Injuries, Including Concussions, Show Scope of Violence at Capitol Riot. *New York Times*. May 7.

Wilson, Reid. (2021). Tens of Thousands of Voters Drop Republican Affiliation after Capitol Riot. *The Hill*. January 27.

THE 2020 CENSUS AND REAPPORTIONMENT

KENNETH C. MARTIS

The Constitution calls for a decennial census of population to allocate seats to the House of Representatives, and, therefore, also to the Electoral College. The first census was held in 1790. In recent decades, the Supreme Court "one person one vote" rulings mandate that even though a state may not have gained or lost a congressional seat, they must redistrict the congressional and state legislature boundaries based upon the new population geography. On April 26, 2021, the Census Bureau announced the state-level results of the 2020 population count. Figure 9.4 illustrates the reapportionment resulting from the 2020 Census. Seven seats were transferred between states. Thirteen states had their House delegation size change; seven states lost seats and six states gained seats.

THE REGIONAL GEOGRAPHY OF REAPPORTIONMENT

Figure 9.4 shows a change in the regional geography of congressional seats and presidential electors. Seats are moving from the North to the South and West. The 2010 reapportionment map shows the exact same trend. In fact, this general movement has been occurring for several decades. A work that analyzes long-term reapportionment change, *The Historical Atlas of State Power in Congress, 1790–1990*, mapped for the first time the reallocation of seats for all censuses since the first in 1790 (Martis and Elmes 1993). The atlas divides American history into four geographic population/reapportionment eras: Original States and New States (1790–1850), Free States and Slave States (1790–1860), Rural and Urban Places (1870–1930), and Sunbelt and Snowbelt (1970–1990). Subsequent counts have shown the Sunbelt-Snowbelt Era extending to the 2000, 2010, and 2020 censuses. Each of the four eras has had a profound effect upon the regional balance of power in the United States, the issues and legislation brought before Congress, and the election of the president. Sunbelt is the term given by demographers and geographers to the combined population in-migration and economic growth along the southern tier of America, namely the South and West regions. The Snowbelt consists of areas and states along the northern tier, especially the Northeast and Midwest.

Figure 9.4 shows that six of the seven states that lost seats in 2020 are in the North. Six states gained electoral power: three are in the South, Texas (+2), Florida (+1), and North Carolina (+1); three are in the West, Colorado (+1), Montana (+1), and Oregon (+1); (Ax 2021; Kondik 2021). Table 9.2 enumerates the total changes for every state in the Sunbelt-Snowbelt Era. Texas (+14), Florida (+13), and California (+9), are by far the biggest winners. In 2000, California gained one additional seat, making it the largest House of Representatives state delegation in American history with fifty-three members. The 2010 census did not change the California number, the first time the state did not increase its delegation since its admittance to the Union in 1850. The 2020 census reported a loss of one seat in California, the first time this has ever occurred. In spite of its first loss, in the 2022 congressional election California will still have by far the largest delegation with fifty-two. Consequently, in the 2024 presidential election, California will have fifty-four electoral votes, one-fifth of the 270 votes needed to elect the president.

The combined congressional losses of the northern states over the last half century are staggering. The last time northern states gained in reapportionment was in 1960: Ohio and Michigan each added one seat because of the post–World War II auto boom; New Jersey gained one mostly from intraregional suburbanization from New York City and Philadelphia; and Maryland gained one, reaping the benefit of the expansion of the federal government in Washington, DC, and the incumbent suburbanization. Since the 1970 census, New York has lost an incredible thirteen seats, Pennsylvania eight, Ohio eight, Illinois seven, and Michigan six.

The combined congressional losses of the Snowbelt states, of course, equal the gains of the Sunbelt states. The biggest gainers in the South are Texas (+14) and Florida (+13). In addition, Georgia (+4) and North Carolina (+3) have made significant increases. Some of the biggest percentage gains have been in the previously sparsely settled West. Arizona increased five seats, going from a

FIGURE 9.4

Apportionment of US House of Representatives 2022

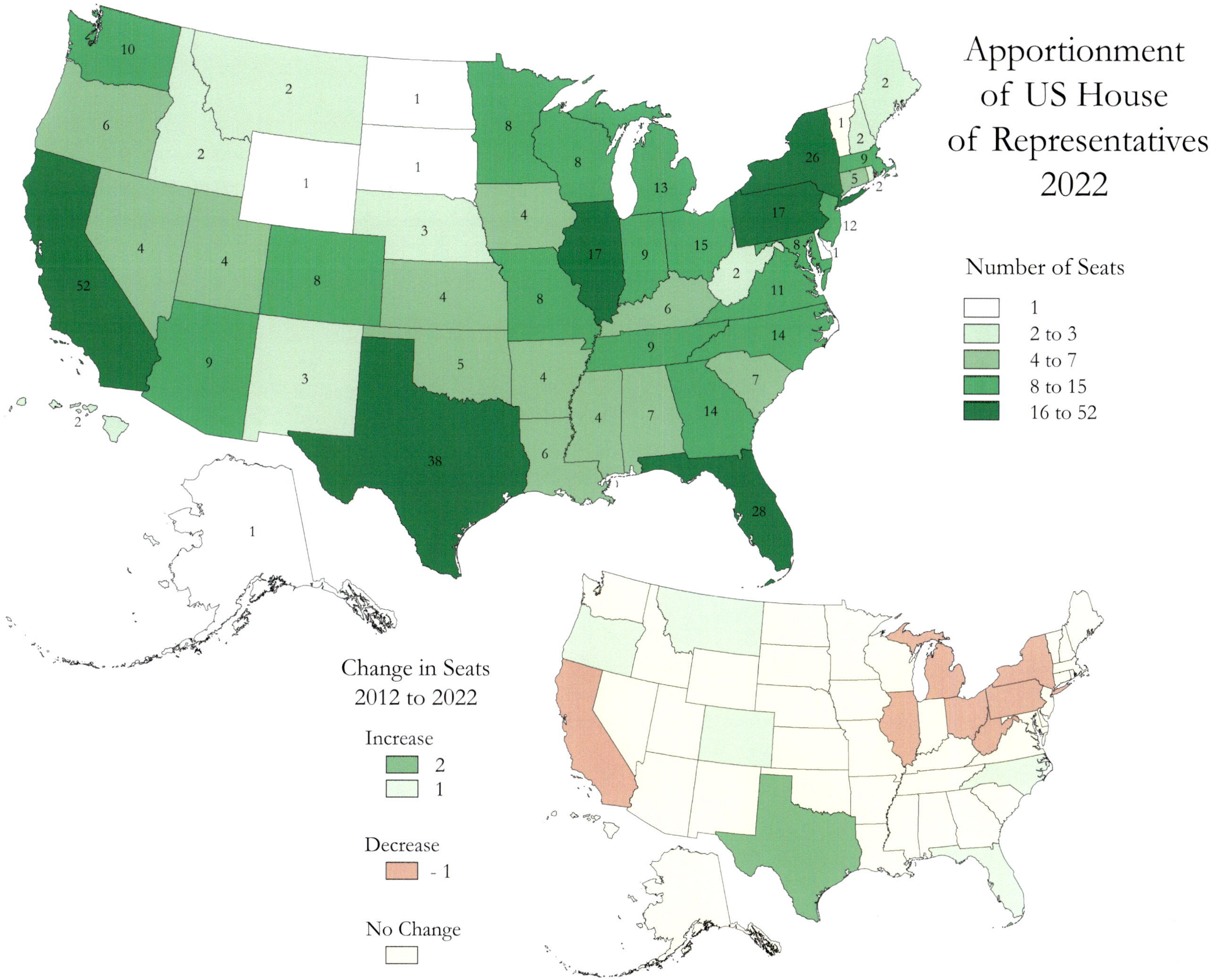

TABLE 9.2. State Reapportionments in the Sunbelt-Snowbelt Era: US House of Representatives, 1972–2030

Elections	AL	AK	AZ	AR	CA	CO	CT	DE	FL	GA	HI	ID	IL	IN	IA	KS	KY	LA	ME	MD	MA	MI	MN
1972-1980	7	1	4	4	43	5	6	1	15	10	2	2	24	11	6	5	7	8	2	8	12	19	8
1982-1990	7	1	5	4	45	6	6	1	19	10	2	2	22	10	6	5	7	8	2	8	11	18	8
1992-2000	7	1	6	4	52	6	6	1	23	11	2	2	20	10	5	4	6	7	2	8	10	16	8
2002-2010	7	1	8	4	53	7	5	1	25	13	2	2	19	9	5	4	6	7	2	8	10	15	8
2012-2020	7	1	9	4	53	7	5	1	27	14	2	2	18	9	4	4	6	6	2	8	9	14	8
2022-2030	7	1	9	4	52	8	5	1	28	14	2	2	17	9	4	4	6	6	2	8	9	13	8
Seats Gain/Loss	0	0	+5	0	+9	+3	-1	0	+13	+4	0	0	-7	-2	-2	-1	-1	-2	0	0	-3	-6	0
% Delegation Gain/Loss	0	0	+125	0	+21	+60	-17	0	+87	+40	0	0	-29	-18	-33	-20	-14	-25	0	0	-25	-32	0

delegation of just four in 1970 to nine in 2022. Nevada has gone from only one House member as recently as 1980 to four in the 2022 election (Overberg and Peterson 2021).

THE POLITICAL GEOGRAPHY OF REAPPORTIONMENT

In the last six censuses, there has been a clear reallocation of regional power in the United States. But has this reallocation translated into political power? Does it benefit the Republicans or Democrats? In the 2020 census, seven states lost one seat; five voted Democrat in the 2020 presidential election (California, New York, Pennsylvania, Michigan, and Illinois) and two Republican (Ohio and West Virginia)—a slight net gain for the Republicans. Six states gained seats; four voted Republican (Texas (+2), Florida, North Carolina, and Montana) and two Democrat (Colorado and Oregon)—another slight gain for the Republicans.

The long-term Sunbelt-Snowbelt trend shows a similar pattern. The big losers were New York (−13/solid D), Pennsylvania (−8/D possible swing), Ohio (−8/R possible swing), Illinois (−7/solid D), and Michigan (−6 D/possible swing). The big gainers were Texas (+14/solid R), Florida (+13/R possible swing), and California (+9/solid D). In the solid states alone, the Democrats have lost eleven congressional seats (electoral votes) and the Republicans have gained fourteen. Of the three largest states, only one, Florida, is historically a battleground. The increase in numbers in swing state Florida is one of the reasons it has played a critical, and sometimes decisive, role in recent presidential elections.

Looking at the wider national political trend, the pattern is more complicated (Chinni 2021). Although Republicans have made great strides controlling the South, these gains have been mitigated by population losses in the Great Plains, Appalachia, and other places. Although the Democrats have recently controlled the Midwest and Northeast, this region is declining in seats/electors, but still has a large population base. In addition, the Democrats have controlled the substantial California vote and other Pacific Coast states and have also made great inroads in some growing Mountain West places like Colorado, Nevada, and New Mexico. The surprising 2020 Democrat presidential and 2021 senatorial runoff wins in Georgia complicates further the historical pattern. Georgia can now be considered a swing state.

A number of political and demographic trends could alter the above alignment of solid or swing categories for both parties. For example, the Hispanic vote is growing in Arizona, Nevada, and North Carolina and they are now swing states rather than safely Republican based on recent election results. The regional maps in chapter 5 show a divided nation: the Northeast, Midwest, and Pacific Coast tending Democrat and the South, Great Plains, and Mountain West tending Republican. In the six presidential elections from 2000 to 2020, the Republicans have won three and the Democrats three. The nation is divided and the growing population and electoral power in the swing states of Florida, Georgia, North Carolina, Nevada, and Arizona will be the key in future elections.

MS	MO	MT	NE	NV	NH	NJ	NM	NY	NC	ND	OH	OK	OR	PA	RI	SC	SD	TN	TX	UT	VT	VA	WA	WV	WI	WY
5	10	2	3	1	2	15	2	39	11	1	23	6	4	25	2	6	2	8	24	2	1	10	7	4	9	1
5	9	2	3	2	2	14	3	34	11	1	21	6	5	23	2	6	1	9	27	3	1	10	8	4	9	1
5	9	1	3	2	2	13	3	31	12	1	19	6	5	21	2	6	1	9	30	3	1	11	9	3	9	1
4	9	1	3	3	2	13	3	29	13	1	18	5	5	19	2	6	1	9	32	3	1	11	9	3	8	1
4	8	1	3	4	2	12	3	27	13	1	16	5	5	18	2	7	1	9	36	4	1	11	10	3	8	1
4	8	2	3	4	2	12	3	26	14	1	15	5	6	17	2	7	1	9	38	4	1	11	10	2	8	1
-1	-2	0	-0	+3	0	-3	+1	-13	+3	0	-8	-1	+2	-8	0	+1	-1	+1	+14	+2	0	+1	+3	-2	-1	0
-20	-20	0	0	+300	0	-20	+50	-33	+27	0	-35	-16	+50	-32	0	+17	-50	+11	+58	+100	0	+10	+43	-50	-11	0

REFERENCES

Ax, Joseph. 2021. U.S. Census Hands More House Seats to Republican Strongholds Texas, Florida. *Reuters*. April 26.

Chinni, Dante. 2021. Census Winners and Losers Paint a Muddled Future for the Parties. *NBC News*. May 2.

Kondik, Kyle. 2021. Census Reapportionment: Seat Trade-Offs Likelier to Benefit Republicans. University of Virginia Center for Politics. April 26.

Martis, Kenneth C., and Gregory Elmes. 1993. *The Historical Atlas of State Power in Congress: 1790–1990*. Washington, DC: Congressional Quarterly Press.

Overberg, Paul, and Kristina Peterson. 2021. Census Shows South and Mountain West Gain Political Power from Population Growth. *Wall Street Journal*. April 26.

10 THE 2020 ELECTION AND BEYOND

THE 2020 ELECTION: LOOKING BEYOND FOR THE UNEXPECTED

STANLEY D. BRUNN

Politics is everything and everything is politics.
—Anonymous

Future historians, political scientists, and geographers will with certainty look at the 2020 US presidential election as an LNO (like none other) phenomenon. This is not only because of the "topsy-turvyness" of the year itself, but what transpired in 2021 and what may lie ahead. As the election year evolved it became evident to the leaders of both political parties and their supporters that there were some unprecedented issues surfacing that those running for this highest elected office would have to confront. Four issues remained front and center: the reporting, spread, and impacts of the COVID-19 virus and the president's handling of the issue; the nationwide TV coverage of the death of African American George Floyd at the hands of a white police officer and the impacts; the construction of a border wall along portions of the US and Mexico border and tandem issues about undocumented refugees and asylum seekers; and the leadership behavior of the American president, his ardent supporters, and those who found him divisive, self-serving, insensitive to race and gender issues, and untrustworthy.

When Americans went to the polls on November 3, voters were faced with some stark differences in political experiences and savvy between President Trump and Democratic presidential candidate Joe Biden, even though both were elderly white males. Public opinion polls in the months preceding the election consistently showed Biden with a steady narrow (about 5 percent) margin; Trump's approval numbers were consistently about 45 percent. While these were national numbers, in a number of crucial or "swing" states, the gaps between the two candidates were narrower. A large percentage of the Democrats supported Biden and a large percentage of Republicans supported Trump. How the independent voters would vote became crucial, even though they were leaning toward Biden.

MAP OF THE FINAL VOTE

Predicting final outcomes in national elections is difficult for two main reasons. In many states, including "swing" states, voters have to wait, sometimes hours and days, because their voting history is one of narrow victories for the winning party. Second, those delays can be prolonged if there are many absentee ballots and/or delays in voting totals, especially from some precincts with a large number of votes, unusually high turnout, or machine problems in counting votes. Declaring a popular vote winner is not difficult in many states where one party has traditionally held sway and the minority party has far fewer votes. The "wait and see" votes are usually not in the very largest states such as California, New York, or Texas, but in states that have grown in population the past decade and have many new voters. Voting analysts take a "wait and see" attitude about how states like Florida, North Carolina, Virginia, Pennsylvania, Michigan, Illinois, and Ohio vote. They care less about Oregon, Kentucky, Arkansas, Kansas, Idaho, Wyoming, Montana, Rhode Island, and Vermont because they have smaller populations and the majorities generally vote consistently for the same party. A related issue is the role of the Electoral College. In a "winner take all" Electoral College, the candidate and party with the most votes have all the College votes, whether the winner wins by one or two votes or one hundred thousand. On election evenings, newscasters will continually update the voting totals of the candidates and the electoral votes for candidates.

While the popular vote totals and electoral vote totals are known on election night or the following morning, the official acknowledgment of the results is not known until the states certify the results on December 14. The Electoral College members are in most cases unknown to the voters; however, they are bound to support the candidate their party selected. The final and official task in certifying the election result is not known until Congress meets in early January to officially certify the next president and vice president.

POSTELECTION AMERICA

The final nationwide results of the election revealed no shocking news; preelection polls predicted the Biden-Harris team would win the popular vote and the Electoral College vote. The Biden-Harris ticket had 81.2 million votes compared to 74.2 million for Trump and Pence. The Democratic team had 51.8 percent of the votes and the Republican candidates 46.8 percent. There were a few surprises to the Biden camp in that he won Georgia and Arizona; Biden also won three states that Trump won by narrow margins in 2016: Wisconsin, Michigan, and Pennsylvania. The major immediate surprise that emerged from the election was that Trump refused to admit defeat and congratulate the winner, as is customary. Instead, he immediately called for recounts in states that he won in 2016 or lost by narrow margins in 2020, believing that recounting the ballots would reveal major errors in the counting and lead him to be declared the 2020 winner. Trump supporters filed many lawsuits in the states that he lost. All lawsuits were dismissed by the courts, but that did not stop him or his ongoing efforts to overturn the results.

One additional event occurred on January 6, 2021, when Congress met to certify the November results. A postelection protest encouraged by Trump included a massive insurrection by Trump supporters who entered the Capitol building seeking to overturn the Electoral College results being counted. The purpose was to create a chaotic setting that would nullify Biden's win and ensure that Trump would be president another four years. The riotous protest was unsuccessful and, following an interruption, the counting continued, and Biden and Harris were declared new president and vice president. This spectacle, visible nationwide and internationally, marked the end of a tumultuous election year and a memory that would remain lasting in the minds of many voters prior to the official inauguration of the forty-sixth president on January 20, 2021, and even longer.

NEW ADMINISTRATION AND A DIVIDED COUNTRY

While the new president and vice president won the popular vote by seven million votes and the Electoral College by a 306 to 232 margin, the branches of Congress were closely split between the Republicans and the Democrats. In the House of Representatives, the Democrats lost seats and now held a narrow lead of 222 to 213 over Republicans. The Senate, which before the November election had more Republicans, was now split fifty-fifty, with any tie-breaking vote cast by Vice President Kamala Harris. As the weeks and months after the inauguration revealed, there was "talk" about bipartisanship by the president and Democrats, but with little success as few Republicans broke party rank and sided with Democrats on legislation. Supporters of both political parties will wait and see what across-the-aisle support there will be in the coming months on infrastructure, expanded voting rights, the documentation and resettlement of refugees, ongoing COVID-19 relief efforts, changing the tax structure to ensure wealthy individuals and corporations pay more, Supreme Court nominations, and foreign policy issues related to Afghanistan, Russia, China, NATO, and Israel/Palestine.

FOUR PENDING POLITICAL ISSUES

The 2021 state and national political calendars do not appear to be a time for quiet contemplation and bipartisanship. Rather there are a number of important issues that will keep the leaders of both political parties alert. Three are headline-grabbing and direct spinoffs of the November election. The first are ongoing (as of July 2021) legal efforts, supported by a large majority of Republicans, to show that the election was rigged and that Trump should be declared winner. These false claims, raised in a number of states, have not been successful. Second, there is also less Republican support for a bipartisan commission to explore the causes and impacts of the January 6 events. Third, there are serious efforts by states that Trump won or narrowly lost to tighten the requirements for those voting, especially states with large numbers of minorities. The thinking is that by tightening the qualifications of those voting for the Democrats in 2020, the impact will be more states voting for Republican officeholders at all levels. Most of the proposed "tightening" efforts include legislation that would restrict the time frame for registration, require photo IDs, and limit mail-in ballots and the number of places where one could vote. Parties are often afraid to let the "non-registered voters vote" because they do not know how they vote (think of the fears of permitting women and racial minorities to vote). The fourth set of issues relate to redistricting, a practice carried out by many states following a decennial census. Each state is allocated X number of districts amounting to 435 in total. Each state, even the very smallest in population, will have at least one member in the US House of Representatives. States "win" or "lose" seats depending on changes in their population numbers (not registered voters) compared to the previous census. If a state gained population, it would carve out new districts; if it lost population, it will still have to carve out new districts. In most states, the redistricting commissions are comprised of political, and partisan, members who seek to preserve their party's position of power, even if that means conjuring up districts with shapes that defy a clear definition. Only a few states have independent nonpartisan commissions. As district boundaries are redrawn and voting restrictions are passed, one can expect lawsuits filed by the Democrats. State courts and even the US Supreme Court will also likely be involved in making some decisions about voting eligibility and the weirdness of some redistricting initiatives.

THINKING AND MOVING AHEAD

Political scientists, historians, and geographers are not immune from examining previous presidential elections from thirty-two, forty-eight, or sixty years ago. The search is for similar state and regional patterns, but also where there are sharp differences in voting results and turnouts. In an examination of county and state vote totals for the two major political parties sixty or sixty-eight years ago, one might look at where the presidential and vice presidential candidates came from, thinking that if they came from two different political cultures, that "balance" would be more likely to bring victory than if they came from the same state or region. That thinking has less meaning today in a country strongly linked by national media reporting and greater national awareness of major issues. The internet, shifts in regional migrations, and national recognition and marketing of consumer products have led to greater national awareness about improvements in infrastructure, women's rights, urban and rural renewal, sustainability, alternative energy sources, and greater minority representation. To be sure, these and others will be on the agenda of those seeking higher political offices in the next fifty years.

The election brought elation and criticism from the American electorate, with some choosing to "stick with the past" versus those "seeking progressive futures." Regardless of which "side" an individual voter or a political party is on, there are likely to be some changes in the political landscape at national and regional levels in the coming decades. The "inklings" were there in 2020 if one looks carefully at the issues raised at local, regional, and national levels.

A number of scenarios, some likely and others less likely, will emerge on the American national scene in the coming decades. These include the following: both national political parties having racial and ethnic minorities leading party tickets; women leading the ticket of one party and men running as vice presidents; Blacks, Hispanics, and Asians playing increasing leadership roles in national parties and high statewide offices; compensation to present generations of past enslaved family members; removal of controversial racial statues and racist or sexist place names; rank voting for those running for major offices (rank the top four); women becoming the leaders of major political parties and in the Senate and House committees, in diplomatic corps, and in military command; more Supreme Court members and more women; new national holidays honoring women; more women on banknotes and stamps; term limits (for Supreme Court justices and for Senators (four terms) and House members (eight terms); abolishing the Electoral College (perhaps by 2048 or before) as it is recognized as grossly unfair and replaced by popular vote; statewide nonpartisan redistricting committees; mobile voter registration where one can vote (only once, of course) anywhere on Election